16th European-African Regional Conference of the International Society for Terrain-Vehicle Systems

Lublin, Poland
11 - 13 October 2023

ISBN: 978-1-7138-8803-1

Printed from e-media with permission by:

Curran Associates, Inc.
57 Morehouse Lane
Red Hook, NY 12571

Some format issues inherent in the e-media version may also appear in this print version.

Printed with permission by Curran Associates, Inc. (2024)

For permission requests, please contact the International Society for Terrain Vehicle Systems at the address below.

International Society for Terrain-Vehicle Systems
c/o Cold Regions Research and Engineering Laboratory
72 Lyme Road Hanover, New Hampshire 03755-1290
USA

http://www.istvs.org

Additional copies of this publication are available from:

Curran Associates, Inc.
57 Morehouse Lane
Red Hook, NY 12571 USA
Phone: 845-758-0400
Fax: 845-758-2634
Email: curran@proceedings.com
Web: www.proceedings.com

ISTVS

Proceedings of the 16th European-African Regional Conference of the ISTVS

October 11-13, 2023
hosted by Lublin University of Technology, Poland
https://2023.istvs.org

Proceedings of the 16th European-African Regional Conference of the ISTVS

ISBN 978-1-942112-55-6 (pdf) [illegible]

Additional copies of this publication are available from https://www.istvs.org/proceedings

Editors

Jarosław Pytka, Editor-in-Chief
Lublin University of Technology, Poland

Paweł Tomiło
Lublin University of Technology, Poland

Massimo Martelli
National Research Council, Italy

Contents

Conference overview

The 16th European-African Regional Conference of the ISTVS is an opportunity to present and disseminate the latest scientific and technological achievements in the field of terramechanics and terrain-vehicle systems with representatives of educational and scientific institutions in the field and professionals from industry, business environment and government agencies. The organizers also welcome the participation of students from relevant undergraduate and postgraduate courses. The intention is to create a congenial atmosphere and comfortable conditions to promote networking, exchange of ideas, and dissemination of knowledge.

The keynotes and technical sessions focus on the following topics:

- soil and other deformable traction substrates, including Martian and lunar regolith, snow, moschus, etc., mechanical characterization for advanced field mobility modeling as well as for minimizing the effects of soil compaction and erosion, with particular emphasis on the impact of weather factors;
- driving systems of vehicles and machines, including innovative concepts of tires, wheels, and tracks for energy saving traction;
- propulsion systems, engines, mechatronic systems, and vetronics to meet the requirements of demanding use cases and increasingly stringent environmental protection standards;
- regolith research into celestial bodies, the Moon, Mars, asteroids for practical solutions in the field of mobility, autonomous, and robotic systems;
- metrology in terramechanics, in particular new solutions for remote autonomous measurements using AI and maintenance-free devices and systems;
- any other topics relevant to the field of terramechanics and terrain-vehicle systems.

Conference Chair

Jarosław Pytka | Lublin University of Technology, Poland
j.pytka@pollub.pl

Scientific Committee

Hosting institution

The Lublin University of Technology is the largest technical university in the Lubelszczyzna region, south-eastern Poland. It consists of six faculties, including the Faculty of Mechanical Engineering, in the building where the 2023 ISTVS Conference is held. The Faculty of Mechanical Engineering boasts a rich tradition and significant achievements in the field of mechanical engineering, technology, material engineering, transport, and automotive technology. In the field of terramechanics, research has been carried out since 1992 and teaching classes in off-road and military vehicles since 2005. This year marks the 70th anniversary of the Lublin University of Technology and the ISTVS conference is an integral part of the celebrations.

Patronage

The ISTVS 2023 Conference received the honorary patronage of the Polish Minister of Education and Science, Prof. Przemysław Czarnek. The Conference was co-financed by the Polish state budget, under the Excellent Science II Program, contract no. KONF/SP/0330/2023/01.

Minister of Education and Science
Republic of Poland

POLITECHNIKA
LUBELSKA
LUBLIN UNIVERSITY
OF TECHNOLOGY

Keynote speakers

Jerzy Lipiec

Vehicular traction effects on agriculture: Soil compaction and soil erosion, crop yield. Implications from the perspective of terramechanics

Professor of agricultural sciences, Institute of Agrophysics, Polish Academy of Sciences, Lublin | www.ipan.lublin.pl

His research activity focuses on the impact of soil compaction, agricultural practices on soil physical properties, plant growth and the environment. He has published more than 100 articles in international scientific journals, 2 books, 24 monographies and book chapters. He coedited *Encyclopedia of Agrophysics* (Springer) and serves or served as co-editor of *International Agrophysics, Acta Agrophysica*, member of the Editorial Board of *Soil and Tillage Research, Acta Agriculturae Scandinavica*. He had several internships with foreign research centres in the Netherlands, USA and Japan and participated in projects within the *Joint Programming Initiative on Agriculture, Food Security and Climate Change* (FACCE-JPI) and *Horizon 2020*.

Andrzej Selenta

The Bekker Project

Born in 1950, graduated in 1975 with a master's degree from the Faculty of Automotive and Construction Machinery of the Warsaw University of Technology, specialization: Heavy Construction Machinery. In 1985, he defended his PhD in Soil Mechanics. In 1989–1994 he was vice-dean for student affairs. For many years he was the dean's representative for international programs: Socrates - Erasmus, Leonardo da Vinci, LLP Erasmus. In 1989–1994 he was the secretary of the Polish section of ISTVS, a society founded by Mieczysław Bekker. In 1991, he co-organized the celebrations related to the naming of M.G. Bekker auditorium in the building of the Faculty of Automotive and Construction Machinery of Warsaw University of Technology. In 1992, he was responsible for the management of M. G. Bekker's archive in his home in Santa Barbara, California. So far, he is responsible for the materials collected in the so-called Bekker archive at the SiMR Faculty of Warsaw University of Technology. Currently retired. Coordinator of *Project Bekker*.

Stelios Skevakis

A novel flexible all-metal wheel for planetary exploration

Director, Production Department, Hellenic Technology of Robotics SA

He is a Mechanical Engineer (School of Engineering, University of Crete, 1996) with MBA (National Technical University of Athens 2012). He has been involved in complex mechanism analysis, design and prototype integration since 2005, with focus on terrestrial locomotion systems. He has joined Hellenic technology of Robotics on 2014 and has been Lead Engineer for a number of ESA projects, among others for the design and manufacturing of a solar panel cleaning manipulator for the Martian environment (DUSTER), the design and integration of all metal flexible wheels for planetary exploration (AWE wheels), as well as the design and integration of a satellite docking system for non-cooperative satellites (PREDAToR). He is currently involved in a high-speed rover development for the lunar surface (RAPID), as well as phase A of the European Moon rover System (EMRS). He has published several conference papers on planetary rovers and wheels. He is co-inventor of 2 International Patents.

Conference program

Pre-conference events – Tuesday, October 10, 2023

10.00 – 15.00 | Pre-Conference Professional Short Course

Novel Methods in Analytical and Experimental Off-Road Vehicle Dynamics
10.00 – 11.30 | Room M107 | Prof. Schalk Els, University of Pretoria, South Africa

Elements of Planetary Rover Design: Innovative Solutions for Optimal Traction on Extraterrestrial Bodies
11.45 – 13.15 | Room M107 | Dr. Lutz Richter, SoftServe, Inc., Munich, Germany

New Trends in Off-Road Vehicle Design
13.30 – 15.00 | Room R302 | Dr. Zbigniew Kiernicki, Lublin University of Technology

14.00 – 17.00 | ISTVS Board of Directors Meeting, Room M107

18.00 – 21.00 | Ice Breaker Welcome Event, Hotel Ibis Styles

1st Conference Day – Wednesday, October 11, 2023

08.00 | Registration

09.00 | Opening ceremony, Auditorium A1

Conference Chair Welcome — Prof. Jarosław Pytka

President's Address — Prof. Corina Sandu

Journal of Terramechanics Editor-in-Chief's Address — Prof. Vladimir Vantsevich

10.00 | Plenary Session, Auditorium A1

Keynote: *Vehicular traction effects on agriculture: Soil compaction and soil erosion, crop yield. Implications from the perspective of terramechanics*
Prof. Jerzy Lipiec, Institute of Agrophysics, Poland

12.00 | Conference photo followed by lunch in Lanchomania

13.00 | Parallel sessions — 3 tracks — Rooms M107/CT302/R302

Autonomous and robotic systems 1
Chair: Dr. Lutz Richter
Room M107

0274 | Deep learning method for IMU-based tracking of Martian rover

0709 | Proposal of swarm rovers' collaborative locomotion with expansion and contraction mechanism for driving a loose slope

0889 | Skidding suppression method using "discrete 4-wheel-drive typed rover" considering steering angle at traveling across loose slope

1453 | Suppressing the reduction of the traveling displacement on loose soil for rovers with the function of a wheel walking

2765 | Obstacle detection vision system enabling autonomous mounding on clearcuts

Soil compaction 1
Chair: Prof. Junya Yamakawa
Room CT302

2177 | Effects of road wheel load, driving speed and track slip upon stress state in sandy soil under tracked vehicle loading

2297 | Traction effects on vertical and horizontal contact stresses and the risk of soil structure deformation

3797 | Disturbed and naturally recovered soil surface as a ground for sub-sequent vehicle mobility operations

4895 | Theoretical analysis and fem simulation of the interfacial forces between tracked vehicle and snow soil

6690 | Comparative evaluation of methods to evaluate penetration resistance for Clegg hammer and cone penetrometer

Off-road vehicle dynamics
Chair: Prof. Schalk Els
Room R302

3859 | Analysis for dynamic characteristics of a tractor cabin during agricultural operations

4141 | Modeling and verification of a full-scale forestry vehicle real-time multi-physics digital twin

5356 | Development and validation of head-feeding combine harvester multi-body dynamics simulation model

6955 | Development of dynamics simulation model for agricultural tractor according to agricultural work

7198 | Analysis of aircraft dynamics while overcoming obstacles on a grass runway

8963 | Analysis of the influence of terrain unevenness and velocity on the ground pressure exerted by demining rollers

14.30 | Break

15.00 | Parallel sessions — 3 tracks — Rooms M107/CT302/R302

Autonomous and robotic systems 2
Chair: Prof. Genya Ishigami
Room M107

6027 | Digital precision planning tool for autonomous forest regeneration of mixed tree species

8151 | Study on estimation of traveling states using strain information on chassis of lunar and planetary exploration rovers

8672 | Trajectory optimization for vegetation override in off-road driving

2145 | Obstacle performance and wheel failure test analysis of Zhurong rover

4976 | Performance evaluation of autonomous driving lateral control simulation model of agricultural tractor

Propulsion systems and engines
Chair: Prof. Rafał Longwic
Room CT302

4816 | Effects of humidity on the emissions of the diesel engines

7477 | Evaluating the pressure performance of DPF filters using engine bench analysis

1166 | Selection of energy saving en-gine mode based on the power delivery and fuel consumption of a 95-kW tractor during agricultural operations

4727 | Analysis of gear durability for agricultural tractor transmission

Driving systems of off-road vehicles and machines
Chair: Massimo Martelli
Room R302

2437 | Improving predictive control methods on off-road vehicles with realistic steering preview inputs

3815 | Data measurement and performance evaluation of an electric all-wheel-drive tractor during plow tillage

3893 | Analysis of load factor and performance evaluation for developing an electric driven tractor

8254 | Characterization of the new US army cold weather all-terrain vehicle (CATV) on winter surfaces

16.30 | Break

17.00 | Workshop: LUT Student Projects on Off-Road and Military Vehicles

2nd Conference Day – Thursday, October 12, 2023

09.00 | Plenary Session, Auditorium A1

Keynote: *A novel flexible all-metal wheel for planetary exploration*
Stelios Skevakis, Hellenic Technology of Robotics, Greece

Keynote: *The Bekker Project*
Dr. Andrzej Selenta, Warsaw University of Technology, Poland

10.00 | Break

10.30 | Parallel sessions — 3 tracks — Rooms M107/CT302/R302

Off-road mobility modeling 1
Chair: Prof. Corina Sandu
Room M107

6211 | Wheel-soil terramechanics model augmentation using machine learning across multiple soil types

1522 | Tractive performance of rigid wheel in granular media using coarse-scale DEM models

6143 | Flexible tire-terrain interaction model for real time simulations

3835 | Effect of steer angle rate upon tyre lateral force generation on two different soils

4508 | Semi-empirical terramechanics model for variable terrain height

Soil characterization 1
Chair: Prof. Peter Kiss
Room CT302

1098 | Simulation of change in support-ing force when imparting vibration by distinct element method

2801 | In Situ Soil Property Estimation for Autonomous Earthmoving Using Physics-Infused Neural Networks

4336 | High-fidelity 3D-DEM simula-tion for wheel mobility in low gravity environment

4456 | Modeling soil-tool interaction of a cultivator sweep using DEM

Metrology in terramechanics
Chair: Prof. Jerzy Józwik
Room R302

1638 | Analysis of tire characteristics on road surface with volcanic ash fall

1779 | The development of a dynamic vegetation override pushbar for autonomous ground vehicles

4595 | Evaluation and comparison of driving performance of a lunar exploration rover wheel in different soils

5552 | Vehicle drawbar test method with improved measurement and control

6861 | Research on airplane performance on a grass airfield

12.00 | Lunch, Lanchomania

13.00 | Parallel sessions — 2 tracks — Rooms M107/R302

Off-road mobility modeling 2
Chair: Prof. Jozsef Koveceses
Room M107

3796 | The issue of assessing the suitability of own simulation models for testing off-road vehicles

6736 | Ride comfort comparison be-tween 4-poster and straight line driving simulations

0448 | Lunar rover discrete element method study and calibration

8859 | An efficient and high-fidelity track model for dynamic simulation of off-road tracked vehicles

4681 | Simplified models of terrain-vehicle interaction for real-time applications

Soil characterization 2
Chair: Tomohiro Watanabe
Room R302

2936 | Moisture content impacts on soil load bearing capacity and its spectral behaviour

3493 | Numerical modeling of a tire on undrained saturated clay using FEM, ALE, and SPH

9992 | Soil shear strength values obtained from its colour

7577 | Implementation of moving loads on ice in NRMM

1805 | Accelerating graph networks for real-time physics simulations

14.30 | Break

15.00 | Parallel sessions — 2 tracks — Rooms M107/R302

Soil compaction 2
Chair: Dr. Loraine ten Damme
Room M107

7523 | Assessing performance of light wheeled vehicles on GRC-1 using 3D scanned footprint measurements

8754 | Characterization of lunar simulant (FJS-1), toyoura sand, and silica sand with physical experiments and dem simulation

9504 | Soil deformation model for efficient simulation of off-road vehicles and compaction force analysis

3507 | Determination of the tractive forces of a tractor based on the recon-struction of tire–soil contact surface obtained by means of photogrammetry

6630 | Real-time measurement of tire sinkage using stereo cameras

Innovative concepts of tires, wheels, and tracks
Chair: Prof. Przemysław Simiński
Room R302

3280 | Comparison of tweels and pneumatic tires on LTATV military vehicle

6059 | Comparison of multi-tracked running gears for light UGVs in terms of terrain obstacles overcoming abilities

7994 | Longitudinal friction coefficient comparison on actual and laboratory surfaces

9268 | Experiments of sinkage characters of wire mesh wheel under different slip ratio

9716 | Non-pneumatic tire for special vehicle – a review of selected functional properties

16.30 | Closing ceremony, Auditorium A1

19.00 | Gala dinner — Hotel Ibis Styles

3rd Conference Day — Technical Tour — Friday, October 13, 2023

Technical Tour of the *Autobox Innovations* facilities in Starachowice

07.30 | Departure from Lublin

10.00 | Arrival at Autobox, Starachowice

10.15 | Welcome by Head Director, Mirosław Kalinowski

10.20 | Presentation of the production facility

11.30 | Transfer to proving ground

12.00 | Test show of vehicles: *Star*, *Honker* — outdoor, in action

13.00 | Break — military pea soup

14.00 | Departure from Starachowice

15.30 | Visit Kazimierz Dolny

16.00 | Dinner at the *U Fryzjera* restaurant, Kazimierz Dolny

18.00 | Departure from Kazimierz Dolny

19.00 | Arrival in Lublin

List of submissions

Full papers

Abstract-only

ISTVS

Conference papers

0274 | Deep learning method for imu-based tracking of Martian rover

Paweł Tomiło [a,*]

[a] *NeuroLab Laboratory of Neuroscience Applications in Management and Marketing, Lublin University of Technology, Lublin, Poland*

* *Corresponding author: p.tomilo@pollub.pl*

ABSTRACT

The double integration method is one of the most used techniques for calculating an object's position in time. As dynamic noise and bias can affect sensor data, filters like the alpha-beta filter or the Kalman filter are frequently utilized. They are in capable of compensating for sensor reading noise. The quality of the sensor has a big impact on how accurate these solutions are. When using inexpensive sensors, a correctly built neural network model can help compensate for noise and determinate the change in the position of the object over time. The article describes the method of tracking an RC vehicle using the low-cost IMU module and software consisting of an artificial neural network model specially developed for this purpose. Base stations that emitted infrared laser scans were utilized as a reference method to confirm the accuracy of the proposed solution and as a means of data collection for training. The proposed model provides acceptable accuracy when compared to the reference method. the tests were carried out on a short section of the road, in the case of longer sections the estimation error may be larger. The developed solution can be used as part of a larger system combined with visual odometry to navigate an autonomous vehicle.

Keywords
Artificial Neural Network
Deep Learning
Inertial measurement unit

1. Introduction

The accurate location of the rover is an important aspect of managing the remote operations of rovers on the surface of Mars. Various methods were used to determine the position of the rovers, such as (Powell et al., n.d.):

- heading sensor and wheel encoder
- inertial measurement unit (IMU);
- visual odometers (VO).

Each of the methods discussed above has drawbacks. In the instance of the wheel encoder, slip and kinematic imprecision are the root causes of the error.

Common inertial measurement unit consists of 3-axis gyroscope, 3-axis accelerometer and 3-axis magnetometer. (Kok et al., 2017). IMU systems are used for many tasks, such as robotics, pedestrian tracking, supporting GPS systems. Drift has a significant impact on state estimation. This drift is the end result of a buildup of several IMU integration errors, which are collectively described as an additive zero-mean Gaussian noise with a slowly varying bias. Due to the rapidly increasing accumulation of drift, estimate that solely relies on IMU measurement integration is not practical for more than a few seconds. The IMU biases severely affecting navigation accuracy (Buchanan et al., n.d.). For position determination, the use of an Extended Kalman Filter (EKF) is a common solution. A precise understanding of the random error model linked to the used sensor is necessary for the EKF approach. The vehicle dynamic model may need to be linearized in order to employ the EKF approach, which might lower estimation accuracy when taking highly nonlinear state errors into account. Additionally, the EKF method's linearization model of the error state is only temporary effective. For this reason, EKF is used when the main method of determining position is unavailable, e.g. when visual odometry is unavailable (Dai et al., 2020).

The use of VO in outdoor settings is a difficult issue. Localization is challenging in outdoor contexts due to a number of elements, such as the fact that terrains are typically not flat, direct sunshine, shadows, and dynamic changes in the environment brought on by wind and sunlight. The primary issues with VO systems relate to the cost of computing, as well as lighting and image circumstances (Aqel et al., 2016).

The article focuses on IMU-based techniques. The primary goal of this article is to develop an artificial neural network model that can infer the location of the Mars rover from IMU signals.

Many methods based on artificial neural networks (ANN) and other methods, including visual odometry, have recently been proposed in the field of IMU-based localization to address the aforementioned issues. For this reason, many publicly available datasets have been created. The sets contain data e.g. from smartphones to determine the location of a person (Chen et al., 2018; Yan et al., 2017), autonomous driving (Geiger et al., 2013; Liao et al., 2021) and drone flight logs ("Flight Review," n.d.). These datasets have been used by many researchers to create a solution that uses IMU indications to locate a given device.

The neural network model proposed by (Zeinali et al., 2022) uses an architecture that uses convolutional layers. They used the Mobile Resnet Block, which consists of successive layers of depth wise convolution 1D, batch normalization, ELU activation function, separable convolution and batch normalization. To the data processed by the described layers, input data is added - a residual connection. The last layer in the block is the ELU activation function. The described neural network model was used to

https://doi.org/10.56884/ZLZU4684
ISBN 978-1-942112-55-6 (pdf)

determine the location of a person using data from a smartphone. A similar approach has been adopted by (Herath et al., 2019). The researchers created three models of neural networks, the first one was based on residual connections, the second one used layered Long short-term memory (LSTM), and the last model used the Temporal Convolution Network (TCN) architecture. TCN architecture uses dilated convolution (Pelletier et al., 2019). The neural network model developed by (AbdulMajuid et al., 2021) uses 4 layers of LSTM to determine the location of the drone's position in 3d space. The artificial neural network model created by (Yusefi et al., 2021) uses monocular visual odometry and an LSTM layer to process the IMU indication. The use of the LSTM layer in the entire model translates into a reduction of the root mean square error of localization in relation to other position estimation methods such as EKF, Uncertain Kalman filter, Mahony and Madgwick algorithms.

The solution described in the article presents a new model of a neural network that can be used to improve the accuracy of determining the position of a Martian rover when visual odometry is not achievable, the proposed solution might be utilized in its place or as a supplement. The model uses LSTM layers in residual connections.

2. Material and methods

A test vehicle and a test site were developed to carry out the tests. In order to estimate the position of the RC vehicle – Martian rover, a measuring station was developed and a neural network model was created. The network architecture created by (AbdulMajuid et al., 2021) was applied as a base model in order to compare the results. The developed models of artificial neural networks were trained on exactly the same PX4 dataset ("Flight Review," n.d.). The PX4 dataset was chosen due to the similar scale and solutions used in drones compared to the target solution.

The selected network model was used in further research. Due to the anomalies in the magnetic field of the planet Mars, accelerometer and gyroscope indications were used to train the model (Langlais et al., 2019). Data collection for the model was carried out on a test site specially developed for this purpose and with the use of a RC car. In order to train the network, appropriate measurements were carried out and a new data set was created to validate the solution.

2.1 Selection of neural network architecture

Location estimation based on the IMU indication, assuming that the true position is known, is a supervised learning issue. Two main layers were used to select the built architecture:

- LSTM layer;
- Convolution layer.

In the LSTM layer, the output depends on 3 main elements: input data at current time step, cell state – current long-term memory, hidden state – output at the previous point in time. In the LSTM layer, gates are used, which are responsible for how information enters the layer, how it is stored in it and how it leaves the layer. For a given timestamp t, the input data to the layer is cell state C_{t-1} and hidden state h_{t-1} and input data x_t. This data is processed by the forget gate, where it is decided which elements of C_{t-1} are useful (for each element of C_{t-1} weights are assigned) based on h_{t-1} and x_t. The next step is called the input gate. This step determines what new information should be added to the cell state. The output from the LSTM layer is newly updated cell state C_t and new hidden state h_t. The output gate is used to assign new weights to h_t by using C_t (Staudemeyer and Morris, 2019).

The convolution layer in the case of timeseries is responsible for the extraction of features from the signal. In order to build a tensor of outputs, it creates a convolution kernel that is convolved with the layer input over a single spatial (or temporal) dimension.

In order to select the optimal model, 4 models of artificial neural networks were developed. Each of them was trained on the exact same subset from the PX4 dataset. The first of the models according to (AbdulMajuid et al., 2021) consists of 4 consecutive layers of LSTM and a layer of densely connected neurons. Another model uses LSTM residual layers. This kind of layer offers a direct route between outputs from neighboring layers. The residual connection executes element-wise addition after applying identity mapping to input. Empirically, it is shown that a neural network with residual connections converges substantially more quickly in training process. (Kim et al., 2017). A model using convolutional architecture was also developed using MobileResnet Block according to (Zeinali et al., 2022). The last of the models was developed as a combination of the LSTM and Conv models by using two separate branches. In the process of training neural network models, a callback was used, whose task was to stop the learning process when the value of the loss function for the validation set did not decrease for 5 epochs, and then to save the optimal weights.

The highest value of the loss function was for the architecture based on convolutional layers, both for the training and validation sets. The smallest value of the loss function is for the model combining LSTM and convolutional layers, however, it was in the case of the model using residual LSTM that the smallest value of the loss function was reached for the validation set. Loss and MAE of neural network models are presented in Table 1.

Table 1
Loss and MAE of neural network models

Type of layer used	Train		Validation	
	Loss	MAE	Loss	MAE
LSTM	0,2023	0,0492	0,3880	0,1442
Residual LSTM	0,1918	0,0458	0,3877	0,1391
Conv	0,5713	0,1757	0,5959	0,2215
LSTM + Conv	0,1925	0,0451	0,4056	0,1568

Finally, the model architecture was chosen that reached the minimum value of the loss function for the validation set – Residual LSTM. The developed model architecture consists of two LSTM layers replacing each other, two LSTM residual layers and one LSTM layer that returns the sequence and one layer of densely connected neurons. The developed architecture of the artificial neural network model is presented in Table 1.

A 12 measurements from the PX4 data set were used to learn the artificial neural network model. Next, the neural network model in the transfer learning process was trained on a new dataset that was created on the test track presented in section 2.2. testing ground. 30 RC vehicle rides made up the set, of which 5 served as test data.

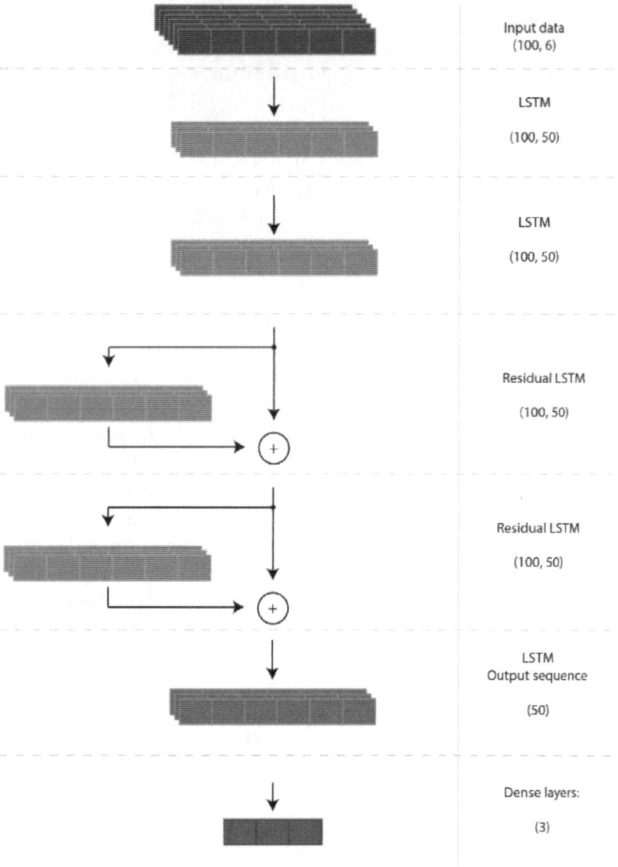

Fig 1. The developed architecture of the artificial neural network model

2.2 Testing ground

In order to carry out the measurements, a test site was prepared. A testing ground was created to carry out the experiment. The grass layer has been removed to create an uneven terrain. The testing ground was prepared so that from the starting point to about halfway along the track there is a decrease in elevation and then an increase. The test track is shown in Figure 2.

Fig. 2. Test track and direction of vehicle movement

Lighthouses were placed at the corners of the test site to determine the position of the vehicle. Lighthouse system is a method of determining a position is 3d space. It uses infrared (IR) light planes rotated by the Lighthouse base stations' drums (Taffanel et al., 2021). This signal is received by the IR receivers on the Lighthouse sensor board mounted on the RC car. Conducted by (Bauer et al., 2021) studies have shown that the lighthouse system is characterized by a repeatability of several millimeters. Accuracy is significantly affected by system calibration, the tilted reference plane, and the changing hierarchy of the system. The effects of bias translate into an increase in positioning error greater than 1 cm.

2.3 RC car and data collection

To conduct the test, a remote-controlled test vehicle was created. A radio frequency controller is used by the user to control the vehicle. The front axle of the car utilizes a wishbone suspension, while the back axle uses leaf springs. The drive shaft connects the drive axle in the rear axle to the drive from the DC motor. The RC controller, which regulates the input voltage to the DC motor and communicates commands to the servo in charge of spinning the wheels, Additionally, the vehicle is fitted with an IMU device that gathers data on acceleration and angular acceleration. Figure 3 displays a photograph and diagram of the created remote-controlled car.

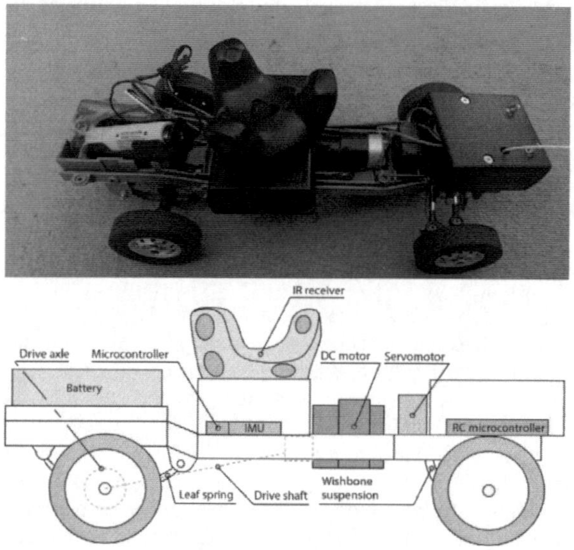

Fig. 3. Developed RC vehicle and its diagram

The IMU is connected to a microcontroller equipped with a WIFI module. It is used for data transmission. The data is sent to the PC computer that serves as the data collection unit via the User Datagram Protocol (UDP). An IR receiver has been installed in the axis of the accelerometer, which receives infrared light to determine the position of the vehicle. These data are also sent to the data processing unit via the RF communication method. Figure 4 shows the data transmission diagram.

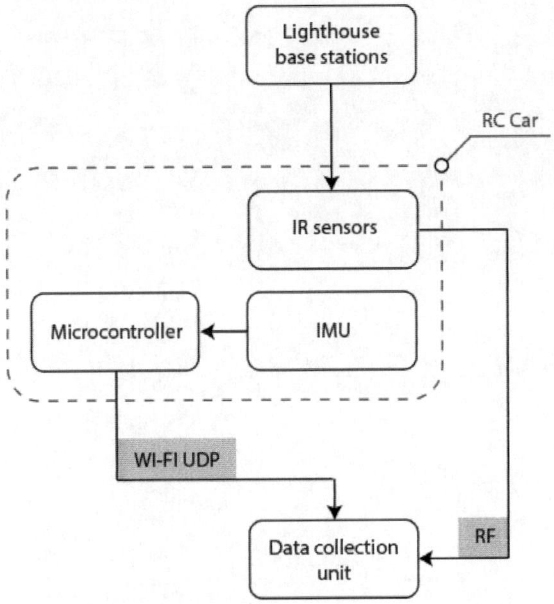

Fig. 4. Data transmission diagram

A series of measurements were carried out to train the neural network model. In each of them, the vehicle followed the route shown in Figure 2.

3. Results

The test of the developed neural network model was carried out on the previously described track. The vehicle was controlled remotely by the user, therefore the routes of the vehicles were different. The average velocity for all measurements was 0.8 m/s. The vehicle position estimation error for the x, y and z axes was calculated on the basis of the mean absolute error. For whole test measurements, the x-axis error is 0.0556 m, the y-axis error is 0.0717 m, and the z-axis error is 0.0213 m. Figure 5 shows the route of the vehicle determined on the basis of the estimation of the artificial neural network model and the lighthouse system.

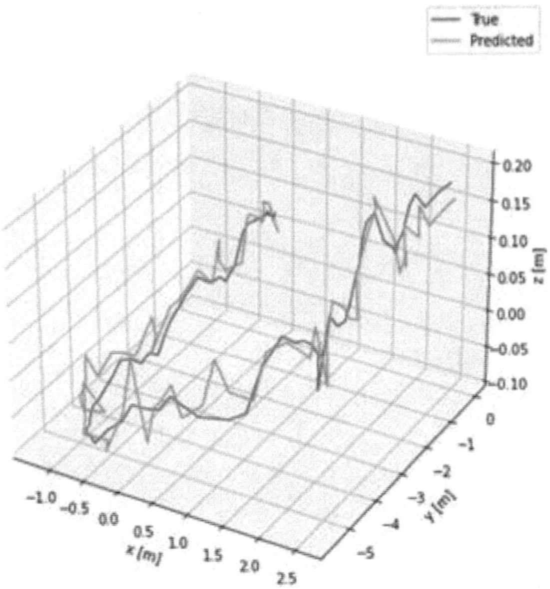

Fig. 5. Route of the vehicle determined on the basis of the estimation of the artificial neural network model and the lighthouse system

4. Conclusions and future work

The use of the developed neural network model for the purpose of estimating the displacement of the vehicle makes it possible to determine the position of the vehicle. The position determination is not accurate and has a moderate error. The tests carried out showed that the largest error occurs for the z axis. This inaccuracy might be brought on by both small-scale and significant impact of surface unevenness. The presented system is characterized by an error appearing for the static state - when the vehicle is not moving but there are various types of vibrations related to the environment. Position estimation based on acceleration data and gyroscope indications is not accurate enough. In addition, the tests were carried out on a short section of the road, in the case of longer sections the estimation error may be larger.

The presented solution can be used as part of a larger system combined with visual odometry to navigate an autonomous vehicle.

Further work on the system will include the creation of neural network models, which, based on a stereo camera, will allow for estimating the depth and determining the displacement of the vehicle based on visual data. The developed model can be used as an element of the optimization algorithm to increase the accuracy of visual odometry system.

5. Declaration of competing interest

The authors declare that they have no known competing financial interests or personal relationships that could have appeared to influence the work reported in this paper.

6. References

AbdulMajuid, A., Mohamady, O., Draz, M., El-bayoumi, G., 2021. GPS-Denied Navigation Using Low-Cost Inertial Sensors and Recurrent Neural Networks.

Aqel, M.O.A., Marhaban, M.H., Saripan, M.I., Ismail, N.B., 2016. Review of visual odometry: types, approaches, challenges, and applications. Springerplus 5. https://doi.org/10.1186/S40064-016-3573-7

Bauer, P., Lienhart, W., Jost, S., 2021. Accuracy Investigation of the Pose Determination of a VR System. Sensors 2021, Vol. 21, Page 1622 21, 1622. https://doi.org/10.3390/S21051622

Buchanan, R., Agrawal, V., Camurri, M., Dellaert, F., Fallon, M., n.d. Deep IMU Bias Inference for Robust Visual-Inertial Odometry with Factor Graphs.

Chen, C., Zhao, P., Lu, C.X., Wang, W., Markham, A., Trigoni, N., 2018. OxIOD: The Dataset for Deep Inertial Odometry.

Dai, H. fa, Bian, H. wei, Wang, R. ying, Ma, H., 2020. An INS/GNSS integrated navigation in GNSS denied environment using recurrent neural network. Defence Technology 16, 334–340. https://doi.org/10.1016/J.DT.2019.08.011

Flight Review [WWW Document], n.d. URL https://review.px4.io/browse (accessed 5.13.23).

Geiger, A., Lenz, P., Stiller, C., Urtasun, R., 2013. Vision meets robotics: The KITTI dataset. International Journal of Robotics Research 32, 1231–1237. https://doi.org/10.1177/0278364913491297/ASSET/IMAGES/LARGE/10.1177_0278364913491297-FIG11.JPEG

Herath, S., Yan, H., Furukawa, Y., 2019. RoNIN: Robust Neural Inertial Navigation in the Wild: Benchmark, Evaluations, and New Methods. Proc IEEE Int Conf Robot Autom 3146–3152. https://doi.org/10.1109/ICRA40945.2020.9196860

Kim, J., El-Khamy, M., Lee, J., 2017. Residual LSTM: Design of a Deep Recurrent Architecture for Distant Speech Recognition. Proceedings of the Annual Conference of the International Speech Communication Association, INTERSPEECH 2017-August, 1591–1595. https://doi.org/10.21437/Interspeech.2017-477

Kok, M., Hol, J.D., Schön, T.B., 2017. Using Inertial Sensors for Position and Orientation Estimation. Foundations and Trends in Signal Processing 11, 1–153. https://doi.org/10.1561/2000000094

Langlais, B., Thébault, E., Houliez, A., Purucker, M.E., Lillis, R.J., 2019. A new model of the crustal magnetic field of Mars using MGS and MAVEN. J Geophys Res Planets 124, 1542. https://doi.org/10.1029/2018JE005854

Liao, Y., Xie, J., Geiger, A., 2021. KITTI-360: A Novel Dataset and Benchmarks for Urban Scene Understanding in 2D and 3D. IEEE Trans Pattern Anal Mach Intell 45, 3292–3310. https://doi.org/10.1109/TPAMI.2022.3179507

Pelletier, C., Webb, G.I., Petitjean, F., 2019. Temporal Convolutional Neural Network for the Classification of Satellite Image Time Series. Remote Sensing 2019, Vol. 11, Page 523 11, 523. https://doi.org/10.3390/RS11050523

Powell, M.W., Crockett, T., Fox, J.M., Joswig, J.C., Norris, J.S., Rabe, K.J., n.d. Targeting and Localization for Mars Rover Operations.

Staudemeyer, R.C., Morris, E.R., 2019. Understanding LSTM-a tutorial into Long Short-Term Memory Recurrent Neural Networks.

Taffanel, A., Rousselot, B., Danielsson, J., McGuire, K., Richardsson, K., Eliasson, M., Antonsson, T., Hönig, W., 2021. Lighthouse Positioning System: Dataset, Accuracy, and Precision for UAV Research.

Yan, H., Shan, Q., Furukawa, Y., 2017. RIDI: Robust IMU Double Integration. Lecture Notes in Computer Science (including subseries Lecture Notes in Artificial Intelligence and Lecture Notes in Bioinformatics) 11217 LNCS, 641–656. https://doi.org/10.1007/978-3-030-01261-8_38

Yusefi, A., Durdu, A., Aslan, M.F., Sungur, C., 2021. LSTM and filter based comparison analysis for indoor global localization in UAVs. IEEE Access 9, 10054–10069. https://doi.org/10.1109/ACCESS.2021.3049896

Zeinali, B., Zandizari, H., Chang, J.M., 2022. IMUNet: Efficient Regression Architecture for IMU Navigation and Positioning.

0448 | Lunar rover discrete element method study and calibration

James Hurrell [a,*], Keisuke Takehana [a], Kentaro Uno [a], Kazuya Yoshida [a]

[a] *Department of Aerospace Engineering, Tohoku University, Sendai, Miyagi, Japan*

**Corresponding author hurrell.astro@dc.tohoku.ac.jp*

ABSTRACT

The Discrete Element Method (DEM) is a high-fidelity simulation tool to study the interaction of granular material. The single-wheel interaction of a lunar rover wheel with Toyoura sand is studied. Traditionally Toyoura sand is considered cohesionless but is measured to have a cohesion similar to Lunar soil. This study aimed to develop a well-calibrated DEM wheel-soil simulation with and without cohesion for an equivalent angle of repose (AoR). 3D DEM single-wheel simulations are performed for Toyoura sand in Earth conditions. Separate calibration for the simulation parameters uses AoR. The Johnson–Kendall–Roberts (JKR) model is used, allowing for particle cohesion modelling. Simulations are kept homogeneous and use spherical particles. A comparison is made between non-cohesive and cohesive simulations. Traction coefficient, wheel sinkage, wheel trace and soil flow under the wheel are examined. The static friction coefficient, rolling friction coefficient and cohesion have the most significant influence on an AoR. The optimisation model determines ideal parameters for simulating AoR that match experimental values. When applied to single-wheel simulations, there is a slight reduction in traction coefficient for cohesive soil. Differences in sinkage, wheel trace and soil motion are negligible. Alongside the parameter calibration performed, the results of the simulations are considered suitable for traction performance. Improvements can be made with soil depth and consideration of void ratio. For a homogenous soil, cohesion may not be critical to the soil response. Future studies aim to use direct comparison to experiment, reduced gravity experiments and lunar simulant results to refine the simulations.

Keywords
Angle of repose
Calibration
Discrete element method
Single-wheel test

Introduction

The focus of extraplanetary terramechanics has been on Lunar and Martian rover systems. Typically, this concerns small, slow-moving autonomous robotic vehicles. Drive systems and suspension are uniquely designed, with a limit to performance due to weight and power requirements. Automation is typically required, with an expected lack of human interaction if issues arise.

An often-used example of the requirement to study wheel-soil interaction is the Spirit and Opportunity rovers which experienced severe mobility issues and highlighted the need to prepare well for extra-planetary missions. Furthermore, given the limits on power availability, maximising and understanding the wheel performance aids in motor selection and can be used to improve rover efficiency (Wong, 2022).

Over the past several years, space agencies and private companies have shown an ever-increasing interest in and usage of the lunar surface, with the private attempts to land by Space IL and ispace. However, the only successful landings have been by Russia, America and China, highlighting the difficulty in exploring this remote environment.

Efforts are being made to explore the effect of lunar gravity on soil behaviour with parabolic flight experiments replicating lunar gravity (Daca et al. 2020; Kobayashi et al. 2010). However, this remains difficult as there is a lack of ability to fully capture or even fully understand the lunar conditions experimentally with limited lunar performance data due to understandable accessibility issues.

This paper summarises the development of a 3D discrete element method (DEM) simulation of wheel soil interaction for lunar study and investigation of lunar conditions. The focus has been to ensure a robust methodology for calibrating soil properties and studying single-wheel interaction with the DEM method.

The DEM method has been widely used in various applications, from pharmaceuticals and mining to agriculture and planetary exploration. Despite granular material composing such a wide range of fields, the base physics understanding of granular material is limited and typically understood through the bulk behaviour (O'Sullivan, 2011).

DEM is considered to have been initiated by Cundall and Strack (1979), detailing modelling a system of spherical particles with contact and subsequent motion constrained by a system of equations.

DEM is one of several methods that can model surface and wheel interaction. This includes continuum approaches and other numerical methods, including DEM. The field of terramechanics is considered to have originated with the works of Bekker in 1956 and subsequent work with the Bekker-Wong model of wheel-soil interaction, a semi-empirical method dedicated to modelling the wheel-soil interaction (Wong, 2022). Other numerical approaches include Finite Element Method (FEM) and Soil Contact Model (SCM), which both involve modifying a meshed surface due to wheel interaction.

Many of these approaches can replicate certain bulk behaviours. Yet, for studying the mechanisms, the interaction between particles and observing particle motion is only possible in a particulate method such as DEM (O'Sullivan, 2011). With the development of high-speed rovers and future human presence on the lunar surface, DEM can be used to study high wheel speed conditions (Rodriguez-Martinez et al., 2019). In addition, studies have found that DEM can better predict wheel performance than Bekker methods (Smith, 2013). The DEM can further be used to

https://doi.org/10.56884/XUAC7188
ISBN 978-1-942112-55-6 (pdf)

study the gravitational influence directly on the soil particles and subsequent motion.

Toyoura sand is the focus of developing the current simulations due to its homogeneity and availability. In the context of soils in Earth-based terramechanics, Toyoura sand is typically considered cohesionless. However, from analysis, there is slight cohesiveness similar to many lunar soil samples and simulants.

Simulations will consider modelling the soil with and without cohesion while maintaining calibration of parameters to see how this affects the wheel-terrain response.

The calibration process developed is supported by that demonstrated by Zhu et al. (2022) and the study of design of experiments (Douglas, 2013). The calibration process has two main steps, screening and optimisation. This is based on the design of experiments methodology, which uses fractional factorial experiment design to get results with a minimum number of simulations. A minimum number of DEM simulations is ideal due to the computational cost of DEM simulations. The screening approach design is a regular two-level factorial design, and the optimisation approach is the Box-Behnken (Douglas, 2013).

While ideally the wheel simulation would be calibrated directly, this is not feasible due to the time taken for a wheel simulation. However, several experiments are used to study soil behaviour and characterise soil properties that can be used to generate the simulation parameters. These include angle of repose (AoR), triaxial or direct shear, cone penetrometer and bevameter experiments. Using these experiments requires a smaller particle count, which means simulations of these experiments can be run much faster than wheel simulations.

Commonly used for DEM simulation as an experiment for calibration of soil properties is the AoR. There are two forms of AoR. Dynamic AoR involves continuously moving material, for example, in a rotating cylinder or drum, while static AoR is a soil pile generated and left to rest.

Three main methods to generate a soil pile for AoR are defined by Beakawi and Baghabra (2018) as the tilting box, fixed funnel and hollow cylinder methods. Static AoR is defined as the angle between the horizontal and maximum slope of its surface that a soil assumes through natural processes (He et al. 2020). In this work, the static AoR is generated by the fixed funnel method and used to calibrate soil properties.

Triaxial testing and the required preparation for a triaxial test are used to obtain data on the soil parameters for simulation.

Furthermore, this is a continuation of work developed in the space robotics lab and aims to further develop the terramechanics simulation capabilities, complementing experimental work. Ono et al. (2020, 2021) most recently developed simulations to study the soil state during wheel motion. Aside from the software and experiments used, the key difference is to develop 3D simulations to improve from 2D, allowing a continuum study of particle motion across the wheel's profile.

It has been found that the overall response of a 2D system can be qualitatively related to the 3D case. Replication of bulk behaviour is observed, as seen in previous work (Ono et al., 2021). However, reducing a 3D problem to 2D loses the contact complexity and displacement of particles. As a result, it may not be possible to draw quantitative comparisons between 2D simulations and 3D experiments (O'Sullivan, 2011).

Simulations are kept to spherical particles and a single particle size with coarse-graining enabled to improve simulation efficiency.

1. Experimental data

1.1 *Angle of repose experiment*

The static AoR generated for this study used the fixed funnel method. A set volume and mass of material is poured through a funnel forming a pile below as shown in Fig. 1. A value is then measured for AoR by Eq. 1, related to the internal friction angle (Beakawi and Baghabra, 2018).

$$AoR = \arctan\left(\frac{AB}{HC}\right) \quad (1)$$

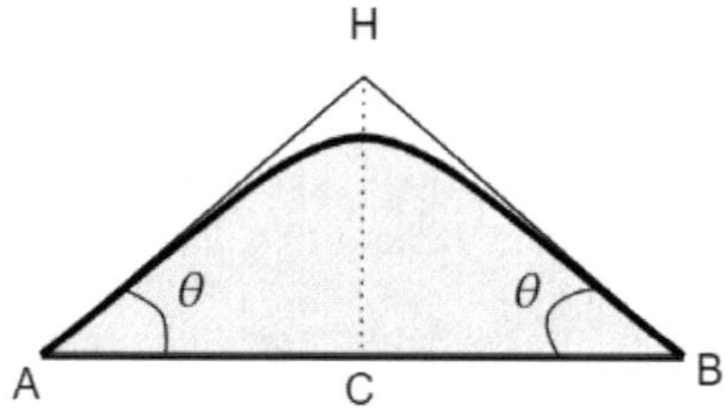

Fig. 1. Angle of repose measurement

1.2 *Single-wheel experiment*

Simulations were based on an experimental testbed for future comparisons. The system allows for change in the load force and setting of slip ratio by controlling angular and horizontal wheel velocity.

The maximum speed of the testbed is 0.36 km/h and the sandbed dimensions are 1.6 m in length, 0.3 m wide and typically 0.12 m depth of material. In comparison, the simulation is set to a sandbed of 0.4 m long, 0.1 m wide and 0.07 m deep. The simulation replicates the experimental area in 25% of length, 33% of width and 58% of depth. Figure 2 shows an overview towards the end of the simulation for 20% slip.

1.3 *Wheel experiment inputs*

Single-wheel experiments can be conducted for a variation of load force, angular velocity, and slip ratios. The settings replicated in the simulation are shown in Table 1.

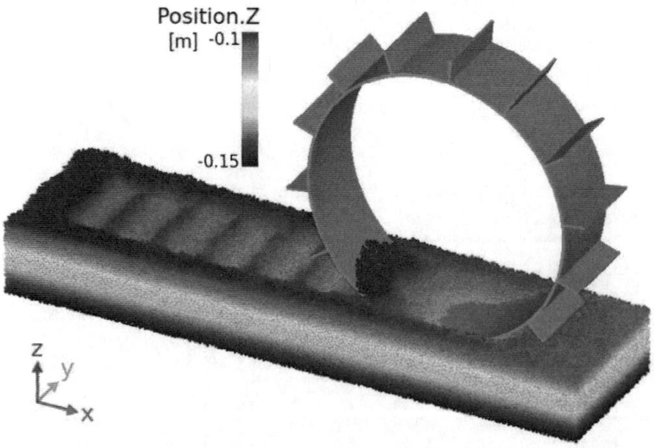

Fig. 2. Single wheel simulation overview for 20% slip, particles coloured by position on the z-axis

Table 1
Single-wheel experimental parameters

Parameter	Value	Units
Wheel Diameter	180	[mm]
Wheel Width	30	[mm]
Wheel Mass	0.1	[kg]
Load Force	25	[N]
Slip Ratio	0,10,20	[%]
Horizontal Velocity	0.02	[m/s]
Angular Velocity	0.222	[rad/s]

The wheel slip can be set at a fixed value. The slip ratio is a value that can characterise the wheel's travelling state. For simplification, slippage of a wheel without grousers is used as defined in Eq. 2:

$$i = \frac{r_w\omega - v_x}{r_w\omega} \tag{2}$$

where r_w is wheel radius, ω is angular velocity and v_x is the translational velocity.

1.4 Wheel experiment measurables

Several aspects of the wheel simulation are measured. Sinkage, traction coefficient, wheel trace, soil motion under the wheel and the void ratio can be considered for the simulated sandbed.

Dynamic sinkage is measured in the experiment and replicated in simulation. This is sinkage measured from when the wheel is at rest given by Eq. 3. Where d is the wheel's vertical centre of mass at rest, s_w is the soilbed height and r_w is the wheel radius.

$$z = d - (s_w - r_w) \tag{3}$$

The force of the particles on the wheel is measured in the simulation and can be split into the force along each axis. The net force in the x direction is given as F_x; dividing this value by the load force, F_z gives the traction coefficient, μ shown in Eq.4.

In the context of wheel performance, the traction coefficient is equivalent to the drawbar pull coefficient, which is the ratio of drawbar pull to rover weight. Drawbar pull is a commonly used value for vehicle performance (He et al., 2020).

$$\mu = \frac{F_x}{F_z} \tag{4}$$

Wheel trace is the track left behind the wheel after motion. The trace for a grousered wheel leaves a distractive pattern that can be measured. The distance between two ruts or depressions in the track is measured and this is given as the wheel trace value.

The void ratio is a measure of the soil condition. If the value is high, the soil is considered in a loose state; if low, then the soil is in a dense state. A minimum and maximum value of void ratio was experimentally measured for Toyoura sand. Bulk density is a similar term that can be used.

$$V = \frac{V_V}{V_S} = \frac{V_T - V_S}{V_S} \tag{5}$$

The void ratio can be calculated as the ratio of the void volume, V_V and void solids, V_S. Void volume can be calculated from the difference in the total volume, V_T and void solids as shown in Eq. 5.

Particle image velocimetry (PIV) is an image analysis technique used to visualise flow in a system. It is typically used in fluid dynamics, but there have been multiple cases of its use in the study of granular material experimentally and from simulated data. In this case, it is used to study the flow of particles, and the soil motion, under the wheel. The result is displacement vectors between images of the general particle motion. This can be visualised as a colour map.

2. Discrete element method simulation

The DEM simulations were developed using Aspherix, a commercial DEM simulation framework and successor to LIGGGHTS, an open-source code previously used in the laboratory's studies.

Initially, a DEM simulation requires the setting of the system geometry, including the boundary conditions, particle locations, contact model and particle properties. A simulation progresses through with a specific timestep where contacting particles are identified and the force between contacts is calculated. The resultant change in momentum and resultant velocities is calculated and the simulation progresses. (O'Sullivan, 2011)

In a wheel simulation, directly stimulating particle size is not feasible as this would require several million particles. Particles are then scaled using coarse-graining and are set to be homogenous as Toyoura sand has a narrow size distribution.

2.1 Discrete element method contact model

The basis of a DEM simulation is the consideration of a contact as a spring dashpot model with dampers and a slider. This can be split into normal, tangential and rolling friction forces modelled in the simulation, with the associated dampening force for normal and tangential force.

Typically used in literature is the Hertz-Mindiln model, which is suitable for non-cohesive soil. A development of this is the Johnson, Kendall, Roberts (JKR) model. This allows for consideration of cohesion within the particle contact as an attractive force.

Force for the JKR model particle contact is given by Eq. 6. If the value of cohesion, ψ is 0, then the JKR model reduces to the Hertz-Mindilin model. Equation 7 gives the particle overlap in the normal.

$$F_{JKR} = \frac{4E^*a^3}{3r^*} - 2\sqrt{2\pi E^*\psi a^3} \tag{6}$$

$$\delta_{JKR} = \frac{a^2}{r^*} - \sqrt{\frac{2\pi\psi a}{E^*}} \tag{7}$$

The equivalent Youngs modulus, E* and equivalent radius r* can be calculated as follows:

$$\frac{1}{E^*} = \frac{1-v_i^2}{E_i} + \frac{1-v_j^2}{E_j} \tag{8}$$

$$\frac{1}{r^*} = \frac{1}{r_i} + \frac{1}{r_j} \tag{9}$$

E and v are the Youngs modulus and Poisson's ratio respectively for two particles.

The tangential force can be calculated, considering tangential overlap, δ_t for Eq. 10:

$$F_t = k_t\delta_t \tag{10}$$

Normal, k_n and tangential, k_t spring stiffness is calculated as:

$$k_n = \frac{4}{3}E^*\sqrt{r^*\delta_n} \tag{11}$$

$$k_t = 8G^*\sqrt{r^*\delta_n} \tag{12}$$

For Eq. 11 and 12, δ_n is equivalent to δ_{JKR} and the equivalent shear modulus, G* is given by Eq. 13:

$$\frac{1}{G^*} = \frac{2(1-v_i)(1+v_i)}{E_i} + \frac{2(1-v_j)(1+v_j)}{E_j} \quad (13)$$

The normal, γ_n and tangential, γ_t dampening force is given as:

$$\gamma_n = -2\sqrt{\frac{5}{6}}\beta\sqrt{S_n m^*} \quad (14)$$

$$\gamma_t = -2\sqrt{\frac{5}{6}}\beta\sqrt{S_t m^*} \quad (15)$$

Where terms β, S and equivalent mass, m* are given by:

$$\beta = \frac{\ln e}{\sqrt{\ln^2 e+\pi^2}} \quad (16)$$

$$S_n = 2E^*\sqrt{r^*\delta_n} \quad (17)$$

$$S_t = 2G^*\sqrt{r^*\delta_n} \quad (18)$$

$$\frac{1}{m^*} = \frac{1}{m_i} + \frac{1}{m_j} \quad (19)$$

e and m are the coefficient of restitution and particle mass, respectively.

Equation 20 gives the resultant force, F, acting on a particle due to a collision, including particle mass, m and gravity, g:

$$F = F_{JKR} + F_t + \gamma_n + \gamma_t + mg \quad (20)$$

Static friction coefficient, μ_s places an upper limit on the tangential force from the Coulomb criterion. This limits tangential force, F_t with normal force, here F_{JKR}, shown in Eq. 21.

$$F_t \leq \mu_s F_{JKR} \quad (21)$$

Rolling friction, μ_r is modelled as a resistive torque that limits the moment of the particle related to the normal force, F_{JKR}

$$M \leq \mu_r {r^*}^2 F_{JKR} \quad (22)$$

3. Simulation calibration

A DEM simulation requires the setting of several parameters. The calibration of these parameters is critical to a valuable and informative simulation. Combinations of parameters can form several solutions to get an appropriate response. As such, it is important to consider the values in the context of the experiment.

Calibration is perhaps avoided in some studies to focus on using DEM itself to generate simulations. This is understandable due to the complexity of DEM simulation and the lack of a standardised approach to validating and calibrating simulations. Despite this, confidence in the accuracy of predictions based on the simulation can only be achieved with careful consideration of the input parameters.

Several efforts have been made to identify experiments and methodologies to perform effective calibration, as Coetzee (2017) reviewed. The two main approaches identified are to measure the parameter value directly at a particle or contact level or measure the bulk behaviour of a particle assembly and reverse calibrate, iteratively changing the parameter values until the bulk response is repeated. Comparison with experimental results is typical to verify the simulation results (Coetzee, 2017).

The approach used in this study is to measure and use several parameter values at a particle level, informing initial parameter values and then using experiments to study the bulk behaviour in simulation and attempt to replicate the response.

Design of experiments is the methodology used to investigate parameter values. It implements the most efficient use of the experiment data by using factorials. As a result, all observations of a result are used to calculate the interaction effects (Douglas, 2013).

A full factorial design considers all possible combinations of parameters and their levels. Due to the large number of runs required to consider all possible combinations, fractional factorial designs are used. This is for screening the parameters, where it is desired to identify the key parameters that influence the result. Subsequent analysis can then be used in optimising the parameters.

3.1 Calibration process

The calibration process first uses results from the AoR and triaxial experiment preparation. This allows for a combination of direct measurement of the simulant particle and contact properties and bulk properties. In preparation for the triaxial experiment, the size distribution, the particle density and minimum and maximum bulk density are measured. The triaxial experiment infers a value of cohesion and internal friction angle.

The next step is calibrating the particle values using the AoR, comparing simulation and experimental values. Several measured particle properties are directly used in the initial simulations.

A screening experiment is where it is desired to identify which parameters significantly affect the experiment response.

For screening, a regular two-level factorial design of resolution IV can be used to identify the effects of parameters and interactions between the parameters.

Box-Behnken is known as a three-level factorial design. The results of a model and associated equation developed from the parameter values and their influence on the response can indicate optimal parameters for the desired value of a response.

4. Results and discussion

4.1 Angle of repose calibration

A soil pile formed for AoR is shown in Fig. 3. The soil pile is formed from 92 grams of Toyoura sand with a funnel sized 12 mm at the discharge point. AoR was generated multiple times, giving 34.6, 34.1, 34.3, 33.6, 33.7 , and 33.4. An average value is determined as 34±1 degrees for Toyoura sand. Measured parameters for Toyoura sand are shown in Table 2. The narrow size distribution of Toyoura sand supports treatment as a homogenous particle size for simulation.

Fig. 3. Static angle of repose generated by the funnel method

Table 2
Measured Toyoura sand parameters

Type	Value	Units
Particle diameter	0.15-0.3	[mm]
Density	2650	[kg/m³]
AoR	34±1	[degrees]
Minimum void ratio	0.596	
Maximum void ratio	0.934	

4.2 Parameter screening results

The screening was performed twice, first without cohesion and then considering cohesion. Both screenings required 16 simulations considering high and low values for 7 or 8 parameters.

The results of considering the effects of parameters on AoR without cohesion are shown in Table 3. Static friction, rolling friction and static friction between the soil and mesh strongly affect the AoR. It is interesting to observe that friction between the soil and mesh is important and seen as the friction between the soil and the plate on which the AoR is formed. When this value is low, forming a soil pile is not possible.

The second screening then included cohesion to see how this could further affect the AoR, summarising the results in Table 4. It was found that cohesion, rolling friction and static friction are the main factors, although the soil-mesh static friction still has an influence. This agrees with prior studies where AoR depends on static friction, rolling friction and cohesion (Coetzee, 2017). As such, cohesion, static friction and rolling friction are used for parameter optimisation.

Table 3
Parameter screening without cohesion

Type	Symbol	Low	High	Contribution
Youngs modulus	E	4	7	0.6
Poisson's Ratio	v	0.1	0.4	0.4
Restitution coefficient	e_1	0.1	0.7	1
Static friction coefficient	μ_s	0.1	0.7	27.8
Rolling friction coefficient	μ_r	0	0.7	17.5
Soil-mesh restitution coefficient	e_2	0.1	0.5	3.1
Soil-mesh static friction coefficient	μ_3	0.1	0.5	23.3
Soil-mesh rolling friction coefficient	μ_4	0	0.7	5.6

Table 4
Parameter screening with cohesion

Type	Symbol	Low	High	Contribution
Youngs modulus	E	5	90	0.4
Cohesion	ψ_1	0	1	34.6
Soil-mesh cohesion	ψ_2	0	1	1.8
Static friction coefficient	μ_s	0.1	0.9	14.4
Rolling friction coefficient	μ_r	0.1	0.9	21.2
Soil-mesh static friction coefficient	μ_3	0.1	0.9	6.6
Soil-mesh rolling friction coefficient	μ_4	0.1	0.9	2.2

4.3 Parameter optimisation results

Box-Behnken is used for finding optimal values for replicating an experimental value of AoR in simulation. It is a three-level factorial design. Low, high and central values are used. The results of the developed model can be used to indicate optimal parameters for the desired value of a response.

Considering three key parameters identified in screening, static friction, rolling friction and cohesion, twelve simulation runs are needed, with an additional 2-5 to study behaviour at the central values.

Originally a value of 1 was used as the limit for cohesion in screening, but this could block the funnel at high static friction and rolling friction. The value of cohesion was reduced to 0.1 and with a value of 0.4 rolling friction, it gave a value of 45 degrees. This was above the 34 degrees expected for Toyoura sand, so these values were used as the upper limit for the optimisation.

The results of the optimisation with parameter values and AoR measured are shown in Table 5. It was noted that 0.7 static friction, 0.2 rolling friction and 0.05 cohesion gives an AoR of 34.2 degrees, which could be used as the non-cohesion values for the AoR and wheel simulation.

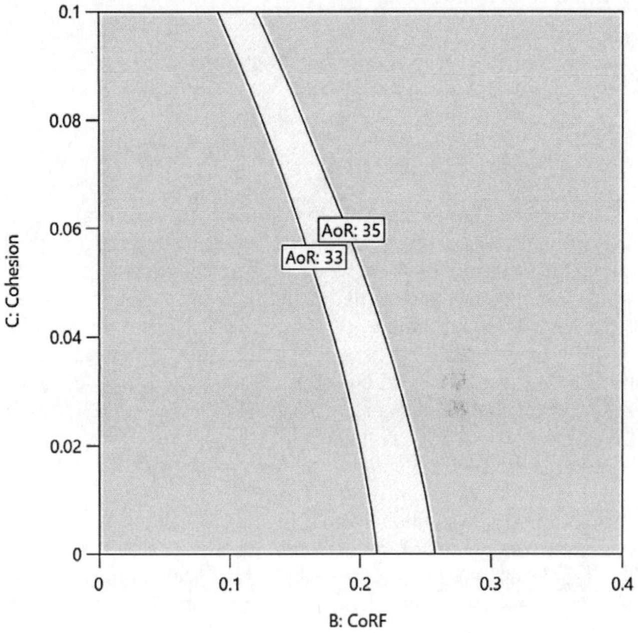

Fig. 4 Coefficient of rolling friction and cohesion modelled relationship

From the analysis of variance, the model's regression equation is given as Eq 23.

$$AoR = 34.72 + 0.0125\mu_s + 10.2\mu_r + 3.54\psi_1 - 1.18\mu_s\mu_r - 1.4\mu_s\psi_1 - 0.125\mu_r\psi_1 + 1.81\mu_s^2 - 3.91\mu_r^2 + 1.26\psi_1^2 \quad (23)$$

The effect of static friction is overshadowed as it has a low coefficient in Eq. 23 compared to rolling friction and cohesion, which have high value coefficients. It can be interpreted that in the 0.5-0.9 range of values, static friction is sufficiently high or that the inclusion of cohesion as part of the contact model results in it being a dominating parameter in the interaction.

Equation 23 can be solved for a set AoR to give parameter values. As a solution can be solved for a range of values static friction was set at a value of 0.68. The value of 0.68 is derived from the internal friction angle, which can be considered equivalent to the AoR value of 34 degrees. The modelled relationship between cohesion and rolling friction can then be examined and is shown in Fig. 4.

Table 5
Parameter optimisation AoR results

Run	μ_s	μ_r	ψ_1	AoR
1	0.7	0.2	0.05	34.6
2	0.9	0.2	0.1	39.2
3	0.7	0.2	0.05	35.1
4	0.7	0.2	0.05	34.6
5	0.5	0.2	0	33.6
6	0.5	0.2	0.1	43.8
7	0.7	0.4	0.1	44.9
8	0.7	0.2	0.05	35.1
9	0.9	0.4	0.05	43.2
10	0.9	0	0.05	23.9
11	0.7	0	0	19
12	0.5	0.4	0.05	43.7
13	0.7	0.4	0	38.4
14	0.7	0.2	0.05	34.2
15	0.7	0	0.1	26
16	0.9	0.2	0	34.6
17	0.5	0	0.05	19.7

The results of the calibration model show the importance and interaction between rolling friction and cohesion in generating the AoR. Using the model at the edges is unsuitable for the Box-Behnken model as it is designed to optimise for parameter values at the centre of the study (Douglas, 2012). This is evident in confirmation checks of parameter values and the resulting AoR summarised in Table 6. When using the model to propose values at the edge of modelled cohesion, 0 and 0.1 in tests 1 and 2, the AoR values are 37.0 and 35.4, respectively.

When tested with a central cohesion value, test 3, a value of 34.2 is found for AoR matching the experimental value. The value for non-cohesion was taken from screening, test 4 and tested with 0.68 static friction and found to match experimental AoR at 33.6.

Tests 3 and 4 values were used in the wheel simulation for cohesive and non-cohesive modelling of Toyoura sand, respectively.

Table 6
Confirmation of angle of repose for the optimisation model

Parameter	Test 1	Test 2	Test 3	Test 4
Static friction	0.68	0.68	0.68	0.68
Rolling friction	0.23	0.106	0.186	0.2
Cohesion	0	0.1	0.05	0
Angle of repose (AoR)	37.0	35.4	34.2	33.6

4.4 Single-wheel simulation results

Following the optimisation, values used for the simulation parameters in the wheel simulations are shown in Table 7. Particle radius is scaled using coarse-graining by 10 to reduce particle count.

In the wheel simulation, to fix the slip ratio, the wheel is set to an angular momentum matching the required angular velocity and the required translation velocity matching the experimental values. The load force was set to 25 Newtons.

The wheel can move on the vertical, z-axis, but no movement is allowed on the y-axis. The wheel then moves with a set motion on the x-axis.

Simulation results are compared over the same time so that the periodic nature of F_x, due to the changing grouser interaction with the soil over time, does not affect the averaged result.

Considering the F_x for 0% slip, the simulation replicates the expected periodic nature of the F_x values due to the grouser interaction. A maximum in F_x is achieved when a new grouser interacts with the soil; the average value is just before the central grouser is vertical and a minimum is found before the rear grouser exits the soil. The period of the motion is approximately 2.1 seconds. Furthermore, the wheel traces or distances between ruts is measured and found to match experimental values.

Figure 5 shows the traction coefficient and sinkage, respectively. The wheel simulation values for the cohesive and non-cohesive are similar, with a slight reduction in traction coefficient for the cohesive simulation. Figure 6 shows the results of PIV for a grouser angle of zero, where the centermost grouser is aligned to the vertical. Soil motion at the centre of the wheel can be compared for soil with and without modelled cohesion. Little difference can be seen visually in the magnitude of velocity or shape of the soil deformation, except at the rear of the wheel. Directly looking at velocity under the wheel finds similar values. Together with the traction and sinkage results, this suggests that cohesion may be not critical to the soil response for Toyoura sand treated as a homogenous soil. Where particles are spherical with homogeneous size and physical properties. However, further investigation of how balancing cohesion with rolling friction for the same AoR should be investigated to see if a trend continues.

An issue found with the simulation was insufficient depth, noted when checking the particle motion under the wheel. Particles would be forced into the bottom of the simulated testbed, which could result in a higher than expected force on the wheel and lower sinkage.

Initial testing of the simulation at 20% slip with sufficient depth to allow for particle motion without the influence of the base of the testbed has a 13% reduction in traction coefficient and a 17% increase in sinkage.

When the void ratio was measured for the simulated testbed, it was found to be at 0.875 compared to the experimental maximum of 0.934. This suggests loose soil in simulation as it is close to the maximum and may need to be reduced to match the actual experimental testbed condition. This can be done by modifying the particle density value or the initial sandbed packing.

Furthermore, decreasing the coarse-graining value from 10 to 6 was found in previous testing to slightly reduce the peak values of F_x. Combining sufficient depth, void ratio and particle scaling could improve the matching of simulation to experiment and will be investigated in future.

Table 7

Single-wheel simulation parameters

Parameter	Symbol	Value	Units
Particle radius (D_{50})	r	0.1	[mm]
Particle radius (scaled 10x)	r	1	[mm]
Particle number	N	236000	
Particle density	ρ	2650	[kg/m³]
Mesh density	ρ	1000	[kg/m³]
Youngs modulus	E	50	[MPa]
Poisson's ratio	v	0.29	
Cohesion	ψ_1	0/0.05	[J/m²]
Soil-mesh cohesion	ψ_2	0	[J/m²]
Static friction coefficient	μ_s	0.68	
Rolling friction coefficient	μ_r	0.2/0.186	
Soil-mesh static friction coefficient	μ_3	0.3	
Soil-mesh rolling friction coefficient	μ_4	0.1	
Coefficient of restitution	e_1	0.4	
Soil-mesh coefficient of restitution	e_2	0.4	

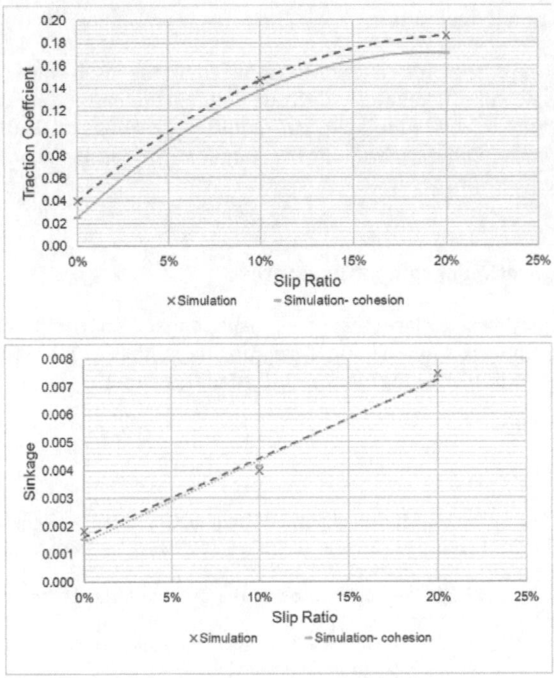

Fig. 5. Wheel simulation results comparing the effect of including cohesion for the same angle of repose, for traction coefficient and sinkage against slip ratio

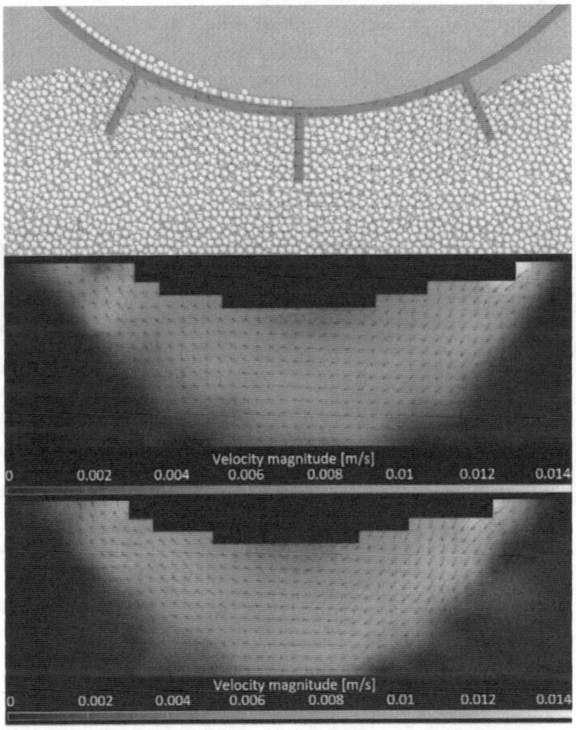

Fig. 6. Particle image velocimetry for soil motion under the wheel at a grouser angle of 0°. Position of the wheel (top) soil with no cohesion modelled (middle) and soil with modelled cohesion (bottom)

5. Conclusions and future work

Parameters calibrated to a simulation are unique to a simulation. They can inform subsequent simulations and similar work but should be re-calibrated to each simulation case. Directly comparing parameters to the experimental values is difficult and simulated values may not represent measurable values. Particle scaling is often used as performing a to-scale simulation is computationally restricted. As the rolling friction and cohesion models are influenced by particle size, these values will be influenced by the scaling and scaling should be kept low.

The calibration method was found to reliably produce parameters to generate an AoR. These parameters are based on a model generated by the design of experiments for optimisation using the Box-Behnken method and the interaction of cohesion, static friction and rolling friction coefficients. Difficulty was found when examining parameter values at the model's edge and the calibration optimises for values at the centre of the study. As such, the selection of suitable parameter ranges is essential.

Improvement to the calibration method could be made by screening for the effect of particle scale changes and changes that can be made to accommodate the void ratio. Changing the particle density is a simple way to modify the void ratio or by changing the initial sandbed packing. The use of size distribution in simulation could further aid in replicating experimental void ratio as could modelling the morphology of the soil particles.

Calibrated soil parameters were then applied to wheel simulations for the same AoR generated with and without cohesion. It was found that sensible results were generated concerning differences due to the slip ratio and the polynomial line fitted to the results at 0, 10 and 20% slip. For 0% slip, the simulation was close to the expected zero traction coefficient but offset. This is understood to be due to the influence of the wheel grousers. When considering soil motion, soil depth was found to be insufficient. With improvements to the simulation to allow sufficient depth for particle motion, it is found for 20% slip the traction coefficient is reduced by 13% and sinkage is increased by 17%.

The difference between simulating wheel motion with and without cohesion, when calibrated to the same value of AoR, was found to be minimal for traction coefficient, sinkage and the soil motion under the wheel. This could suggest that cohesion is not critical to the soil response, for the cases of Toyoura sand, which can be modelled as spherical particles with homogeneous size and physical properties. Many studies ignore the cohesion value for Toyoura sand (Ono et al., 2021; Wong, 2022. However, for more cohesive soils the lower traction coefficient and minor changes to soil flow morphology may increase and will be investigated.

Difficulty in using DEM to replicate experimental data is shown. Sensible wheel response can be simulated but the experiment must be clearly characterised and well understood to be suitably replicated.

AoR while an easy experiment to produce can be difficult to measure and have variations depending on the technique used to produce the AoR. In addition, the results will be independent of the sandbed condition. This highlights the difficulty of using AoR to calibrate wheel experiments.

AoR alone may be insufficient to fully calibrate the parameters for the wheel simulation. Future work will investigate using the cone penetrometer experiment with the calibration process as developed here. This may be better suited for calibrating a wheel simulation than the AoR, as cone penetrometer and wheel experiments can be performed in the same testbed. The soil condition, primarily void ratio, will then be considered. Ideally, a bevameter experiment would be used and should be considered in future work as this would better represent the wheel motion and shearing of the soil.

To understand the lunar regolith parameters, cone penetrometer experiments were performed with the Lunokhod rover, experiments during Apollo and analysed particle size distribution, chemical composition, geophysical properties and particle shape from returned samples (Grant et al., 1991). These values are then what lunar regolith simulants are typically based on depending on the desired use case. A lunar regolith simulant, FJS-1 the Fuji Japanese Simulant, is available in the lab. Furthermore, there is potential to consider an in situ value of AoR from rover wheel traces, craters or boot prints on the lunar surface.

Following corrections to the simulated soil bed depth, void ratio and particle scaling, the influence of lunar gravity and comparison with lunar regolith simulant experimental data can be investigated. It will be interesting to explore the effect of cohesion in the case of a heterogeneous soil such as lunar regolith and the potential difference to a homogenous simulant.

6. Nomenclature

a	Contact radius	[m]
d	Wheel vertical centre of mass position	[m]
e	Coefficient of restitution	-
E	Young's modulus	[Pa]
F_x	Net force in the direction of travel	[N]
F_z	Load force	[N]
F	Particle force	[N]
G	Shear modulus	[Pa]
g	Gravity	[m/s^2]
i	Slip ratio	-
k	Spring stiffness	-
m	Mass	[kg]
r	Particle radius	[mm]
r_w	Wheel radius	[m]
S_w	Soilbed height	[m]
V	Void ratio	-
V_x	Translational velocity	[m/s]
z	Sinkage	[m]
γ	Dampening force	[N]
δ	Particle overlap	[m]
μ	Traction coefficient	-
μ_s	Static friction coefficient	-
μ_r	Rolling friction coefficient	-
ν	Poisson's ratio	-
ρ	Density	[kg/m^3]
ψ	Cohesion	[J/m^2]
ω	Angular velocity	[rad/s]

7. Acknowledgements

The author would like to acknowledge the assistance of students Keisuke Takehana and Kizaki Shino in understanding the experiment to be simulated. In addition, thanks to Isabel Casasbuenas Cabezas and Prof. Kamura for assistance in measuring soil properties.

8. Declaration of competing interest

The authors declare that they have no known competing financial interests or personal relationships that could have appeared to influence the work reported in this paper.

9. References

Beakawi Al-Hashemi, H. M., & Baghabra Al-Amoudi, O. S., 2018. A review of the angle of repose of granular materials. In *Powder Technology* (Vol. 330, pp. 397–417). Elsevier B.V.

Coetzee, C. J. 2017. Review: Calibration of the discrete element method. Powder Technology, 310, 104–142.

Cundall, P. A., & Strack, O. D. L. 1979. A discrete numerical model for granular assemblies. *Géotechnique*, *29*(1), 47–65.

Daca, A., Tremblay, D., & Skonieczny, K., 2021. The Relationship between Cone Penetration Resistance and Wheel-Soil Interactions in Lunar Gravity. *IEEE Aerospace Conference Proceedings*, *2021-March*.

Douglas, M., 2012. Design and Analysis of Experiments, eight ed., Wiley

Grant H. Heiken, David T. Vaniman, & Bevan M. French. 1991. *Lunar Sourcebook*. Cambridge University Press.

He, R., Sandu, C., Mousavi, H., Shenvi, M. N., Braun, K., Kruger, R., & Els, P. S. 2020. Updated Standards of the International Society for Terrain-Vehicle Systems. *Journal of Terramechanics*, *91*,

Kobayashi, T., Fujiwara, Y., Yamakawa, J., Yasufuku, N., & Omine, K., 2010. Mobility performance of a rigid wheel in low gravity environments. Journal of Terramechanics (4), 261–274.

O'Sullivan, C., 2011. Particulate Discrete Element Modelling- A Geomechanics Perspective

Ono, S., Namikawa, S., & Yoshida, K., 2020. Analysis of Soil Deformation and Wheel Traction on Loose Terrain Using PIV. *IEEE Aerospace Conference Proceedings*.

Ono, S., Lichtenheldt, R., & Yoshida, K., 2021. Soil flow analysis for planetary rovers based on particle image velocimetry and discrete element method. *Proceedings of the 20th International and 9th Americas Conference of the International Society for Terrain-Vehicle Systems, ISTVS 2021*.

Rodríguez-Martínez, D., van Winnendael, M., & Yoshida, K., 2019. High-speed mobility on planetary surfaces: A technical review. *Journal of Field Robotics*, *36*(8), 1436–1455.

Smith, W., Melanz, D., Senatore, C., Iagnemma, K., & Peng, H., 2013. Comparison of DEM and traditional modeling methods for simulating steady-state wheel-terrain interaction for small vehicles. *Proceedings of the 7th Americas Regional Conference of the ISTVS*, *56*, 1–16.

Zhu, J., Zou, M., Liu, Y., Gao, K., Su, B., & Qi, Y., 2022. Measurement and calibration of DEM parameters of lunar soil simulant. *Acta Astronautica*, *191*, 169–177.

Wong, J.Y., 2022. Mechanics of Vehicle–Terrain Interaction. In Theory of Ground Vehicles, J.Y. Wong (Ed.). pp. 77-219

0709 | Proposal of swarm rovers' collaborative locomotion with expansion and contraction mechanism for driving a loose slope

Kanta Fujimura [a,*], Daisuke Fujiwara [b], Kojiro Iizuka [a]

[a] *Graduate School of Engineering and Science, Systems Engineering and Science, Shibaura Institute of Technology, Saitama, Japan*

[b] *Department of Mechanical and Electrical Engineering, Suwa University of Science, Nagano, Japan*

* *Corresponding author: mf22111@shibaura-it.ac.jp*

ABSTRACT

NASA, China, and the Soviet Union have performed missions to the Moon since the 20th century. Large-sized rovers have been used for the exploration of the Moon's surface. However, some organizations, such as JAXA and NASA, plan to explore the Moon with a lot of small sized rovers called swarm rovers. Swarm rovers will make the exploration more efficient. Two-wheeled rovers have been proposed for lunar swarm rovers, but a wheel-typed vehicle has a risk of slipping due to loose ground on the Moon. To solve this problem, the authors propose collaborative locomotion of two-wheeled swarm rovers for crossing loose slope. In this locomotion, multiple rovers connect to each other. When one rover moves, the other rovers stop, and their wheels are fixed relative to the ground. Thus, the locomotion uses the traction of locked wheels (non-rolling wheels). The authors considered the traction force of the locomotion theoretically, and the consideration suggested that multiplying the number of connected rovers and the locked wheels makes the driving performance higher. To confirm the effect of the proposed locomotion, this paper developed test bed rovers. The rover has magnets and a linear actuator; therefore, they can connect to each other and perform locomotion like an inchworm. Furthermore, the authors implemented experiments of driving loose slope with 6 test bed rovers to evaluate the locomotion's performance. The results suggested that the proposed locomotion could cross higher loose slope than the normal traveling. Also, the amount of slip suppression increased as the number of connected rovers multiplied. However, once the number of rovers exceeded a certain number, the amount of slip suppression did not increase. In summary, this paper argues that swarm rovers can suppress a slip on loose slope via collaborative locomotion using a drawbar pull force of locked wheels.

Keywords
Swarm robot
Loose ground
Loose slope
Lunar exploration
Rover
Collaborative locomotion

1. Introduction

NASA, China, and the Soviet Union have been carrying out missions to the Moon since the 20th century. Large-sized rovers have been used for the exploration of the Moon's surface. However, some organizations, such as JAXA and NASA, plan to explore the Moon with a lot of small-sized rovers called swarm rovers (ispace, 2021; JAXA, 2021; NASA, 2021a, 2021b). Swarm rovers will make the exploration more efficient. Two-wheeled rovers are planned for lunar swarm rovers. On the other hand, the lunar surface is covered with rugged terrain and a granular material called regolith. In addition, some slopes of regolith are quite steep, which slope angels are about 20 to 30 degrees. Therefore, wheel-typed vehicles have a risk of slipping and being stuck due to increasing traction load on those high loose slope on the Moon. It is difficult for wheeled rovers to cross on over 15 degrees loose slopes, which is showed in section 3.1. Fortunately, the collaborative locomotion of swarm rovers can solve this stucking problem. For instance, Ozkan-Aydin et al. (2021) say that multiple leg-typed robots which connect to each other have higher driving performance on an offload terrain and a bumpy road than an individual robot. However, the knowledge to appraise the driving performance of two-wheeled rovers' collaborative locomotion on loose grounds are not enough. This paper proposes the collaborative locomotion of two-wheeled swarm rovers for crossing loose slopes. Specifically, the locomotion uses a drawbar pull force of locked wheels. The locomotion has the possibility of increasing a drawbar pull force and the amount of slip suppression as the number of rovers multiplies. Furthermore, we implement experiments of driving loose slopes with test bed rovers to evaluate the proposed locomotion's performance.

2. Method

2.1 Proposed locomotion

Previous studies proposed a traveling locomotion that uses traction of locked (non-rolling) wheels (Andrade et al., 1998; Kemurdjian et al., 1992; Moreland et al., 2011). Moreland et al. (2011) reported that the drawbar pull force of the locomotion exceeds it of the normal driving. Andrade et al. (1998) indicated the specification of the push-pull locomotion rover with 6-wheel. The study indicated that the maximum climbing angle of Marsokhod rover closes to 30-35 degrees. In this study, the authors applied the locomotion to swarm rovers' collaborative locomotion. Figure 1 shows an image of the proposed locomotion. Two-wheeled rovers connect to each other with the connection unit which is compliant in the driving direction. When one rover moves, the other rovers stop. The wheels of the stopping rovers are fixed relative to the ground. Thus, the locomotion uses the traction of locked wheels. In Figure 1-a, only the head rover moves, and the other rovers stop. In Figure 1-b, only the middle rover moves, and the others retain the current state. The locomotion has possibility of increasing the drawbar pull force as the number of rovers and the locked wheels multiplied without limiting the number of rovers. Moreover, various rovers' formations that can deal with traversing, getting over objects, etc. can be considered. The connecting mechanism must expand and contract to push up and pull up other rovers. The link must be compliant in the longitudinal direction. The authors developed test bed rovers that can perform the proposed locomotion. Figure 2 and Table 1 show the specifications of the test bed rovers. The rovers can connect to each other using magnets attached to the front and tail. Moreover, the rovers are equipped with a linear actuator; therefore, they can connect to each other and perform locomotion like an inchworm. Figure 3 shows images of

https://doi.org/10.56884/FVWG4188
ISBN 978-1-942112-55-6 (pdf)

the rovers performing the proposed locomotion. Three rovers connect and perform the proposed movement. In Figure 3-a, only the head rover moves. The other rovers stop, and their locked wheels are fixed relative to the ground. In Figure 3-b, only the middle rover moves, and the others stop. In Figure 3-c, only the end rover moves, and the others stop. Circuitry and batteries are external and not included in the weight of the rover.

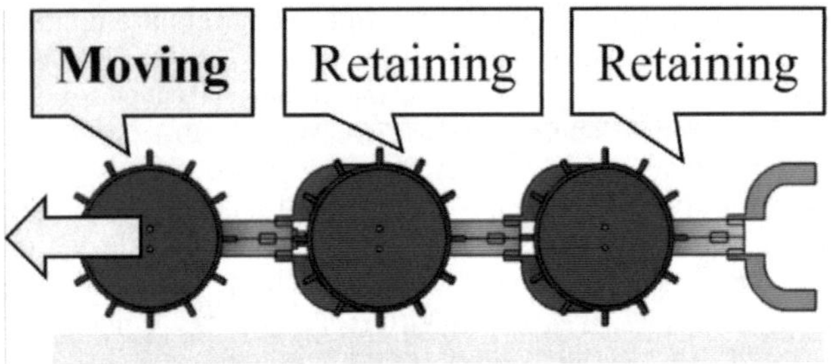

(a) Three rovers connect and perform the proposed collaborative movement. In this picture, only the head rover moves. The other rovers stop, and their wheels are fixed relative to the ground.

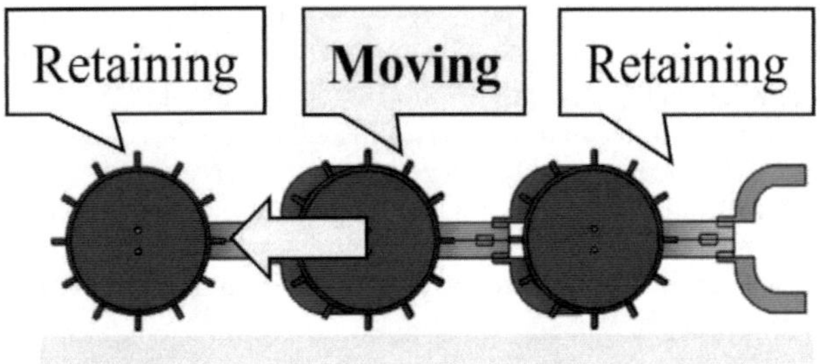

(b) Only the middle rover moves, and the others retain the current state.

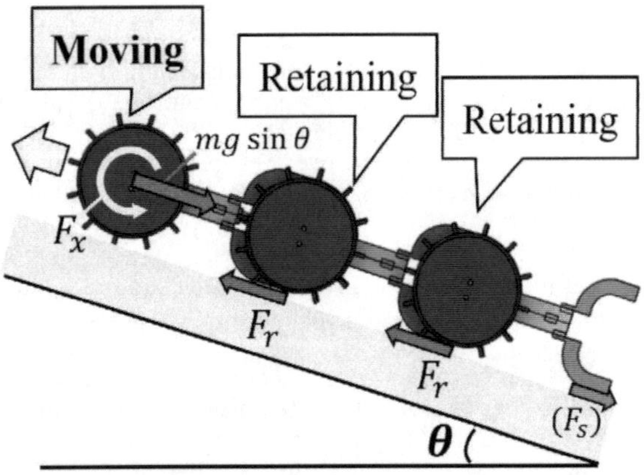

(c) Rovers cross on slope with the proposed locomotion. In this picture, only the head rover moves. The other rovers stop.

Fig. 1. Proposed collaborative locomotion

Table 1.
Specification of the test robot

Parameter	Value	Unit
Body size	W 177 × L 165 × H 100	[mm]
Weight	440	[g]
Wheel diameter	88	[mm]
Wheel lug's height	9	[mm]
Number of lugs	12	[-]

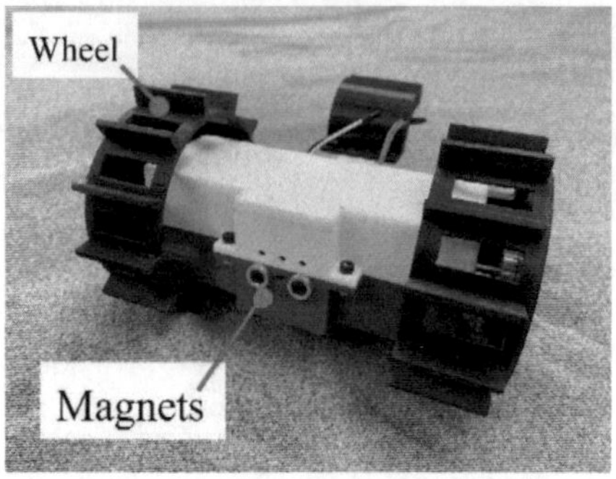

(a) Overview of the test robot. The two-wheeled robot has magnets (attached to the front and the back part of the robot) and a linear actuator.

(b) Description of the linear actuator expanding.

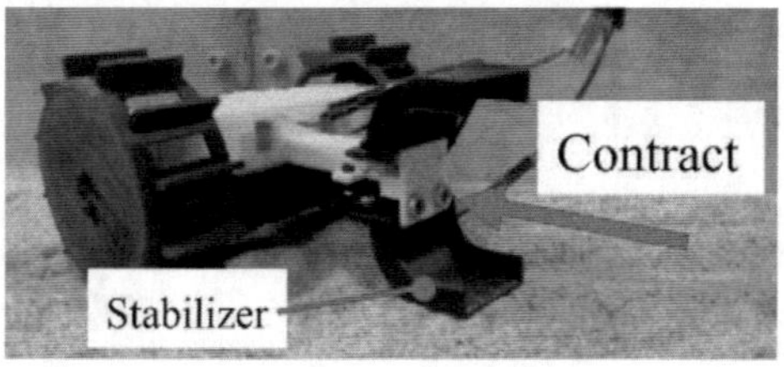

(c) Description of the linear actuator contracting.

Fig. 2. Proposed test robots

2.2 Preliminary Simulation

In the first place, the authors consider how slope angle just connecting two-wheeled rovers can cross on loose slope theoretically. Traction force of one wheel driving on a loose ground F_x is as follows:

$$F_x = rb\int_{-\theta_r}^{\theta_f}\{\tau_x(\theta)\cos\theta - \sigma_x(\theta)\sin\theta\}d\theta \tag{1}$$

where, r: Radius of the wheel, b: Width of the wheel, σx: Stress working in a loose ground, τx: Shear stress between a surface of the wheel and a loose ground (Muro, 1993). Traction force when multiple rovers are connected is as follows:

$$F_n = 2F_x \times n \tag{2}$$

n is the number of rovers. Then traction load Mn, which is arisen by its gravity on a slope, is given by Eq (3).

$$M_n = n \times mg\sin\theta + F_s \tag{3}$$

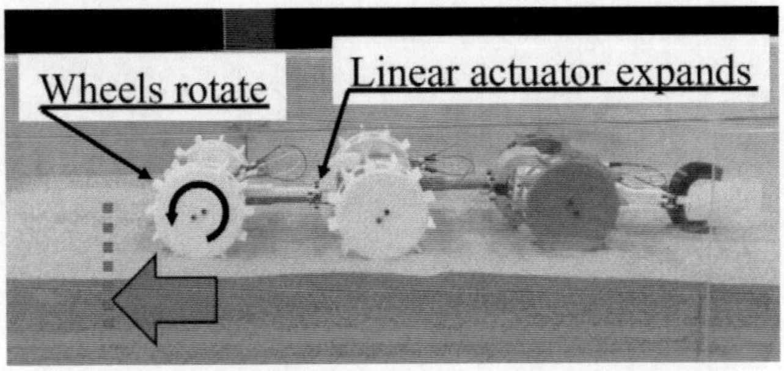

(a) Three rovers connect and perform the proposed movement. In this picture, only the head rover moves. The other rovers stop, and their wheels are fixed relative to the ground.

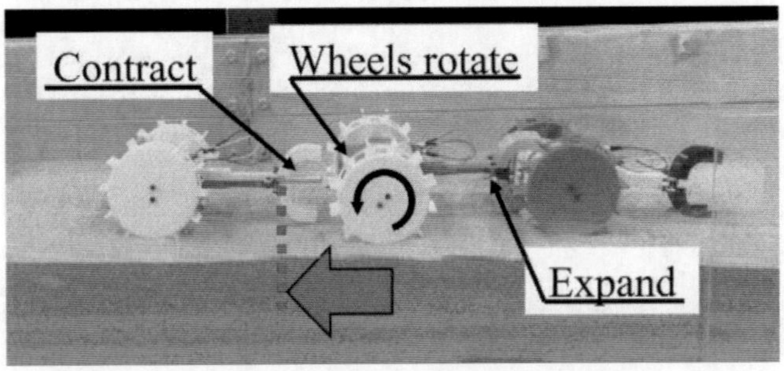

(b) Only the middle rover moves, and the others stop.

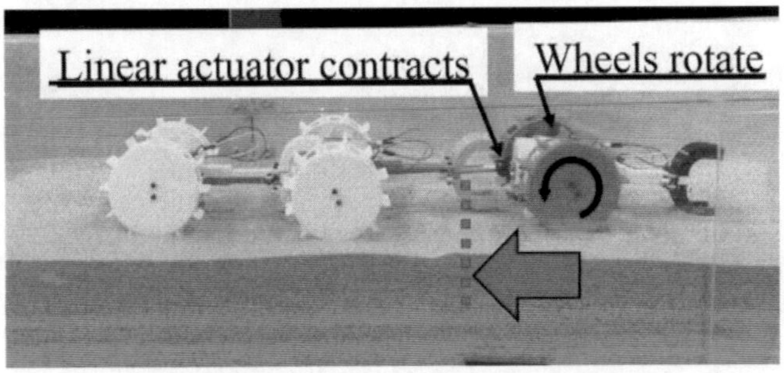

(c) Only the end rover moves, and the others stop.

Fig. 3. Picture of proposed locomotion with three rovers

Fs is a load of stabilizer which means a tail of a rover (shown in Fig.2-c). This paper supposes that stabilizers other than the end rover do not touch the ground. When traction force *Fn* exceeds traction load *Mn*, the vehicle can drive up on a slope. The function is as follows:

$$F_n \geq M_n \tag{4}$$

When the equal sign is valid in Eq (4), the authors organize Eq (4) as Eq (5). θ is a slope angle on which rovers can drive.

$$\theta = \sin^{-1}\left(\frac{2F_x}{mg} - \frac{F_s}{nmg}\right) \tag{5}$$

Herein, this paper considers the swarm locomotion using traction of locked wheels. First, we consider the supporting force (drawbar pull force) of a locked wheel based on knowledge of soil mechanics. Figure 4 shows an image of forces that work on the wheel that moves while removing the ground. There are three forces in Fig. 4. F_{wx1} is the shear force generated when the ground under the wheel is sheared. F_{wx2} is the resistance force evolved from the wall of the ground made by the wheel sinking or the ground being removed. $F_{wx3}$is the friction force works between the side surface of the wheel and the ground. The drawbar pull force of a locked wheel F_{wx} is given by Eq (6).

$$F_{wx} = F_{wx1} + F_{wx2} + F_{wx3} \tag{6}$$

In this paper, the authors ignore F_{wx3} because the force is much smaller than the other two forces. Thus, we defined that the drawbar pull force F_{wx} consists of F_{wx1} and F_{wx2}. Next, this paper state details of F_{wx1} and F_{wx2}. When rigid objects settle into a loose ground and push the ground, shear stress work between the bottom surface of the objects and the ground (Sakai et al., 2010; Sato et al., 2007; Yasukawa et al., 2021). Shear stress gradually increases when shear displacement is applied under arbitrary vertical stress. Then, the maximum shear stress τ_{max} is given by Eq (7).

$$\tau_{max} = c + \sigma \tan\varphi \tag{7}$$

Equation (7) is known as the Coulomb equation, where, c: Cohesion of soil, σ: Vertical stress working on the sheared surface, φ: Angle of internal friction. The maximum shear stress is not immediately available at the beginning of the contact area. However, it will be reached asymptotically, according to the Janosi and Hanamoto (1961) equation:

$$\tau = \tau_{max}\{1 - \exp(-J/K)\} \tag{8}$$

where J is the shear displacement and K is the soil shear deformation modulus. F_{wx1} is shear stress τ multiplied by grounding area between the wheel and ground A, and is as follows:

$$F_{w1} = \tau \times A \tag{9}$$

When a board with unit width moves in ground, passive earth pressure works on the board. Coulomb (1776) considered a sandy soil mass without cohesion that is bounded by a wall with a slip surface as shown in Fig. 5 (Sakai et al., 2010; Sato et al., 2007; Yasukawa et al., 2021). He assumed that the soil mass moves along the slip surface because of the movement of the wall. Earth pressure on the wall surface was calculated using an equilibrium approach where the force from the weight of soil mass *W*, reaction force to prevent the mass from sliding off *R*, and reaction force from the board *Pp*. The passive earth pressure *Pp* is given by Eq (10).

$$P_P = \frac{1}{2}\gamma_t H^2 K_P \tag{10}$$

$$K_P = \frac{\sin^2(\beta+\varphi)}{\sin^2\beta\sin(\beta-\delta)}\left\{1 - \sqrt{\frac{\sin(\varphi+\delta)\sin(\varphi+i)}{\sin(\beta-\delta)\sin(\beta-i)}}\right\}^{-2} \tag{11}$$

where, γ_t is the unit weight of the soil. *H* is a settlement of the wheel in ground. *Kp* is called Coulomb's passive earth pressure coefficient. When a locked wheel moves in ground, the soil mass As which is shown in Fig. 4 moves with the wheel. This state is observed with PIV (Particle Image Velocimetry) in our previous experiment. Thus, this paper defined that *Pp* is the passive earth pressure that is received from the sand mass pushed by the mass As as a single object with the wheel. F_{wx2} is soil pressure *Pp* multiplied by the breadth of the board *B*. Its formula is as follows:

$$F_{w2} = P_P \times B \tag{12}$$

When rovers drive on a loose slope, the traction load generated by the rovers' weight works on the rover itself. Therefore, the drawbar pull force of a locked two-wheeled rover *Fr* is as follows:

$$F_r = 2F_{wx} - mg\sin\theta \tag{13}$$

Then we consider the situation when multiple rovers are connecting to each other shown in Fig. 1. Then traction force *Fn*, when one rover moves and the others stop, is as follows. *n* is the number of rovers.

$$F_n = 2F_x + (n-1)F_r \quad (14)$$

Traction load *Mn*, which is generated by its gravity on a slope, is given by Eq (15). *Fs* is a load of stabilizer. We suppose that stabilizers other than the end rover do not touch the ground. Therefore, *Fs* works only when the end rover moves.

$$M_n = \begin{cases} 1 \times mg\sin\theta + F_s & \text{(The end rover moves)} \\ 1 \times mg\sin\theta & \text{(The others move)} \end{cases} \quad (15)$$

When traction force *Fn* exceeds traction load *Mn*, the vehicle can drive up a slope.

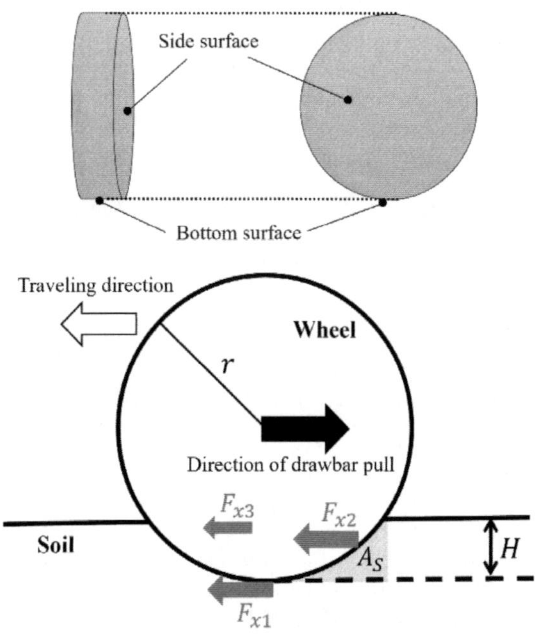

Fig. 4. Force working on the locked wheel

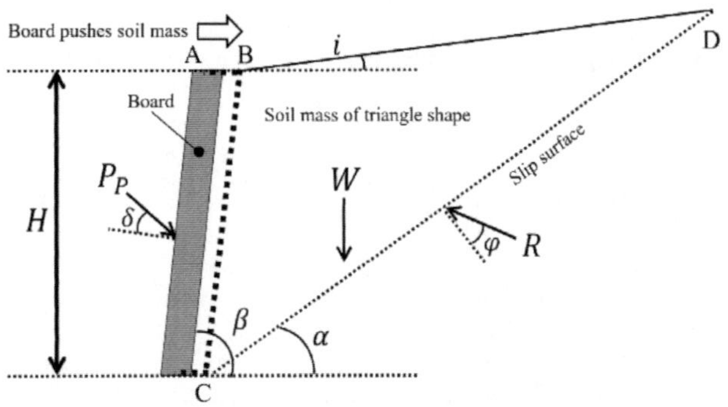

Fig. 5. Assumptions for Coulomb's earth pressure and the force triangle

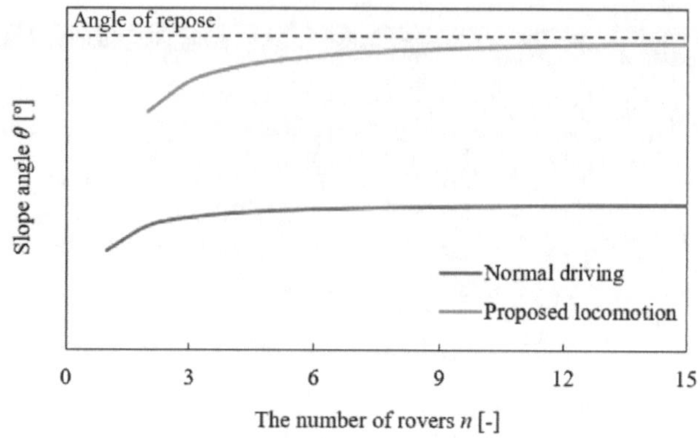

Fig. 6. Slope angle at which rovers can drive

The authors plotted the value of θ that means a slope angle on which rovers can drive with the normal driving and the proposed locomotion on Fig. 6, respectively. The traction force *Fx* and the drawbar pull force F_{wx} are constant to compare under similar conditions. Moreland et al. (2011) implemented experiments to compare the drawbar pull force of the push-pull locomotion with the traction force. The mass of the experimental vehicle was 400 kg. The vehicle had rigid wheels (71 cm in diameter and 25 cm in width) and drove on GRC-1 lunar soil strength simulant. The result indicated that the drawbar pull force is larger than the traction force. This paper supposed that the ratio of the drawbar pull force F_{wx} from the traction force *Fx* is 2.0. Figure 6 suggests that the increasing trend of the proposed locomotion tends to be larger than that of the normal driving as the number of rovers multiplies.

3. Driving Experiment

In this chapter, the authors implemented two driving experiments. These experiments are described below.

1. Experiment to compare the performance of the proposed locomotion and the normal traveling.
2. Experiment to investigate the relationship between the traveling displacement and the number of connected rovers.

3.1 Experiment to compare the performance of the proposed locomotion and the normal traveling.

3.1.1 Purpose and Method

The authors implemented driving experiments on a loose slope using the test robots shown in section 2.1 to compare the proposed locomotion and the normal driving mode. The length, height, and width of the experimental field are 2300 mm, 150 mm, and 500 mm, respectively (Fig. 7). Table 2 shows the experimental conditions. The slope angles were from 5° to 25°. Driving methods were the normal and the proposed locomotion. The number of connected robots was 1 to 3. The driving distance was 450 mm, and the authors used a motion capture system to measure the traveled distance. When the robots did not move more than 10 seconds, we concluded that the robots were stuck. When the robots drove the set distance without being stuck, we concluded that they could travel on the angle.

3.1.2 Results and Discussion

Table 3 shows the experimental results. The results said that the individual robot was stuck on the 10 degrees slope, and connected robots which drove normally were stuck on the 20 degrees slope. However, the robots could cross the 25 degrees slope with the proposed locomotion. The results mean that the proposed locomotion has higher driving performance on a loose slope than normal driving.

3.2 Experiment to investigate the relationship between the traveling displacement and the number of connected rovers.

3.2.1 Purpose and Method

The authors implemented driving experiments on a loose slope to investigate the relationship between the number of connected rovers and the driving performance. During the experiments, we used test bed rovers shown in Fig. 8. Table 4 summarizes the specifications of the test bed rover. The rovers have almost the same specifications as the rovers of Chapter 2.4, but their weights and sizes are slightly larger than the rovers of Chapter 2.4 because circuitry and batteries are included in the rovers. Figure 9 indicates the overview of the experiment that uses six rovers. Table 5 shows the experimental conditions. The slope angles were 0° and 25°. The number of connected rovers was two to six. The rovers drove on the proposed locomotion, and they drove five sequences. One sequence means the period where the head rover starts moving and the end rover finishes moving. The authors used a motion capture system to measure the traveled distance and calculated the average value of traveled distances. After that, we calculated R, which means a ratio of traveled distance of a sequence on 25° slope from it on 0° slope. The formula is as follows:

$$R = X_{25}/X_0 \times 100 \quad (16)$$

where, X_0: distance of one sequence movement on 0°, X_{25}: distance of one sequence movement on 25°.

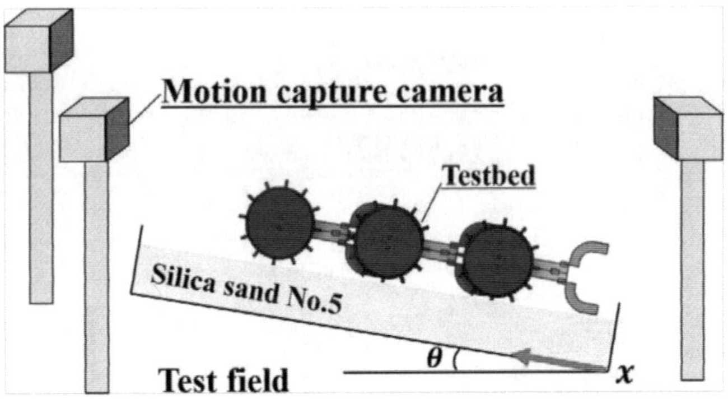

Fig. 7. Test field for experiments using test robots

Table 2
Conditions of experiment to compare the performance of the proposed locomotion and the normal traveling

Item [Unit]	Value
Ground condition	Silica sand No.5
Slope angle [°]	0, 5, 10, 15, 20, 25
Driving distance [mm]	450
Trial	5
Driving pattern	Normal: 1, 2 and 3 robots Proposed: 2 and 3 robots

Table 3
Results of normal and proposed traveling. "Normal" means the robots drove on the experimental conditions with normal driving. "Proposed" means the rovers performed the proposed collaborative locomotion. Circle-marks and X-marks mean: ○: Driving, ×: Stuck

Slope/ Robots	1, Normal	2, Normal	3, Normal	2, Proposed	3, Proposed
0°	○	○	○	○	○
5°	○	○	○	○	○
10°	×	○	○	○	○
15°	×	○	○	○	○
20°	×	×	×	○	○
25°	×	×	×	○	○

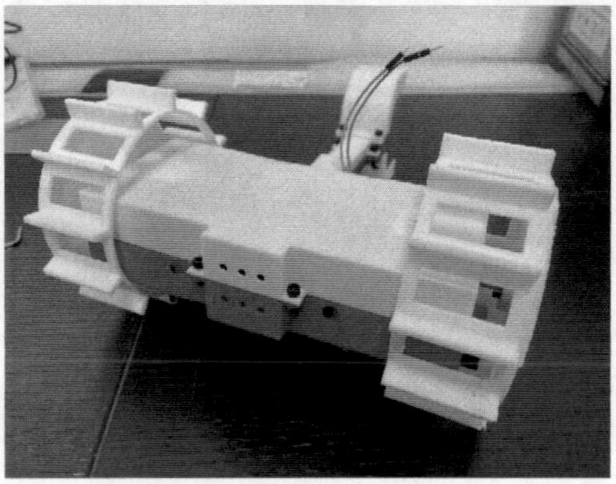

Fig. 8. Test rover used the experiment in section 3.2

Table 4
Specification of the test rover

Parameter	Value	Unit
Body size	W 210 × L 220 × H 110	[mm]
Weight	491	[g]
Wheel diameter	98	[mm]
Wheel lug height	10	[mm]
Number of lugs	12	[-]

3.2.2 Results and Discussion

Figure 10 shows the experimental results. When the number of robots was two, the ratio R was 57.9%. When the number of robots was three, the ratio was 82.2%. When the number of robots was four, the ratio was 84.0%. It means that the proposed rovers can get more than 80% of driving distance on 25 degrees silica sand No.5 slope from 0 degrees if there are three rovers and more. Comparing three and two rovers, the proposed locomotion with three rovers suppressed a slip more considerably than two rovers. On the other hand, the ratio R between three rovers and more had little difference. The result suggests that the amount of slip suppression increased as the number of connected rovers multiplied. However, once the number of rovers exceeded a certain number, the amount of slip suppression did not increase even if the number of rovers multiplied.

Table 5
Conditions of the experiment to investigate the relationship between the traveling displacement and the number of the connected rovers

Item [Unit]	Value
Ground condition	Silica sand No.5
Slope angle [°]	0, 25
Number of sequences	5
Trial	5
Driving pattern	Proposed
Number of robots	2 to 6 robots

Fig. 9. Picture of 6 test rovers.

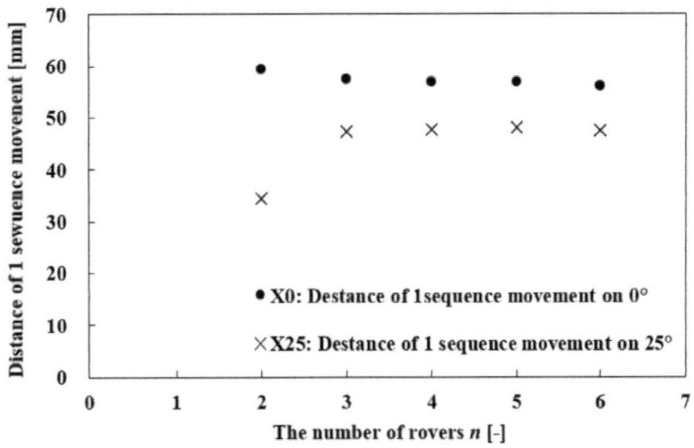

(a) Distance of 1 sequence movement

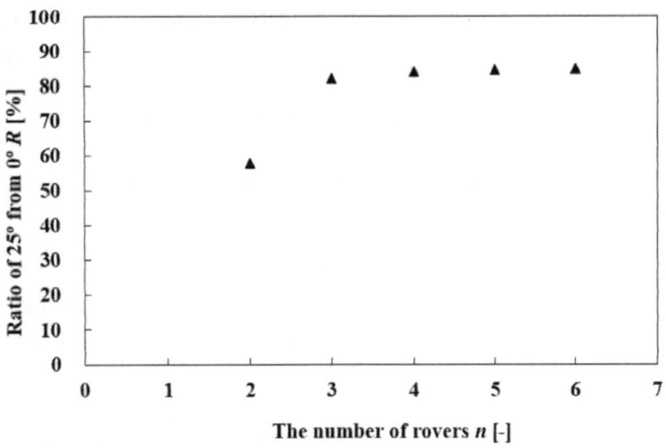

(b) Ratio of 25 degrees to 0 degree

Fig. 10. Results of traveling displacements and *R*

4. Conclusions and future work

This paper proposed collaborative locomotion of two-wheeled swarm rovers for crossing loose slopes to solve the problem of slippery due to loose ground. Then, the authors considered the swarm locomotion using traction of locked wheels theoretically. After that, we developed test bed rovers that perform the proposed locomotion and implemented driving experiments on loose slopes with the rovers. The results suggested that the proposed locomotion could cross higher loose slopes than the normal traveling. Also, the amount of slip suppression increased as the number of connected rovers multiplied. However, once the number of rovers exceeded a certain number, the amount of slip suppression did not increase even if the number of rovers multiplied.

In a future study, the authors plan to implement experiments to investigate the driving performance by varying the number of robots moving and the others stopping. In addition, we plan to implement driving experiments on 32 degrees slope close to the angle of repose and to appraise the traversability over loose sloped terrains.

5. Nomenclature

A	Contact area	[m^2]
b	Width of a wheel	[m]
c	Cohesion of soil	[kN/m^2]
Fn	Traction force when multiple rovers are connected	[N]
Fr	Drawbar pull force of a locked two-wheeled rover	[N]
Fs	Load of stabilizer	[N]
F_{wx}	Drawbar pull force of a locked wheel	[N]
F_{wx1}	Shear force generated when the ground under the wheel is sheared	[N]
F_{wx2}	Resistance force evolved from the wall of the ground made by the wheel sinking	[N]
F_{wx3}	Friction force works between the side surface of the wheel and the ground	[N]
Fx	Traction force of one wheel driving on a loose ground	[N]
g	Gravitational acceleration	[m/s^2]
H	Settlement of the wheel in ground	[m]
J	Shear displacement	[m]
K	Soil shear deformation modulus	[m]
Kp	Coulomb's passive earth pressure coefficient	[-]
m	Mass	[kg]
Mn	Traction load	[N]
n	The number of rovers	[-]
Pp	Passive earth pressure	[N/m^2]
R	Ratio of X25 from X0	[%]
r	Radius	[m]
X_0	Distance of one sequence movement on 0°	[mm]
X_{25}	Distance of one sequence movement on 25°	[mm]
γt	Unit weight of the soil	[kN/m^3]
θ	Slope angle	[°]
σ	Vertical stress	[kN/m^2]
σx	Stress working in a loose ground	[kN/m^2]
τ	Shear stress	[kN/m^2]
τx	Shear stress	[kN/m^2]

τ_{max}	Maximum shear stress	[kN/m^2]
φ	Angle of internal friction	[rad]

6. Acknowledgements

This work was supported by JSPS KAKENHI Grant Number JP21K14105.

7. Declaration of competing interest

The authors declare that they have no known competing financial interests or personal relationships that could have appeared to influence the work reported in this paper.

8. References

Mettam, G.R., Adams, L.B., 2009. How to prepare an electronic version of your article, in: Jones, B.S., Smith, R.Z. (Eds.), Introduction to the Electronic Age. E-Publishing, Inc., New York, pp. 281–304.

Andrade, G., Amar, F.B., Bidaud, P., Chatila, R., 1998. Modeling Robot-Soil Interaction for Planetary Rover Motion Control. Proceedings of the 1998 IEEE/RSJ, International Conference on Intelligent Robots and Systems, Innovations in Theory, Practice and Applications (1998), pp. 576-581.

ispace, 2021. Contract signed for "Lunar transportation, operation, and data acquisition of a small robotic system. https://ispace-inc.com/jpn/news/?p=1903 (accessed 25 March 2023).

Janosi, Z., Hanamoto, B., 1961, The analytical determination of drawbar pull as a function of slip, for tracked vehicles in deformable soils, Proceedings of the 1st Intl. Conference Mech. Soil-Vehicle Systems, Turin, Italy (1961)

JAXA, 2021. Data Acquisition on the Lunar Surface with a Transformable Lunar Robot, Assisting Development of the Crewed Pressurized Rover. https://global.jaxa.jp/press/2021/05/20210527-1_e.html (accessed 27 March 2023).

Kemurdjian, A., Gromov, V., Mishkinyuk, V., Kucherenko, V., Sologub, P., 1992. Small Marsokhod Configuration, Proceedings 1992 IEEE International Conference on Robotics and Automation.

Moreland, S., Skonieczny, K., Wettergreen, D., Asnani, V., Creager, C., Oravec,H., 2011. Inching Locomotion for Planetary Rover Mobility, 2011 Aerospace Conference.

Muro, T., 1993. Teramechanics, GIHODO SHUPPAN, Japan.

NASA, 2021a. Cooperative Autonomous Distributed Robotic Exploration. https://www.nasa.gov/directorates/spacetech/game_changing_development/projects/CADRE (accessed 27 March 2023).

NASA, 2021b. CADRE of Mini Rovers Navigate Simulated Lunar Terrain. https://www.nasa.gov/feature/glenn/2021/cadre-of-mini-rovers-navigate-simulated-lunar-terrain (accessed 27 March 2023).

Ozkan-Aydin, Y., Goldman, D.I., 2021. Self-reconfigurable multilegged robot swarms collectively accomplish challenging terradynamic tasks, Science Robotics, vol. 6, No. 56.

Sakai, T., Kathuyama, K., Hossain, M.Z., Pyrak-Nolte, L.J., 2010. Soil Mechanics (2) -Fundamental Properties-, CORONA PUBLISHING, Japan.

Sato, K., Nishimura, T., Sakurai, S., Kobayashi, Y., Kiku, T., Suami, K., 2007. Soil engineering from the basics, Asakura Publishing, Japan.

Yasukawa, I., Imanishi, K., Tateishi, Y., 2021. Soil mechanics, third ed, Ohmsha, Japan.

0889 | Skidding suppression method using "discrete 4-wheel-drive typed rover" considering steering angle at traveling across loose slope

Kojiro Iizuka [a,*], Tatsuhiko Suga [a], Takaaki Sakata [a], Daisuke Fujiwara [b]

[a] *Department of Machinery and Control Systems, Shibaura Institute of Technology, Saitama, Japan*

[b] *Department of Mechanical and Aeronautical Engineering, Suwa University of Science, Nagano, Japan*

* *Corresponding author: iizuka@shibaura-it.ac.jp*

ABSTRACT

Wheeled rovers, which are capable of wide-area and detailed exploration, are widely used for lunar and planetary exploration to expand human activity areas. The Moon and planets have many steep slopes such as craters, and their surfaces are loose soil ground with accumulated particles called regolith. When a rover traverses such loose soil slopes, there is a risk that the rover may deviate from the target path because it is getting to skid in the downward direction of the slope. Therefore, it is necessary to suppress or decrease the skidding. A discrete 4-wheel-drive typed rover has been focused on to solve this problem in this study. Discrete 4-wheel-drive typed rover can be moved by a chassis unit and a wheel independently like a legged rover. This proposed method was more effective in suppressing skidding than the conventional wheel-driven method, but skidding was observed during wheel motion due to the lack of force on the upper side of the slope. This study is focused on "the steering angle", which can generate force on the upper side of the slope, to further suppress skidding. Experiments were conducted using the test bed to measure the amount of skidding across an 800 mm slope. The results showed that the amount of skidding decreased as the steering angle is increased. Moreover, the steering angle to the upper side of the slope is effective in suppressing skidding. Thus, in addition, a discrete driving method, to add "the steering angle" has indicated effective suppression of skidding because a steered wheel generates forces on the upper side of the slope.

Keywords
Lunar exploration rovers
Push pull locomotion
Discrete 4-wheel-drive typed rovers (Push pull locomotion)
Steering angle

1. Introduction

The Moon and planets have many steep slopes such as craters, and their surfaces are loose soil ground with accumulated particles called regolith. The hugely successful MER Mars rovers SPIRIT and OPPORTUNITY during their multi-year missions on Mars both experienced repeated incidents of immobilization in weak terrain (fig.1)(NASA, 2023). The wheels of the MER sank into loose soil because of the rotation and movement of the wheel. This means that it is so difficult for wheels or rovers to get effective driving force from the soil. The loose surface is easy to destroy by given forces some effect like the movement of wheels (fig.2)(Iizuka, 2005). There are a lot of ideas to traverse this state like the surface covered with loose soil. The effective method is traction control. Considering the wheels' rotation speed is critical without making loose soil destroy. If the wheels rotate quickly and move, the surface of the loose soil is easy to destroy. In others' ideas, we know that focusing on the wheel's form is effective. Especially, there is a lot of research treated with the grousers which had considered number, length, and width (Guo et al., 2020, Sutoh et al., 2012). These can be thought of in a way that includes terramechanics. This means increasing the shear force by design. Many hardware and wheel geometry studies consider hill climbing in the straight direction. Of course, Intelligence approaches such as path planning consider various ground movements. However, it is so tricky for a wheeled typed rover to travel in a lateral direction on loose soil with a steep slope. During path planning and traveling according to the path, the rover may deviate from the path due to the physical balance between the left and right sides of the rover. In fact, when a rover exits a crater or the edge of a crater, it is likely to travel laterally on the slope. The normal scenario is to correct the path by taking measures such as the difference in rotation of the left and right wheels and the change in posture direction.

Fig. 1. The poor condition of rovers during traveling loose soil (NASA, 2023)

Fig. 2. The loose surface state destroyed by rotating and moving of wheel (Iizuka, 2005)

https://doi.org/10.56884/NRZZ6629
ISBN 978-1-942112-55-6

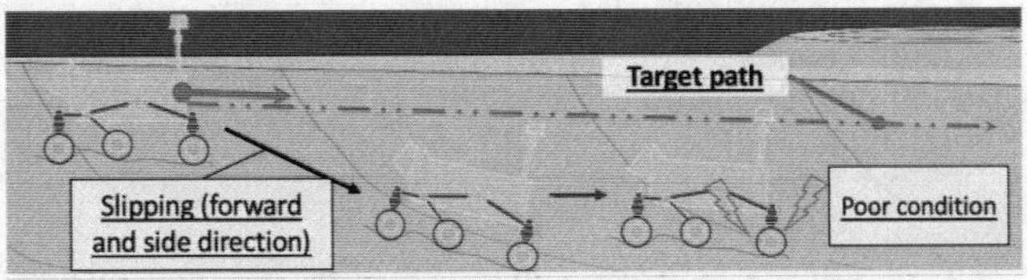

Fig. 3. Slipping behavior to forward and side direction affecting to rovers at traveling across loose soil with slope

If such measures are not taken, slippage occurs in the forward and lateral directions with respect to the planned path, and the travel path of the rover deviates significantly from the planned path (fig.3). In addition, the rovers traveling on the loose ground possibly cause the deterioration of the traveling state. Our research group has focused on the bearing force of locked wheels based on the knowledge of terramechanics. When a locked wheel is towed, it shows a high resistive force. Therefore, a mechanism that can actively change the position of the chassis has been installed inside the vehicle body. The mechanism can move the front and rear wheels alternately (Push Pull Locomotion). This method, "Push Pull Locomotion," has been proven in driving tests to significantly increase the ability to climb slopes. Furthermore, the amount of skidding was greatly reduced in the test of running horizontally (laterally direction) on a loose slope. However, it is necessary to eliminate the skidding to travel accurately on the planned path. Therefore, this study proposes a "Push Pull Locomotion" that considers the direction of the wheels (steering angle), utilizing the effect of the Push Pull Locomotion. In the traveling tests, the test bed rover with steering angle is used to perform lateral travel on a loose slope. From the results of driving tests, the effect of skidding suppression is experimentally verified.

The paper is organized as follows. In the next section, we describe State of the robot traveling laterally on a loose slope. In Section 3, we describe mechanics model considerations at traveling laterally on the loose slope. Section 4 and 5 describe experiments using testbed rover with “Push Pull Locomotion with steering angle wheels”, and the conclusion is reported in section 6.

2. State of the robot traveling laterally on a loose slope

As the rover moves laterally across the soft slope ground, the rover tends to drop to the valley side (downward) of the slope because of gravity. If the rover continues to travel, the attitude of the rover is changed downward (in the yaw direction). In addition, there is slippage between the wheels and the loose soil on which the rover is traveling. This means that the rover is forced to skid. As shown in fig. 4, after the rover moves several meters laterally, the amount of skidding becomes large and the rover deviates significantly from the planned path. Preventing "skidding", which occurs when a rover travels laterally on a loose slope, is a fundamental issue, and there are many methods and research to deal with this problem.

2.1 Research on reducing "skidding" conditions

Ishigami et al. proposed a unique applied Terramechanics model including side forces related to wheels using the single-wheel testing machine and the four-wheel rover and conducted a study in which the mathematical model follows the trajectory of the rover traveling laterally on a loose slope (Ishigami et al., 2007). Inotsume et al. proposed a applied Terramechanics model of the wheels tilted in the roll direction toward the upper part of the slope (Inotsume et al., 2013).

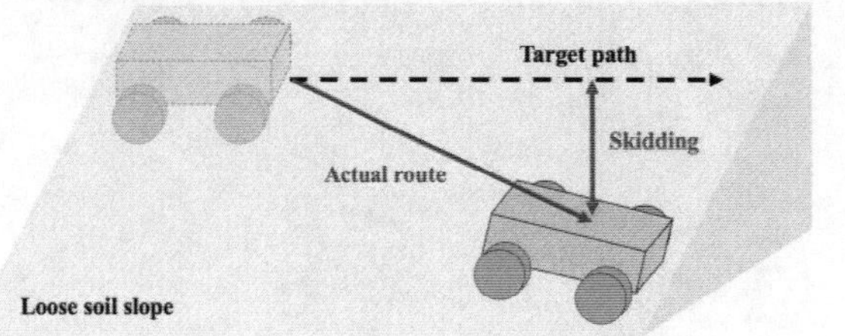

Fig. 4. "Skidding" condition that occurs when a rover travels laterally on loose soil with a slope

Moreland proposed the rover in which the wheelbase (each of the front and rear parts) between the chassis of the wheels and the center of the vehicle body can rotate left and right in the pitch direction (Moreland et al., 2011). This means that the height of the body of the rover from the ground to the left and right can be changed actively. This posture of the rover can position the center of gravity of the vehicle on the mountain's side, thus preventing the rover from skidding. Some research groups such as these methods have proposed several models for following the planned path (terramechanics) and means to reduce the amount of skidding. However, it has not been possible to eliminate skidding during lateral movement of loose slope slopes yet. This is fundamentally because the continuous motion of the wheel is causing the destruction of the loose soil. The continuous rotating motion of the wheels provides the rover with traction from the loose soil. At the same time, however, it also means that the surface of the loose soil with a slope itself is being destroyed (fig. 2). The fact that a rover moves laterally on a loose slope means that the surface of the loose soil on which the rover is traveling will be destroyed simultaneously in the direction of travel and the direction of the bottom of the slope. Therefore, we investigate the suppression of skidding by moving the four chassis-wheel units of a rover independently of each other and using the resistance force from the soil obtained by the locked wheels.

2.2 Push Pull Locomotion

In the previous section, it was mentioned that the continuous rotational motion of the wheel destroys the surface of loose soil and causes side skidding. Therefore, “Push Pull Locomotion”, which is a discontinuous locomotion, was focused on. As shown in fig. 5, “Push-pull locomotion” is a method of moving by using a contraction mechanism of the body, which is divided into a supporting side and an operating side like a scale insect. Slope "climbing" tests with a rover equipped with push-pull locomotive have demonstrated this effect (Fujiwara et al., 2019). Furthermore, our research group has proposed “Advanced Push Pull Locomotion”, which takes advantage of the collapse-prone characteristics (shear characteristics) of loose ground. It uses this property to actively sink the wheel and take advantage of the resisting force of soil. The method of traveling horizontally on the loose soil with a slope has been studied by utilizing the function of this push-pull locomotion.

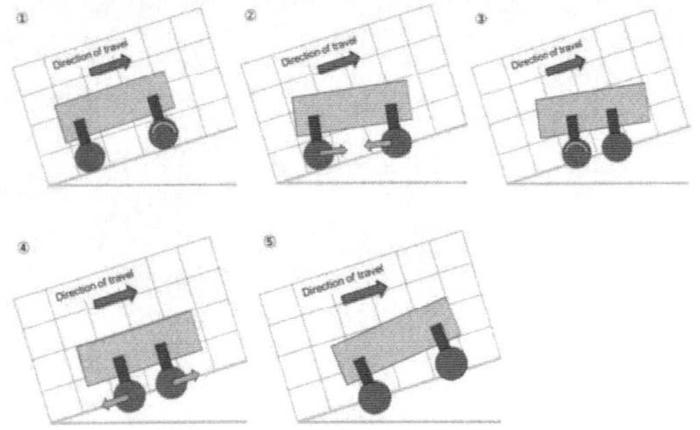

Fig. 5. Push Pull Locomotion using by using a contraction mechanism

2.3 Discrete 4-Wheel-Drive typed rover

As shown in fig. 6, our research group has been studied lateral movement on a soft slope using a test bed with four wheels and their chassis each capable of actively changing its positional relationship with the vehicle body (Push Pull Locomotion function). This update was necessary because the load distribution of the wheels distributed to the left and right is different when the rover moves laterally on a slope.

The traveling test of this unique "Discrete 4-Wheel-Drive typed testbed" on the loose soil with a slope moving laterally showed that the amount of skidding was significantly reduced compared to that of a normal rover without "a function of Push Pull Locomotion. However, it is not possible to eliminate the amount of skidding. Therefore, a basically mechanics model of the lateral movement has been considered to verify the effective elements for the skidding.

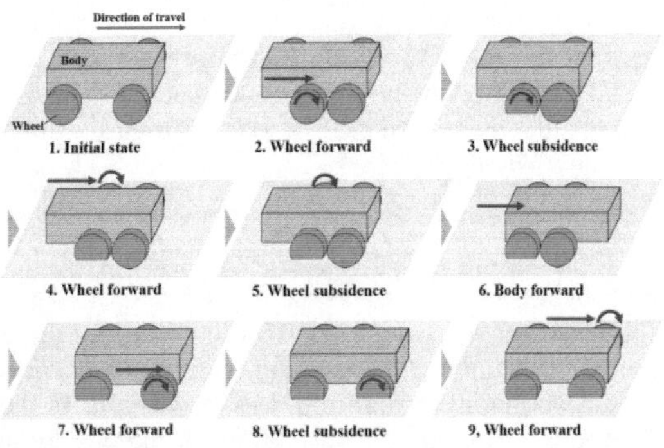

Fig. 6. "Discrete 4-Wheel-Drive typed rover" with four wheels and their chassis each capable of actively changing its positional relationship with the vehicle body

3. Mechanics model considerations at traveling laterally on the loose slope

3.1 Lateral traveling on slope without "push-pull locomotion" (version of Discrete 4-Wheel-Drive)

First, fig. 7 shows the state of the force when the rover is at rest on the slope. The orientation of the rover is facing horizontally on the slope. The downward gravity component and the ground reaction force loaded to each wheel are a balancing state on a slope. The reaction forces from the ground above and below the slope are $2F_{yu}$ and $2F_{yl}$, respectively, and the combined force F_y is shown in eq. (1). The forces on each of the four wheels are shown in fig. 7(b). Equation (2) shows the balance between the gravity of the rover toward the down-slope direction and the combined force received from the ground.

$$F_y = 2F_{yu} + 2F_{yl} \tag{1}$$
$$Mg\sin\theta = F_y \tag{2}$$

Figure 8 shows a schematic diagram of the rover immediately after it starts moving. When the rover starts moving, the ground collapses due to the rotation of the wheel, and the reaction forces F_{yu} and F_{yl} from the ground decrease as shown in fig. 8(a). The reduced F_{yu} and F_{yl} are denoted as F'_{yu} and F'_{yl}, respectively, and the reaction force $F'y$ from the ground is shown in eq. (3). The force applied to the rover in the direction of the slope is as shown in eq. (4).

$$F'_y = 2F'_{yu} + 2F'_{yl} \tag{3}$$
$$Mg\sin\theta \geqq F'_y \tag{4}$$

The force F_{yt} applied to the rover in the slope direction is expressed by Equation (5). As shown in fig. 8(b), the driving force F_x in the
direction of travel is shown in eq. (6).

$$F_{yt} = Mg\sin\theta - F'_y \tag{5}$$
$$F_x = 2F_{xu} + 2F_{xl} \tag{6}$$

The combined force F_v applied to the rover is downward on the slope. This combined force to the down-slope side causes the rover to skidding when it moves laterally down the slope.

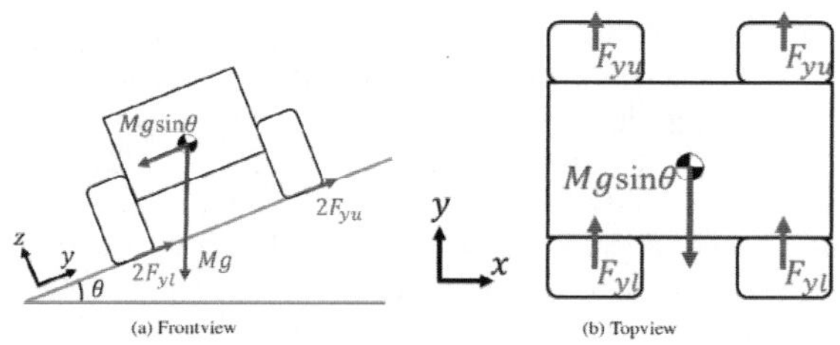

(a) Front view (b) Top view

Fig. 7 The forces generated when the rover moves laterally on the slope

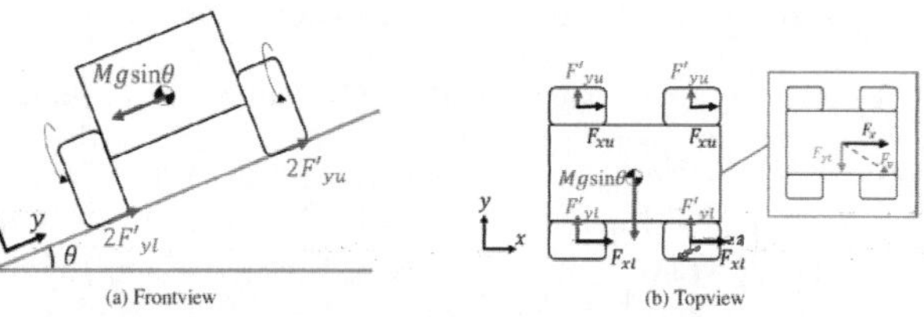

(a) Front view (b) Top view

Fig. 8. Schematic diagram of the rover immediately after it starts moving

3.2 Lateral traveling on slope with "push-pull locomotion" (version of Discrete 4-Wheel-Drive)

Next, the proposed method (Discrete 4-Wheel-Drive method) is examined with respect to the force applied to the rover when it moves laterally on the slope. When the rover is stationary on the slope, the force relationship is the same as that of a normal rover (fig. 7(b)). As shown in fig. 9, immediately after traveling, the wheels move one by one, and the driving force appears on one wheel. Figure 9 shows the case when the rear wheels on the down-slope side of the rover are driven. The reaction force F''_y (force in the slope direction) obtained from the ground can be expressed by eq.7.

$$F''_y = 2F_{yu} + F_{yl} + F'_{yl} \quad (7)$$

The static (locked) wheels have a larger support force than the dynamic wheels due to the lack of fluidization of soil. The proposed method (Discrete 4-Wheel-Drive method) has three stationary wheels. Therefore, the integrated force of the supporting wheel is larger than the force of the moving wheel in the direction of the slope (down direction). In other words, the idea is as shown in eq. (8).

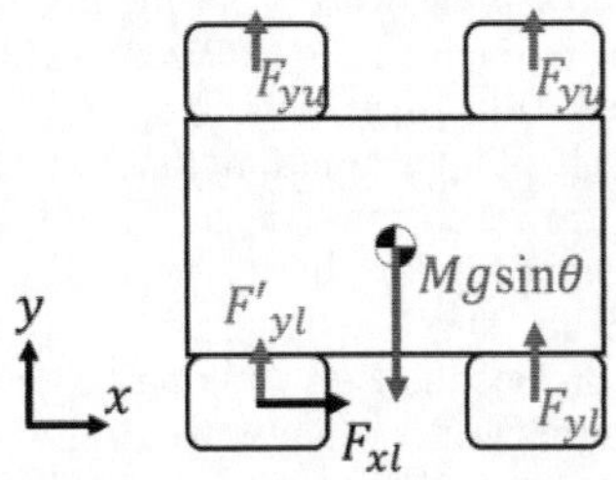

Fig. 9. A case when the rear wheels on the down-slope side of the rover are driven

Fig. 10. The attitude of the rover tendency when the wheels start moving for the first time.

$$F''_y > F'_y \quad (8)$$

The traveling by the proposed method maintains a higher support force than the normal traveling, and the skidding can be significantly suppressed compared to the normal traveling. However, previous studies have confirmed that skidding also occurs at higher slopes, and skidding has not yet been entirely suppressed (Inaba et al., 2021). In particular, as shown in fig. 10, the attitude of the rover tends to change when the wheels start moving for the first time.

3.3 Consideration of Steering Angle to prevent skidding

In order to completely eliminate the skidding, it is proposed to add a "steer angle" to "Discrete 4-Wheel-Drive method". By adding steering angles to the wheels, the driving force can be directed to the upper side of the slope as shown in fig. 11. In the case of traveling on loose soil, since the wheels are buried, when the wheels are traveling laterally on the slope, the resistance of soil acting on the sides of the wheels to prevent them from slipping down the slope (side force) occurs. However, this side force decreases when the wheels move. This is because the movement of the wheels moves the soil. Therefore, it was considered that the force of the upward component of the slope, which occurs when the wheel has a steering angle, is effective in assisting the reduction of side forces. Figure 12 shows the force when a steering angle is added to the proposed method. By adding a steering angle β on the upper side of the slope, the upper component of driving force $F_{xl}\cos\beta$ acts on the upper side of the slope. The combined force using this component can be expressed by eq. 9.

$$F_y^{steer} = 2F_{yu} + F_{yl} + F'_{yl} + F_{xl}\cos\beta \quad (9)$$

Since the force toward the upper side of the slope has the relationship shown in eq. 10, it is considered that skidding during wheel motion can be suppressed by adding a steering angle. From the above,

$$F_y^{steer} > F''_y \quad (10)$$

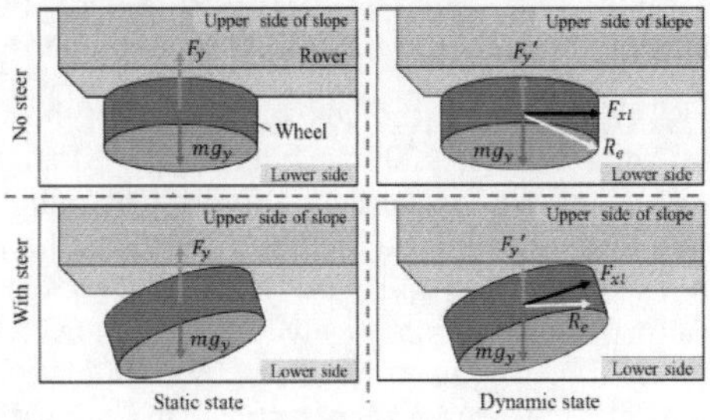

Fig. 11. Force relationship on one wheel; static vs dynamic (left and right), with steer vs without steering (up and down)

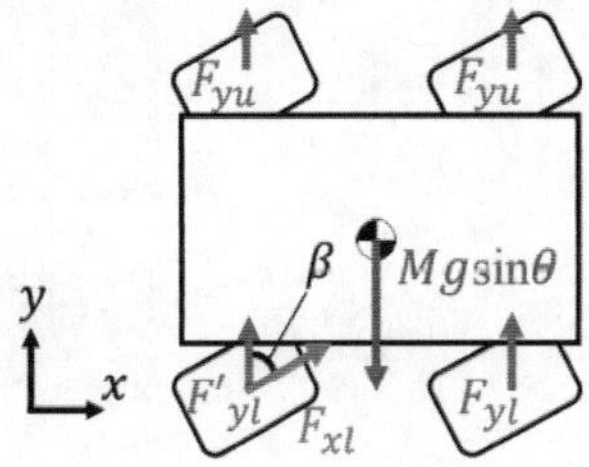

Fig. 12. Force when steering angle is added to the proposed method

the effect of the steering angle on the suppression of skidding is experimentally verified.

4. Traveling experiments of lateral movement on a loose soil with a slope

4.1 Proposed "Discrete 4-Wheel-Drive typed rover"

A testbed with "Push Pull Locomotion" capability is described. Figure 13 shows the test bed rover used for running tests. The size is 450mm wide, 581mm long, 315mm high, and weighs 150N. In addition, as shown in fig. 13(b), the mounting position of the body

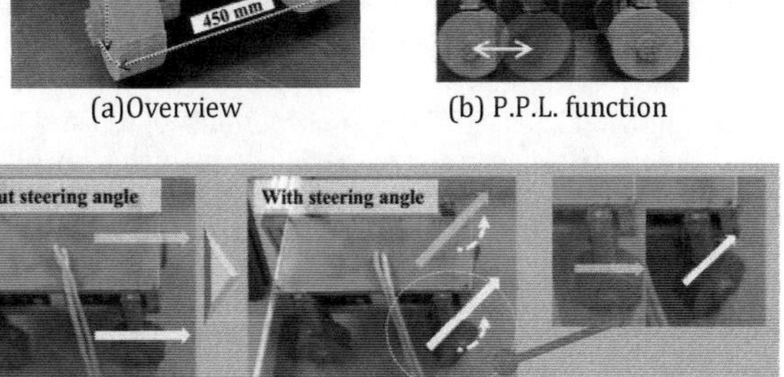

(a)Overview (b) P.P.L. function

(c) State of without steering angle and without one

Fig. 13. Testbed for traveling test on slope

and chassis can be changed actively. The stroke length is MAX 162.5mm. The testbed is equipped with four DC motors for wheel drive, four DC motors for chassis movement, and four DC motors for steering. Each motor is driven by a microcomputer (Arduino), and the parent microcomputer controls all motors. The size of the wheel is Φ160mm, width 70mm.

4.2 Environment for traveling test on loose soil

Figure 14(a)-(b) shows the soil bin used for the traveling experiments. The soil bin is 1500mm wide and 2000mm long and is covered with sand with a depth of 150mm. The outside of the soil bin is composed of a strong aluminum frame, and the slope can be made by using a jack. Figure 14(c) shows the experimental situation. In order to obtain highly accurate movement and trajectory data during the traversing, three motion capture cameras are used. Video cameras are installed in front of the rover, on top of the rover, and at an oblique angle from the upper part of the rover so that the entire rover can be imaged.

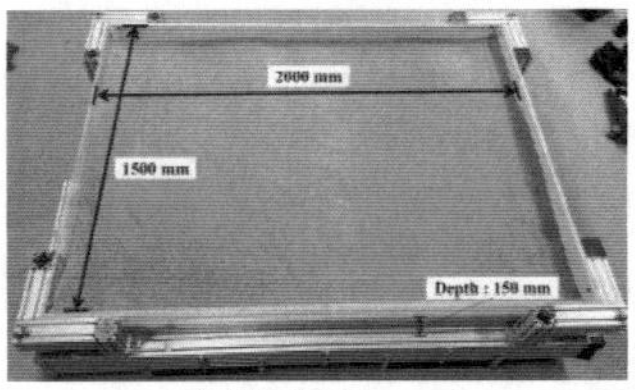

(a)Soil bin

(b) Side view of soil bin

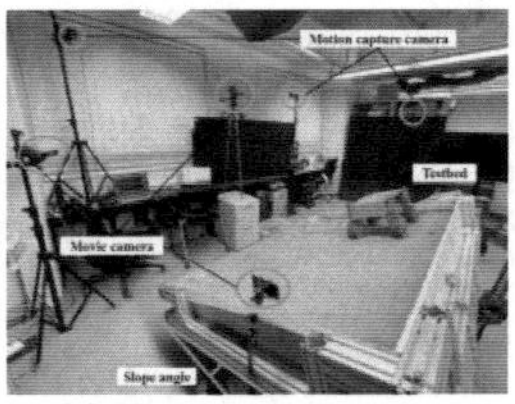

(c) Measurement state

Fig. 14. Experimental environment

4.3 Environment setting

This section describes the traveling experiments. The following four types of traveling patterns are used, and the amount of skidding is observed from the traveling trajectory for comparison.

- normal traveling (without "push pull locomotion")
- "Push pull locomotion"
- normal traveling with steering wheels (front)
- "Push pull locomotion" with steering wheels (front)

These comparisons reveal the effects of the push pull locomotion compared to the normal lateral movement, and the effects of the push pull locomotion on the lateral movement when the steering wheel is used. The experimental slopes are 15 and 20 degrees. These trials' angles are the angle at which a normal rover without push-pull locomotion is clearly skidding sideways.

5. Results and Discussion

Figure 15 shows traveling experiments of lateral movement on loose soil with a slope at an obliquity of 15 degrees. Figure 15(a) shows the experiment in normal mode (without push-pull locomotion). The rover's posture is changing, and skidding is also observed. The skidding is observed when the rover travels to the edge of the soil bin. Figure 15(b) shows the traveling condition of the rover equipped with the "Push Pull Locomotion" function. In this traveling, the amount of skidding has been reduced compared to that in the normal mode (without Push Pull Locomotion). It is effective to move the four wheels individually. However, the rover still has skidding. Figure 15(c) shows state of lateral movement when the steering angle of the wheels is set to 30 degrees. The attitude of the rover has recovered, but the downward slip is not controlled because of the lateral slip. Figure 15(d) shows state of lateral movement of the rover with steering wheels (front, 30 degrees) and "push pull locomotion function" The amount of skidding of this traveling method has been reduced greatly reduced compared to the other three types of traveling. The rover with this method has successfully traveled in the direction of lateral movement on loose soil with a slope without skidding, even under the influence of gravity. Figures 16-19 show the trajectories of the four typed traveling. These movement trajectory data were obtained by motion capture, and each traveling experiment was conducted five times. These figures show the x-axis in the direction of translation and the y-axis in the direction of slope. A smaller value on the y-axis indicates a larger amount of skidding. In the results of normal traveling (without "push pull locomotion" function) shown in fig. 16, the rover is skidding from the initial stage of traveling, and it is skidding about 500mm when it moves 800mm.

(a) normal traveling (without "push pull locomotion")

(b) "Push pull locomotion"

(c) normal traveling with steering wheels (front wheels)

(d) "Push pull locomotion" with steering wheels (front)

Fig. 15. Lateral traveling experiments on slope of loose soil (slope 15°)

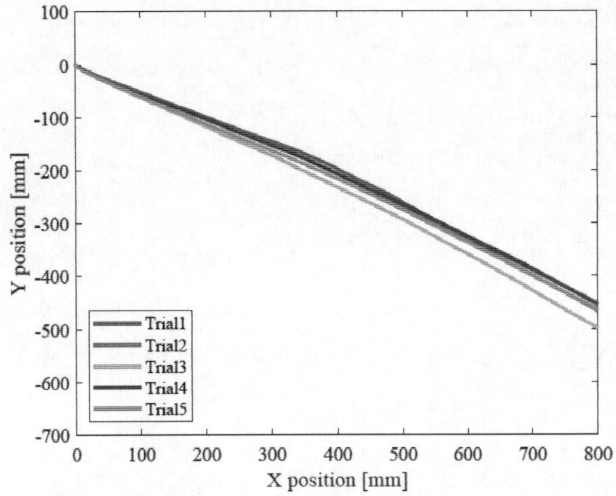

Fig. 16. Traveling trajectory of the normal rover (without "Push Pull Locomotion" function)

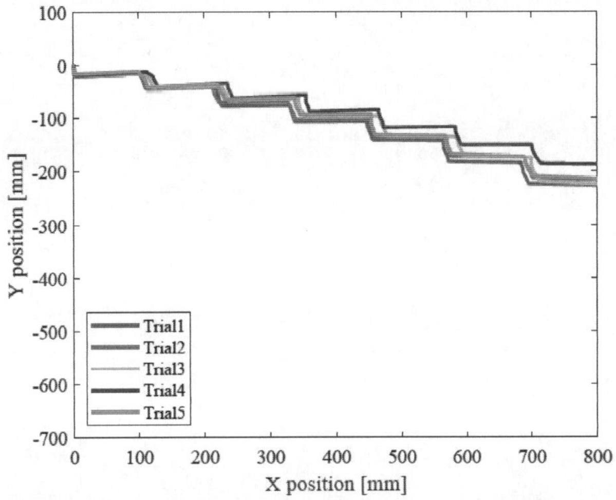

Fig. 17. Traveling trajectory of the rover with "Push Pull Locomotion" function

In the result of the rover using the "Push Pull Locomotion" function shown in fig. 17, the skidding of the rover is suppressed in the initial stage of traveling, but the rover still has suppressed in the initial stage of traveling, but the rover still has occurred skidding when the chassis and wheels move in stages. This result shows the effect of about 60% reduction of skidding compared result shows the effect of about 60% reduction of skidding compared to the normal traveling (without "Push Pull Locomotion" function). As shown in fig. 18, the traveling experiments of normal traveling with the steering wheels (front) show a decrease in the amount of skidding compared to that of normal traveling without the steering angle.

However, the rover changed its attitude significantly to the mountain side (yaw direction) due to the steering of the front wheels. Finally, the rover (with steering wheels without "Push Pull Locomotion" function) had occurred skidding about 300 mm at 800 mm travel. Figure 19 shows the results of traveling experiments of the rover with the steering wheels (front) and "push pull locomotion" function. There was no significant skidding at the beginning and during the traveling, and the proposed method succeeded in traveling without any skidding at all. The front wheels of the rover have steering angle, so the rover slightly had changed its attitude to the mountain side. Since the measurement location is the center of the markers attached to each corner of the vehicle body, the rover had skidding of 50 mm when it traveled 800 mm. However, fig. 15(d) shows that the rover almost did not have skidding at the position of the marker in the left front of the rover.

6. Conclusion and future works

This study proposed the traveling method for the rover with "Push Pull Locomotion" function and wheels with steering angle for traveling laterally on loose soil with a steep slope without skidding. In the traveling experiments, the rover equipped with the "push-pull locomotion" function and having a steering angle on the front wheels significantly reduced the amount of skidding compared to a normal rover (without the "push-pull locomotion" function). Furthermore, the proposed method is superior to the rover with the "push-pull locomotion" function (no steering angle). Finally, the proposed method was able to virtually eliminate sideslip during lateral movement on loose soils on steep slopes. However, two actuators are mounted to actively change the steering angle of the front wheels. Basically, four motors are added to realize the push-pull locomotion function. This means that the system consumes much more power than a normal rover. Therefore, to reduce the number of actuators, we will design the robot with an awareness of mechanical design. In particular, we plan to design a structure in which the beam between the wheels rotates passively according to the difference in the number of revolutions of the left and right wheels. This will allow the beam to rotate in the yaw direction and provide a steering angle equivalent to that of the wheels. And then, in the future, this study plans to update the proposed mechanical model for skidding movement to a model that incorporates subsidence and side force considered more Terramechanics-wise.

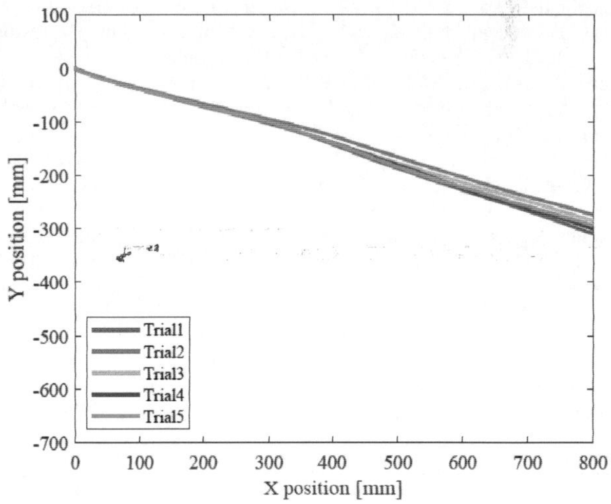

Fig. 18. Traveling trajectory of the normal rover with steering wheels (front) (without "Push Pull Locomotion" function)

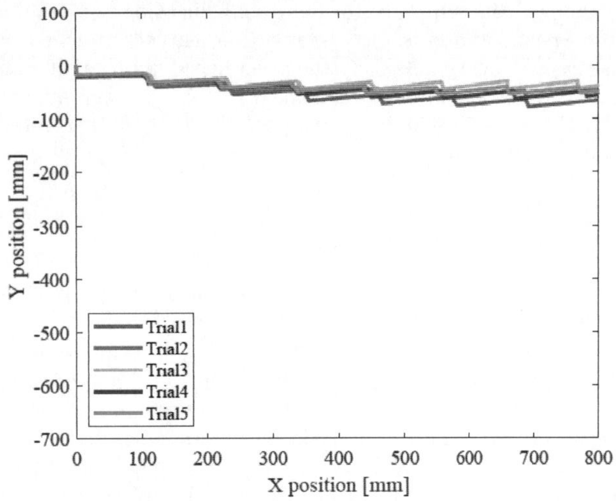

Fig. 19. Traveling trajectory of the rover with steering wheels(front) and "Push Pull Locomotion" function

7. Declaration of competing interest

The authors declare that they have no known competing financial interests or personal relationships that could have appeared to influence the work reported in this paper.

8. References

NASA, Spirit and Opportunity, https://www.nasa.gov/mission_pages/mer/freespirit.html (accessed 29 May 2023).

K. Iizuka, H. Kanamori, T. Kubota, 2005, A Study of Terramechanics for Movement on Planetary Surface, Robotics Symposia, Vol.10, pp.33-38, 2005 (in Japanese)

Junlong Guo, Weihua Li, Liang Ding, Tianyou Guo, Haibo Gao, Bo Huang, Zo ngquan Deng, 2020, High-slip wheel-terrain contact modelling for grouser-wheeled planetary rovers traversing on sandy terrains, Mechanism and Machine Theory, Vol.153, 104032.

M. Sutoh, J. Yusa, T. Ito, K. Nagatani, K. Yoshida, 2012, Traveling performance evaluation of planetary rovers on loose soil J. Field Robot., 29, pp. 648-662

Ishigami, G., Miwa, A., Nagatani, K., Yoshida, K., 2007, Terramechanics-Based Model for Steering Maneuver of Planetary Exploration Rovers on Loose Soil, J. Field Robot, (24), 233-250.

Inotsume, H.; Sutoh, M.; Nagaoka, K.; Nagatani, K.; Yoshida, 2013, K. Modeling, analysis, and control of an actively reconfigurable planetary rover for traversing slopes covered with loose soil. J. Field Robot., 30, 875–896.

Moreland, S., Skonieczny, K., Wettergreen, D., Asnani, V., Creager, C., Oravec, H., 2011. Inching Locomotion for Planetary Rover Mobility, 2011 Aerospace Conference.

Fujiwara, D., Iizuka, K., Asami, D., Kawamura, T., Suzuki S., 2019, Study on Traveling Performance for Variable Wheel-Base Robot Using Subsidence Effect, Int. J. Mech. Eng. Robot. Res., 8(2), 233-238.

Inaba, K., Iizuka, K., 2021, Proposal of a planetary exploration rover with a wheel using sinking performance for crossing slope (in Japanese), Journal of the Robotics Society of Japan, vol.39, No.10, pp.958-965.

1098 | Simulation of change in supporting force when imparting vibration by distinct element method

Tomohiro Watanabe [a,*], Ryoma Higashiyama [b], Kojiro Iizuka [c], Dai Watanabe [c]

[a] *Department of Engineering, Mechanical Engineering Program, Niigata University, Niigata City, Niigata Prefecture, Japan*

[b] *Graduate School of Systems Engineering and Science, Shibaura Institute of Technology, Saitama City, Saitama Prefecture, Japan*

[c] *Department of Science and Engineering, Machinery and Control Systems, Shibaura Institute of Technology, Saitama City, Saitama Prefecture, Japan*

* *Corresponding author: watanabe@eng.niigata-u.ac.jp*

ABSTRACT

Robots with leg mechanisms have been investigated for planetary exploration as exploration rovers with outstanding running performance. The legged rovers are efficient in moving on rough terrain because the leg mechanism has a high degree of freedom. However, the surfaces of planets (For example, Mars and the Moon, etc.) are not only rough but also loose. Owing to the inherent characteristics of loose ground, legged rovers mostly slip because the motion of their legs deforms the surface of loose ground.

To solve this problem, a walking method was proposed for preventing legged rovers from slipping. Our previous study confirmed that the shear strength of the ground and sinkage of the rover legs increased when imparting vibration. An increase in these parameters cause an increase in the supporting force, which is the resistance force exerted on the legs of the rovers by the ground. The running performance of a legged rover is related to the supporting force. The walking method was proposed based on the effects that vibration changes the ground parameters. The proposed method's effectiveness was shown by an experiment in which a legged testbed walked on loose ground.

To improve the running performance of legged rover, the supporting force must be estimated to select a suitable vibration that can efficiently increase the supporting force. In this study, the supporting force when imparting vibration to the ground was simulated by the discrete element method (DEM). The simulated result was similar to the actual change in the supporting force when imparting vibration qualitatively. This result offers valuable information for understanding how imparting vibrations to increase the supporting force.

This paper is based on a previous study (Higashiyama et al., 2023). This study presents new experimental results from a previous study and maintains the originality.

Keywords
Loose soil
Supporting force
Vibration
Terramechanics
Discrete element method

1. Introduction

Many missions are undertaken to explore the planets for human development. Perseverance, the rover developed by NASA, reached Mars in 2021 and investigated the ground environment and sources of life (NASA, 2023). Chang'e-5, the rover developed by China, reached the Moon in 2020 and succeeded in the sample return mission (Qiao et al., 2021).

In these missions, exploration robots, known as rovers, are used to move on the surfaces of planets or other extraterrestrial bodies. Many of the rovers were wheeled vehicles. However, the running performance of the current rovers must be improved to expand the area that can be explored. Recently, robots with leg mechanisms have been investigated as exploration rovers with outstanding running performance (Komizunai et al., 2010; Yeomans et al., 2013). However, the surfaces of Mars and the Moon are loose. The ground on the surface is covered with loose soil, which is referred to as regolith. Moreover, the loose ground is uneven with many slopes because of craters and is easily deformed by an external force. A walking method was proposed for legged rovers that reduces the distance they slip on loose ground using vibration (Watanabe et al., 2020a). The effectiveness of the proposed walking method was evaluated by conducting an experiment in which a legged testbed walked on loose ground.

In the proposed method, the supporting force, increased by vibration, is important. Understanding the relationship between the supporting force and vibration parameters is essential to clarify the mechanism whereby imparting vibrations increases the supporting force. This relationship offers valuable information as a design guideline for the walking motion.

The running performance of a legged rover is considered to be related to the supporting force, which is the resistance force exerted on the legs of the rover by the ground. In a previous study, this relationship was investigated (Watanabe et al., 2020b). In the next step, the supporting force must be estimated to select a suitable vibration that can efficiently increase the supporting force for improving the running performance of rover. Estimating the supporting force offers valuable information for understanding how imparting vibrations increases the supporting force. In this study, the supporting force when imparting vibration was simulated to the ground by the discrete element method (DEM). The changes in the ground when imparting vibration are complex because the ground particles receive compaction, shearing, and friction. The DEM method is suitable for estimating these changes because it can reproduce the movement of each particle.

In this study, the experiment in which the rod was dragged on the ground after imparting vibration from the rod to the ground was simulated. The supporting force was measured when the rod was dragged on the ground. Moreover, the simulated results and experimental results were compared.

2. Experimental method for measuring support force

2.1 Explanation of increasing supporting force by vibration

Figure 1 shows the change in supporting force caused when the leg is dragged on the ground. The supporting force increases immediately after dragging the leg because the movement of the leg compacts the ground. After the supporting force reaches a peak value, it decreases because the ground crumbles. The peak value of the supporting force is increased by vibration (Fig. 1). This factor is considered that the ground is compacted, and the density of the ground is increased by vibration.

https://doi.org/10.56884/HKEZ2827
ISBN 978-1-942112-55-6 (pdf)

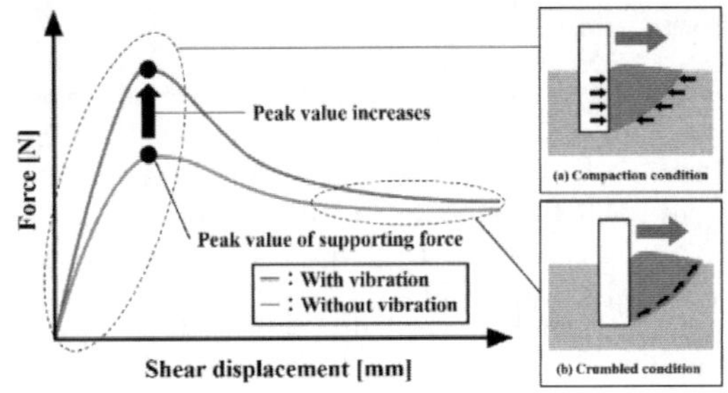

Fig. 1. Supporting force vs shear displacement (Higashiyama et al., 2023)

2.2 *Experimental method*

For comparison with the results of the DEM simulation, the rod emulated the rover's leg was dragged on the ground, and the supporting force was measured. Figure 2 shows the experimental setup that consists of a soil tank, rod, and force sensor. Figure 3 illustrates the flow of the experiment. First, the ground was mixed and flattened. Then, the rods were placed on the ground. The sinkage of the rod were 3cm, 5 cm, and 7cm. Next, the vibration was generated for 30 s. Subsequently, the rod was dragged after the vibration ceased. The force sensor measured the supporting force value. The rod was moved at a speed of 8.87 mm/s and dragged for 5 seconds. The soil tank had a length, width, and height of 309, 439, and 300 mm, respectively. Figure 2 (b) shows the size of the rod, which is a quadrangular prism. The bottom end of the rod is square, with sides of 20 mm, and the rod was 200 mm long. The ground material consisted of Silica No. 5. This experimental machine was mounted on a vibration generator (Wave Maker 05), which propagates vibrations to the rod. The frequency of the vibration was set to 10 Hz and 30 Hz. The amplitude of the vibration was set to 1 mm. The direction of the vibrations was the same as that of the dragging direction of the rod. Five trials were performed for each experimental condition. These values of sinkage and vibration parameters were referred to a previous study (Watanabe et al., 2020b). Table 1 presents the experimental conditions.

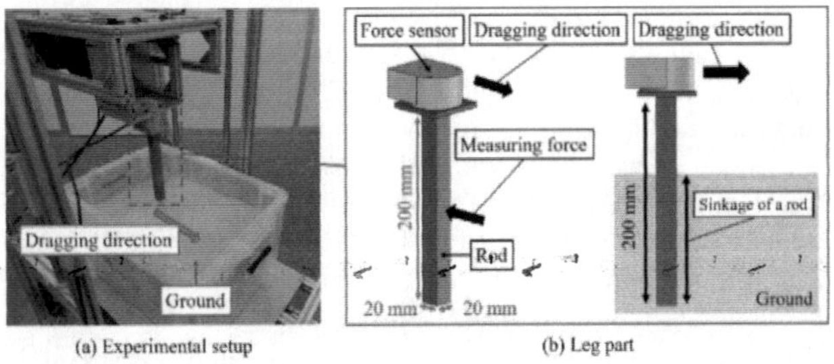

Fig. 2. Experimental setup for measuring supporting force (Higashiyama et al., 2023)

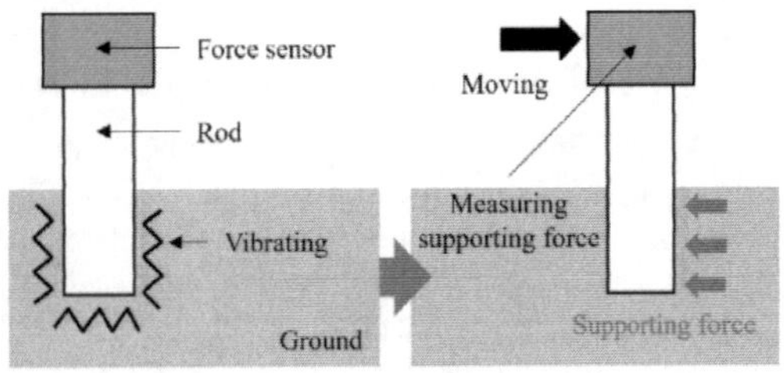

Fig. 3. Experimental flow for measuring supporting force (Higashiyama et al., 2023)

Table 1
Conditions for measuring supporting force

Item	Conditions (value)
Number of trials	5
Sinkage	3 cm, 5 cm, 7cm
Vibration time	30 s
Traction speed	8.87 mm/s
Traction time	5 s
Kinds of sand	Silica No. 5
Vibration generator	Wave Maker 05
Vibration amplitude	1 mm
Vibration frequency	10 Hz, 30 Hz
Force sensor	PFS055YA251U6

3. Simulated method for measuring supporting force

3.1 *Explanation of DEM simulation model*

In the DEM method, contact forces that are elastic repulsion and friction are modeled, and the particle's movement is calculated from these contact forces based on the equation of motion. Figure 4 shows the DEM model which is used in this paper. In the normal direction, the spring and dashpot connect parallel to each particle (Fig. 4 (a)). In the tangential direction, the friction slider is in the model additionally to express the friction (Fig. 4 (b)). The movement of particles can be calculated in Eq. 1 and Eq. 2. K_i is the number of particles that touch particle i.

$$m_i \frac{dv_i}{dt} = \sum_{j-1}^{k_i} \left(F_{n_{ij}} + F_{t_{ij}}\right) + m_i g \tag{1}$$

$$I_i \frac{d\omega_i}{dt} = \sum_{j-1}^{k_i} \left(r_i F_{t_{ij}}\right) + R_i \tag{2}$$

n and t mean normal direction and tangential direction. Where t is time, m is a particle's mass, v is a particle's velocity, ω is a particle's angular velocity, g is gravity acceleration, and I is a moment of inertia. R_r is rotational resistance which expresses rolling friction between particles. $F_{n_{ij}}$ is a normal force between particle i and j. $F_{t_{ij}}$ is a tangential force between particle i and j. $F_{n_{ij}}$ and $F_{t_{ij}}$ is calculated in Eq. 3 and Eq. 4.

$$F_{n_{ij}} = K_n u_n + \eta_n \frac{du_n}{dt} \tag{3}$$

$$F_{t_{ij}} = \min\{\mu_p F_n, K_t(u_t + r\varphi_r) + \eta_n(\frac{du_t}{dt} + r\frac{d\varphi_n}{dt})\} \tag{4}$$

K is a spring constant, η is a damping coefficient of dashpot, u is relative displacement, μ_p is relative angular displacement, m is the coefficient of friction between particles. Rotational resistance R_r is calculated in Eq. 5.

$$R_r = -\mu_r b |F_n| \frac{\varphi_r}{|\varphi_r|} \tag{5}$$

μ_r is the coefficient of rolling friction between particles, and b is the radius of the particle. The spring constant between particles in the normal direction K_n is calculated in Eq. 6.

$$K_n = \frac{4}{3\pi}\left(\frac{1}{\delta_i + \delta_j}\right)\sqrt{\frac{r_i r_j}{r_i + r_j}} \tag{6}$$

$$\delta_i = \frac{1 - v_i^2}{E_i} \tag{7}$$

$$\delta_j = \frac{1 - v_j^2}{E_j} \tag{8}$$

E is Young's modulus of a particle and a wall, and v is Poisson's ratio. K_n, the spring constant when a particle and a wall touch, is calculated in Eq. 9.

$$K_n = \frac{4}{3\pi}\sqrt{r_i}\frac{1}{\delta_i+\delta_w} \tag{9}$$

$$\delta_w = \frac{1-\nu_w^2}{E_w\pi} \tag{10}$$

The spring constant between particles in the tangential direction K_t is calculated in Eq. 11.

$$K_t = \frac{K_2}{2(1+\nu_i)} \tag{11}$$

The viscosity of dashpot η_{ij} is calculated in Eq. 12.

$$\eta_{ij} = 2\sqrt{\frac{2m_i m_j}{m_i+m_j}K_m} \tag{12}$$

The parameters of particles on the DEM simulation were set to fit the angle of repose, measured in the experiment, and simulated in the DEM method. This decision method of particle parameters is referred from the study by Nakano et al. (Nakano et al., 2012). Table 2 shows particle parameters on the DEM simulation.

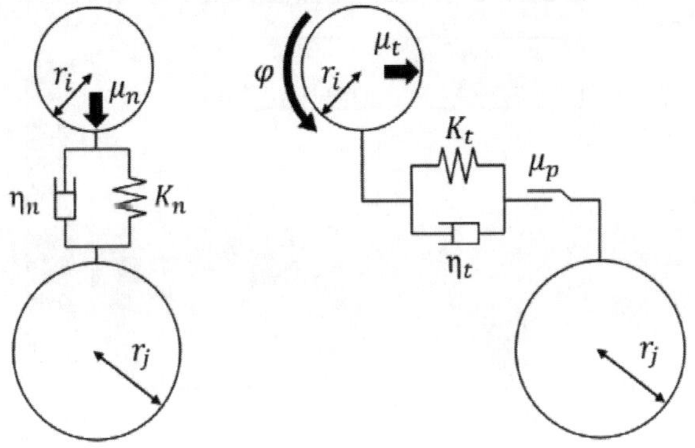

(a) Normal direction (b) Tangent direction

Fig. 4. Simulation model (Higashiyama et al., 2023)

Table 2
Simulation parameter

Parameter	Symbol	Unit	Value
Normal damping coefficient	η_n	-	0.1
Tangential damping coefficient	η_t	-	0.1
Friction coefficient (Particle to particle)	-	-	25
Rolling friction coefficient (Particle to particle)	-	-	0.01
Normal spring coefficient	K_n	N/mm	0.01
Tangential spring coefficient	K_t	N/mm	0.0029
Friction coefficient (Particle to wall)	-	-	0.1
Rolling friction coefficient (Particle to wall)	-	-	0.01
Density	ρ	g/cm³	1.43
Young's modulus	E	MPa	10000
Poisson's ratio	ν	-	0

3.2 Simulated method

The simulation, which was the same as the previous section experiment, was conducted. Figure 5 illustrates the environment and flow of the simulation. First, the rods were placed on the ground. The sinkage of the rod was 5 cm. The simulation was conducted in the case without and with vibration. In the case with vibration, the vibration was generated for 30 s. Subsequently, the rod was dragged after the vibration ceased. The rod was moved at 8.87 mm/s and dragged for 5 seconds. The rod size is the same as in Fig. 2 (b). The shape of the vibration wave is a triangular wave to imitate a sine wave. Table 3 presents the simulated conditions.

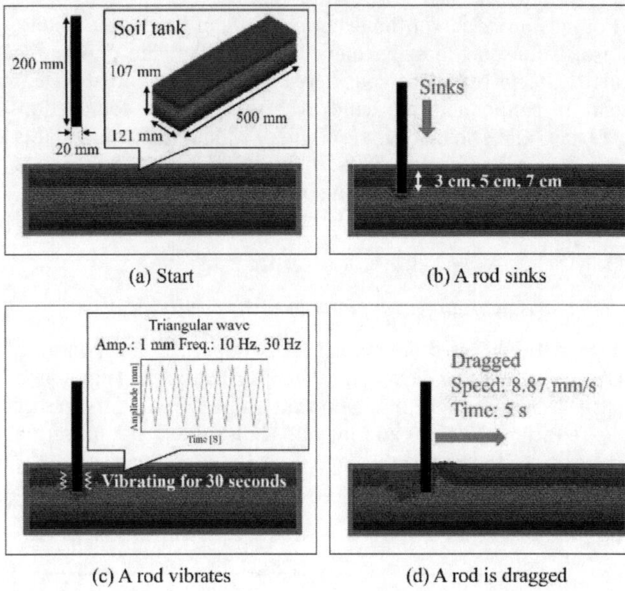

Fig. 5. Simulation flow for measuring supporting force (Higashiyama et al., 2023)

Table 3
Condition for simulating

Item	Conditions (value)
Sinkage	3 cm, 5 cm, 7 cm
Vibration time	30 s
Traction speed	8.87 mm/s
Traction time	5 s
Traction wave	Triangular wave
Vibration amplitude	1 mm
Vibration frequency	10 Hz, 30 Hz

4. Comparison of experimental and simulated results

4.1 Variation in the supporting force against the shear displacement

The variation in the supporting force against the shear displacement is shown in Fig. 6 (without vibration, Sinkage: 5 cm) and Fig. 7 (with vibration, Amp.: 1 mm, Freq.: 30 Hz, Sinkage: 5cm). These figures show the experimental and simulated results. The variation in the supporting force against the shear displacement exhibited two types of shapes. The type of variation shown in Fig. 6 was observed in certain experiments without vibration. In this type, the supporting force initially increased, whereas after, it converged. The type of variation in Fig. 7 was observed in all experimental results when vibration was imparted and in some experimental results without vibration. In this type, the supporting force initially increased, after which it decreased and converged. The variations in the supporting force against the shear displacement in the experiment and the simulation were similar.

4.2 Comparison of supporting force in each sinkage of a rod

Figure 8 and Figure 9 compare the peak value of supporting force in each sinkage condition. Figure 8 shows the experimental results. This graph shows the standard errors calculated from the five experimental trials. Figure 9 shows the simulated results. In both graphs, the larger sinkage of rod, the more peak value increased. In all conditions, the peak value with vibration is larger than without vibration.

In the comparison of the experimental and simulated results, the peak value in the experiment is larger than the one in the simulation. The reason is considered that the sand in the DEM simulation cannot express sand's characteristic of compaction. Expressing sand's characteristic of compaction is important in this study because Vibration makes the sand compacted. It is considered that particle's parameters in the DEM simulation are needed to adjust for expressing sand's characteristic of compaction.

4.3 Comparison of supporting force in each vibration

Figure 10 shows the peak value of supporting force in each vibration condition and compare the experimental and simulated result. The bars of experimental results show the standard errors calculated from the five experimental trials. The larger vibration used, the more the peak value increased. In the comparison of the experimental and simulated results, the peak value in the experiment is larger than the one in the simulation.

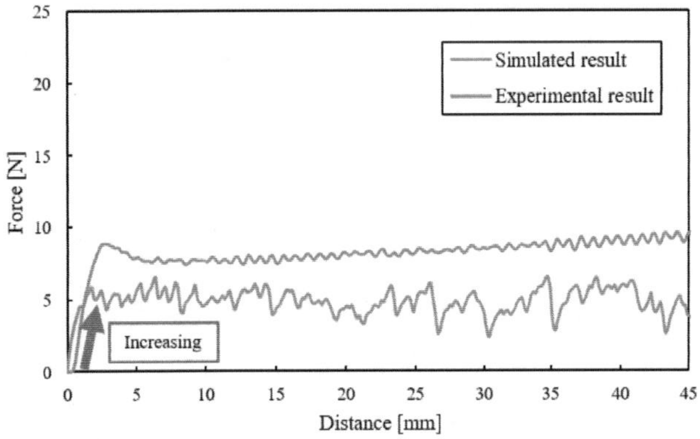

Fig. 6. Supporting force vs distance (without vib., Sinkage.: 5 cm, (Higashiyama et al., 2023)

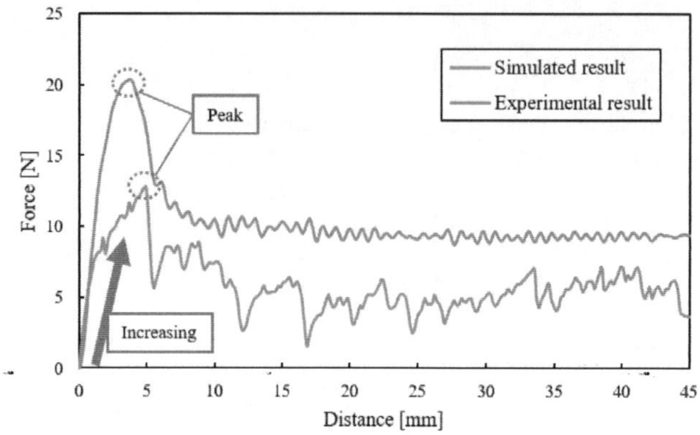

Fig. 7. Supporting force vs distance (Vibration's amplitude is 1mm. Vibration's frequency is 30 Hz. Sinkage of a rod is 5 cm, (Higashiyama et al., 2023)

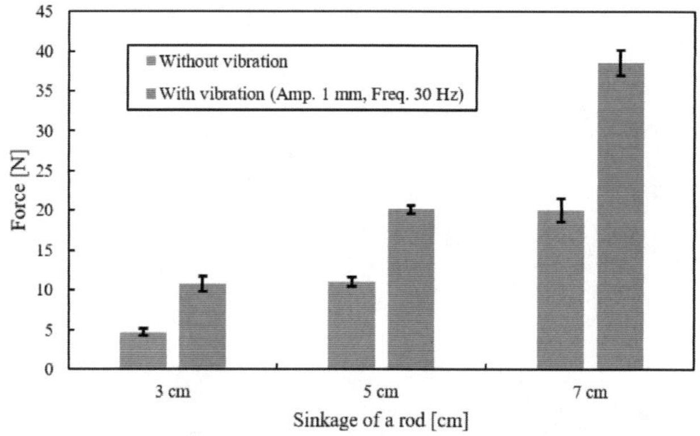

Fig. 8. Comparison of supporting force in experiments (Vibration's amplitude is 1mm. Vibration's frequency is 30 Hz. Sinkage of a rod are 3 cm, 5 cm and 7 cm)

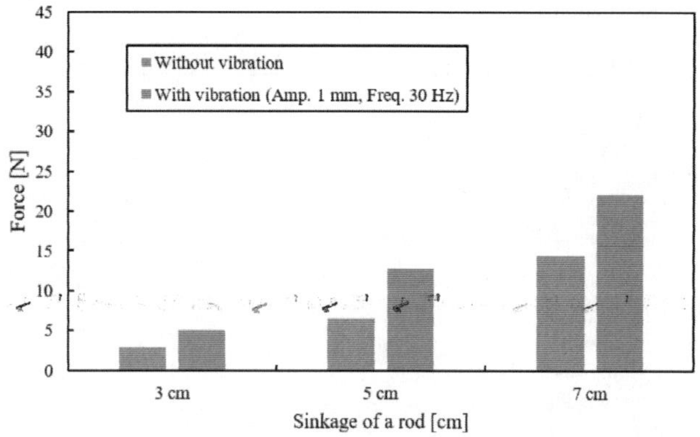

Fig. 9. Comparison of supporting force in simulation (Vibration's amplitude is 1mm. Vibration's frequency is 30 Hz. Sinkage of a rod are 3 cm, 5 cm and 7 cm)

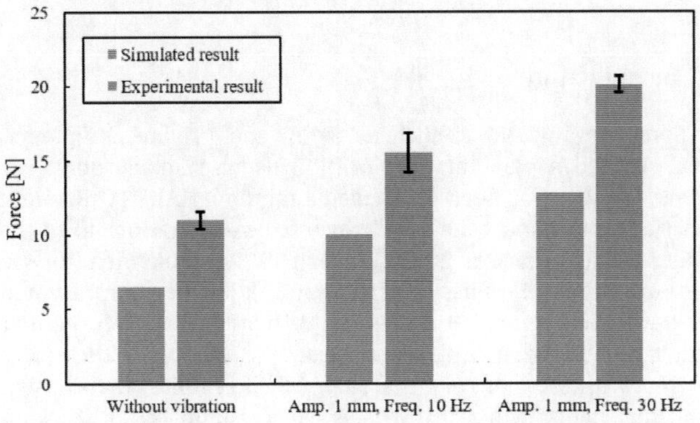

Fig. 10. Comparison of supporting force in each vibration (Vibration's amplitude is 1mm. Vibration's frequencies are 10 Hz and 30 Hz. Sinkage of a rod is 5 cm)

5. Conclusions and future work

In this study, the change in the supporting force by vibration was simulated in the DEM method. The experiment in which the rod was dragged on the ground after imparting vibration from the rod to the ground was simulated. The supporting force was measured when the rod was dragged on the ground. In the comparison of the experimental and simulated results, the following four common points were confirmed.

1. The variations in the supporting force against the shear displacement in the experiment and the simulation were similar.
2. The peak value with vibration is larger than without vibration.
3. The larger sinkage of a rod, the more the peak value increased.
4. The larger vibration used, the more the peak value increased.

From these experimental results, it has become clear that the supporting force can be simulated qualitatively in the DEM method.

In this study, the experimental conditions were limited because there are the initial stages of this study. In a future study, the simulation will be carried out using more experimental conditions (kinds of sand, sinkage of a rod, and vibration parameters) for evaluation in detail. The peak value in the experiment is larger than the one in the simulation. The reason is considered that the sand in the DEM simulation cannot express sand's characteristic of compaction. It is considered that particle's parameters in the DEM simulation are needed to adjust for expressing sand's characteristic of compaction. Therefore, the experimental method for selecting particle's parameters will be proposed to express sand's characteristic of compaction. Finally, the DEM simulator will be improved to fit these values.

6. Acknowledgements

This study was supported by JSPS KAKENHI (Grant Number JP 21J11944). Moreover, this work was supported by REFEC (Research Foundation for the Electrotechnology of Chubu).

7. Declaration of competing interest

The authors declare that they have no competing interests.

8. References

Higashiyama R., Watanabe T., Iizuka K., Watanabe D., 2023. Simulation of distinct element method to analyses of increased supporting force by imparting vibration for improving walking performance of lunar exploration legged rover, in: Nagoya, Japan, JSME annual Conference on Robotics and Mechatronics.

Komizunai S., Konno A., Abiko S., and Uchiyama M., 2010. Development of a static sinkage model for a biped robot on loose soil, in: Sendai, Japan, 2010 IEEE/SICE International Symposium on System Integration.

Nakano M., Abe T., Kano J., and Kunimoto K., 2012. DEM analysis on size segregation in feed bed of sintering machine, ISIJ Int., 52, 1559-1554.

NASA, MARS 2020 MISSION PERSEVERANCE ROVER, https://mars.nasa.gov/mars2020/mission/overview/ (accessed 19 May 2023).

Qiao, L., Chen, J., Xu, L., Wan, S., Cao, H., Li, B., and Ling, Z., 2021. Geology of the Chang'e-5 landing site: Constraints on the sources of samples returned from a young nearside mare, Icarus, 164.

Watanabe, T., Iizuka, K., 2020a. Proposal of walking to reduce slipping behavior using compaction effect of loose soil caused by a propagation of vibration for small light lunar planetary exploration rovers with legs, Tra. JMSE, 86.

Watanabe, T., Iizuka, K., 2020b. Study on Connection between Kind of Vibration and Influence of Ground for Leg Typed Rovers, Int. J. Mech. Eng. Rob. Res., 9.

Yeomans B., Saaj C. M., and Winnendael M. V., 2013. Walking planetary rovers – experimental analysis and modelling of leg thrust in loose granular soils, J. Ter., 50, 107-120.

1453 | Suppressing the reduction of the traveling displacement on loose soil for rovers with function of a wheel-walking

Daisuke Fujiwara [a,*], Qingze He [b], Kojiro Iizuka [c]

[a] *Department of Mechanical and Electrical Engineering, Suwa University of Science, Nagano, Japan*

[b] *Graduate School of Engineering and Science, Systems Engineering and Science, Shibaura Institute of Technology, Saitama, Japan*

[c] *Department of Systems Engineering and Science, Machinery and Control Systems, Shibaura Institute of Technology, Saitama, Japan*

* *Corresponding author: fujiwara@rs.sus.ac.jp*

ABSTRACT

The rovers with cylindrical wheels have a risk of failing to move in the loose soil area, such as the surface of the lunar/planet. The movement using the resistive force of the locked wheel, for example, wheeled walking, or push-pull locomotion, can reduce the risk. In the locomotion, the rover fixes a wheel to the ground and pushes or pulls the other wheel. Many studies have developed these kinds of rovers, for example, Scarab, Marsokhod, and ExoTeR. Furthermore, our previous studies confirmed that the resistive force increases as the sinkage of the locked wheel increases. This effect suppresses the reducing the traveling distance. However, the problem is that increasing the traction load reduces the amount of traveling distance, especially in loose soil with a high slope. To address this problem, this paper focuses on the relationship between wheel rotation and pushing or pulling velocity. Particularly, this paper changes the amount of wheel slip during the push and pulling sequence. To confirm the effect, this paper firstly measures the drawbar pull force acting on the locked wheel. This paper secondly performs a traveling experiment on loose soil using a small test bed rover. The traveling methods are wheel-walking, wheel-walking with large sinkage, and rotational wheel traveling. The slope angles are 15, 20, 25, 30, and 32. The wheel is a lug wheel. The experimental results indicated that the method using the large slip during the push or pull scheme suppressed the reduction of the displacement.

Keywords
Wheel-walking
Push-pull locomotion
Planetary rover
Slip effect
Resistive force

1. Introduction

Some organizations, such as JAXA, and private companies, have planned a planetary exploration using a small robot. For example, ispace has been proceeded a mission, HAKUTO-R, and it has tried to transport the small rover SORA-Q and Rashid to the Lunar surface (ispace, 2023). Also, JAXA has planned to use SORA-Q in the SLIM mission (JAXA, 2023). Furthermore, Dymon has developed a small rover, Yaoki (Dymon, 2023), for Lunar exploration. Planetary surface, for example, the Lunar surface, is covered with a lot of rough terrain, called regolith; therefore, a small rover with cylindrical wheels has a risk of getting stuck on the rough surface.

To solve this problem, previous studies have developed some wheel-walking robots (Fig. 1). In this locomotion, the rover locked one wheel relative to the ground, and the wheel supports the other wheel locomotion. In a previous study, Creager et al. (2015) developed a push-pull locomotion rover, Scarab (Fig. 1 ①). This study indicated that the drawbar pull force of the push-pull locomotion rover was higher than that of the normal traveling. Kermurdjian et al. (1992) and Andrade et al. (1998) indicated the concept of the 6-wheel rover, Marsokhod (Fig. 1 ②). The paper indicated that its specification of the possible angles for traveling loose soil with a slope was 30-35 degrees. Furthermore, our previous study (Fujiwara et al., 2019) confirmed that an increase in sinkage of the locked wheel contributed to an increase the resistive force of the wheel. The authors confirmed that this effect could improve allowable traveling slope angles on rough terrain.

In that locomotion, the supporting of the locked wheel is an important factor for supporting the other moving wheel. Therefore, we have investigated the relationship between sinkage and resistive force in a previous study, whereas the investigation of the other driving wheel's phenomenon has not been enough. Also, the problem is that increasing the traction load reduces the amount of traveling distance, especially in loose soil with a high slope. According to terramechanics, the drawbar pull force of the rotational wheel increase with an increase in slip ratio. Then, the drawbar pull force reaches a certain maximum at an intermediate slip ratio (Ishigami et al., 2007, Bekker, 1969). This paper considers that this relationship can apply to the driving wheel of the wheel-walking on rough terrain. For example, when the drawbar pull force of the rotational wheel is improved by increasing the slip ratio, the force of the rotational wheel makes up for the resistive force of the locked wheel. This leads to suppressing the reduction of the traveling displacement in the wheel-walking rover.

Therefore, the objective is to suppress the reduction of the traveling displacement for the wheel-walking rover using the effect of an increasing drawbar pull force of the rotational wheel.

First, to confirm this effect, this paper validates the relationship between drawbar pull force and slip ratio using a test bed rover with a force sensor. Second, this paper confirms the traveling displacement of the rover on rough terrain at each slope angle. Finally, we evaluate its effect.

https://doi.org/10.56884/WXAI8535
ISBN 978-1-942112-55-6 (pdf)

Fig. 1. Concept of wheel-walking rovers

2. Method

Figure 2 shows the locomotion that this paper mainly focuses on. In this locomotion, the rover locks one pair of wheels relative to the ground and moves the other pair of wheels using the resistive force of the locked wheels. During the wheel rotation of the other pair of wheels, the rotational wheel generates drawbar pull force. According to terramechanics, the drawbar pull force F_x of the rotational wheel is

$$F_x = rb\int_{\theta_r}^{\theta_f}\{\tau_x(\theta)\cos\theta - \sigma(\theta)\sin\theta\}d\theta \qquad (1)$$

where r and b are a wheel radius and width. Also, θ_r and θ_f are the contact angle between a wheel and soil. $\tau_x(\theta)$ is the shear stress of the wheel. $\sigma(\theta)$ is the normal stress of the wheel. Previous study indicated that $\sigma(\theta)$ is changed depending on the slip ratio (Ishigami et al., 2007). The slip ratio is

$$s = 1 - v_w/r\omega \qquad (2)$$

where v_w is the traveling speed of the rover when the rover moves in a normal rotational mode. In the driving mode, the slip ratio is $0 < s < 1$. The slip ratio becomes large when the wheel slip on the loose soil. In the case of wheel-walking mode, v_w is regarded as the speed of linear actuator v_l. The speed of the pair of driving wheel cannot exceed speed of the v_l. The slip ratio becomes large when the $r\omega$ is large compared with v_l.

Meanwhile, the passive earth force F_p acts on the locked wheel when the slope resistance acts on a rover's body during climbing. Our previous study (Fujiwara et al., 2019) confirmed that the resistive force F_p increased as an increase in sinkage h and normal force. When we use this effect, the rover sinks each locked wheel before step 1 and step2. This locomotion is a wheel-walking locomotion with the large sinkage.

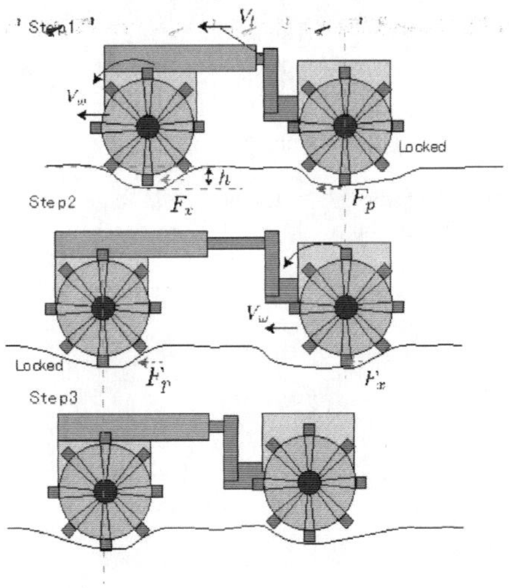

Fig. 2. Wheel-walking locomotion of our study

While the 4-wheel rover pulls or pushes the wheel on the loose soil with a high slope, the sum of the F_x and F_p has to be more than the slope resistance to climb.

$$2F_x + 2F_p \geq mg\sin\theta \qquad (3)$$

where m is the mass of a rover, θ is a slope angle.

In summary, the forces F_x and F_p are an important factor for the wheel-walking. Also, this paper assumes that the force F_x can be increased when the rover intentionally slips the wheel.

3. Experiment

This paper performed the experiments for measuring the drawbar pull force of the rotational wheel during the wheel walking and evaluated traveling performance at different wheel rotational speeds. Table 1 summarizes the conditions.

3.1 Measuring the force acting on locked wheel during extending/shrinking wheelbase

To measure forces acting on the locked wheel during changing the wheelbase length, a wheel unit is attached under the slider parts (Fig. 3). The slider parts locked the wheel unit in a longitudinal direction, but the wheel unit freely moves to a vertical direction. The force sensor which is set between the linear actuator and body measures the push/pull force generated by the rotational wheel while extending/shrinking the wheelbase. Each mass of the wheel unit is 0.78 kg. When the rover extends its wheelbase, we measure the pull force. Meanwhile, we measure the push force when the rover shrinks its wheelbase. The moving speed of the linear actuator is 20 mm/s (constant). However, the wheel rotational speed is 5-26 rpm, which means the slip ratio is 0.23-0.85. The wheel width and diameter are 26 and 100 mm (including lag's height), respectively.

Table 1
Experimental conditions

Description [Units]	Value
Wheel width b[mm]	26
Wheel diameter including lug's height [mm]	100
Number of lugs [-]	18
Mass of one wheel unit [kg]	0.78
Rotational wheel speed [rpm]	5-26
Linear actuator speed v_l[mm/s]	20
Slip ratio s[-]	0.28-0.85
Soil [-]	Silica sand No.5
Soil density [g/cm^3]	1.43
Slope angle θ_d[°]	15-32

※When the rover can move enough with a low wheel rotational speed on a low slope angle (e.g. 15°), the rotational speed was only 5 rpm. The traveling experiment of the push-pull mode with large sinkage were only performed at the slope angle 32° (5 and 26 rpm).

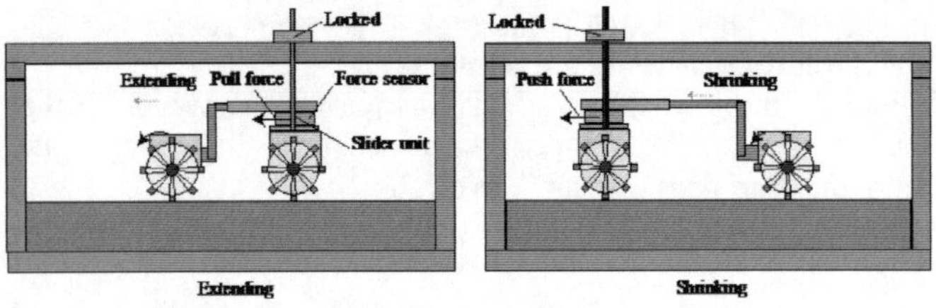

Fig. 3. Equipment for measuring push/pull force

3.2 Traveling experiment on the rough terrain with slope

To confirm the amount of slip of the wheel-walking rover, this paper develops a test bed rover as indicated in Fig. 4. The rover length, width, and height are approximately 283, 174, and 107 mm, respectively. The mass is 1.56 kg (including battery and circuit). The wheel dimensions are the same as the one used in section 3.1. The slope angles are 15, 20, 25, 30, and 32. The locomotion modes are a normal traveling, wheel-walking (push-pull) locomotion, wheel-walking locomotion with large sinkage. The rover travels toward the longitudinal direction by 3-4 sequences at different wheel rotational speeds (5-26 rpm) in the case of the wheel-walking locomotion. Meanwhile, the rotational speed of the normal traveling is only 5 rpm. To confirm the amount of slip, the motion capture system measures the displacement. The positions of the marker are on the top of each wheel unit and the center of the linear actuator. The plate smooths the soil surface before the experiment. The evaluation metric is traveling reduction s_d :

$$s_d = 1 - X_a/X_i \tag{4}$$

where X_a is the actual distance of the total sequences or traveling, and X_i is the ideal distance of the total sequences or traveling. Therefore, s_d becomes large when the rover's displacement is small.

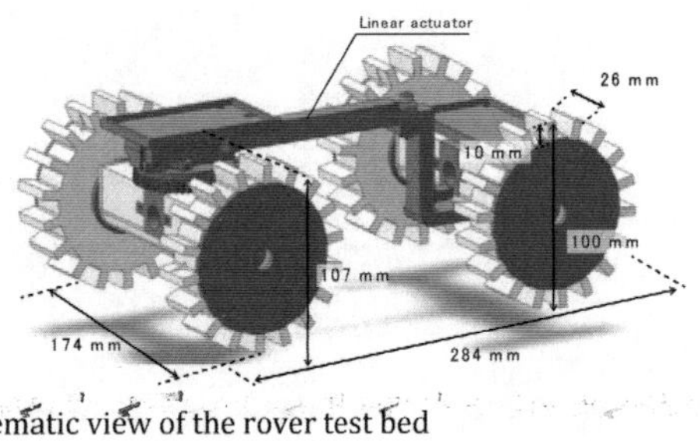

Fig. 4. Schematic view of the rover test bed

4. Result and Discussion

4.1 The force of the locked wheel during extending /shrinking

The force acted on the locked wheel unit during extending or shrinking the other wheel unit (Fig. 5). First, the force rapidly increased, then the force trend became declining. The phenomenon of the rapid increase could be considered as a cause of the acceleration of the wheel. Also, the wheel sunk into the soil during wheel rotation; therefore, the direction of the push/pull force changed, and this could decrease the force of the longitudinal factor. The results indicated that the force at the high slip condition was larger than that of the low slip condition.

To compare the force at each condition, this paper got the maximum value of the force. From Fig. 6, the force gradually increased as increased slip ratio. The trends of these results corresponded to the relationship between the slip ratio and drawbar pull force.

4.2 Amount of displacement of the test bed rover

When the test bed rover climbed on loose soil with a slope while extending and shrinking its wheelbase (Fig. 7), the movement trajectory was a stepwise increase (Fig. 8).

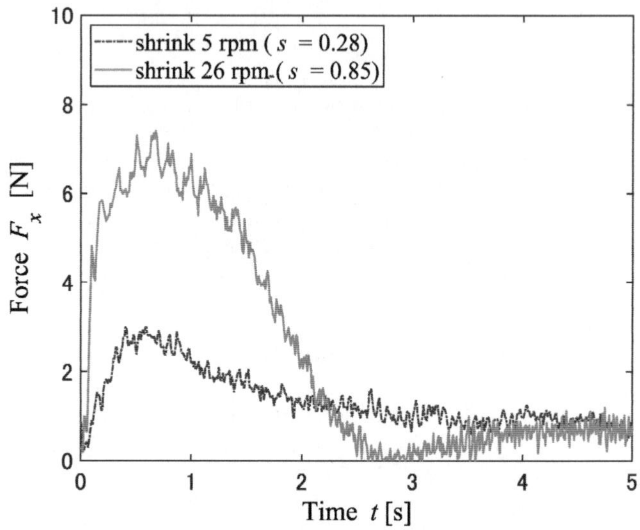

(a) Pull force during extending wheelbase

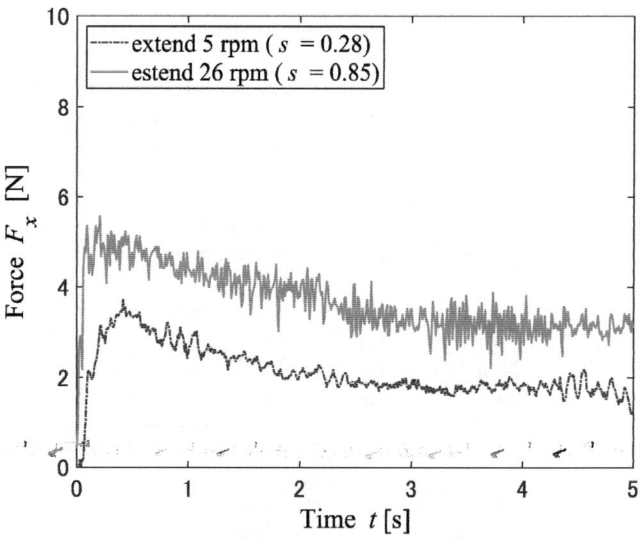

(b) Push force during shrinking

Fig. 5. The data is the typical data. The blue and red line mean the force in the case of low and high slip conditions

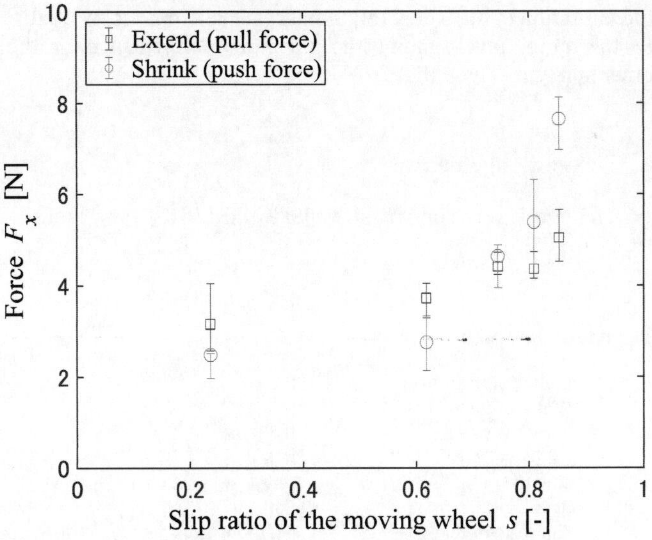

Fig. 6. The resistive force at different slip ratio. The plotted data is an average of five trials. The data is the maximum value of the force at each condition. Each bar means maximum and minimum values at each experimental condition

Fig. 7. Rover climbing the loose soil with slope

Both push-pull modes suppressed the reduction of displacement. The slip ratio of the normal traveling mode was large when the slope over 25° (Fig. 9). On the other hands, the slip ratio s_d of the traveling decreased as increased the wheel slip ratio s (Fig. 10).

When the wheel slip s is low (0.28) at the surface with the slope angle 32°, the slip ratio of the push-pull mode with large sinkage is lower than that of the normal push-pull mode. Meanwhile, the slip ratio s_d of both push-pull modes are approximately the same values (the slip ratio of the push-pull with large sinkage is slightly low) when the wheel slip s is high (0.85). This was because we confirmed that the sinkage at the high slip condition was larger than that of the low slip condition. This suggested that the resistive force could be high at the high slip condition in addition to an increase of the drawbar pull force of the wheel. These results suggested that both an increase of the drawbar pull force and the resistive force are important for traveling.

In summary, the results suggested that increasing the drawbar pull force of the rotational wheel could contribute to suppressing the reduction of the slip for walking locomotion on loose soil. Although this rover is 4-wheel and uses a relatively small wheel, the traveling experiment suggested that the rover could climb the high slope close to the angle of repose.

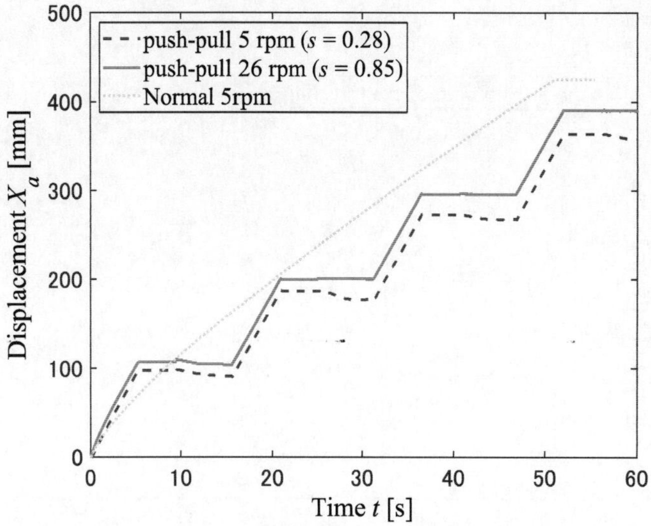

(a) Slope angle 20°

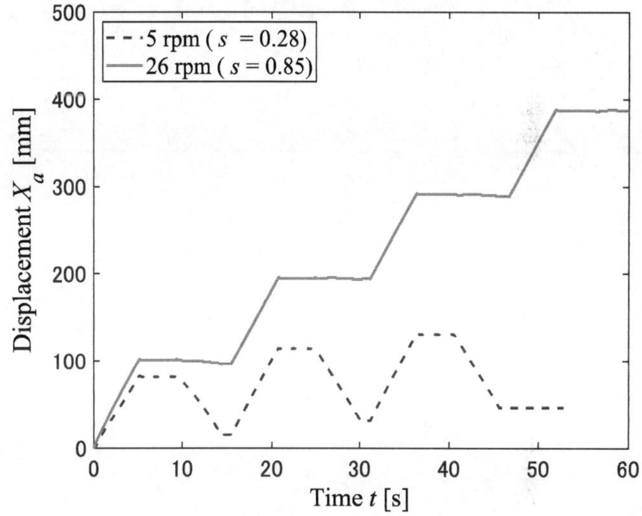

(b) Slope angle 32°; push-pull mode

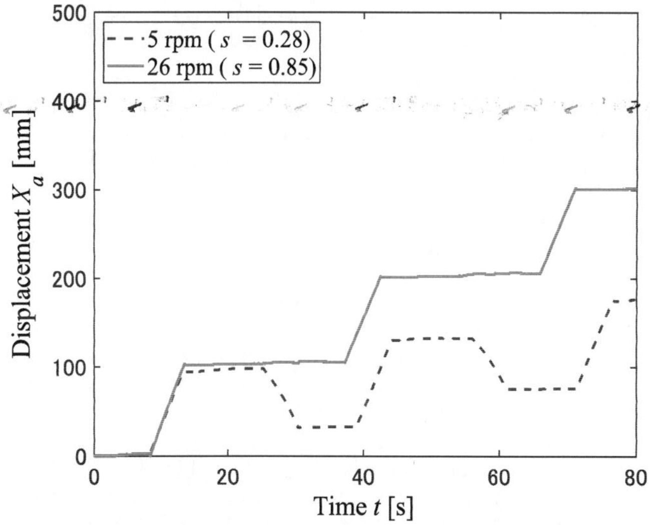

(c) Slope angle 32°; push-pull locomotion with large sinkage

Fig. 8. One example of the displacement trajectory

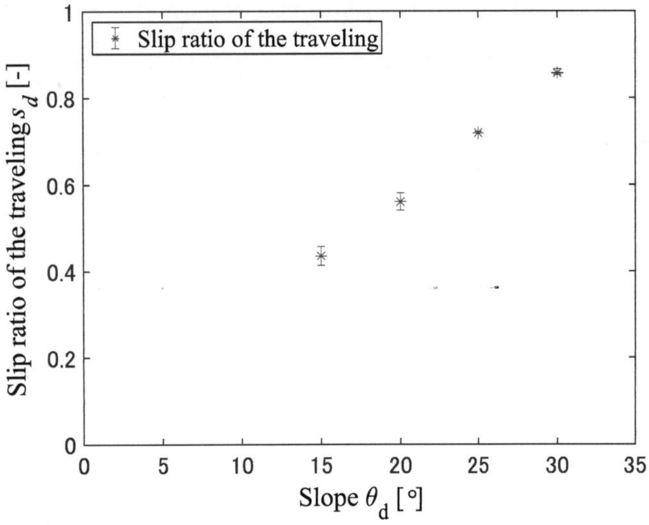

Fig. 9. Slip ratio s_d at different slope angles in the case of the normal traveling. Each plotted data is an average of the five trials. Each bar means the maximum and minimum values

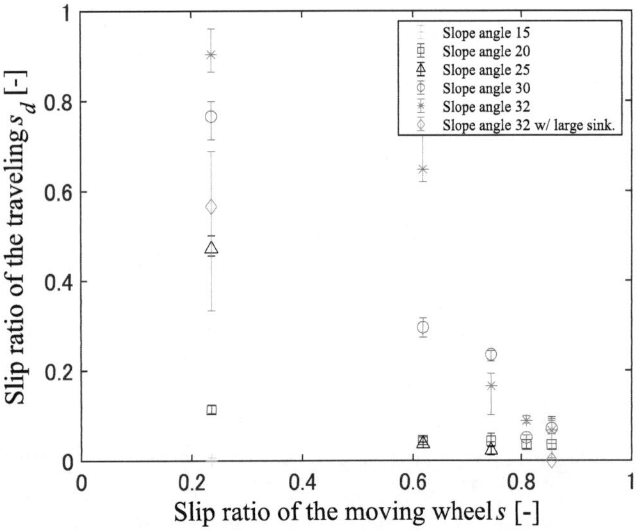

Fig. 10. Slip ratio s vs slip ratio s_d at different slope angles in the case of the push-pull and push-pull with large sinkage. Each data is an average of the five trials. Each bar means the maximum and minimum values

5. Conclusion and future works

To suppress the slip of the wheel-walking rover, this paper especially focused on the drawbar pull force of the driving wheel. Then, this paper increased the drawbar pull force by controlling the amount of slip. Measuring the force of the locked wheel indicated that increasing the slip of the rotational wheel contributes to improving the drawbar pull force. The climbing experiment on loose soil with slope also indicated that the improvement of the drawbar pull force of the rotational wheel during extending/shrinking its wheelbase could contribute to suppressing the reduction of the rover's displacement.

In future works, we will consider the force F_x and F_p at the push-pull mode theoretically. Also, the authors will investigate the limitation of the wheel rotational speed and optimal slip ratio. Furthermore, investigating the traveling performance of the other soil type is essential.

6. Acknowledgements

This work was supported by JSPS KAKENHI Grant Number JP21K14105.

7. Nomenclature

r	Radius of the wheel	[mm]
b	Width of the wheel	[mm]
F_x	Drawbar pull force	[N]
F_p	Resistive force	[N]
h	Sinkage	[mm]
m	Rover's mass	[kg]
s	Slip ratio of the wheel	[-]
s_d	Slip ratio of the rover	[-]
v_l	Velocity of the linear actuator	[mm/s]
v_w	Velocity of the wheel	[mm/s]
X_a	Actual distance	[mm]
X_i	Ideal distance	[mm]
θ	Wheel angle	[°]
θ_d	Slope angle	[°]
θ_r	Contact angle	[°]
θ_f	Exit angle	[°]
τ_x	Shear stress	[N/mm^2]
σ	Normal stress	[N/mm^2]
ω	Angular velocity	[rad/s]

8. Declaration of competing interest

The authors declare that they have no known competing financial interests or personal relationships that could have appeared to influence the work reported in this paper.

9. References

Andrade, G., Ben Amar, F., Bidaud, P., Chatila, R., 1998, Modeling Robot-Soil Interaction for Planetary Rover Motion Control, Proc. 1998 IEEE/RSJ International Conference on Intelligent Robots and Systems. Innovations in Theory, Practice and Applications.

Bekker, M.G., 1969, Introduction to terrain-vehicle systems, University of Michigan Press.

Creager C., Johnson, K., Plant, M., Moreland S., Skonieczny, K., 2015, Push-pull locomotion for vehicle extrication, J. Terramechanics, 57, 71-81.

Dymon, Yaoki, 2023, https://dymon.co.jp/, (accessed on 29 May 2023).

Ishigami, G., Miwa, A., Nagatani, K., Yoshida, K., 2007, Terramechanics-Based Model for Steering Maneuver of Planetary Exploration Rovers on Loose Soil, J. Field Robot, (24), 233-250.

Ispace, HAKUTO-R Mission 1, 2023, https://ispace-inc.com/jpn/m1, (accessed on 29 May 2023).

JAXA, Slim Mission, 2023, https://www.isas.jaxa.jp/home/slim/SLIM/index.html, (accessed on 29 May 2023).

Kermurdjian, A., Gromov, V., Mishkinyuk, V., Kucherenko, V., Sologub, P., 1992, Small Marsokhod Configuration, Proc. 1992 IEEE International Conference on Robotics and Automation.

Fujiwara, D., Iizuka, K., Asami, D., Kawamura, T., Suzuki S., 2019, Study on Traveling Performance for Variable Wheel-Base Robot Using Subsidence Effect, Int. J. Mech. Eng. Robot. Res., 8(2), 233-238.

1522 | Tractive performance of rigid wheel in granular media using coarse-scale DEM models

Bohumir Jelinek [a,*], Angela Card [a], George L. Mason [a], Thomas James [a], Karl Grebner [a], Thomas Skorupa [b], Jody D. Priddy [c]

[a] *Center for Advanced Vehicular Systems, Mississippi State University, Mississippi State, Mississippi, United States*

[b] *U.S. Army Combat Capabilities Development Command Ground Vehicle Systems Center, Warren, Michigan, United States*

[c] *U.S. Army Engineer Research and Development Center, Geotechnical and Structures Laboratory, Vicksburg, Mississippi, United States*

* *Corresponding author: bj48@cavs.msstate.edu*

ABSTRACT

Understanding the interaction between wheel and granular media in variable loading conditions is critical for prediction of net traction of wheeled and tracked vehicles in off-road environments. The discrete element method (DEM) is routinely used for modeling vehicle performance, but although simulations seem realistic, the method's accuracy is often not fully established.

In this work, the DEM modeling accuracy is assessed by the comparison of ten DEM soil models with laboratory soil-bin measurements of the net traction, gross traction, and sinkage of a wheel operating in sand. Laboratory soil-bin measurements, serving as a reference for DEM simulations, were taken from physical experiments (Shinone et al., 2010) examining a 165/60R13 wheel with constant circumferential velocity of 97.6 mm/s and vertical contact load of 980 N operating in powered conditions under slips in the range of -5.9% to 54.8%. The set of ten DEM models representing particle-particle and particle-geometry interactions was a subset of the Generic EDEM Material Model (GEMM) database from Altair®'s EDEM™ software package, choosing the best match to the bulk density and angle of repose for dry sand. Parameters for particle-wheel interactions were left identical to GEMM parameters for particle-geometry interactions.

Given the large particle size and no additional calibration of the DEM model, identifying a single GEMM material model with reasonable similarity for the net traction and sinkage is encouraging. The largest discrepancy between DEM and the lab experiment was observed at negative slips. A need for smaller particle size and further calibration are suggested for a future effort.

Keywords
Discrete element method
Net traction
Gross traction
Sinkage
Slip

1. Introduction

Prediction of vehicle mobility for vehicles in off-road environment presents a challenge for agriculture, construction, military, and space exploration. Suspension and powertrain systems are usually characterized well, but the accuracy of mobility models suffers from hard-to-predict terrain response. The subgrade materials that a vehicle operates on may include rock, gravel, sand, silt, clay soil, or their mixtures with various degrees of saturation. Even relatively homogeneous subgrade is hard to model numerically at the continuum level due to large deformation and material non-linearity, as well as at the particle level due to the complex geometry of soil particles, variability in contact mechanics, and high number of degrees of freedom involved. Predictions of mobility using numerical models must therefore compromise between sufficiently detailed representation and capabilities of high-performance computing.

Multiple studies have used DEM to characterize the interaction of granular subgrades with rigid and flexible wheels. The bulk of DEM work has focused on predicting the mobility of rovers on Earth, the Moon, and Mars in regolith sediments (Nakashima et al. 2010; Knuth et al. 2012; Nakashima and Kobayashi, 2014; Johnson et al., 2015, 2017) exploring the effects of gravity on sinkage and motion resistance of a rigid rover wheel. Tractive and steering performance of off-road wheels on dry sand was analyzed by Du et al. 2017a, 2017b, 2018. These studies examined the effects of lug type, intersection of lug bars, and central angle, with the objective to improve performance prediction for straight and steered performance. The scope of off-road mobility modeling was extended beyond rigid wheels by introducing finite element pneumatic tire model (Recuero et al., 2017). Vertical stress under the wheels of agricultural machinery obtained from DEM using Yade (Šmilauer et al., 2015) was compared with pseudo-analytical continuum model and field measurements field measurements (De Pue et al., 2020, 2019; De Pue and Cornelis, 2019).

Yang et al., (2020) performed DEM-FEM (Finite Element Method) simulations of flexible multi-layer tire on gravel at 10%, 20%, and 30% slip and showed a reasonable agreement of sinkage, gross tractive effort, and tractive force with experiment. These models, however, require a careful calibration (Coetzee, 2017; Zeng et al., 2020a, 2020b), and, in many cases, non-spherical particles are used to better match the experiments (Coetzee, 2020).

In this work, a laboratory experiment to predict tractive performance of a wheel in sand conducted by Shinone et al., 2010 is compared with numerical simulations deploying ten different out-of-the-box coarse-grain DEM models selected to mimic sand-like macroscopic properties. The goal of the study is to evaluate capability of coarse-grain DEM models to predict the net traction, gross traction, and sinkage of a smooth wheel operating in sand under braked and powered conditions.

2. DEM methodology

A particle bed to simulate wheel-soil interactions was generated using particle-factory functionality of the Altair®'s EDEM™ software. The particle bed was assembled from five identical material blocks repeated along the longitudinal direction. Only the particles in a single block were settled under periodic boundary conditions to lower the simulation time needed to populate the bed with particles. A smooth-surface cylindrical wheel without tread was constructed from triangular facets. Figure 1 illustrates the wheel and enclosing box while Table 1

https://doi.org/10.56884/HITS7216
ISBN 978-1-942112-55-6 (pdf)

specifies their dimensions. Figure 1 illustrates the wheel and enclosing box while Table 1 specifies their dimensions.

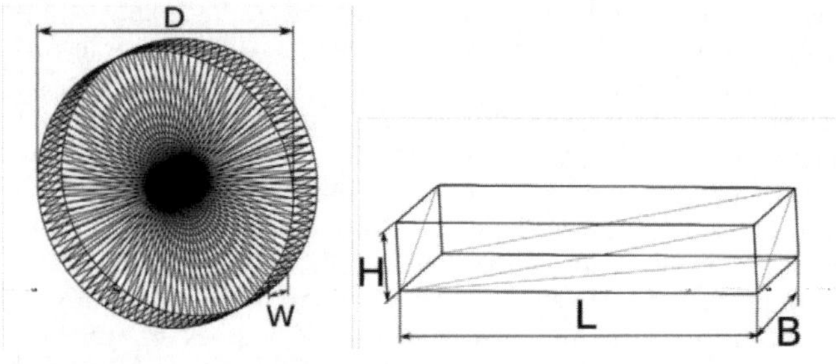

Fig. 1. Triangular facet representation of the wheel and enclosing box

Table 1
Wheel and Box Dimensions

Property	Symbol	Units	Value
Wheel diameter	*D*	cm	53.49
Wheel width	*W*	cm	16.99
Wheel weight	F_G	N	980.33
Box length	*L*	cm	301.5
Box width	*B*	cm	48.0
Box depth	*H*	cm	60.5
Soil depth	*h*	cm	60
Number of particles	*N*	-	97152

Ten different DEM soil models with sand-like macroscopic response were evaluated. Models were chosen from Altair®'s EDEM™ GEMM database based on the following requirements: a bulk density range of 1500-2000 kg/m^3 and an angle of repose of 34 degrees. To represent homogeneous soil with a degree of randomness, particle sizes were scaled to produce a normal distribution. Blocks were duplicated within particle beds and then settled to negligible kinetic energies. To control particle bed variation between simulations, only particle properties were adjusted with Altair's EDEM.PY Python package. Table 2 and Table 3 present GEMM particle properties.

Table 2
Particle Properties

Mat. #	1507	1301	1311	1615	1706
ν	0.25	0.25	0.25	0.25	0.25
ρ	3300	3300	3300	3300	3300
E	10	10	10	10	10
e^p	0.35	0.15	0.55	0.55	0.35
μ_s^p	0.92	0.68	0.68	1.04	1.16
μ_r^p	0	0	0	0	0
γ^p	4.5	0	0	18	0
e^g	0.5	0.5	0.5	0.5	0.5
μ_s^g	0.45	0.45	0.45	0.45	0.45
μ_r^g	0.15	0.15	0.15	0.15	0.15
γ^g	4.5	0	0	18	0

Table 3
Particle Properties

Mat. #	1111	1501	1308	1105	1636
ν	0.25	0.25	0.25	0.25	0.25
ρ	3300	3300	3300	3300	3300
G	10	10	10	10	10
e^p	0.55	0.35	0.35	0.15	0.75
μ_s^p	0.44	0.92	0.68	0.44	1.04
μ_r^p	0	0	0	0	0.05
γ^p	0	18	9	18	0
e^g	0.5	0.5	0.5	0.5	0.5
μ_s^g	0.45	0.45	0.45	0.45	0.45
μ_r^g	0.15	0.15	0.15	0.15	0.15
γ^g	0	18	9	18	0

*a superscript of *p* implies particle-particle interaction while *g* is a particle-geometry interaction

All particles were a multi-spherical shape. Individual spheres had a radius of 9.5mm with a fixed overlap. Figure 2 visualizes the multi-sphere particles.

Fig. 2. Multi-spherical particle shape

The Hertz-Mindlin (no slip) and Hertz-Mindlin with JKR models were used to represent DEM contact models. Hertz-Mindlin (Mindlin and Deresiewicz, 1953) is the default model used in EDEM due to its accurate and efficient force calculation (1993). The Hertz-Mindlin JKR model was used for particles with a non-zero JKR coefficient. For Hertz-Mindlin, the normal force is a function of normal overlap δ_n:

$$F_x = \frac{4}{3}E^*\sqrt{R^*}\delta_n^{\frac{3}{2}} \quad (1)$$

where the E^* is the equivalent Young's Modulus and R^* is the equivalent radius. They are defined as:

$$\frac{1}{R^*} = \frac{1}{R_i} + \frac{1}{R_j} \quad (2)$$

$$\frac{1}{E^*} = \frac{(1-v_i^2)}{E_i} + \frac{(1-v_j^2)}{E_j} \quad (3)$$

with E^i, v^i, R^i, and E^j, v^j, R^j being the Young's Modulus, Poisson ratio, and radius of each sphere in contact. Additionally, there is a damping force, $F_n{}^d$, given by:

$$\boldsymbol{F_n^d} = -2\sqrt{\frac{5}{6}}\beta\sqrt{S_n m^*}\,\boldsymbol{v_n^{rel}} \quad (4)$$

$$m^* = \left(\frac{1}{m_j} + \frac{1}{m_i}\right)^{-1} \quad (5)$$

$$\beta = \frac{-\ln e}{\sqrt{ln^2 e + \pi^2}} \quad (6)$$

$$S_n = 2E^* \sqrt{R^* \delta_n} \quad (7)$$

where m* is the equivalent mass, $\boldsymbol{v}_n^{rel}$ is the normal component of the relative velocity, β, and S_n (the normal stiffness). The constant e is the coefficient of restitution, the tangential force, F_t, depends on the tangential overlap δ_t and the tangential stiffness S_t.

$$F_t = -S_t \delta_t \quad (8)$$

$$S_t = 8G^* \sqrt{R^* \delta_n} \quad (9)$$

Here, G^* is the equivalent shear modulus. Additionally, tangential damping is defined as:

$$\boldsymbol{F}_t^d = -2\sqrt{\frac{5}{6}}\beta\sqrt{S_t m^*}\boldsymbol{v}_t^{rel} \quad (10)$$

where $\boldsymbol{v}_t^{rel}$ is the relative tangential velocity. The tangential force is limited by Coulomb friction, $\mu_s F_n$, where μ_s is the coefficient of static friction.

The Hertz-Mindlin contact model with JKR cohesion (Johnson et al., 1971) allows modelling of cohesive materials. It calculates the normal force as a function of overlap δ_n and the surface energy parameter γ as follows:

$$F_{JKR} = -4\sqrt{\pi\gamma E^*} a^{3/2} + \frac{4E^*}{3R^*} a^3, \quad (11)$$

$$\delta_n = \frac{a^2}{R^*} - \sqrt{\frac{4\pi\gamma a}{E^*}} \quad (12)$$

where E^* is the equivalent Young modulus and R^* is the equivalent radius from Eq. (2). Wheel slip and circumferential velocity were used as an initial condition to each system. Wheel slip is defined as:

$$i = 1 - \frac{v_f}{\omega r} \quad (13)$$

Equation 13 defines i as wheel slip, v_f as the wheel forward velocity, ω as wheel angular displacement, and r as wheel radius. Post-processing values such as net traction and gross traction were calculated using Eq. 13 and Eq. 14, respectively. Sinkage amount was based on the initial position of the suspended wheel.

$$F^x = \sum_j F_j^x \quad (14)$$

$$T = \sum_j F_j^t d_j \quad (15)$$

F^x is the forward component of the contact force, F^t is the wheel-surface-tangential component of the contact force, D is the wheel diameter and d_j is distance from the contact point to the wheel axis. F^t has negative/positive sign if the force acts to increase/decrease wheel rotation.

3. Results

With a load of 980 N, the wheel was placed -1.3 meters away from the center of the bed. By coupling a driving program to EDEM, the wheel moved at a constant circumferential velocity of 97.6 millimeters per second to the final position of 1.3 meters. With one degree of freedom, the wheel only moved freely along the surface's normal axis (allowing sinkage). Figure 3 shows the initial position of the system.

Fig. 3. Initial position: GEMM 1636, 45.8% slippage

Figure 4 shows a final state of a system with sinkage.

Fig. 4. Final position: GEMM 1636, 45.8% slippage

For each particle, wheel slippage ranged from -5.9% to 54.8%. Net traction, gross traction, and wheel height were plotted at each time step. Figures 5, 6, and 7 show plots of these values, for GEMM 1636 at 45.8% wheel slippage, over the simulation span. Initial data outliers are due to the wheel dropping above the surface.

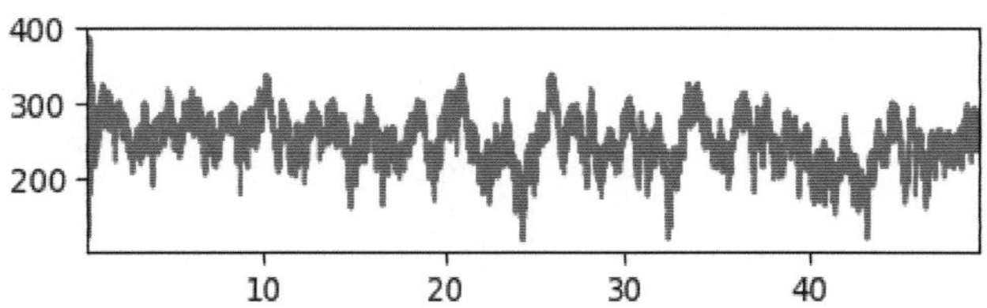

Fig. 5. Net traction (N) v. time (s): GEMM 1636, 45.8% slippage

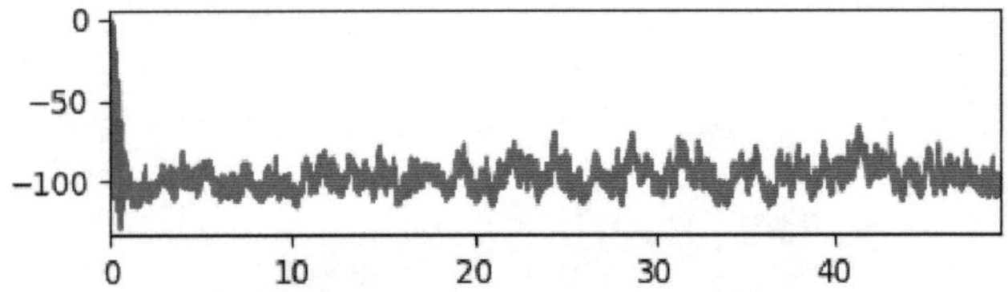

Fig. 6. Gross traction (Nm) v. time (s): GEMM 1636, 45.8% slippage

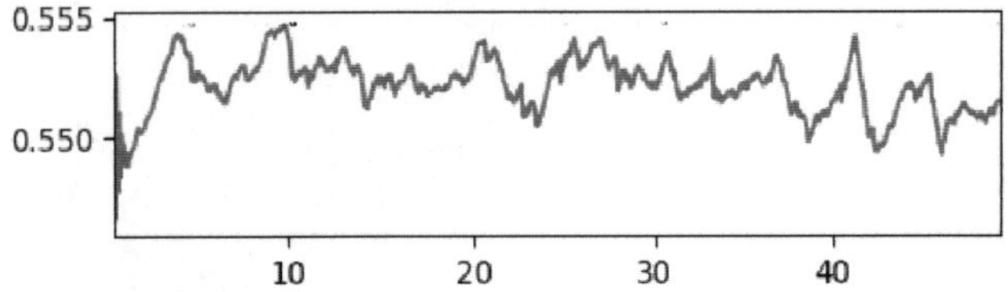

Fig. 7. Center of mass height (m) v. time (s): GEMM 1636, 45.8% slippage

Figures 8 and 9 show average net and gross traction. Data outside the interval of -1 meter to 1 meter was excluded.

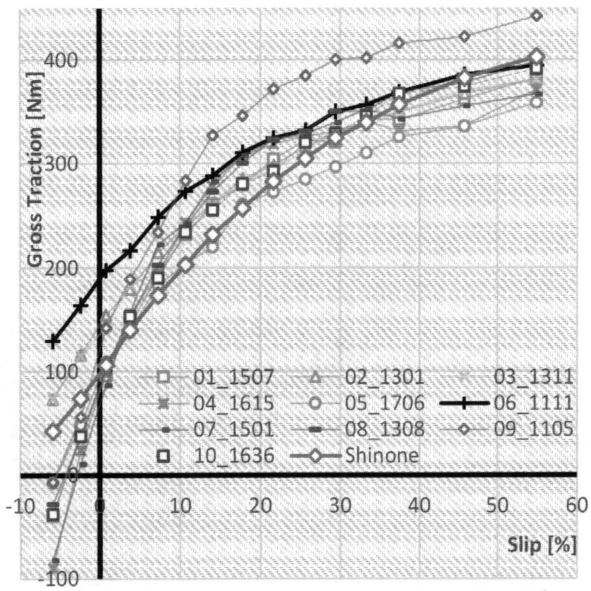

Fig. 8. Gross tractive effort

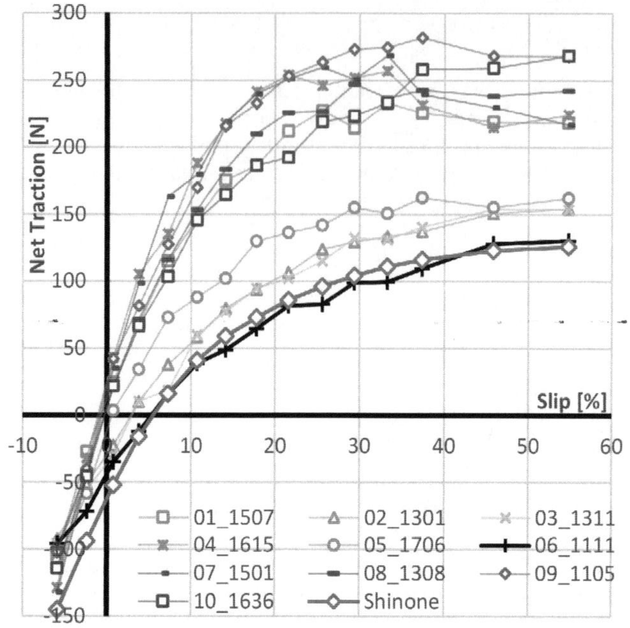

Fig. 9. Net tractive effort (drawbar pull)

Figure 10 shows wheel sinkage based on the lowest point of the wheel contact surface and the uncompressed height of the particle bed.

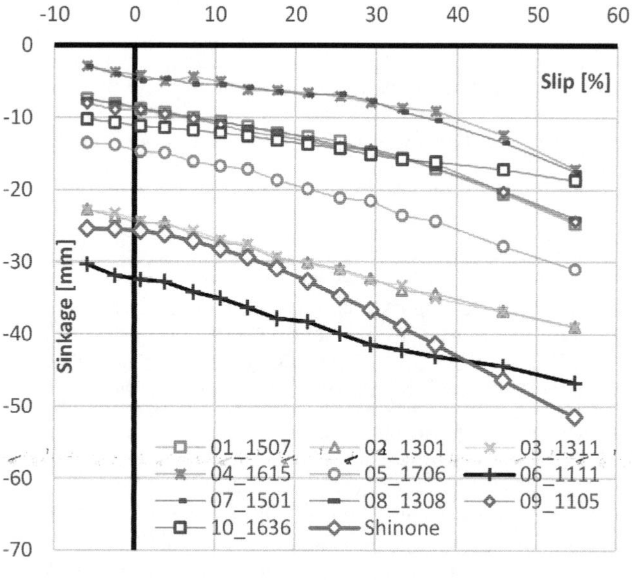

Fig. 10. Sinkage

4. Conclusions and future work

Out of ten materials selected from Altair®'s EDEM™ Generic Material Model (GEMM) database to match macroscopic properties of dry sand, GEMM material number 1111 provided the best agreement with lab test results reported by Shinone et al., 2010. The simulation results are similar to the experiment. Net traction was the most accurate model except that slightly lower values occurred at negative slips. Sinkage was overestimated by up to 25% for slips lower than 40% and underestimated by up to 10%

for slips higher than 40%. The gross traction agrees well with DEM results for a slip larger than 40%, while at lower positive and negative slips the gross traction is overestimated by up to 200% at the slip of -5.9%.

5. Nomenclature

a	Contact radius	[m]
B	Box width	[m]
D	Wheel diameter	[m]
d_j	distance from the j-th particle wheel contact point to the wheel axis	[m]
e	Coefficient of restitution	[]
E	Young's modulus	[m]
F^d	Damping force	[m]
F_G	Wheel weight	[N]
F_r	Rolling friction force	[N]
F_s	Static friction force	[N]
F_x	Forward component of force on wheel	[N]
G	Shear modulus	[MPa]
H	Box depth	[m]
h	Soil depth	[m]
i	Wheel slip	[]
L	Box length	[m]
m	mass	[kg]
N	Number of particles	[]
R	Hertz-Mindlin radius	[m]
S	Stiffness	[N/m]
T	Gross tractive effort	[Nm]
v_f	Forward component of wheel velocity	[m/s]
v_t^{rel}	Relative tangential velocity	[m/s]
W	Wheel width	[m]
α	Contact overlap	[m]
β	Relative velocity	[m/s]
γ	JKR surface energy parameter	[J/m²]
δ	Overlap	[m]
μ	Coefficient of friction	[]
ν	Poisson's ratio	[]
ρ	Bulk Density	[kg/m³]
ω	Angular displacement	[1/s]

6. Acknowledgements

This material is based upon research conducted under contract W912HZ-22-C0004 with the U.S. Army Engineer Research and Development Center (ERDC). The views and conclusions contained herein are those of the authors and should not be interpreted as necessarily representing the official policies or endorsements, either expressed or implied, of ERDC or the U.S. Government.

7. Declaration of competing interest

The authors declare that they have no known competing financial interests or personal relationships that could have appeared to influence the work reported in this paper.

8. References

Coetzee, C., 2020. Calibration of the discrete element method: Strategies for spherical and non-spherical particles. Powder Technol. 364, 851–878.

Coetzee, C.J., 2017. Review: Calibration of the discrete element method. Powder Technol. 310, 104–142.

De Pue, J., Cornelis, W.M., 2019. DEM simulation of stress transmission under agricultural traffic Part 1: Comparison with continuum model and parametric study. Soil Tillage Res. 195, 104408.

De Pue, J., Di Emidio, G., Verastegui Flores, R.D., Bezuijen, A., Cornelis, W.M., 2019. Calibration of DEM material parameters to simulate stress-strain behaviour of unsaturated soils during uniaxial compression. Soil Tillage Res. 194, 104303.

De Pue, J., Lamandé, M., Schjønning, P., Cornelis, W.M., 2020. DEM simulation of stress transmission under agricultural traffic Part 3: Evaluation with field experiment. Soil Tillage Res. 200, 104606.

Du, Y., Gao, J., Jiang, L., Zhang, Y., 2018. Development and numerical validation of an improved prediction model for wheel-soil interaction under multiple operating conditions. J. Terramechanics 79, 1–21.

Du, Y., Gao, J., Jiang, L., Zhang, Y., 2017a. Numerical analysis of lug effects on tractive performance of off-road wheel by DEM. J. Braz. Soc. Mech. Sci. Eng. 39, 1977–1987.

Du, Y., Gao, J., Jiang, L., Zhang, Y., 2017b. Numerical analysis on tractive performance of off-road wheel steering on sand using discrete element method. J. Terramechanics 71, 25–43.

Johnson, J.B., Duvoy, P.X., Kulchitsky, A.V., Creager, C., Moore, J., 2017. Analysis of Mars Exploration Rover wheel mobility processes and the limitations of classical terramechanics models using discrete element method simulations. J. Terramechanics 73, 61–71.

Johnson, J.B., Kulchitsky, A.V., Duvoy, P., Iagnemma, K., Senatore, C., Arvidson, R.E., Moore, J., 2015. Discrete element method simulations of Mars Exploration Rover wheel performance. J. Terramechanics 62, 31–40.

Johnson, K.L., Kendall, K., Roberts, A.D., 1971. Surface energy and the contact of elastic solids. Proc. R. Soc. Lond. Math. Phys. Sci. 324, 301–313.

Knuth, M.A., Johnson, J.B., Hopkins, M.A., Sullivan, R.J., Moore, J.M., 2012. Discrete element modeling of a Mars Exploration Rover wheel in granular material. J. Terramechanics 49, 27–36.

Mindlin, R.D., Deresiewicz, H., 1953. Elastic Spheres in Contact Under Varying Oblique Forces. J. Appl. Mech. 20, 327–344.

Nakashima, H., Fujii, H., Oida, A., Momozu, M., Kanamori, H., Aoki, S., Yokoyama, T., Shimizu, H., Miyasaka, J., Ohdoi, K., 2010. Discrete element method analysis of single wheel performance for a small lunar rover on sloped terrain. J. Terramechanics 47, 307–321.

Nakashima, H., Kobayashi, T., 2014. Effects of gravity on rigid rover wheel sinkage and motion resistance assessed using two-dimensional discrete element method. J. Terramechanics 53, 37–45.

Recuero, A., Serban, R., Peterson, B., Sugiyama, H., Jayakumar, P., Negrut, D., 2017. A high-fidelity approach for vehicle mobility simulation: Nonlinear finite element tires operating on granular material. J. Terramechanics 72, 39–54.

Shinone, H., Nakashima, H., Takatsu, Y., Kasetani, T., Matsukawa, H., Shimizu, H., Miyasaka, J., Ohdoi, K., 2010. Experimental Analysis of Tread Pattern Effects on Tire Tractive Performance on Sand using an Indoor Traction Measurement System with Forced-slip Mechanism. Eng. Agric. Environ. Food 3, 61–66.

Šmilauer, V., others, 2015. Yade Documentation 2nd ed. The Yade Project.

Yang, P., Zang, M., Zeng, H., 2020. DEM–FEM simulation of tire–sand interaction based on improved contact model. Comput. Part. Mech. 7, 629–643.

Zeng, H., Xu, W., Zang, M., Yang, P., 2020a. Calibration of DEM-FEM model parameters for traction performance analysis of an off-road tire on gravel terrain. Powder Technol. 362, 350–361.

Zeng, H., Xu, W., Zang, M., Yang, P., Guo, X., 2020b. Calibration and validation of DEM-FEM model parameters using upscaled particles based on physical experiments and simulations. Adv. Powder Technol. 31, 3947–3959.

1638 | Analysis of tire characteristics on road surface with volcanic ash fall

Junya Yamakawa [a,*], Ryosuke Eto [a]

[a] Department of Mechanical Engineering, National Defense Academy, Yokosuka, Kanagawa, Japan

* Corresponding author: yamakawa@nda.ac.jp

ABSTRACT

In order to investigate the impact of volcanic disasters on vehicle driving, data was collected from driving an instrumented vehicle on asphalt roads covered with volcanic ash. This presentation will discuss the results obtained by analyzing the data collected on the simulated volcanic ash road. The tire side force generated during vehicle turning is largely due to the sideslip angle, but it is also affected by the camber angle as well as the vertical load. Since the side force data during driving is collected while the values of not only the sideslip angle but also the vertical load and camber angle are changing, it is necessary to determine the relationship between these variables and the side force. Therefore, we analyzed the side force data using Gaussian process regression, which allows regression from multiple variables. Since Gaussian process regression takes time to find appropriate values for the hyperparameters of the kernel, this report mainly describes the modeling with neural networks.

Keywords
Volcanic ash
Asphalt road
Side force
Sideslip angle
Camber angle
Neural network
Gaussian process regression

1. Introduction

The Japanese archipelago has many active volcanoes. Mt. Fuji, the highest volcano at 3776 m above sea level, has erupted repeatedly in the past with a cycle of about 100 years. If an eruption were to occur, ash fall of 2 - 10 cm is expected in the Tokyo metropolitan area, which is about 100 km away, and ash fall of 50 cm is expected in the area near Mt. Fuji. In October 2021, an event was held in Yamanashi Prefecture, where Mt. Fuji is located, to educate citizens about volcanic eruption preparedness by driving on asphalt-paved roads covered with volcanic ash to experience how their cars would be affected by the ash fall caused by an eruption.

In our previous report (Yamakawa et al., 2022), we described the results of the analysis of data collected by the instrumented vehicle driving paths set up at the event site. In this report, we present a study on how to process the data collected on curved roads covered with volcanic ash, which is expected to be more slippery than asphalt road surfaces.

2. Experiment

2.1 Experiment fields

The road the vehicle run was a 4 m wide, 30 m radius, flat or sloping 2.5% curved road, covered with a 3 cm thick layer of volcanic ash. The volcanic ash used was excavated from volcanic ash deposited by past eruptions and classified as fine grained, as shown in Fig. 1. The weather was generally clear throughout the experiment period, and the data were collected under nearly dry conditions.

Fig. 1. Fine volcanic ash spread on asphalt road surface

2.2 Instrumented vehicle

Figure 2 shows a Mitsubishi Motors Model 73 light-duty truck used in the driving experiment, and its main specifications are shown in Table 1. The vehicle is 4-wheel drive, but it can be switched to rear 2-wheel drive depending on the purpose. The tires are YOKOHAMA GEOLANDAR M/T+ for off-road use, and the instruments are mounted on the front two wheels. This device measures triaxial forces applied to the tires from the road surface, ground speed by laser Doppler velocimetry, and tire attitude angle. Ground speed is measured in two directions on the running surface, so the tire's lateral slip angle can be determined from the tire's orientation and direction of travel. The tire pressure was set to 180 kPa, the standard pressure for size 218/85R 18LT tires.

https://doi.org/10.56884/ABIS2469
ISBN 978-1-942112-55-6 (pdf)

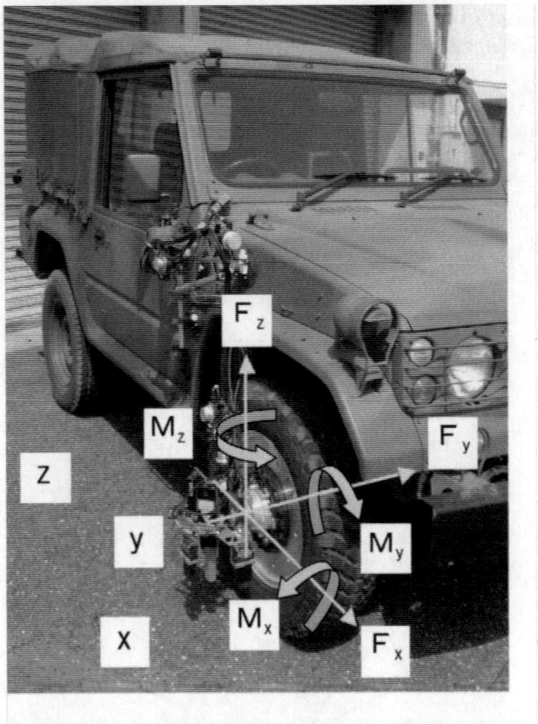

Fig. 2. The instrumented vehicle and coordinate systems

Table 1
Main specification of the instrumented vehicle

Type	Value [-]
Length / Width / Height	4140 mm / 1765 mm / 1970 mm
Wheelbase	2420 mm
Tread (front/rear)	1485 mm / 1500 mm
Minimum Ground Height	245 mm
Vehicle Weight	1940 kg
Engine	4M40 diesel, turbo
Cylinder / Displacement	In-line 4-cylinder/2835 cc
Maximum Power	92 kW (4000 rpm)
Maximum Torque	294 Nm (2000 rpm)

2.3 Side force and related variables

Since the ground speed is measured in the longitudinal and lateral directions near the tire, the sideslip angle β can be expressed by the following equation based on the direction of the tire and the direction of travel (Figure 3).

$$\tan \beta = V_y / V_x \quad (1)$$

where (V_x, V_y) is the ground speed near the tire in the x- and y-directions, respectively. The inclination of the tire γ from the vertical plane shown in Fig. 3 is called the camber angle. During a turn, not only the side force is generated by the sideslip angle, but also the side force called camber thrust is generated by the camber angle, and the side force measured by the instrument is the sum of these forces. The vertical load of the tire on the outer side of the turn increases and decreases on the inner side, so the effect of the vertical load must be included in the consideration.

2.4 Conditions for experiment

To clarify the relationship between the sideslip angle of the tire and the side force, the vehicle was driven in two directions on curved roads at a speed of 10 to 30 km/h, and data was obtained for right and left turns. On a gradient road, the right turn is downhill and the left turn is uphill, causing a difference in the vertical load between the front and rear wheels, but this is not a particular problem for deriving a model of the side force characteristics.

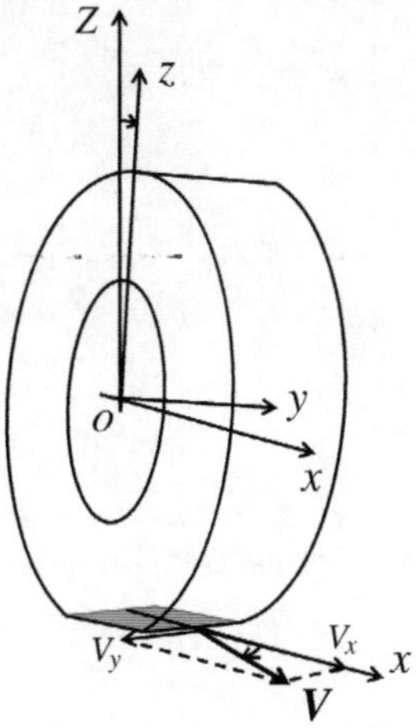

Fig. 3. Tire travel velocity, sideslip angle and camber angle

3. Experimental results and models

3.1 Side force and sideslip angle

The data acquired by the instrumented vehicle provides information on the characteristics of side force on an ash fall road surface. Figure 4 shows the relationship between side force and sideslip angle when four wheels are driven. Figure 5 shows the relationship when rear wheels are driven (front wheels are rolling). The figures superimpose the results of all experiments conducted on flat and gradient roads. During turning, a side force called camber thrust is generated by the camber angle, and the measured side force is the sum of the side force due to the sideslip angle. In addition, the centrifugal force generated changes the vertical load of the tire. Because the vertical load and camber angle vary between the outer and inner wheels, the side force values obtained are not under the same conditions. Therefore, in Figs 4 and 5, changes in vertical load from -4000 N to 4000 N, which have a large effect on side force, are indicated by color shading based on a vertical load of 5600 N for straight run. The color is darker when the increase in vertical load is large, and is closer to white when the decrease in vertical load is large. The left wheel is red and the right wheel is blue.

In general, the side force decreases as the longitudinal slip ratio increases (Abe, 2008). Therefore, the side force differs between the four-wheel drive condition, in which longitudinal slip occurs in the front wheels and exerts longitudinal force, and the rear-wheel drive condition, in which the front wheels roll and no longitudinal slip occurs. However, a comparison of Figs 4 and 5 shows close agreement. In this experiment, the four-wheel drive system exhibited almost no longitudinal slip in the front wheels because the vehicle was running at a nearly constant speed.

The side force is divided into two at the right wheel's sideslip angle of 6 to 12° in Fig. 4 and at the left wheel's side slip angle of -8 to -4° in Fig. 5 due to the influence of the vertical load. This is due to the fact that the large sideslip occurs when the vehicle speed exceeds 30 km/h, resulting in significant vertical load changes on the inner and outer wheels.

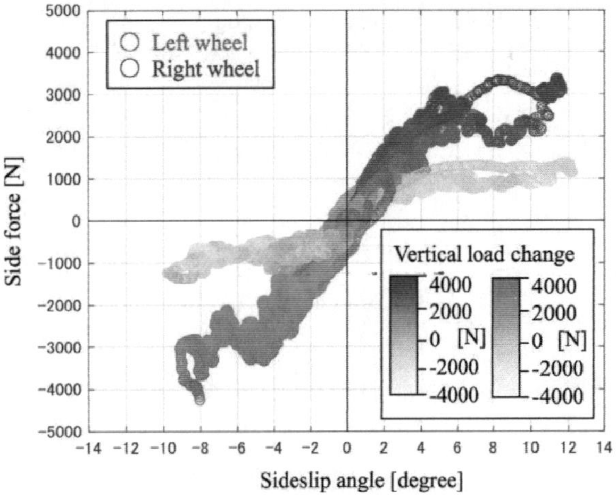

Fig. 4. Relationship between side force and sideslip angle in 4-wheel drive

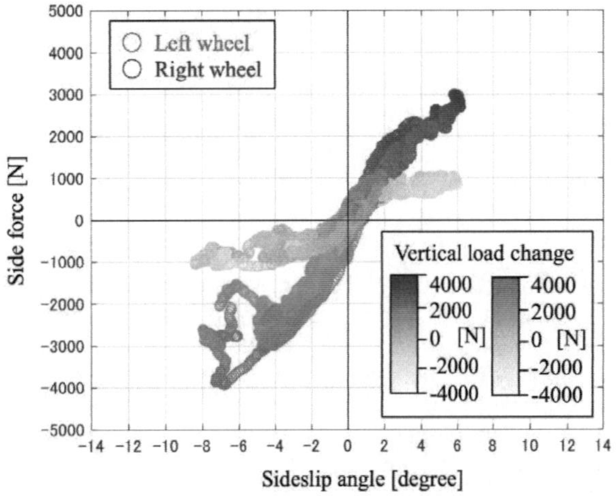

Fig. 5. Relationship between side force and sideslip angle in rear-2-wheel drive

3.2 Regression model

Side force includes the effects of vertical load and camber angle. Considering these effects, this section discusses the influence of volcanic ash on the relationship between side force and sideslip angle during rear-wheel drive. For this purpose, a regression model is generated with parameters related to the sideslip angle, vertical load, and camber angle as inputs. Previous studies (Mizuno et al., 1999, Aota et al., 2021) have used magic formula models or Gaussian processes in their regression models. The Magic Formula requires empirical knowledge of the range of parameters to be determined, and it is difficult to identify parameters from the results of this experiment, for which there are few data. In the Gaussian process regression model, the Gram matrix becomes large when the number of data is large, and parameter identification is time-consuming. Therefore, this study mainly describes regression models using neural networks.

3.3 Neural network model

A neural network (NN) is a functional model with a complex shape composed of activation functions and weight parameters that control their behavior (Suyama, 2019). A schematic diagram of the forward propagating NN used in this study is shown in Fig. 6. The input layers can be thought of as the sideslip angle, camber angle, and vertical load change, and the output as the side force. In our previous work, we have used this input/output to generate Gaussian process regression models (Aota et al., 2021). Since the Gaussian process regression model is identical to the NN model with one hidden layer and an infinite number of units (Suyama, 2019), we also set the hidden layer to 1 and the number of units to a sufficiently large value in this study. The tanh function was used as the activation function, and the Deep Learning Toolbox of Matlab, a numerical analysis software from MathWorks, was used to train the NN model.

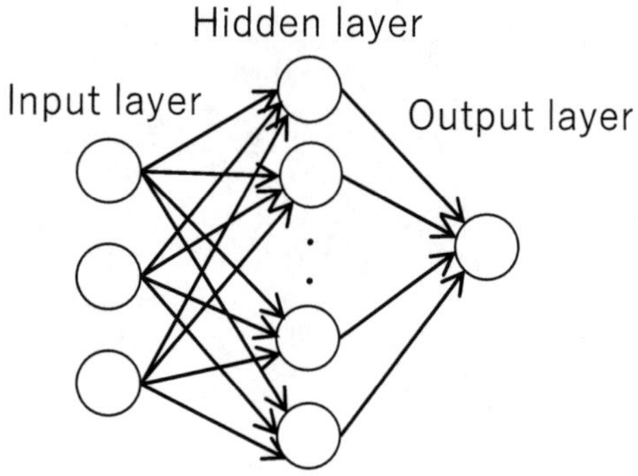

Fig. 6. Schematic diagram of a neural network

To confirm the data used for input, Fig. 7 shows the relationship between the camber angle and the sideslip angle in rear-wheel drive. Figure 8 shows the relationship between the change from the standard vertical load and the sideslip angle when the rear wheels are driven. The vertical load has a wide enough distribution, but the camber angle is generally small, ranging from 0 to 1º. In the previous report (Yamakawa et al., 2022), we modeled by Gaussian process regression, using a kernel function called ARD (Bishop, 2006), where the value of the hyperparameter indicates the magnitude of the association to input-output. The analysis results show that the hyperparameters related to the camber angle are much smaller than those related to the vertical load change and the side slip angle, indicating that the camber angle has a small effect on the side force. Therefore, in the regression model developed in this study, the camber angle is kept constant and two inputs, the sideslip angle and the vertical load change, are used.

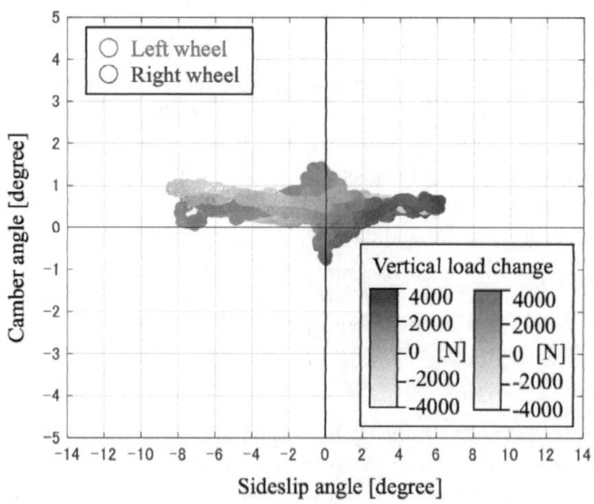

Fig. 7. Relationship between camber angle and sideslip angle in rear-wheel drive

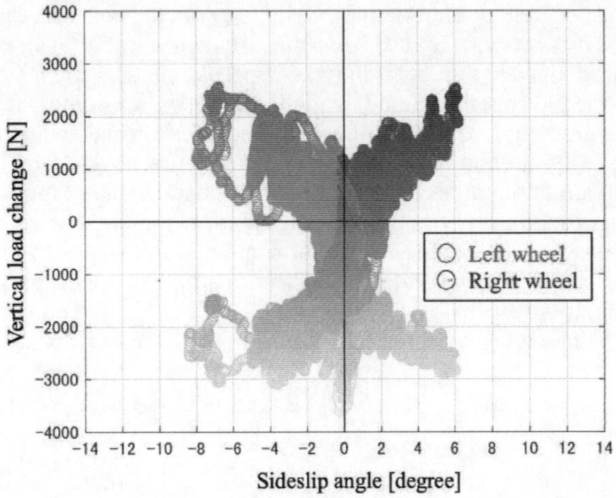

Fig. 8. Relationship between vertical load change and sideslip angle in rear-wheel drive

3.4 Validation of the NN regression model

Overlearning is a problem in regression models using neural networks. In this study, the NN model is trained with a large number of random initial parameters and generalized by averaging the outputs. This method can also be used to evaluate the accuracy of the predictions based on the standard deviation of the predictions. The 43192 sets of experimental data points were randomly split 9:1, with 90% of the data being training data and the remaining 10% being validation data.

Figure 9 shows the root mean square error (RMSE) of the validation data and predictions when 100 NN models were created. In this study, the number of units was set to 20, which is approximately where the error converges.

Figure 10 shows the side force predicted by the NN model when the number of units is 20. A 2000 N decrease from the standard vertical load is shown in blue, a standard load in black, and an increase of 2000 N in red. The solid line shows the mean value of the prediction, and the dashed line shows the range when the standard deviation is added or subtracted. Roughly 68% of the data falls within this range. Since there is little difference in predictions between 40 and 100 NN models, we decided to use the number 40 models.

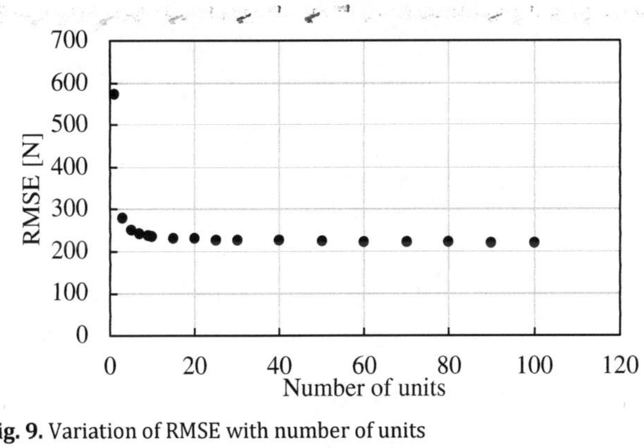

Fig. 9. Variation of RMSE with number of units

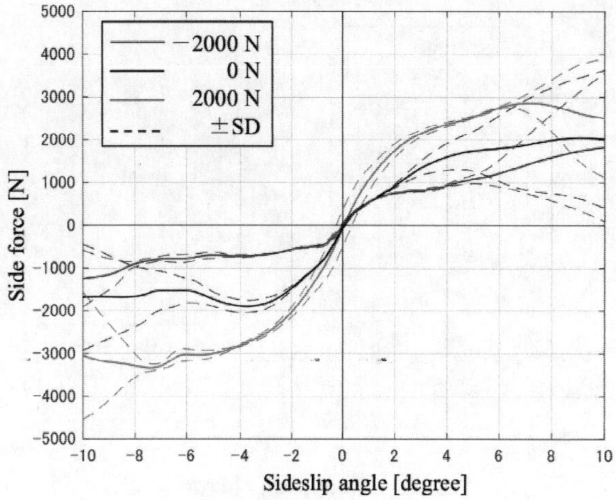

(a) Number of models: 40

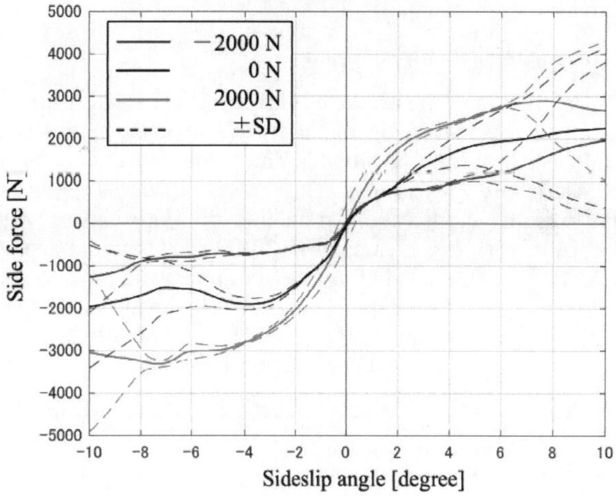

(b) Number of models: 100

Fig. 10. Predicted side force on volcanic ash road surface

Figure 11 shows a comparison of the experimental and predicted values, and it can be seen that the predicted values are in good agreement with the experimental values. There are many data in the range of -8 ~ 4º sideslip angles when the vertical load change is -2000 N, in the range of -4 ~ 2º when there is no vertical load change, and in the range of -8 ~ -2 º and 4 ~ 6º for 2000 N. Therefore, the standard deviations are small and highly accurate in this range, and the standard deviations are large and less accurate outside the range.

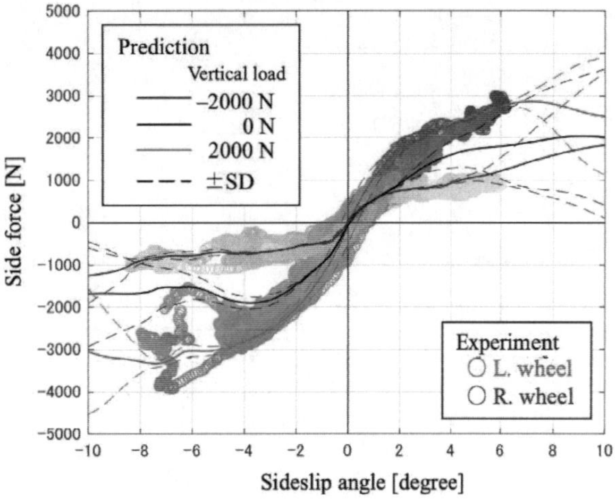

Fig. 11. Comparison of experimental data and predicted values on volcanic ash

3.5 Comparison with asphalt road

Figure 12 shows the predicted side force on asphalt using the Gaussian process regression model and the predicted side force on volcanic ash using the NN model. The solid colors indicate the difference in the vertical load. The Gaussian process regression model uses the model and parameters developed in previous studies (Aota et al., 2021, Yamakawa, 2021) and for a camber angle of 0.5º. On asphalt, the side force reaches its maximum value at a sideslip angle of 6~8º, while on volcanic ash, the increase in side force slows down and approaches its maximum value at 2~6º. The maximum value is also smaller than that of asphalt.

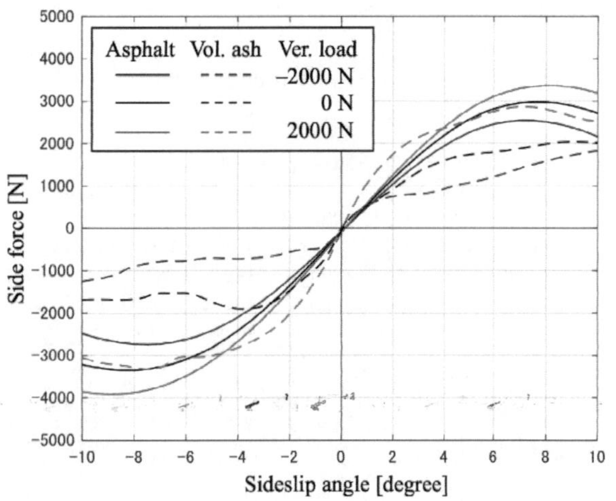

Fig. 12. Comparison of side forces between asphalt and volcanic ash roads

When the vertical load changes -2000 N for volcanic ash, the side force continues to increase at sideslip angles of 4º or greater. In this range, the accuracy is low due to the paucity of experimental data, and it is considered to be saturated or reduced in practice. When the vertical load change is 0 or -2000 N, there is no difference in cornering power between asphalt and volcanic ash, but when the vertical load change is 2000 N, cornering power is greater for volcanic ash. This is thought to be due to the fact that the tire sinks and the force on the sides of the tire from the volcanic ash increases.

3.6 Gauss process regression model

The calculation of side force on asphalt shown in Fig. 12 was performed using Gaussian process regression, and the same Gaussian process was used for the data on volcanic ash road surfaces. However, in obtaining the hyperparameters, the former used the Metropolis method, while the latter used a combination of the gradient and Newton-Raphson methods (Yamakawa et al., 2022). This method searches for hyperparameters to maximize the likelihood of the model relative to experimental values. Since the gradient method is computationally time-consuming, the values obtained after a certain amount of computation are used as the initial values for the Newton-Raphson method to speed up the computation.

The following kernel was used to model the Gaussian process.

$$k(\boldsymbol{x}_m, \boldsymbol{x}_n) = \eta_0 exp\left\{-\frac{1}{2}(\boldsymbol{x}_m - \boldsymbol{x}_n)^{\mathrm{T}}\Sigma_1(\boldsymbol{x}_m - \boldsymbol{x}_n)\right\} + \boldsymbol{x}_m^{\mathrm{T}}\Sigma_2\boldsymbol{x}_n \quad (2)$$

$$\Sigma_1 = \begin{bmatrix} \eta_1 & 0 & 0 \\ 0 & \eta_2 & 0 \\ 0 & 0 & \eta_3 \end{bmatrix};\ \Sigma_2 = \begin{bmatrix} \eta_4 & 0 & 0 \\ 0 & \eta_5 & 0 \\ 0 & 0 & \eta_6 \end{bmatrix} \quad (3)$$

where $\boldsymbol{\eta}$ is hyperparameters, values indicating how much the input data $\boldsymbol{x}^{\mathrm{T}} = (\beta, \gamma, \Delta F_Z)$ affects the output lateral force. The m and n in the equation indicate the number of the data.

Modeling using Gaussian processes has one hidden layer in a neural network model, corresponding to an infinite number of units (Neal, 1996). Figure 13 shows the model processed with data combining one left turn and one right turn each. The model assumes a camber angle of 0.5º. The number of data is randomly divided into four parts, 1469, as training data, which differs from the neural network shown earlier in terms of the number of data processed. The model is slightly different because the effect of the camber angle is included in the model, but the same characteristics can be seen.

In a neural network, 40 simulations are performed to obtain the mean and standard deviation of the simulations in order to suppress overlearning. In regression by Gaussian process, the mean and standard deviation of the model are obtained simultaneously. The dashed line in Fig. 13 shows the standard deviation, indicating that the standard deviation is small where there is a large number of data and large where there is little data. This is also the case with the model created using a neural network.

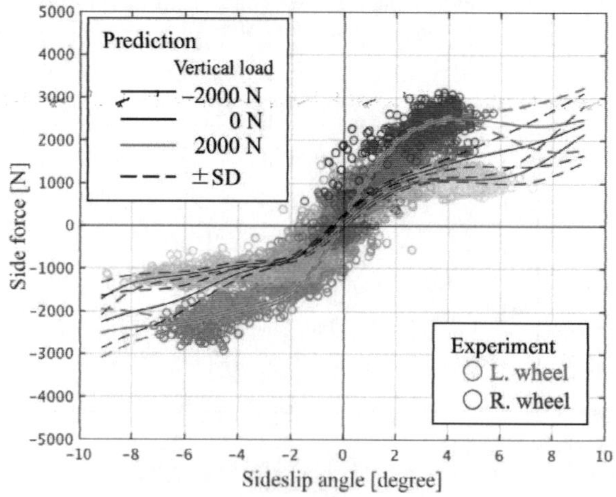

Fig. 13. Model based on Gaussian regression

4. Conclusions

Models of side force were constructed by processing data obtained from tires during turning on asphalt covered with volcanic ash by an instrumented vehicle. Although the data was obtained for a single tire from an off-road vehicle, the characteristics of the tire during turning were confirmed. The proposed method can be used to model the side force for different data.

The cornering power, which is the coefficient of side force generation near zero sideslip angle (slope, the increment of side force with respect to the sideslip angle), changes significantly with the vertical load in running experiments on 3 cm thick volcanic ash. On asphalt surfaces, cornering power is not significantly affected by the vertical load, but on volcanic ash surfaces, cornering power increases with the vertical load. Compared to an asphalt surface, there is little difference in cornering power at the standard vertical load, but it increases by a factor of about 1.5 when the vertical load is increased by 2000 N. When driving on volcanic ash surfaces, the side force saturates at a relatively small 3-5º sideslip angle, but the maximum side force is sufficiently greater for asphalt surfaces.

5. Acknowledgements

This research was supported by JSPS Kakenhi Grant-in-Aid for Scientific Research JP20K04357.

6. References

Abe, M., 2008. Automotive vehicle dynamics, Theory and application, 2nd ed., Tokyo Denki University Press. (in Japanese)

Aota, K., Yamakawa, J., Eto, R., 2021. Estimation of Forces on Vehicle Tires Running on Actual Ground by Gaussian Process Regression, Terramechanics, 41, pp. 45-50. (in Japanese)

Bishop, C. M., 2006. Pattern Pecognition and Machine Learning, Springer Science+Business Media, LLC.

Mizuno, M., Takahashi, T., Hada, M., 1999. Magic Formula Tire Model Using the Measured Data of a Vehicle Running on Actual Roads, Toyota Central R&D Review, Vol.34, No.3, pp.39-46. (in Japanese)

Neal, R. M., 1996. Bayesian Learning for Neural Networks, Number 118 in Lecture Notes in Statistics. Springer-Verlag.

Suyama, A., 2019. Bayesian Deep Learning, Kodan-sha, pp. 1-41, pp. 187-212. (in Japanese)

Yamakawa, J., Aota, K., Eto, R., 2021. Acquisition of Interaction Characteristics between Tire and Ground by Gaussian Process Regression. Proceedings of the ISTVS 20th International Conference, Online.

Yamakawa, J., Eto, R., Ichikado, Y., Yoshimoto, M., Nishizawa, T., Kubo, T., Yamada, H., 2022. Research on Vehicle Running Performance on Paved Roads Covered with Falling Volcanic Ash. Proceedings of the 11th Asia-Pacific Regional Conference of the ISTVS, Online.

1805 | Accelerating graph networks for real time physics simulation

Mohammad Jasim Usmani [a], Krzysztof Skonieczny [a,*]

[a] Department of Electrical and Computer Engineering, Concordia University, Montreal, Quebec, Canada

* Corresponding author: mohammadjasim.usmani@mail.concordia.ca, krzysztof.skonieczny@concordia.ca

ABSTRACT

Physics-based simulations are crucial for analyzing interactions between physical objects and are very commonly used in areas such as terramechanics. The discrete element method is a high-fidelity way to model the interactions between the soil particles, however, it is very computationally intensive. One promising alternative is to use Graph Network architecture which uses graph neural networks to learn and approximate the dynamics of the physical system fed into it. Even though this is faster, due to the high dimensionality of the dataset, the inference is performed on a very large graph and rollouts are not in real-time. To accelerate the simulation, one way developed recently in our laboratory is to use dimensionality reduction techniques such as principal components analysis to identify the modes in the dataset that carry the highest variance. These modes are then fed into the Graph Network along with the rigid body to train the model. Although the reduced-order dataset results in a much smaller graph than the original dataset, the graph fed into the network is still naively fully connected. This research aims to further accelerate this subspace framework by identifying a partial graph for the network without compromising either the performance or training. This identification of the partial modes is done by Neural Relational Inference (NRI). The NRI model is based on a variational autoencoder where the encoder can extract a partial graph that can still perform inference on the GNS without loss in accuracy. We demonstrate the effectiveness of our approach on blade cutting as well as wheel simulation datasets and show that this framework has reduced the inference time for excavation blade-driven granular flow from 0.35 seconds per second of simulation to 0.17 seconds and from 0.38 seconds to 0.2 seconds for wheel driven granular flow, an approximately 2x and 1.8x speed up respectively.

Keywords
Graph Neural Networks
Neural Relational Inference
Principal Component Analysis
Real Time Simulation

1. Introduction

Simulating granular flow and physics-based simulations are generally computationally intensive and expensive. These simulations necessitate a prior understanding of the underlying physics of the environment and objects involved, posing challenges in developing real-time simulators with both accuracy and efficiency. Specifically, modeling situations like robot-terrain interaction and soil interaction with excavators is particularly difficult. Real-time simulators are crucial for visualizing the interaction of granular particles with various objects. However, achieving computational efficiency remains a significant obstacle.

To address this challenge, a novel approach based on Graph Neural Networks was introduced by Sanchez-Gonzalez et al. (2020). They employed a graph network architecture proposed by Battaglia et al. (2018) to learn the underlying dynamics of various physics simulations involving substances like sand, water, and "goop". The proposed Graph Network Simulator resulted in accurate rollout performance as well as generalization. However, this was still far from real-time rollout. To tackle the problem of real-time simulation for wheel-terrain interaction as well as excavation blade-driven granular simulation interaction, Haeri & Skonieczny (2021) proposed to reduce the granular data to reduced PCA modes and fed these to the Graph Network along with the particles representing an interacting rigid body. A similar approach was previously proposed by Holden et al. (2019) for using PCA modes and training neural networks on subspace dataset to generate real-time simulations of different objects for video games. Using subspace data instead of full space data significantly decreased the space complexity which reduced the graph size significantly. As the rollout time is proportional to the number of edges in the adjacency matrix, this reduction led to a 700x speed up in simulation time. However, the adjacency matrix of the graph was assumed fully connected.

This research aims at identifying a sparser adjacency matrix of the graph that retains key connections. By reducing the number of edges, the graph would facilitate faster inference time. Neural Relational Inference (NRI), an unsupervised variational autoencoder model proposed by Kipf et al. (2018) takes in the particle trajectories as input and gives a partial graph indicating the most important edges. The main contribution of this paper is accelerating the accurate modeling of rigid body-driven granular flows by utilizing a partial graph instead of a fully connected one. The Neural Relational Inference model is employed to identify the most optimal partial graph that reduces inference time without sacrificing accuracy.

The paper's structure is as follows: the next section describes the existing framework of the graph network simulator, specifically formulated for simulating rigid body-driven interactions in granular flows. Subsequently, the paper discusses the Neural Relational Inference (NRI) model and its implementation for the case of excavation and wheel-soil interaction. Finally, the paper concludes by comparing the performance of the Graph Network Simulator with a fully connected graph against the NRI-generated partial graph.

2. Graph Network Simulator for Rigid Body driven granular flows

The Graph Network Simulator (GNS) utilizes graph neural networks to learn physics and extract underlying dynamics from the training data. In this context, the granular particles can be represented as nodes in a graph, while the edges symbolize the

https://doi.org/10.56884/PFFS7901
ISBN 978-1-942112-55-6 (pdf)

interactions between these particles. The GNS employs the message-passing framework introduced by Gilmer et al. (2017) to learn and update the features associated with the nodes and edges. The GNS has three modules which are the encoder, processor, and decoder to embed, aggregate, and update the node features as well as the edge features. The encoder's role is to represent the initial states of the particles as a latent graph, with the particle positions serving as node features and the displacements between sender and receiver nodes encoded as edge features. Graph construction is accomplished using the nearest neighbor algorithm, creating connections between sender and receiver nodes based on proximity. The processor uses the interaction network as proposed by (Gilmer et al., 2017) to perform the message passing and update the initial graph by the encoder. The decoder takes the updated graph from the processor to further decode the dynamics by predicting the position of the particles. To promote generalization and avoid error accumulation, random-walk noise is added to the input training data to ensure that the data distribution in the input is like the one in the rollout. The noise is sampled from a normal distribution with zero mean and 0.00067 as the standard deviation.

This framework was extended for the case of rigid body driven granular flows (Haeri & Skonieczny, 2021). The dataset for training the GNS was generated using high-fidelity Non-local Granular Fluidity Material Point Method (NGF-MPM) simulations (Haeri & Skonieczny, 2022). The generated dataset is three-dimensional data of the granular particles and rigid body particle positions along with the force acting on the rigid body. The excavation dataset contains 250 examples for a single soil type with a flat plate cutting soil in a ramp-horizontal-ramp motion or a continuous curved motion at different angles, depths, and speeds. The wheel dataset consists of 243 examples for three different wheel sizes with varying slip, internal friction, load, and gravity. The GNS trained on the excavation dataset was focused on mainly predicting the position of the granular particles and the force on the rigid body while the GNS trained on the wheel dataset also had to predict the position of the particles representing the rigid body which made it more complex.

The following changes were adapted to the GNS for accurate and real time modelling of granular flow:

1. The dataset fed in the network includes the position of the rigid body (R_i), the interacting force on the rigid body (F_i), and, nominally, the positions of the granular particles (S_i). To make the simulation real-time, the dimensionality of the granular particle position data was reduced using Principal Component Analysis (PCA) to 8 modes (S'_i).
2. The GNS was adapted to predict not only particle positions but also the forces acting on the rigid body.
3. The loss function was adapted to include the mean squared error of the predicted and ground truth interaction force on the rigid body.
4. Instead of using the nearest neighbor algorithm for graph construction in the encoder, a fully connected undirected graph was used for message passing and learning.

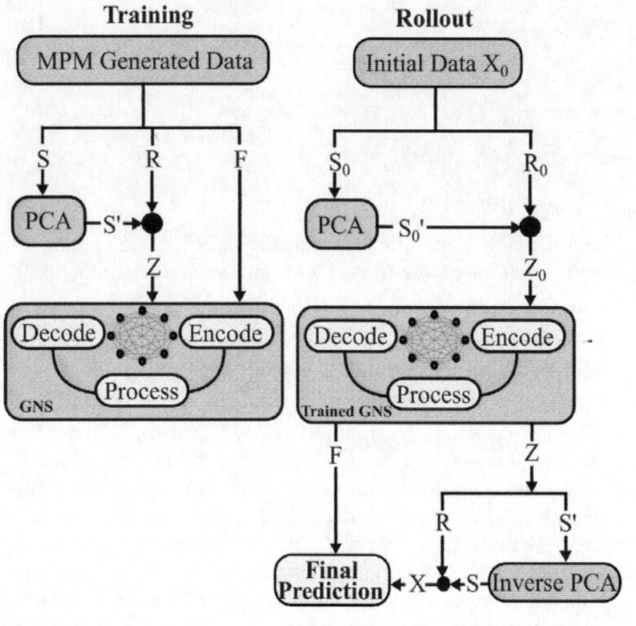

Fig. 1. GNS formulation for training and rollout for accelerated rigid body driven granular flow where S refers to the full space granular particle position, S' refers to the PCA reduced granular particle positions, R refers to the particles representing rigid body, F refers to the force acting on the rigid body.

Through their formulation, the authors successfully achieved accurate real-time modeling of excavation and wheel-soil interactions. Nevertheless, the graph construction approach employed in the encoder module was a simplistic fully connected scheme. This approach incurs a time complexity of $O(N^{2.718})$ when updating the edge features, where N represents the number of nodes in the graph. Consequently, employing a smaller graph would decrease the time complexity and enhance inference speed. This raises the question of identifying the optimal partial graph configuration that minimizes the number of edges, consequently reducing the rollout time.

3. Neural Relational Inference

The Neural Relational Inference (NRI) model is an unsupervised learning model based on the variational autoencoder architecture where the latent space encodes the pairwise interactions between the edges of the senders and receivers (Kipf et al., 2018). It is able to learn and encode the dynamics from the trajectory dataset. The NRI uses graph neural networks to represent and learn the interaction between the nodes by message-passing. Here the NRI trains an encoder that models the physical interactions between the input trajectories. . The role of the decoder is to predict future trajectories as well as reconstruct the ground truth trajectories fed into the model.

The NRI tries to optimize the evidence lower bound:

$$\mathcal{L} = E_{q_\phi(z|x)}[\log p_\theta(x \mid z)] - \mathrm{KL}[q_\phi(z \mid x) | p_\theta(z)] \quad (1)$$

where $q_\phi(z \mid x)$ is the discrete distribution learned by the encoder, $\log p_\theta(x \mid z)$ is the predicted trajectory by the decoder, $p_\theta(z)$ is the prior and KL refers to the KL divergence. The encoder tries to predict the pairwise relationship between the nodes in the graph with a fully connected graph as input. The following message-passing operations take place.

$$h_j^1 = f_{\text{emb}}(x_j) \tag{2}$$

$$v \to e: \quad h_{(i,j)}^1 = f_e^1([h_i^1, h_j^1]) \tag{3}$$

$$e \to v: \quad h_j^2 = f_v^1(\textstyle\sum_{i \neq j} h_{(i,j)}^1) \tag{4}$$

$$v \to e: \quad h_{(i,j)}^2 = f_e^2([h_i^2, h_j^2]) \tag{5}$$

$$q_\phi(z \mid x) = \text{softmax}(h_{(i,j)}^2) \tag{6}$$

Equations 2,3,4 and 5 mathematically describe the messaging passing operations where h_i^1 and h_j^1 refer to the node embedding of the node v_i and v_j in layer 1 respectively. h_i^2 and h_j^2 refers to the node embedding of the node v_i and v_j in layer in 2 respectively. $h_{(i,j)}^1$ and $h_{(i,j)}^2$ refers to the edge embedding of node v_i and v_j in the first layer and second layer respectively. $q_\phi(\mathbf{z} \mid \mathbf{x})$ in Eq. 6 represents the predicted probability of the latent space given the trajectory of the particles and f refers to the neural network. However, sampling from the encoder is not possible as the distribution is discrete. Instead, the NRI model uses the Gumbel-softmax trick to sample from this distribution where g refers to a sample from the Gumbel distribution and τ softmax temperature which decides how smooth the samples will be. The decoder aims to predict the future trajectory given the latent input graph z where each edge type is represented as a neural network.

$$z_{ij} = \text{softmax}\left(\left(h_{(i,j)}^2 + g\right)/\tau\right) \tag{7}$$

The decoder can be mathematically described as follows:

$$v \to e: \quad \tilde{h}_{(i,j)}^t = \textstyle\sum_k z_{ij,k} \tilde{f}_e^k([x_i^t, x_j^t]) \tag{8}$$

$$e \to v: \quad \mu_j^{t+1} = x_j^t + \tilde{f}_v(\textstyle\sum_{i \neq j} \tilde{h}_{(i,j)}^t) \tag{9}$$

$$p(x_j^{t+1} \mid x^t, z) = \mathcal{N}(\mu_j^{t+1}, \sigma^2 I) \tag{10}$$

In Equations 8, 9, and 10, Kipf et al. (2018) provides mathematical descriptions of the message passing operations within the proposed model. In particular, $\tilde{h}_{(i,j)}^t$ represents the edge embedding, while μ_j^{t+1} denotes the mean of the predicted trajectory. The fundamental concept underlying the NRI model is to acquire a comprehensive understanding of the physical dynamics exhibited by trajectories based on a dataset of previous trajectories. This understanding enables the model to accurately predict future timesteps given an initial condition. Since the NRI model is built upon the framework of a variational autoencoder, the bottleneck encodes pairwise interactions, thereby providing valuable insights into relative influence within the dataset.

4. Accelerating Graph Network Simulator for rigid body driven granular flows using NRI

In the framework of the Graph Network Simulator for rigid body driven granular flows, our group's prior work used a fully connected graph of PCA modes. To make the GNS computationally efficient while maintaining its accuracy, the NRI model is used in this work to identify a partial connection between the extracted PCA modes. NRI was applied to both the excavation dataset as well as the wheel dataset. In the excavation example, only PCA modes (reduced-order representation of granular particle positions) and the force on the center of mass of the excavation blade were predicted, as the rigid body (flat plate blade) followed a preset trajectory. In the wheel example, both the PCA modes (representing granular particle positions) and the wheel position were predicted along with the force on the center of mass of the wheel.

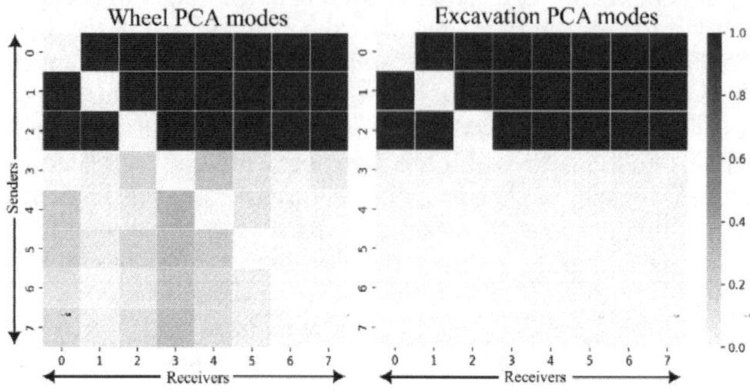

Fig. 2. Output from NRI: partial graph representing the interaction between PCA modes for wheel and excavation datasets.

The NRI model was fed the positions and the velocities (differences between subsequent pairs of position data) of the excavation dataset and the wheel dataset. The encoder learns an interaction network representing a partial graph for the PCA modes of the datasets as shown in Fig. 2. The senders refer to the nodes that send messages using the message passing process to the receiver nodes. A dark square indicates an edge with important message passing, while lighter squares indicate little to no information being passed. A partial graph can thus be constructed with just the edges corresponding to dark squares. The partial graph framework is split between four parts where the graph has edges between senders and receivers. The senders and receivers define the interaction between PCA modes to PCA modes, rigid body to rigid body, rigid body to PCA modes, and PCA modes to rigid body. The partial graph formulations for each of these interactions are explained below, and are summarized by Fig. 3 and Fig. 4 for wheel and excavation datasets, respectively.

PCA modes to PCA modes: The interaction between PCA modes is predicted by the NRI model, as described above. It is worth noting that the NRI prediction solely captures the interaction between the PCA modes, disregarding the interaction of stationary granular particles within the PCA mode. To address this limitation, an additional mode is introduced to represent the stationary granular particles (compare Fig. 2 and 4).

Rigid Body to Rigid Body: When considering the interaction between particles representing rigid bodies, message passing occurs only if the motion of the rigid body is also predicted by the network. In the excavation case, where the rigid body motion is not predicted, no message passing occurs between the rigid body particles, resulting in the absence of edges in the adjacency graph to represent these interactions.

However, in the case of the wheel, where the network predicts the motion of the rigid body, message passing becomes necessary between the particles representing the rigid body. To determine the appropriate rigid body edges, various trials were conducted. For the wheel scenario, it was determined that establishing edge connections from each rigid body particle to its consecutive rigid body particles yielded satisfactory results.

Rigid Body to PCA modes: The message-passing from the rigid-body to the PCA modes occurs in both the excavation example as well as the wheel. This part is crucial for rigid body driven granular flow, so all the particles representing the rigid body have an edge in the adjacency matrix for all the connections with all the PCA modes. This enables the GNS to learn the relative motion of the PCA modes with respect to the rigid body.

PCA modes to Rigid Body: The message-passing in this case occurs only when the positions of the particles representing the rigid body are also predicted. For this case, the GNS needs to learn the motion of the particles representing the rigid body relative to the PCA modes. Messaging passing is necessary between the granular particles and the wheel, leading to an undirected graph that enables the model to learn the dynamics of the wheel motion. Therefore, there is an edge in the adjacency matrix for all the connections between PCA modes to the rigid body in the case of the wheel but no edge in the case of excavation.

In the excavation example, the position of the particles representing the blade is pre-defined and is not predicted by the network. Consequently, there is no messaging passing required from the granular particles to the blade, resulting in a directed graph.

Thus, the final partial graph for the wheel driven granular flow (see Fig. 3) and excavation driven granular flow (see Fig. 4) is illustrated below. Recall that in prior work a fully connected undirected graph was used, which would correspond to such figures with all squares filled in.

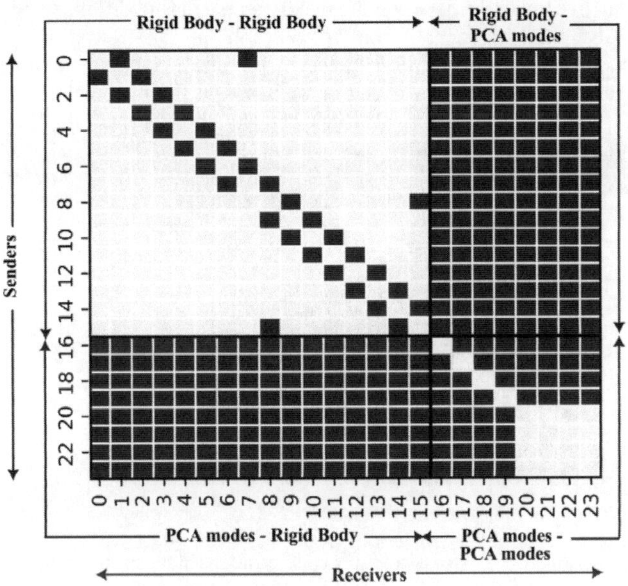

Fig. 3. Partial graph for wheel driven granular flow.

The optimized partial graph formulation resulted in a significant reduction in the number of edges within the adjacency graph. Specifically, for excavation-driven granular flow, the number of edges decreased from 784 to 188, while for wheel-driven granular flow, it decreased from 576 to 332. These reduced partial graphs successfully replicated the desired outcomes while offering faster rollout times, primarily attributed to the decreased number of edges present in the adjacency matrix. However, in the case of predicting the node positions of the wheel, a discrepancy was observed where the particles within the rigid body failed to maintain rigid relative spacing. This issue is addressed in the subsequent sub-section through the utilization of least square fitting techniques.

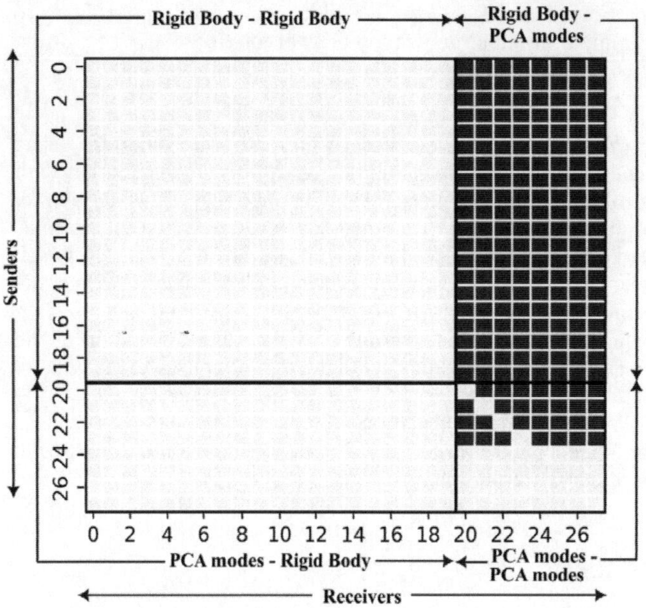

Fig. 4. Partial graph for excavation driven granular flow.

4.1 Rigid Body Correction by Least Square Fitting of 3D points

In the case of the wheel, predicting the position of the particles representing the rigid body introduces a slight error in the wheel position. This error propagates to incorrect slip predictions and eventually inaccurate predictions for the granular particles. To mitigate the accumulation of errors during the rollout, it is essential to correct the position of the particles representing the rigid body, i.e. maintaining the wheel's rigid shape. To address this issue, a formulation is presented to minimize the discrepancy between the particles representing the wheel before and after rotation. This involves finding an appropriate rotation matrix and translation vector that minimize the error. The problem can be formally described as a minimization problem, with the objective of minimizing the discrepancy between the pre-rotation and post-rotation positions of the wheel particles.

$$\min_{R\in\Omega,t} \sum_{i=1}^{n} |Rx_i + t - y_i|^2 \quad (11)$$

Here n is the number of rigid body particles, R is the rotation matrix and t is the translation vector. x_i and y_i are the particle positions before wheel rotation and after rotation respectively. This is adapted from Söderkvist & Wedin (1994) who formulated the process as shown in algorithm 1 below.

Algorithm 1: Rigid Body Correction by Least Square fitting (Söderkvist & Wedin, 1994)

Input: $x_1, x_2, \dots, x_n, y_1, y_2, \dots, y_n$
Output: R, t
$\bar{x} = \frac{1}{n}\sum_{i=1}^{n} x_i, \bar{y} = \frac{1}{n}\sum_{i=1}^{n} y_i$
$A = [x_1 - \bar{x}, \dots, x_n - \bar{x}]$, $B = [y_1 - \bar{y}, \dots, y_n - \bar{y}]$
$C = BA^T$
$USV^T = C$
$R = U \times \text{diag}(1, 1, det(UV^T)) \times V^T$
$t = \bar{y} - R\bar{x}$

This algorithm does not explicitly correct the particle positions of the rigid body but rather extracts a rotation matrix and a translation vector to rotate the rigid body particles from before the rotation. This algorithm is added along with the rollout to minimize any error in the rigid body prediction.

4.2 Training Noise

The training noise added to the data was generated using random-walk, initially sampled from a normal distribution with zero mean and standard deviation 0.00067. This addition of random noise in the training data is due to several reasons. Random noise leads to data augmentation in every step which promotes generalization and prevents overfitting. However, the standard deviation of the PCA modes was in the order of 10^{-5}, which is lower than the order of the standard deviation of the noise. This led to training data that was randomized to the point of data corruption, rather than helping in generalization. To tackle this, the standard deviation of the normal distribution was modified from 6.7×10^{-4} to 6.7×10^{-6}. This reduced the error in predictions during rollouts.

5. Results

To generate the rollout, the graph network simulator is given the initial positions, and the GNS then iteratively predicts the particle positions and force acting on the rigid body.

Runtime Efficiency: Table 1 shows the runtime efficiency of the MPM method, full space dataset, subspace dataset with a fully connected graph, and subspace dataset with a partially connected graph (this work). The partial graph has a faster training time as well as rollout time compared to the other methods and is 2x faster than the subspace dataset with a fully connected graph.

Table 1
Run time comparison of rigid body driven simulation using Material Point Method (MPM), full space GNS (FS), subspace GNS with fully connected graph (FCG) (Haeri & Skonieczny, 2021) and subspace GNS with partially connected graph (PCG).

Dataset	Method	Training [sec/step]		Rollout [sec/sec]		Simulation [sec/sec]
		GPU[1]	GPU[2]	CPU[1]	GPU[1]	CPU[2]
	MPM	-	-	-	-	270
Excavation	FS	OOM	1.000	-	OOM	-
	FCG	0.016	0.064	0.350	0.440	-
	PCG	**0.01**	-	**0.15**	**0.206**	
	MPM	-	-	-	-	240
Wheel	FS	OOM	0.500	-	OOM	-
	FCG	0.017	0.068	0.380	0.460	-
	PCG	**0.011**	-	**0.197**	**0.217**	-

Prediction Accuracy: In order to compare the accuracy of the models, the visualization of the particles and forces of the simulation are compared with the ground truth and the prediction in the x-y plane for the excavation dataset at timesteps 1 second, 3 second, and 5 second mark (see Fig. 5). The histogram (see Fig. 6)

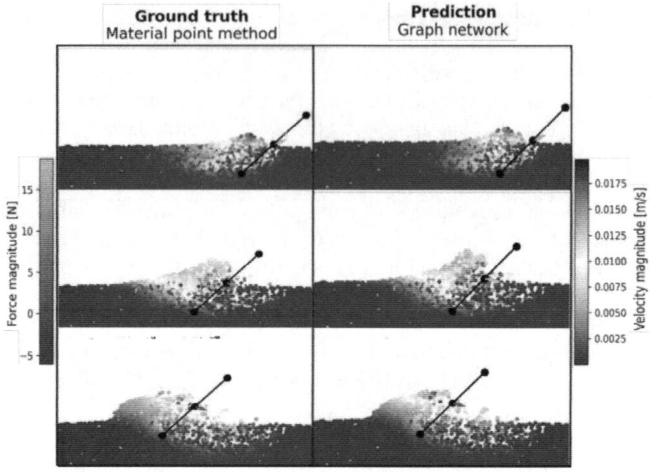

Fig. 5. Ground Truth and Prediction of the excavation driven granular flow at time 1 second, 3 second and 5 second.

quantifies the error between the predicted particle positions from the GNS with the partially connected graph and the ground truth from MPM for the excavation dataset. In the prediction, most of the particles have a mean squared error between the order of 10^{-8} and 10^{-6} (units of mm²) which reflects the accuracy of the simulation.

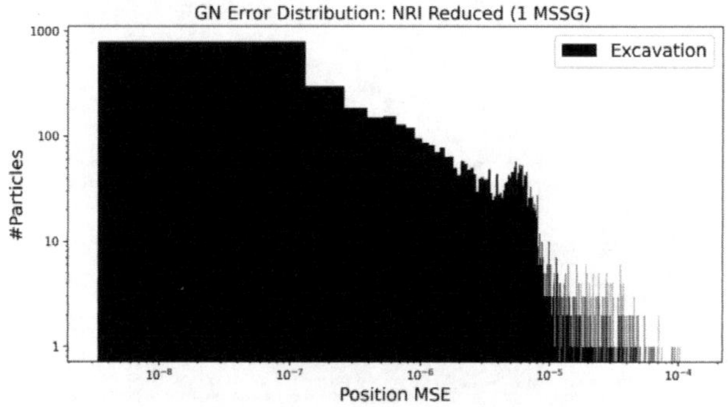

Fig. 6. Histogram of error between particle positions of the ground truth and prediction of the excavation driven granular flow.

In the context of the wheel dataset, a comparison is made between the visualization of particles and forces obtained from the simulation, the ground truth, and the predictions in the x-y plane. When the simulation does not incorporate the rigid body correction, the particles representing the rigid body exhibit incorrect predictions. Consequently, this leads to an amplified error in the position of the granular particles. It is important to note that the positions of the granular particles are dependent on the positions of the particles representing the rigid body. Therefore, any inaccuracies in the prediction of the rigid body positions result in increased errors in the positions of the granular particles (see Fig. 7).

When the rigid body motion is corrected, both the rigid body particles as well as the granular particles produce accurate prediction with respect to the ground truth. The visualization between the ground truth and the prediction for the case when rigid motion is corrected as well as when the rigid body motion is not corrected is shown in the figure below (see Fig. 8).

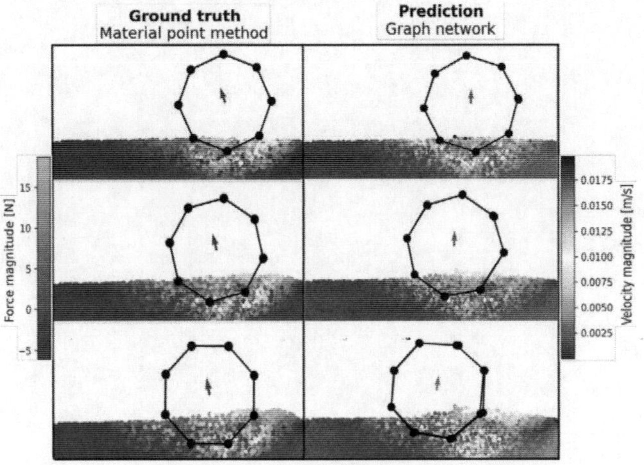

Fig. 7. Ground Truth and Prediction of the Wheel driven granular flow without rigid body correction at time 1 second, 3 second and 5 second.

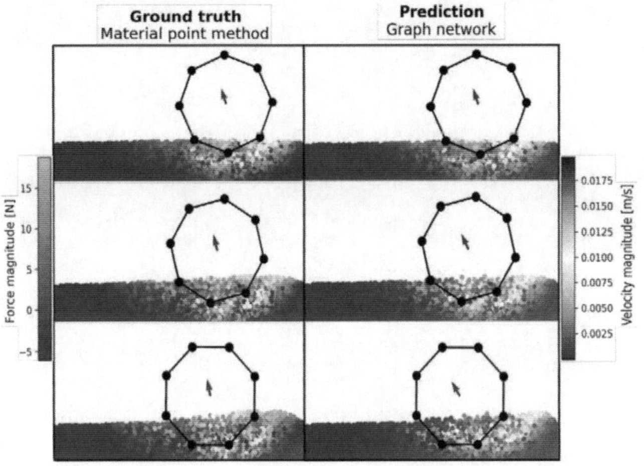

Fig. 8. Ground Truth and Prediction of the Wheel driven granular flow after applying rigid body correction at time 1 second, 3 second and 5 second.

The histogram quantifies the error between the predicted particle positions from the GNS with the partially connected graph and the ground truth from MPM for the wheel dataset. In the prediction, most of the particles have a mean squared error between the order of 10^{-8} and 10^{-6} which reflects the accuracy of the simulation. However, there is higher error in the case when there is no rigid body correction applied in the rollout (see Fig. 9).

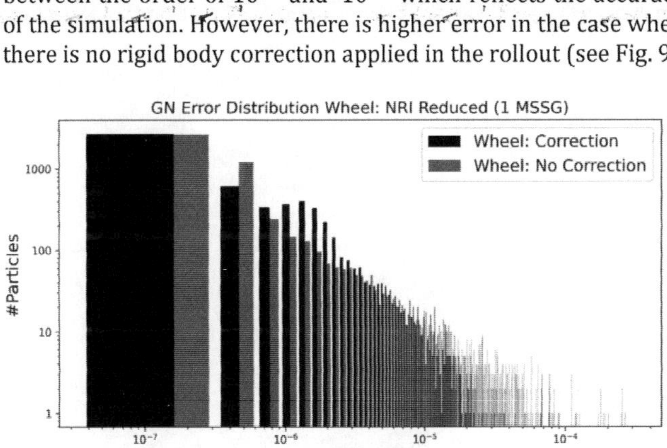

Fig. 9. Histogram of error between particle positions of the ground truth and prediction of the wheel driven granular flow.

6. Conclusion

This research aims to enhance the computational efficiency of utilizing graph neural networks within the graph network architecture, particularly for physics-based simulations involving rigid body-driven granular flows. While graph networks have demonstrated effectiveness in learning dynamics from datasets, they also impose significant computational requirements. To address this challenge, a partial graph framework is proposed in this study, which reduces the computational cost by reducing the number of nodes in the graph. The generation of the partial interaction graph between the Principal Component Analysis (PCA) modes is achieved through the employment of the neural relational inference (NRI) model which extracts the most important relationship between the PCA modes.

The experimental findings indicate that the utilization of the partial graph framework for training and inference does not compromise the model's performance or the accuracy of the rollouts. Moreover, it successfully reduces the rollout time. Specifically, the partial graph approach led to a 2x improvement in rollout time for the excavation blade-driven granular flow example and a 1.8x improvement for the wheel-driven granular flow example.

Although the NRI-based partial graph framework has expedited rollout time, alternative architectures such as MLP mixer for time series (Chen et al., 2023) and graph MLP mixer (He et al., 2022) have also recently demonstrated promising outcomes. As the MLP mixer architecture solely consists of MLP (Multi-Layer Perceptron) without graph operations, it is expected to generate fast inferences. This idea can be explored in future work to investigate even faster rollouts for rigid body driven granular flows.

7. Acknowledgements

This research was supported by Natural Sciences and Engineering Research Council of Canada along with industrial support and collaboration with CM Labs Simulation Inc.

8. References

Battaglia, P. W., Hamrick, J. B., Bapst, V., Sanchez-Gonzalez, A., Zambaldi, V., Malinowski, M., Tacchetti, A., Raposo, D., Santoro, A., Faulkner, R., Gulcehre, C., Song, F., Ballard, A., Gilmer, J., Dahl, G., Vaswani, A., Allen, K., Nash, C., Langston, V., ... Pascanu, R. (2018). Relational inductive biases, deep learning, and graph networks. https://arxiv.org/abs/1806.01261v3

Chen, S.-A., Li, C.-L., Yoder, N., Cloud, G., Research, A. I., Arik, S. O., & Pfister, T. (2023). TSMixer: An all-MLP Architecture for Time Series Forecasting. https://arxiv.org/abs/2303.06053v2

Gilmer, J., Schoenholz, S. S., Riley, P. F., Vinyals, O., & Dahl, G. E. (2017). Neural message passing for quantum chemistry. 34th International Conference on Machine Learning, ICML 2017, 3.

Haeri, A., & Skonieczny, K. (2021). Subspace Graph Physics: Real-Time Rigid Body-Driven Granular Flow Simulation. https://arxiv.org/abs/2111.10206v1

Haeri, A., & Skonieczny, K. (2022). Three-dimensionsal granular flow continuum modeling via material point method with hyperelastic nonlocal granular fluidity. Computer Methods in Applied Mechanics and Engineering, 394. https://doi.org/10.1016/j.cma.2022.114904

He, X., Hooi, B., Laurent, T., Perold, A., LeCun, Y., & Bresson, X. (2022). A Generalization of ViT/MLP-Mixer to Graphs. https://arxiv.org/abs/2212.13350v1

Holden, D., Duong, B. C., Datta, S., & Nowrouzezahrai, D. (2019). Subspace neural physics: Fast data-driven interactive simulation. Proceedings -

SCA 2019: ACM SIGGRAPH / Eurographics Symposium on Computer Animation. https://doi.org/10.1145/3309486.3340245

Kipf, T., Fetaya, E., Wang, K. C., Welling, M., & Zemel, R. (2018). Neural relational inference for Interacting systems. 35th International Conference on Machine Learning, ICML 2018, 6.

Sanchez-Gonzalez, A., Godwin, J., Pfaff, T., Ying, R., Leskovec, J., & Battaglia, P. W. (2020). Learning to simulate complex physics with graph networks. 37th International Conference on Machine Learning, ICML 2020, PartF168147-11.

Söderkvist, I., & Wedin, P. Å. (1994). On condition numbers and algorithms for determining a rigid body movement. BIT, 34(3). https://doi.org/10.1007/BF01935651

2145 | Obstacle performance and wheel failure test analysis of Zhurong Rover

Hongtao Cao [a], Meng Zou [a,*], Jianqiao Li [a], Baofeng Yuan [b], Dong Pab [b]

[a] Key Lab for Bionics Engineering of Education Ministry, Jilin University

[b] Institute of Spacecraft System Engineering, CAST

* Corresponding author: zoumeng@jlu.edu.cn

ABSTRACT

A large number of rocks are distributed on the surface of Mars, and the wheels have potential risks such as puncture and fracture, which pose a threat to the trafficability of the Mars rover. Therefore, it is of great significance to carry out the experimental analysis of the obstacle performance and failure mechanism of the Zhurong Rover wheel.

A ground test model of the Zhurong Rover was used as the ground simulation vehicle, and the stone height and driving speed are taken as variables. The wheel drive mechanism power, current and wheel obstacle resistance are taken as the test indexes, and the obstacle performance test of the Mars rover is carried out. The test results show that the variation trend of average power and current is obvious, and there will be obvious slope peak when crossing the obstacle, which can judge when the Mars rover crosses the obstacle and the state of crossing the obstacle. When the wheel crosses the obstacle, the vehicle power, current and wheel resistance are positively correlated with the obstacle height and driving speed.

In order to verify the strength and damage resistance of the wheel, the extreme driving conditions were set to test the failure modes such as wheel damage, puncture and fracture. The Zhurong Rover wheel was shown to have excellent damage resistance.

The analysis results verify the obstacle surmounting performance of the Mars rover and the strength and damage resiliency of the wheels.

Keywords
Zhurong rover
Climbing obstacle capability
Power
Current
Resistance

1. Introduction

There are a lot of rocks on the surface of Mars(e.g. Yingst, Cropper et al. 2016), And the distribution of rock is more dense than at the surface of the moon(e.g. M., P. et al. 2008). The harsh driving environment puts forward higher requirements for the mobility performance of the Mars rover. The main risks affecting the mobility of Mars rover are wheel breakage, wheel sinkage, and wheel inability to travel due to line failures. The wheels of the U.S. Mars rover Curiosity are riddled with punctures in their rims due to rock erosion, with holes and perforations accounting for more than half of the wheels(Rankin, Maimone et al. 2021).

Therefore, sufficient ground test verification work is very important for the Mars rover. The ' Apollo ' project in the United States established a space environment simulation laboratory (SESL) at the Johnson Space Center in Houston, which permitted simulation experiments at reduced vehicle weight and in vacuum conditions, for testing the US Apollo LRV at system level. In addition, an outdoor lunar terrain test site was established to train astronauts and test the performance of the lunar rover(Arvidson and R. 2002). The ' Lunokhod ' rover program of the former Soviet Union also carried out ground tests such as mobility test, teleoperation test, detector navigation and obstacle avoidance test during the development process (Gulick, Morris et al. 2001). China has established a field verification wear technology system for the Chang 'e-3 lunar rover and provided a reference for the development of the Zhurong Rover (Dang, Zhang et al. 2022).

In this paper, verification tests on the ground movement performance and of the climbing obstacle capability of the Zhurong Mars Rover are reported. The relationship between obstacle height, driving speed and vehicle current, power and resistance is compared and analyzed, which provides a basis for judging the trafficability of Mars rover. Experimental verification under extreme conditions was carried out to study the strength and damage resistance of the Mars rover wheels.

2. Tests and Methods

2.1 Test equipment

Figure 1a shows the Zhurong rover ground test model used in the mobility tests reported here. The mass of the ground test model is 90 kg, which is about 3/8 of the mass of the whole rover. This means, the weight on Earth of the ground test models approximates Zhurong's weight on Mars which is considered an important condition for representative mobility tests, because of the influence of vehicle weight on sinkage, motion resistance, and drawbar pull. The Mars rover mobility system is in the form of a six-wheel drive main and secondary rocker suspension, with the right side of the front end shown in Fig.1a named the +Y side and the left side the -Y side. The wheels of the Mars rover shown in Fig.1b are 300mm in diameter, 200mm in width, 80cm between the front and rear wheels, and 130cm between the left and right wheels; there are 20 5mm wheel grousers evenly distributed on the wheels to assist in driving.

https://doi.org/10.56884/PJHH7857
ISBN 978-1-942112-55-6 (pdf)

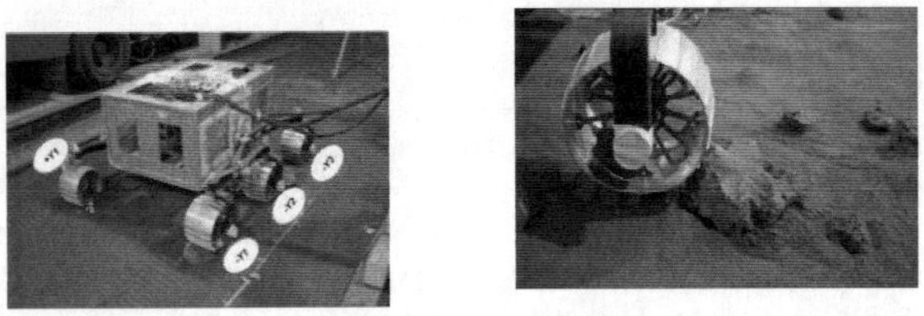

a. Mars Rover Mobility System b. Wheels in the Crossing

Fig. 1. Diagram of test vehicle and wheels

2.2 Test site

The experiments were conducted at the Key Lab for Bionics Engineering of Education Ministry of Jilin University, where the Mars simulants JLU Mars-1(D=40μm), Mars-2(D=200μm) and Mars-3(D=700) were laid (D means the median particle size of soil.). The distribution map of the test area is shown in Fig.2. In the Rocky sand drift area, the obstacle resistance test bench shown in fig.2b is set up. Fig.2c, d shows the ground of self-made slate and sharp-angle stone, which is used to verify the mobility of the vehicle in extreme driving environment.

As shown in Fig.2 b1-4. The obstacle resistance test platform is buried under the simulated fire soil, and the obstacle is exposed on the surface. When the test vehicle crosses the obstacle, the force sensor under the obstacle will collect the load on the wheel. The multi-directional force sensor is installed in the test bench to obtain the wheel pressure, the driving direction of the vehicle to drive the friction force, and the force between the simulated force module and the contact wheel is obtained.

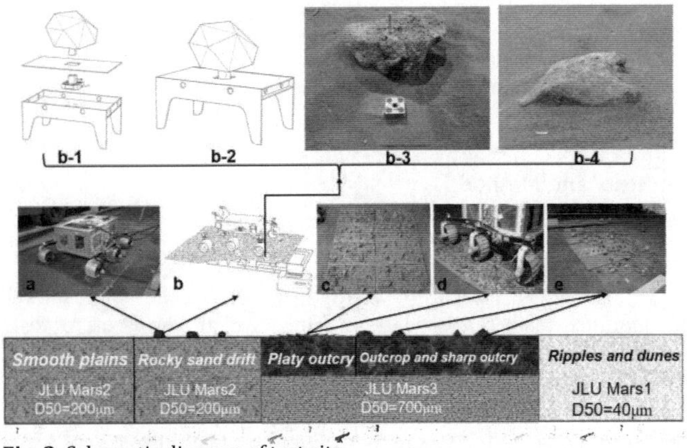

Fig. 2. Schematic diagram of test site

2.3 Test scheme

The controller is connected with the ground simulation vehicle through the cable, issuing instructions such as forward, backward and steering, and recording the power and current data of each wheel in real time. The multi-directional force sensor at the obstacle resistance test bench can record the force of the wheel when the wheel crosses the obstacle.

(1) Climbing obstacle capability

In view of the complex environment on the surface of Mars, the speed of the Mars rover is generally low, and three speed conditions of 50m/h,100m/h and 200m/h were used in the tests.

There will be obstacles such as stones of different sizes in the complex environment on the surface of Mars. When the height of the stones is low, in order to reduce the driving distance of the Mars rover, the Mars rover can go straight for vertical obstacle crossing. The permissible obstacle height is set in the range of 50-200 mm by design of the mobility system, which can ensure that the Mars rover can cross the obstacle vertically under the premise of safety.

In the obstacle surmounting performance analysis test, the obstacle surmounting performance of the whole vehicle is verified when the wheel passes through one obstacle on one side, and the corresponding data obtained are analyzed and calculated to explore the obstacle surmounting law.

(2) Obstacle anti-piercing ability test

In order to simulate the strength and anti-damage ability of the Mars rover wheel and suspension under extreme conditions, the hard stone plate + sharp-angle stone ground was processed and prepared. A total of 12 pieces were laid in the form of 3 × 4 in the soil tank, each of which was about 95-99 cm long, about 50 cm wide and about 6 cm thick. The overall area after laying is about 300cm × 200cm. The maximum height of the hornstone is about 8cm, and there are about 4-6 large hornstones per slab, accounting for 10 % -20 % of the total area.

Fig. 3. The Mars rover travels on sharp slate.

(3) Drop-on-obstacle test

In this test case, a steel spike is buried in front of the obstacle to test the ability of the wheel to resist damage after the wheel falls off the stone. A three-way force sensor is set below the stone to detect the force of the stone on the wheel. As shown in Fi.4.

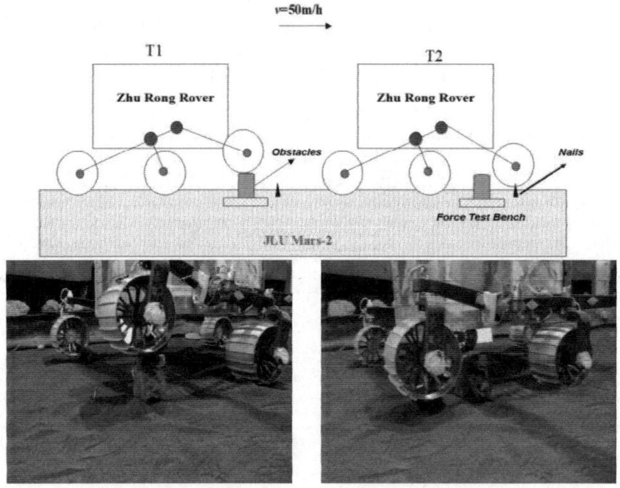

Fig. 4. Mars rover obstacle crossing drop test

3. Test results and analysis

3.1 Climbing obstacle capability analysis

Fig.5 and Fig.6 show the measured data of the total current and overall electrical input power of the six-wheel drive of the Zhurong ground test model when it crosses the 200mm stone at a speed of

50m/h. The six peaks in the figure are the real-time situation of power current when the Mars rover wheels cross obstacles sequentially. In the diagram, it can be seen that the wheel will produce obvious peaks when it crosses the obstacle, which can judge when the Mars rover crosses the obstacle and the obstacle state.

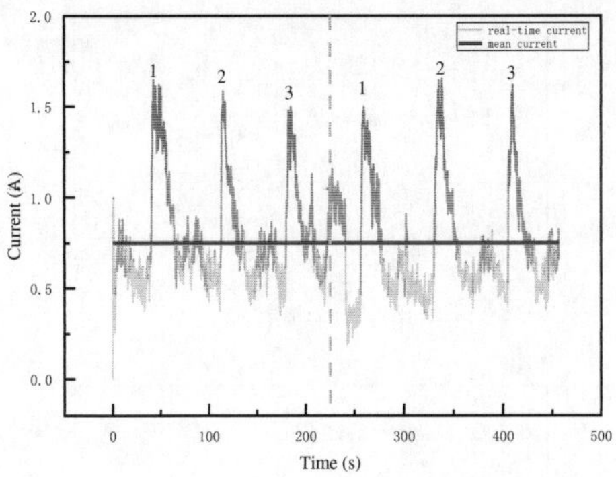

Fig. 5. Obstacle real-time total current of all six-wheel drives

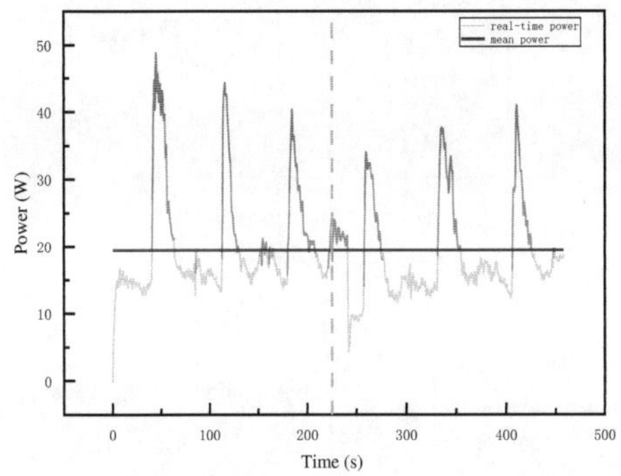

Fig. 6. Obstacle real-time total power of all six-wheel drives

Fig.7 and Fig.8 are shown as the changes of the current and power of the Mars rover when it crosses obstacles at different heights at different driving speeds. It can be seen from the diagram that the average current and average power of the whole vehicle show a linear increasing trend.

For a speed of 50m/h, if obstacle height increases from 90mm to 200mm, the average power increases by 31.4%, and the average current increases by 24.3%. The maximum current and maximum power also show a linear increasing trend. When crossing the 90mm obstacle, the maximum current is 1.3A and the maximum power is 23.6W. The maximum current increases to 1.76A and the maximum power increases to 49.9W when crossing the 200mm obstacle.

For a speed of 100m / h, if obstacle height increases from 90mm to 200mm, the average power increases by 27.3 %, and the average current increases by 23.6 %. The maximum current and maximum power also show a linear increasing trend. When crossing the 90mm obstacle, the maximum current is 1.32A and the maximum power is 36.5W. The maximum current increases to 1.84 A and the maximum power increases to 65.2 W when crossing the 200 mm obstacle.

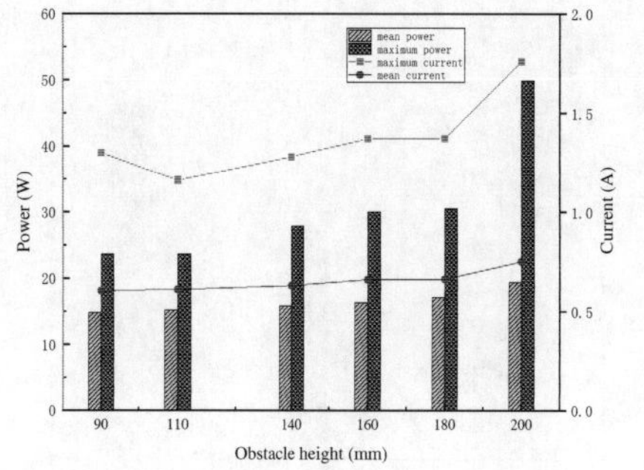

Fig. 7. The change of current power at 50m/h

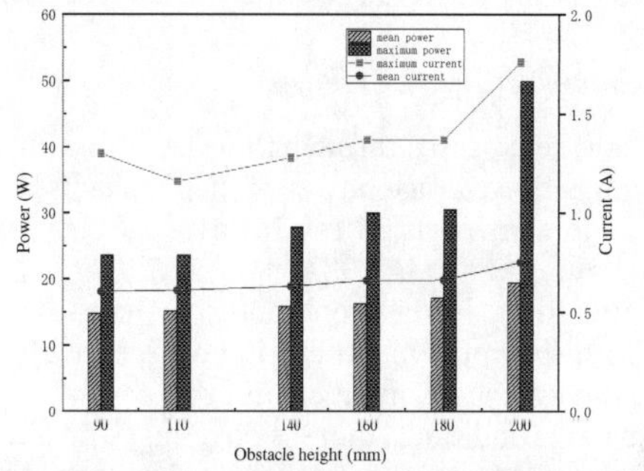

Fig. 8. The change of current power at 100m/h

When the wheel of Zhurong Rover crosses the obstacle, the change of the obstacle resistance of the wheel with height is shown in Fig.9, The force shown in the diagram is obtained from the resultant force measured by the three-way force sensor. It is concluded that with the increase of obstacle height, the resistance of the wheel also increases linearly. In addition, the greater the speed, the greater the kinetic energy, which will lead to the increase of the impact force on the wheel. The maximum wheel resistance in the test is 587.2 N.

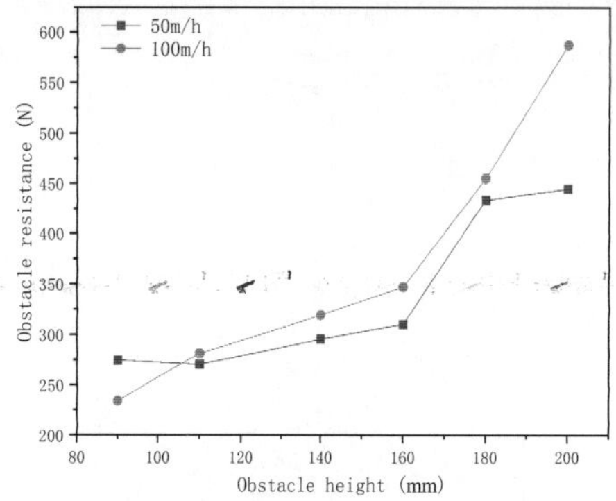

Fig. 9. The change of obstacle resistance

Under the height of 100mm obstacle, the change of driving parameters of the test vehicle with speed is compared and analyzed. It can be seen from the figure that the average power and average current of the vehicle show an increasing trend. At 100m/h and 200m/h, the average power increases by 66.20% and 181.93% compared with that at 50m/h, and the average current increases by 31.14% and 51.83%.

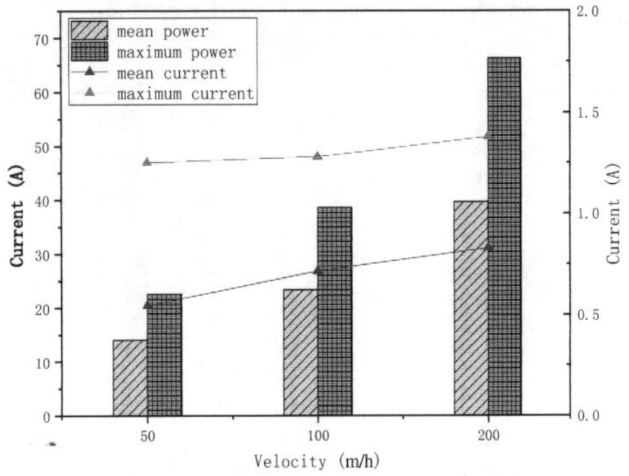

Fig. 10. The change of current and power under different driving speeds

3.2 Wheel failure verification analysis

When the wheel passes through the hard stone pavement at a speed of 50m/h, the average power of the mechanism is 18.455w and the maximum power is 53.1w. The average current is 0.663A and the maximum current is 1.93A.

Due to the strict driving conditions, the current and power fluctuations are obvious, but there is no excessive current and power, and the average value is within the normal range. Under extreme working conditions, there is no damage or fracture of the wheel, which verifies the good driving performance of the vehicle.

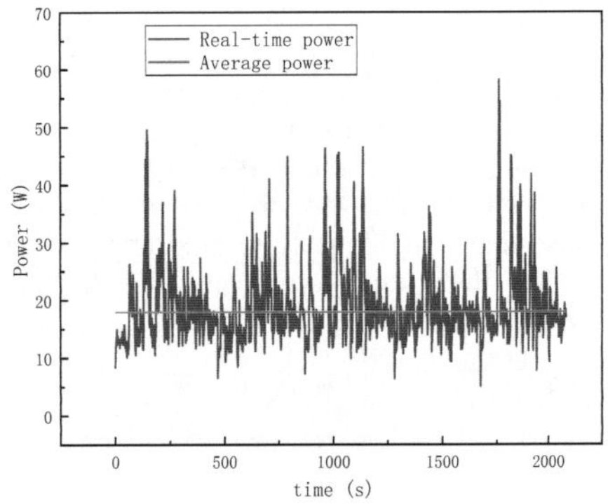

a. power diagram

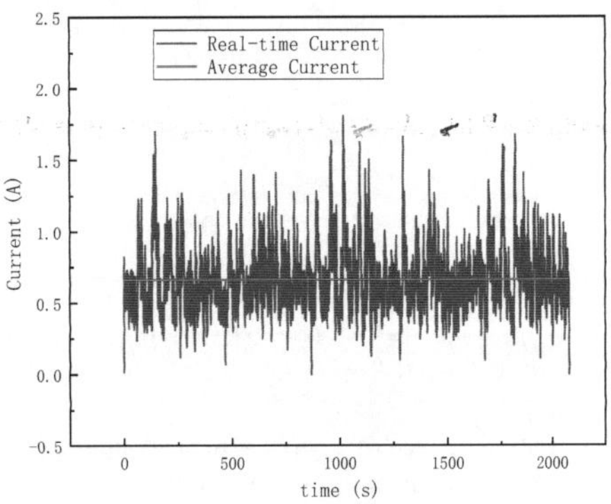

b. current diagram

Fig. 11. Power current diagram of hard stone plate test condition of all six-wheel drives

In order to further verify the impact resistance of the wheel, we carried out a 200mm obstacle drop test. The measured data of the Zhurong Rover test is shown in Fig,12. After the test, no obvious damage marks were observed on the wheels, but the scratches and debris falling shown in Fig.13 appeared.

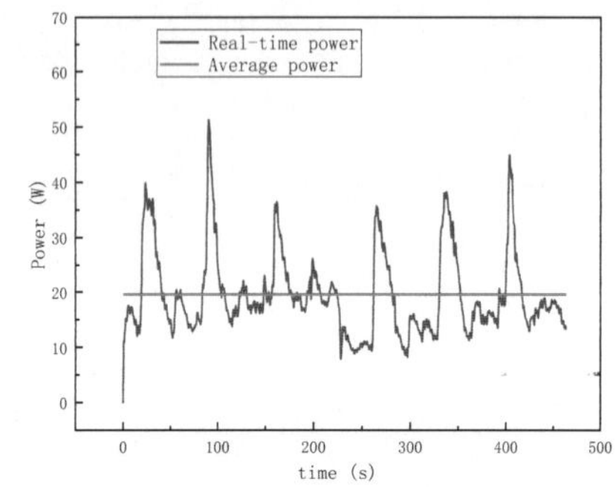

a. power diagram

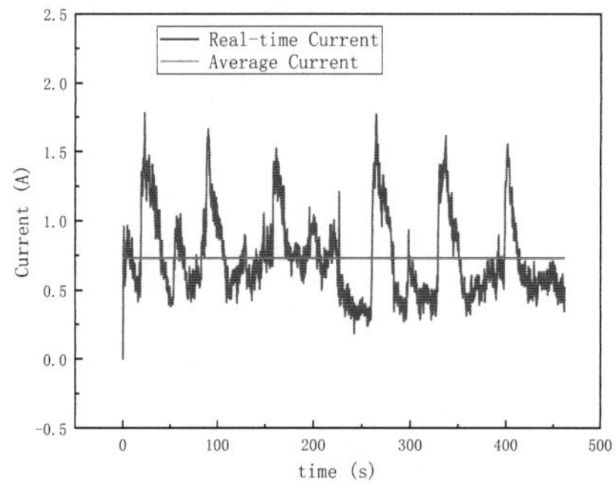

b. current diagram

Fig. 12. Obstacle drop test power current diagram

Fig. 13. Scratches and debris falling situation

3.3 Result analyze

In this section, the vertical obstacle surmounting ability of Zhurong rover is analyzed through experiments, and the variation rules of vehicle power, current and force under different working conditions are analyzed. The analysis shows that the current and power will produce obvious peak curve when the wheel crosses the obstacle, which can judge when the Mars rover crosses the obstacle and the state of crossing the obstacle. With the increase of obstacle height, the average current and average power of the vehicle show a positive correlation trend. The maximum current in the test is 1.84A, and the maximum power increases to 65.2W. As the obstacle height increases, the resistance of the wheel also increases linearly. In the test, the wheel is subjected to a maximum resistance of 587.2N. In addition, it is found that the change of driving speed has a significant effect on the power current. The average power at 100m/h and 200m/h increases by 66.20% and 181.93% compared with that at 50m/h, and the average current increases by 31.14% and 51.83%.

In order to simulate the extremely harsh Mars driving environment and further verify the strength and damage resiliency of the Mars rover wheel, the ability of the wheel to resist puncture damage was verified by the test of the stone sharp corner stone condition and the obstacle drop condition. The test results show that the Mars rover can cross the working condition of the slate corner stone, by observing the appearance of the wheel, it is concluded that the obstacle-crossing drop test does not cause damage to the wheels of the Mars rover, only slight scratches and debris fall.

4. Analysis and Discussion of Obstacle Theory

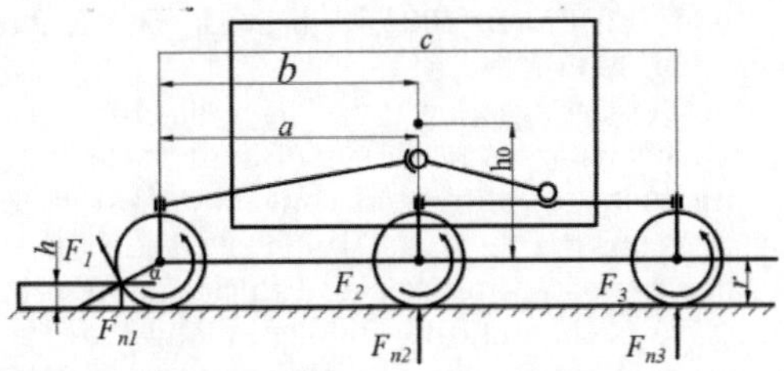

Fig. 14. Force analysis of wheel vertical obstacle crossing

Figure 14 shows the schematic diagram of the force acting on the vehicle when the front wheel crosses the obstacle. The force balance equation is:

$$\begin{aligned} &F_{n1}\cos\alpha - F_1\sin\alpha - F_2 - F_3 = 0 \\ &F_{n1}\sin\alpha + F_1\cos\alpha + F_{n2} + F_{n3} - W = 0 \\ &F_1 r + F_2 r + F_3 r + bW - F_{n2}a - F_{n3}c = 0 \end{aligned} \qquad (1)$$

in the form: $\sum_{i=1}^{3} F_i = \mu F_{ni}$, $\sin\alpha = 1 - \frac{h}{r}$, a=b=c/2.

b is the horizontal distance of the center of gravity of the front wheel track, c is the horizontal distance between the front wheel and the rear wheel, a is the horizontal distance between the front wheel and the middle wheel, r is the wheel radius, h is the vertical height of the obstacle, μ is the rolling resistance coefficient, W is half of the vehicle weight, Fi is the reaction force of the ground to the vehicle, and Fni is the propulsion force at the contact point between the wheel and the ground.

The Mars rover is driven by six driving motors. When a single wheel or several wheels of the Mars rover encounters stones, it is not the obstacle-crossing wheels that increase the output torque alone, but the six wheels work together to complete the obstacle-crossing. The obstacle surmounting performance of six-wheeled vehicle with independent suspension is affected by the force of each wheel and the position of the center of gravity. Because the front and rear wheel track and the height of the center of gravity of the Zhurong Mars rover can be adjusted via its active suspension , and the wheel spines distributed on the outer circle of the Mars rover wheel provide better grip for the wheel obstacle surmounting, which also contributes to the improvement of the obstacle surmounting ability. The test found that the final obstacle surmounting ability h / D is higher than 0.5.

5. Conclusions

In view of the complex terrain environment on Mars, this paper conducts experimental research on the obstacle-crossing performance of the Zhurong Rover using a ground test rover model, and verifies the obstacle-crossing ability of the wheels. When the wheels cross the obstacle, the current and power will produce an obvious peak curve, which can judge when the Mars rover crosses the obstacle and the obstacle-crossing state. The results of various evaluation indicators can provide a basis for judging the driving environment of the on-orbit Mars rover. Due to the scaling of the weight of the test rover, it is meaningful to correct the ground test data through the real data in orbit, so as to provide a more reliable driving scheme for on-orbit operation. The vertical obstacle crossing test shows that the obstacle crossing ability h / D of the Mars rover is higher than 0.5, and the vehicle current, power and resistance are positively correlated with the obstacle height. In further tests, the resiliency against damage of the Mars rover wheels was tested by slate and obstacle crossing drop tests. The test verifies that the Mars rover wheel has good strength and damage resistance.

6. Acknowledgements

This work was supported by the National Natural Science Foundation of China (grant numbers 52075217, 51775233) and Key Funding of Science and Technology Department of Jilin Province (grant numbers 20200401144GX).

7. Declaration of competing interest

The authors declare that they have no known competing financial interests or personal relationships that could have appeared to influence the work reported in this paper.

8. References

Arvidson, E. R., 2002. FIDO prototype Mars rover field trials, Black Rock Summit, Nevada, as test of the ability of robotic mobility systems to conduct field science. J. Geophys. Res. 107(E11): 8002.

Dang, Z. J., Zhang, J., Chen, B., 2022. Field Validation of Egress Process for Planetary Rover." Proc. 11th Asia-Pacific Regional Conference of the ISTVS.

Gulick, V. C., Morris, R.L., Ruzon, M.A., Roush, T. L., 2001. Autonomous image analyses during the 1999 Marsokhod rover field test. J. Geophys. Res., Planets 106(E4): 7745.

Golombek, M. P., Huertas, A., Marlow, J., McGrane, B., 2008. Size-frequency distributions of rocks on the northern plains of Mars with special reference to Phoenix landing surfaces. J. Geophys. Res., Planets 113, E00A09.

Rankin, A., Maimone, M., Biesiadecki, J., Patel, N., Levine, D., Toupet, O., 2021. Mars curiosity rover mobility trends during the first 7 years. J. Field Robotics, 20, 0040, R3.

Yingst, R. A., Cropper, K., Gupta, S., Kah, L.C., Williams, R.M.E., Blank, J., Calef III, F., Hamilton, V.F., Lewis, K., Shechet, J., McBride, M., Bridges, N., Martinez Friaz, J., Newsom, H., 2016. Characteristics of pebble and cobble-sized clasts along the Curiosity rover traverse from sol 100 to 750: Terrain types, potential sources, and transport mechanisms. Icarus: 280, 72-92.

2177 | Effects of road wheel load, driving speed and track slip upon stress state in sandy soil under tracked vehicle loading

Michał Kuszneruk [a,*], Paweł Tomiło [b], Jarosław Pytka [a], Dawid Tatarynow [a], Rafał Longwic [a]

[a] *Faculty of Mechanical Engineering, Lublin University of Technology, Lublin, Poland*

[b] *Faculty of Management, Lublin University of Technology, Lublin, Poland*

* *Corresponding author: m.kuszneruk@pollub.pl*

ABSTRACT

This study analyses experimental data of stress state under loads of running tracked vehicles in sandy soil measured with the SST (Stress State Transducer) in field experiments. Based on measured soil pressures, soil stresses in the orthogonal and octahedral planes were determined at the depth of 20 cm beneath the test vehicles' right track. Three different tracked vehicles of masses—44.3, 12.13, and 9.7 Mg—were used in the experiment. Effects of road wheel loads distribution, driving velocity, and track slip on soil stress state were determined and analyzed. The field experiment were compared with analytical predictions based on Boussinesq theory as well as the Mean Maximum Pressure (MMP) and Nominal Ground Pressure (NGP) models. A model of an artificial neural network was developed, which allowed the detection of the moment when the road wheel of the tracked vehicle is above the stress state transducer (SST) sensor.

It was concluded that the vehicle maneuver or drive mode (rolling, driving, velocity) affected soil stress state significantly. This is because of a varying wheel load vector during different drive modes. Finally, use of a flat value of wheel load in modelling and simulations of soil stress distribution may lead to erroneous results, without taking into account the wheel mode effect.

Keywords
Tracked vehicle
Soil stress state
Stress state transducer
Measurement
Sandy soil
Neural network model

1. Introduction

Tracked undercarriages have the advantages of minimizing vehicle impact and soil, what is understood as less structural changes in the soil due to the significant load from the road wheels, mainly less soil compaction and less disturbances in soil water-air relations. The measure of the vehicle load on soil is the contact pressure, so in this sense, a tracked undercarriage system can be much more advantageous than a wheeled one. Another advantage of tracked undercarriage is a better vehicle mobility, higher ability to overcome low-load-bearing terrain, which also comes down to lower contact pressure, which is characteristic of the tracked chassis (Althoff and Thien, 2005, Dąbrowski, 2001, Rybansky, 2020, Shoop et al, 2005).

Tracked undercarriages have also been used in agricultural vehicles, originally as an integral part of the vehicle, but now as separate track units retrofitted on conventional tractors. With tracked suspension, the vehicle's weight is distributed over a greater surface compared to wheeled vehicles and the resulting contact pressure is theoretically uniform along the track. This increases the maximum traction since the shear strength of the soil is used over a greater area (Arvidsson et al., 2011). Tracked tractors cause less compaction, understood as an increase of bulk density than wheeled tractors of a similar weight (Brown et al., 1992; Kinney et al., 1992). Arvidsson et al (2011) have compared wheeled and tracked tractors. They aimed to compare soil compaction by means of bulk density, penetration resistance and saturated hydraulic conductivity, for tracks, single and dual wheels mounted on the same tractor type. Measurements were made on two clay soils (Eutric Cambisols) in Sweden, using an 85 kW tractor with a total weight of 7.7 Mg. The rubber track system consisted of four tracks mounted on the conventional wheel axles of the tractor. The measured stresses were similar for the tracks and dual wheels at all depths studied (15, 30 and 50 cm), but were considerably higher for the single wheels at all depths. The stress distribution in the driving direction was relatively even along the front and rear tracks, which is an advantage compared with a long single track, which often has an uneven longitudinal stress distribution. Bulk density and penetration resistance were consistently highest and saturated hydraulic conductivity lowest after wheeling with single wheels, while there were no statistically significant differences between tracks and dual wheels. Ansorge and Godwin (2010) showed that the deformation under a track with a load of 12 Mg was approximately the same as that for a single wheel of approximately 4 Mg.

However, contrary to theoretical predictions, the pressure distribution under the track belt is not uniform, even with no dynamical factors. This is due to the concentration of stresses at points under the centers of the road wheels. The maximum stress under a track can be several times higher than the average stress, calculated as tractor weight divided by track contact area, as a result of non-uniform stress distribution, especially in the longitudinal but also in the lateral direction (Keller et al., 2002). High peak stresses occur under the wheels and rollers and under the track's links with cleats (Dąbrowski, 2001).

Many studies have shown that wheel-soil or track-soil interactions have been used to explain soil stress (van den Akker, 2004). Compared to contact pressure, soil stress—especially a complete stress state—gives more precise information on a dynamic soil response to wheel loads. Also, a number of researchers deal with in-situ measurements of soil stresses (Arvidsson and Keller, 2007; Blackwell et al., 1989; Harris and Bakker, 1994; Horn et al., 1989, 1992; Keller et al., 2002; Keller, 2005; Kirby, 1999; Lamandé and Schjønning, 2011a, 2011b,

https://doi.org/10.56884/AMYL6665
ISBN 978-1-942112-55-6(pdf)

2011c; Lamande et al., 2015; Pytka and Dąbrowski, 2001; Pytka and Konstankiewicz, 2002; Pytka et al., 2004, 2006; Pytka, 2005, 2009, 2010, 2013; Schjønning and Lamande, 2010; Wiermann et al. 1999, Wiermann et al. 2000). Problems with obtaining precise and reliable soil stress data in field experiments are mostly caused by the fact that we do not have a reference value of soil stress, either measured in laboratory conditions or calculated with a general law (Pytka, 2009, 2013; Lamande et al., 2015).

With the prime aim of obtaining a true values of soil stresses under vehicle loading, this paper is focused on (1) determining and analyzing soil stress state under loads of tracked vehicles considering the following factors of different loads (road wheel load), different forward velocity, and track slip; and (2) comparing them with those calculated with the use of chosen analytical methods. This will answer the questions of whether the soil stress distribution under tracked suspension is uniform and of how the soil stress changes due to various driving action (drive modes, i.e. free rolling, driving at track's slip, increasing forward speed).

2. Materials and methods

2.1 Soil stress determination

Field measurements were performed with a stress state transducer (SST). Figure 1 shows the SST used in this study: a schematic of the geometry in upper drawing and the device partially installed in soil, lower photo. This device, originally developed for soil stress measurements by Nichols et al. (1987), consists of six pressure transducers, oriented so that a complete stress state at center of the SST can be determined based on measured pressures. The stress transducers installed in the SST are strain gauge type with a 20-mm external diameter and an active sensing area diameter of 15 mm. Sensing membranes were machined from aluminum and are 0.8 mm thick (Pytka, 2009). Each transducer has its signal amplifier installed close to the strain gauge to minimize noises. The transducers were calibrated in air. Other details can be found in Pytka (2009, 2013). In the present study, the single SST was installed at a depth of about 15 cm in the centerline of the right track. The installation was simple, involving excavation, positioning of the SST and refilling the soil. It was assumed that this method of installation does not influence the accuracy of measurements significantly in a sandy soil.

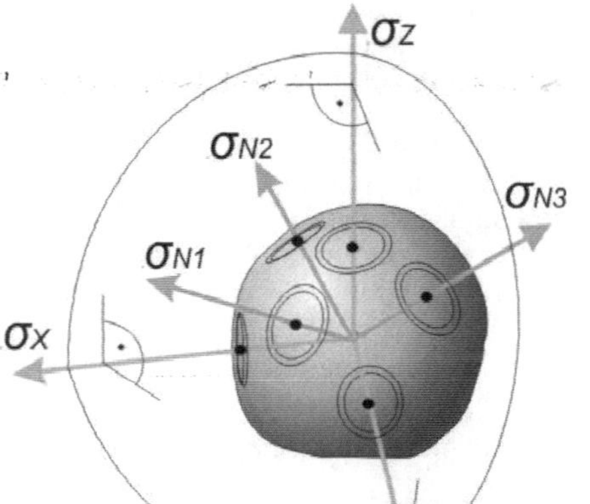

Fig. 1. Stress state transducer used in this study – a schematic of the device and the SST positioned at the measurement point in sandy soil

During installation, a fixed reference point was marked and a distance between that point and the SST was determined. This helped identify the location of the SST after refilling the soil (fig. 2). A portable computer with a data acquisition board recorded soil pressures. Based on the measured soil pressures, a complete soil stress state was calculated using an algorithm described by Nichols et al. (1987) and Harris and Bakker (1994). The complete soil stress state consists of principal stresses: σ_1, σ_2, σ_3, and stresses acting on the octahedral plane: normal stress σ_{OCT} and shear stress τ_{OCT}.

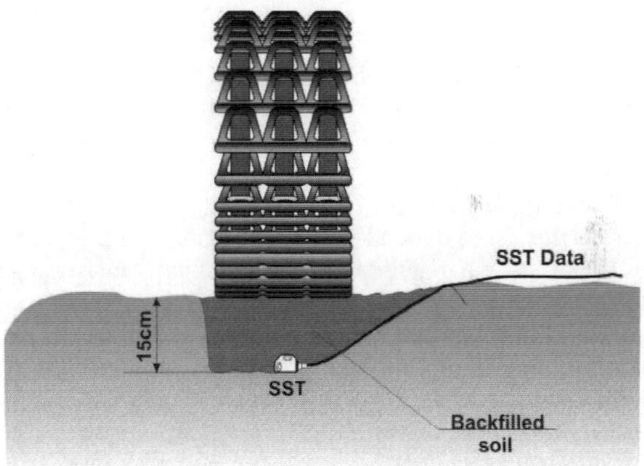

Fig. 2. Installation of the SST in soil

2.2 Soil characterization and field procedures

The measurements were performed in a podzolic sand, which is characterized in Table 1. The measurement site was located on the proving grounds of the 1. Warsaw Armored Brigade in Wesoła, near Warsaw, Poland. The first column in the Table 1 describes the soil's texture, the second column gives Wong's model coefficients. Cohesion, angle of internal friction, and Frohlich's coefficient of stress concentration are included in the third column. The soil had a bulk density of 1.72 g/cm^3 and a moisture content between 5 and 7%. For each test variant, as described in 2.3, at least five repetitions (five test runs) were performed. After each pass, the SST was replaced and the soil at the measuring point and about 10 meters before and after the point was manually homogenized. This let the vehicle run smoothly and allowed all the measurements to be conducted in similar soil conditions. A chalk powder was sprayed over the soil or the location was marked to guide the driver. The soil pressure was saved to a portable, industrial class computer. The data was then recalculated into soil stress components using the methods described in Nichols et al. (1987) and Harris and Bakker (1994).

Results from each individual test run were analyzed by searching for maximum (local) values associated with subsequent road wheels. Then, the maximum values found for each road wheel were averaged and the standard deviation was calculated.

Table 1
Sandy soil characteristics

1	2	3
Sand – 82%	S_{max} = 103[kN/m²]	C = 0.7 [kN/m²]
Clay – 11%	K_w = 0.51[cm]	φ = 30 º
Silt – 7%		ν=3

2.3 Test vehicles

Three military tracked vehicles were selected for the tests; the vehicles weighed 44.3 Mg, 12.13 Mg, and 9.7 Mg (Tables 2 and 3). The three test vehicles have steel tracks. Each track runs around a series of road wheels, causing potential variations in loads as a vehicle moves over a given spot. In order to obtain data to determine the wheel load - soil stress relationship, the vehicles were driven with a velocity of 5 km/h. In such conditions, dynamic effects are small and the mass distribution thorough the road wheels is the most significant factor affecting soil stress. The effect of vehicle speed was investigated for vehicle B (12.13 Mg), which was driven at three velocities: 5, 10, and 20 km/h. Vehicle A (44.3 Mg) was used to determine stress-strain relationship and the effect of rubber pads on soil stress. The effect of track slipping was investigated for the lighter vehicles, B (12.13 Mg) and C (9.7 Mg). The vehicles were stopped by increasing the braking force on an additional vehicle connected behind the test vehicle as described in Pytka et al. (2006) and Pytka (2013). The second vehicle, attached via a dynamometer, generates the braking force, and the increase in this force depends on the operator - the driver who applies the brake, what generates a nearly linearly increasing slip from 0 to 100%; a 100% slip occurs for a fully stopped braking vehicle. The slip of the track belt was determined indirectly, based on the simultaneous measurement of the forward speed and the rotational speed of the sprocket drive wheel. At a given tracks slip, the test vehicle passed over the SST sensor (Dąbrowski, 2001, Pytka and Dąbrowski, 2001).

Figure 3 shows details track undercarriage of the test vehicle, A (44.3 Mg).

Table 2
Technical data of the vehicles used in the experiment

Parameter	Test vehicle		
Total mass, Mg	44.3	12.13	9.7
Total contact area, m² (l×b)	5.22 (4.5×0.58)	1.830(3.545× 0.258)	2.18 (4.22x0.25)
Single element co. area, m² (l_1×b_1)	0.008 (0.138×0.58)	0.036 (0.142×0.258)	0.043 (0.168x0.25)
Number of wheels / wheel's diameter, m	6 / 0.74	6 / 0.70	6 / 0.68
Calculated contact pressure, kPa	90.2	66.2	44.3

Table 3
Mass distribution over the road wheels of the test vehicles

Road wheel	Vehicle mass distribution, Mg/kPa		
	44.3 Mg	12.13 Mg	9.7 Mg
1	3.200/392	1.211/393	1.040/286
2	3.637/446	1.010/328	1.130/311
3	3.910/479	1.070/347	1.170/322
4	3.810/467	0.990/321	1.147/315
5	3.940/483	0.896/291	1.090/300
6	3.643/446	0.831/270	0.840/231

Fig. 3. A side view of the track with road wheels of the vehicle A (upper photo) and a front view of the track with details of a single track element (lower photo)

Table 3 includes mass distribution over the road wheels of the test vehicles. The road wheel is an element of the tracked suspension system and performs the function of transferring the vertical load to the track belt. Basically, it's not driven. In this work, the vehicle mass distribution over the road wheels have been performed by means of weighing individual load wheels after unfastening (removing) the track belt, using an on-ramp scale.

2.4 Methods for contact pressure and soil stress calculation

In this work, three models for predicting contact pressure and soil stress have been used:

- the nominal ground pressure (NGP), expressed in the following equation:

$$NGP = \frac{G}{2rmb} \quad (1)$$

where: m - number of axles, b - tire width, r – road wheel radius

- Rowland's MMP procedure for tracked vehicles on fine-grained soils:

$$MMP = \frac{1.26\,G}{mb\sqrt{ld}} \quad (2)$$

where: d – wheel diameter, l – the length of the contact surface of the track with the ground

- Boussinesq model was used to calculate vertical soil stress:

$$\sigma_Z = \frac{3}{2}\frac{G}{\pi z^2}\cos^5\beta \quad (3)$$

where z – the depth of, β are the polar coordinates of the point at which the stress is determined.

Averaged values of peak maximum measured stresses have been compared with calculated stresses, obtained with the use of the three above models. The data used for comparison includes contact pressure determined for the total track contact area, contact pressures determined separately for each road wheel, peak mean maximum σ_1 soil stresses determined by averaging the data for all six road wheels, predicted soil stresses calculated from the Boussinesq' theory and the NGP and MMP models averaged for all six road wheels.

3. Results

Figure 4 shows a sample complete soil stress state from a single pass of one of the test vehicles. Five curves (σ_1, σ_2, σ_3, σ_{OCT}, and τ_{OCT}) are shown as functions of time. The shapes of the curves are very similar for the three test vehicles, the peak maximum values differ.

3.1 Principal stresses

The three principal stresses reach peak values six times, which is a consequence of the six road wheels of the test vehicles. The soil stress curves tell us also about stress concentration under the wheels as seen by the peak stresses that represent loading of each road wheel. The stress relaxation between the road wheels suggests non-homogeneous load distribution under the track. With a tracked undercarriage, the total load from the vehicle's weight is theoretically distributed under the much larger total contact area of the track. However, in the field, most of vehicle's weight is distributed on the small contact area of a single track element under a road wheel. This causes soil stress concentration under the wheels. This is obviously dependent on track tension force, which can be controlled within a limited range. Keller et al (2002) have measured stresses under a Class Challenger 2-65E rubber-tracked tractor. They conducted two experiments, in which the vertical soil stress was measured during ploughing. They concluded that the vertical stress was found to be much higher under the rear than under the front part of the track, but it depended on the height of the point of application of the draught force. Moreover, the vertical stress was concentrated under the supporting rollers and wheel as well as under the center line of the track. The relationship between maximum stress and average stress was 3.8. A conclusion from their research was that using tracks instead of wheels may reduce the risk of soil compaction but only when the tractor is well balanced.

Table 4
Soil stress data for the test vehicle A

Load/contact pressure, Mg/kPa	Soil stress, kPa *		
	σ_1	σ_{OCT}	τ_{OCT}
3.2/392	392.3/61.4	49.36/18.1	244.9/51.1
3.637/446	455.86/38.5	103.33/5.3	329.6/18.7
3.910/479	473.7/15.6	85.45/12.9	321.9/6.1
3.810/467	433.4/49.7	104.38/5.4	291.33/37.9
3.940/483	432.93/8.2	107.36/5.6	290.73/10.2
3.643/446	478.3/74.7	103.54/5.5	312.06/41.1

(*) Average value/standard deviation

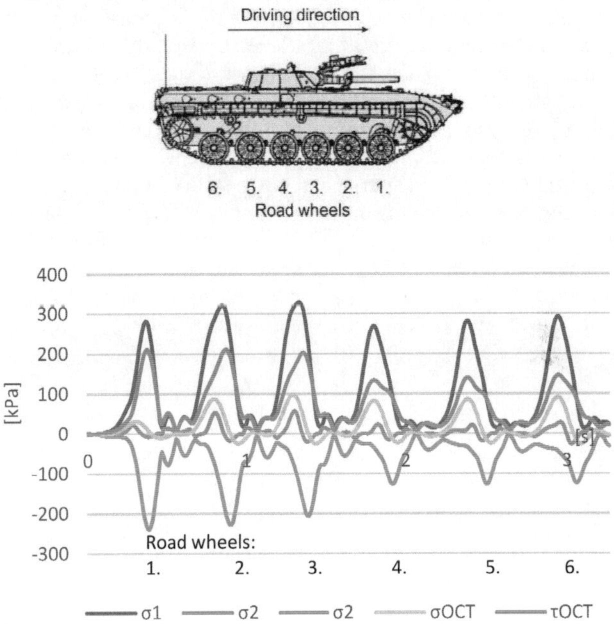

Fig. 4. Soil stress state components, determined from pressures measured of the test vehicle B (12.13 Mg). A sideview of the test vehicle with road wheels numbering

Analyzing the mutual proportions between the individual components of the soil stress state, we notice that the major stress σ_1 is the highest stress state component and reflects to the vertical load, whereas the σ_3 stress reaches negative values: the σ_3 vector is in the opposite direction from traction. Values of the minor stress σ_2, which reflects to the lateral force, are about 20% of σ_1 values on average, which may indicate that the soil stress state is two-dimensional in this case.

3.2 Octahedral soil stresses

The octahedral shear stress surpasses the octahedral normal stress. The ratio of the maximum value of the octahedral shear stress to the value of the octahedral normal stress reaches the highest value of 5.57 for the first road wheel, to decrease for the subsequent wheels to the value of 1.51 for the last wheel. The domination of shear over compression would suggest the shearing and structural damage in soil under the wheels are dominant and have greater importance than volumetric deformations (soil compaction). Taking into account the above values of the τ_{OCT}/σ_{OCT} ratio, it can be seen that shearing occurs most intensively under the first road wheel, while under the next wheels it is weaker and the extent of volumetric changes increases. One of the definitions of the soil compaction process is the filling of soil pores by solid particles of soil. In the case of sand, this mechanism occurs to a lesser extent. Rather, it should be expected a continuous change in the mutual arrangement of individual particles until a stable arrangement is obtained, at which the soil skeleton transfers the load without significant deflection. This is one possible explanation for why octahedral shear stress is higher than normal stress.

3.3 *Vehicle mass distribution over roadwheels*

In general, the relationship between contact pressure and the measured stress in the soil for individual road wheels has a linear character with a degressive initial phase. This relationship, constructed for all vehicles is presented in Figure 4. It was found, among others, that the standard deviation of the measured values of stress in the soil decreases for the successive road wheel and reaches the lowest values for the last of the road wheels (Pytka, 2013). Table 4 collects soil stress data with average values and standard deviations for the vehicle A (44.3 Mg). On the one hand, there may be an effect caused by the successive wheel pass and soil compaction - then the dispersion of the measured stress in the soil is smaller due to the more homogeneous soil structure (Cottrell, 1964, Horn and Lebert, 1994). But perhaps the effect of the measurement method may also be important, namely soil compaction improves the contact between the soil and the sensor membrane and makes the results more reproducible.

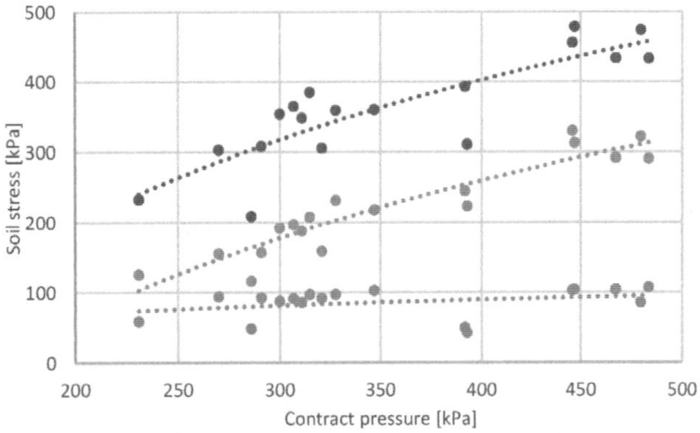

Fig. 5. A relationship between contact pressure and measured soil stresses for individual road wheels of the three test vehicles. Points in blue: major stress σ_1, gray: τ_{OCT}, orange: σ_{OCT}.

3.4 *Effects of vehicle speed*

Figure 6 shows a comparison of the major principal stress σ_1 for three different driving speeds: 5, 10 and 20 km/h, obtained for the test vehicle B (12.13 Mg). It can be seen that the peak values decrease by 30-50% for higher speed, but the shape of the stress course changes, namely it is more stable, without stress relaxation, seen from the drops of the curves between individual road wheels. Different intensity of soil stress relaxation can be explained by the three-phase effect of the soil medium: both elasticity and viscosity of soil material are time or velocity dependent. The so-called hardening effect, which occurs at higher loading rate, results in higher elasticity and consequently reduces mechanical stress (Cottrell, 1964, Mirsayapov and Koroleva, 2020).

This is an interesting issue from the point of view of future research, especially when taking into account the influence of soil water content and porosity and, consequently, the influence of the viscosity of the liquid and gaseous phases.

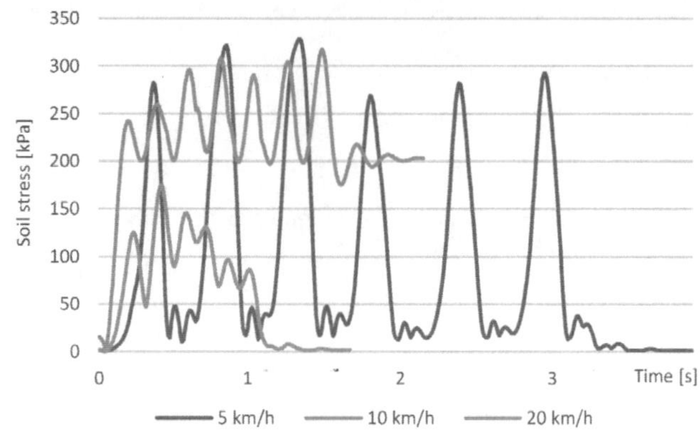

Fig. 6. Effect of speed on major stress σ_1: orange line – 5km/h, blue – 10km/h, green – 20km/h

3.5 *Effects of track slipping*

During pull tests, a triggering phenomenon occurs changes in the distribution of vehicle mass and contact pressure, namely, the vertical component of the braking force may cause the front wheels of the test vehicle to lift. This happens when the hook on the test vehicle is fixed below the height of the center of gravity, consequently dynamic effects combined with unevenness of the road surface cause decentering of the braking force and a vertical component of the braking force occurs, which raises or lowers the test vehicle. Then, the outline of soil stress under the track has a typical, triangle-like shape. The graph in Fig. 7 shows this phenomena, increasing peak values of stresses for subsequent road wheels of the test vehicle B (12.13 Mg). In general, this effect of external braking occurs at a relative low percentage slip (0-30%) and influences the interactions between soil and track.

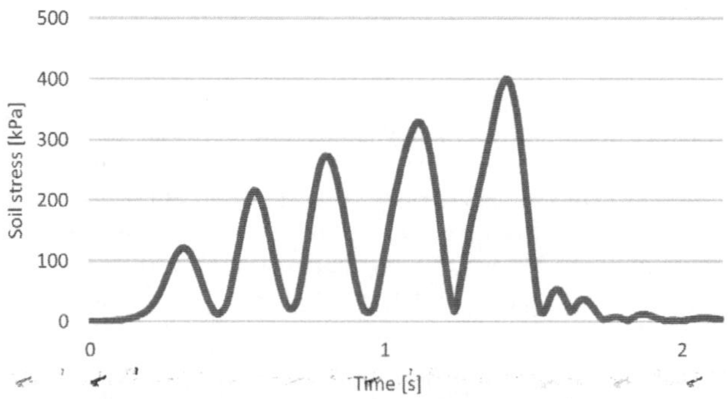

Fig. 7. A "triangle-shaped" σ_1 soil stress curve obtained during a ride with a low slip (10%)

The graphs in Figure 8 show time courses of σ_1 soil stress for test vehicle C (9.7 Mg) ridding with track slipping. This vehicle has its hook at the C.G. height, so no vertical component of the braking force occurred. For the higher values of braking force, stress curves become "saw-shaped", which can be explained due to the effect of vibrations of the vehicle at high braking force with a maximum power (when the braking force increases, the driver of the test vehicle counteracts it by setting the maximum power). These vibrations are transmitted into the soil and affect measured soil stresses.

In general, no effect of slip on the maximum stress values in the soil was found.

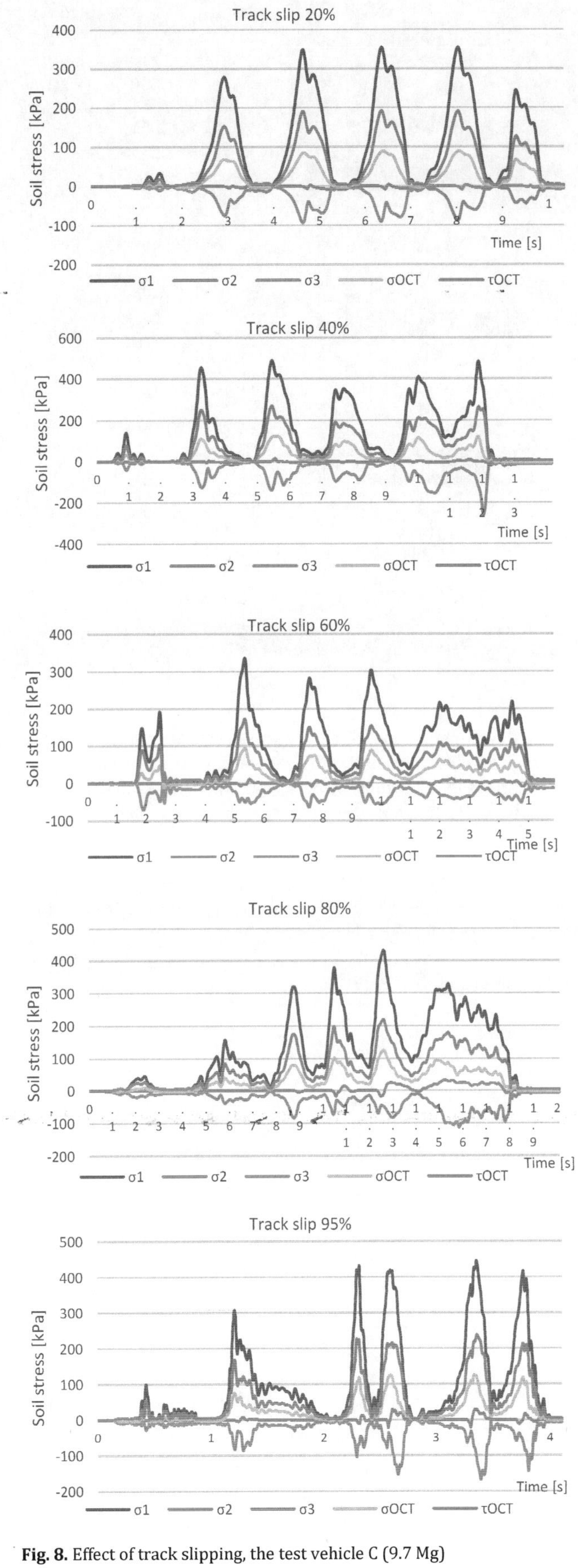

Fig. 8. Effect of track slipping, the test vehicle C (9.7 Mg)

4. Comparison between measured and calculated soil stresses

The measured peak stresses of σ_1 and the roadwheels' contact pressure were of similar magnitudes (Figure 9). Measured peak maximum stress σ_1 (bar B) and the contact pressure for the roadwheels (bar C) are almost identical and always reach the highest value for the three vehicles. Predictions from the three analytical methods (bars D, E, and F) fall within the range of about 40 - 60% of measured peak maximum stress, this tendency is also valid for the three vehicles, moreover the MMP (bar F) predictions reach the highest value of the three analytical pressures, the NGP (bar E) was always the lowest here. The lowest value among the comparison belongs to the contact pressure determined by relating the vehicle's weight to the total contact area of the track (bar A in Fig. 9), this measure reaches about 20% of the measured peak maximum stress σ_1, and, as before, this tendency is valid for all the three test vehicles.

Averaged peak mean maximum σ_1 soil stress represent averaged contact pressure under roadwheels. This is confirmed by measurement results as well as theoretically. On the other hand, unexpectedly, the values calculated with the MMP model differ significantly from both contact pressure under track and averaged contact pressure under roadwheels - they assume intermediate values. This contradicts the idea of the MMP model, which was assumed to give results consistent with the stress vertical stress measured at a depth of approximately 20 cm (Hetherington and Littleton, 1987, Hetherington and White, 2002).

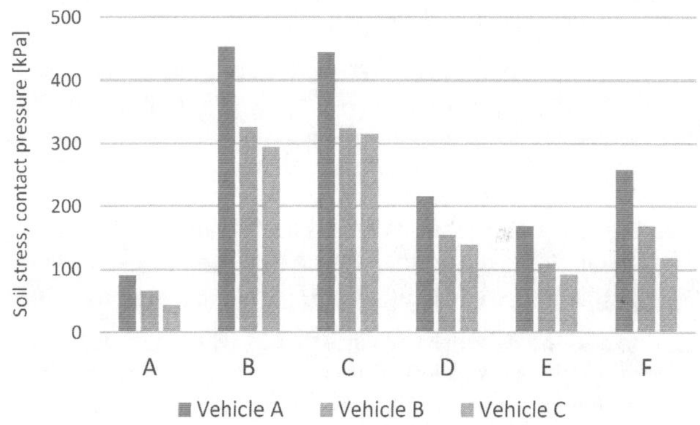

Fig. 9. A comparison of loads and measured as well as calculated stresses. A – contact pressure under the track, B – averaged contact pressure under roadwheels, C – averaged peak mean maximum σ_1 soil stress, D – soil stress calculated from the Boussinesq equation, E, F – soil stress obtained from the NGP and MMP models, averaged for all six road wheels

5. Neural network model for soil stress propagation under tracked vehicle loading

In order to determine the moment in time when a given vehicle wheel passed over the sensor, an artificial neural network model will be used. The collected data is weakly-labeled, therefore, based on the assumptions, the interval in which the vehicle wheel was above the sensor was determined. Additionally 2 auxiliary classes have also been used, finally the data is classified into one of 3 classes, these are:

- Class 1, when stresses are minimal or absent,
- Class 2, stress build-up or drop;
- Class 3, proximity to maximum value.

5.1 Feature selection

The soil stress state under tracked vehicle loading consists of 5 timeseries. An analysis was carried out using the Spearman correlation coefficient to determine the relationship between the timeseries. Figure 10 shows the correlation matrix.

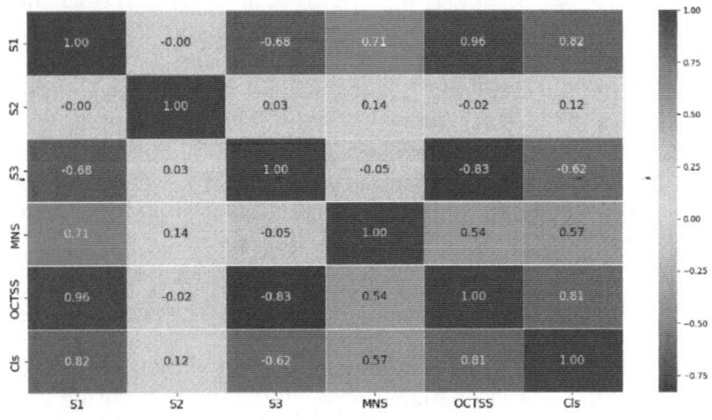

Fig. 10. Correlation matrix

Soil octahedral shear stress τ_{OCT} shows a positive correlation with major stress σ_1, similarly to mean normal octahedral stress σ_{OCT}. The minor stress σ_3 is moderately negatively correlated with σ_1. Due to the pouring correlation, the σ_1, σ_2, and σ_{OCT} stresses timeseries were not used as input data to the artificial neural network model.

5.2 Artificial neural network model

The developed neural network model uses the (Long Short-Term Memory) layer. This kind of layer has the capacity to add or remove data from the gates controlled state cell (Staudemeyer & Morris, 2019). The developed artificial neural network model consists of a single LSTM layer with the number of units of 10. In the training phase, a dropout layer was used, which randomly sets the input values for given units as 0, with rate equal to r_d. The layer scales the remaining values by $\frac{1}{1-r_d}$. This solution allows to reduce the tendency of the network to overfitting (Srivastava et al., 2014). The dropout layer was followed by a layer of densely connected neurons with the number of neurons equal to 10. The output layer is also a layer of densely connected neurons with the number of neurons equal to the number of classes - 3. The model has 663 parameters, so it can be used in tinyML solutions. Architecture of the artificial neural network model is shown in Figure 11.

In the learning process, a training and validation set was used in the proportions of 4:1. The categorical cross-entropy was adopted as the loss function, and binary accuracy (the frequency with which the class predicted by the model corresponds to the true class) was used as an additional metric. The ADAM algorithm was used as an optimizer (Kingma & Ba, 2014). The network reached the minimum value of the loss function for the validation set of 0.2887, and for the training set of 0.3512. The accuracy achieved for the training set was 0.8784, and for the validation set 0.8432.

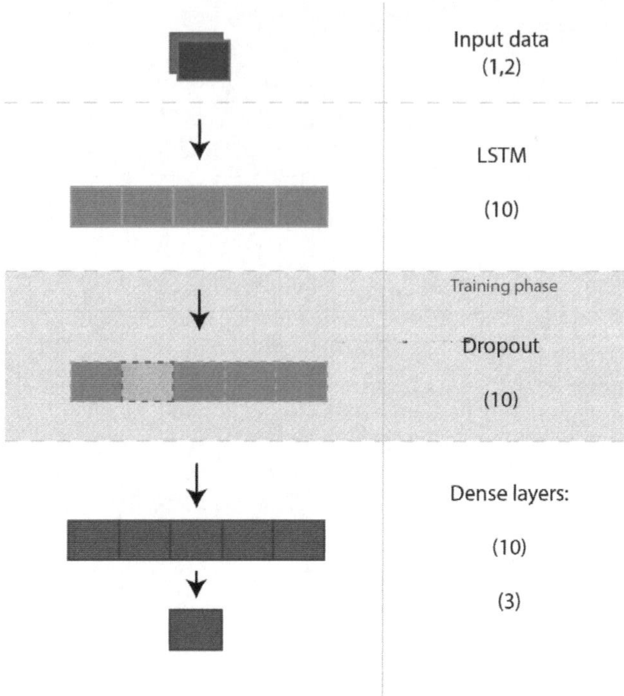

Fig. 11. Architecture of the artificial neural network model

5.3 Results

The developed model of the neural network was tested on 3 time courses of stresses under the wheels of the three test vehicles In the case of the test vehicle (12.13 × 10³ kg), a new measurement was made, which had not previously been used as a set for learning the neural network model.

In the case of the test vehicle (44.3 × 10³ kg), the first wheel generates significantly less stress than the other wheels and there is an increase in tension between the last two wheels relative to the other wheels. Figure 12 shows the timeseries of the variability of stresses under the wheels for the test vehicle (44.3 × 10³ kg) and the predictions of the network.

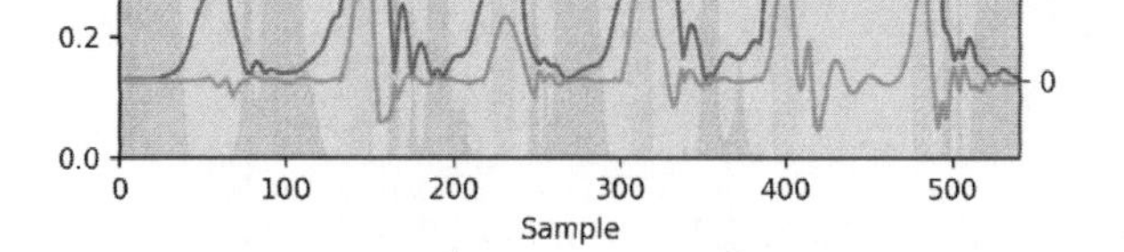

Fig. 12. Prediction of the neural network model for the test vehicle (44.3 × 10³ kg)

For the test vehicle (9.7 × 10³ kg), there are lower stresses for the first and last wheels, but in this case their value is not significantly different from the other wheels. Figure 13 shows the timeseries of the variability of stresses under the wheels for the test vehicle C and the predictions of the network.

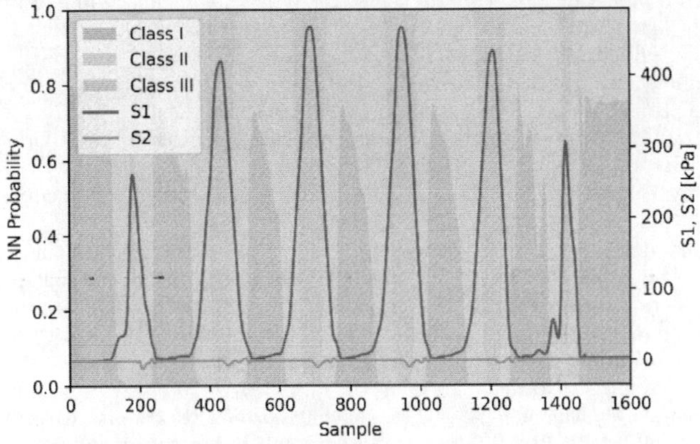

Fig. 13. Prediction of the neural network model for the test vehicle (9.7 × 10^3 kg)

The test vehicle B is characterized by the similar stresses on all wheels. There are significant differences between the 1,2,3 and 4,5,6 wheels. Figure 14 shows the timeseries of the variability of stresses under the wheels for the test vehicle (9.7 × 10^3 kg) and the predictions of the network.

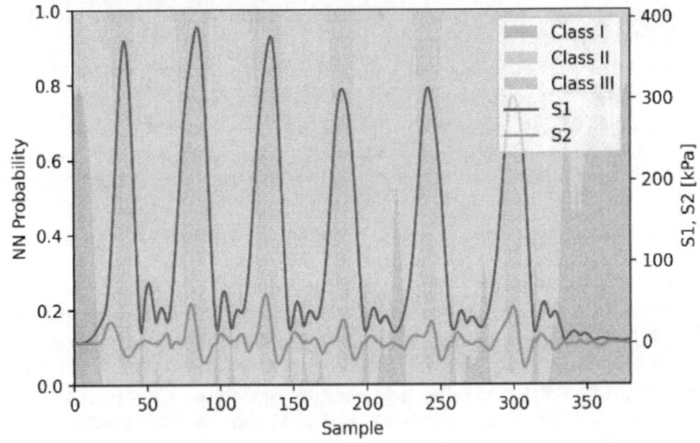

Fig. 14. Prediction of the neural network model for the test vehicle B

5.4 Resume

The wheel trafficking the sensor can be detected by the neural network model that has been constructed. With rising stress levels, the likelihood of class III (closeness to maximum value) occurrence rises. The model accurately predicts the moment the wheel contacts the sensor. The likelihood of the class III for the first wheel for the test vehicle (44.3 × 10^3 kg) is low, but not zero. This implies that it is possible to improve the accuracy for various scenarios by expanding the data set with further measurements.

Class III serves as an indicator of when the value increases until it reaches a maximum value and then decreases in value. This corresponds to the moment the vehicle wheel runs over the sensor, and therefore the first moment of occurrence of class III corresponds to the moment when the wheel runs over the sensor and lasts until the class is changed to class II, which proves that the wheel has left the sensor area. On this basis, it is possible to count the number of vehicle wheels, and thus it allows for estimation of vehicle position relative to the sensor using an appropriate threshold of inference probability of the developed artificial neural network model. For inference, a model using LSTM layers was used to determine a given class in real time.

6. Conclusions and future work

Soil stress state under loading of three tracked vehicles has been determined in this study. Based on the analysis of soil stress state components σ_1, σ_2, σ_3, σ_{OCT}, and τ_{OCT}, the following has been concluded:

1. The soil stress distribution along the track line was non-uniform, affected by vehicle mass distribution and by loading of next wheels under the track. Stress relaxation between the road wheels was pronounced, reaching approximately 90% of the peak stress indicated under a wheel.
2. Faster speeds resulted in decreasing peak stress values and less stress relaxation. For the highest speed, 20km/h, stress relaxation dropped to approximately 20%.
3. The effect of track slip was significant but irregular. In most cases, octahedral shear stress τ_{OCT} increased at track slip contrary to major stress σ_1 which decreased. A logical explanation of that soil stresses changed at different track slip is that increasing slip resulted in changes of vehicle mass distribution and wheel load.
4. The comparison of the major stress σ_1 with theoretical stress data has given diverse results for the Boussinesq model: stress values calculated with ν=3 are approximately 50% smaller than measured peak maximum stresses. Similarly, a comparison of experimental results with MMP and NGP values showed under-estimation of analytical peak maximum stresses.

The final conclusion is that the vehicle maneuver or drive mode (rolling, driving, various speed) affected soil stress state significantly. This is because of a varying wheel load vector during different drive modes. Finally, use of a flat value of wheel load in modelling and simulations of soil stress distribution may lead to erroneous results.

A neural network model with LSTM layer was developed in order to detect the moment a roadwheel is over the SST sensor. The predictive performance of the model was good and it can be used for further analysis of experimental data, also in development of algorithms that could be used in vehicle on-board computers.

In future work, the authors plan to study the impact of the track tension force on the state of stress in the soil. In addition, the analysis using the neural network model will be extended, mainly in order to predict stresses occurring under the load of a tracked vehicle, moving at different speeds and with track slipping.

7. Nomenclature

b_1	Width of a single track element	[m]
l	Length of track belt in contact with soil	[m]
l_1	Length of a single track element	[m]
σ_1	Other components	[N]
σ_2	Peak tire lateral force	[N]
σ_3	Average tire lateral force	[N]
σ_{OCT}	Tire lateral peak force time	[s]
τ_{OCT}	Steer angle rate	[deg/s]

8. Declaration of competing interest

The authors declare that they have no known competing financial interests or personal relationships that could have appeared to influence the work reported in this paper.

9. References

Althoff, P.S., Thien, S.J., 2005. Impact of M1A1 main battle tank disturbance on soil quality, invertebrates, and vegetation characteristics. J. Terramech. 42, 159–176

Akker van den, J.J.H., 2004. SOCOMO, a soil compaction model to calculate soil stresses and the subsoil carrying capacity. Soil Till. Res. 79, 113-127

Ansorge, D., Godwin, R.J., 2010. Soil density increases resulting from alternative tire and rubber track configurations in laboratory and field conditions. In: Dedousis, A.P., Bartzanas, T. (Eds.), Soil Engineering, Soil Biology 20. Springer Verlag, Berlin Heidelberg, pp. 81–90.

Arvidsson, J., Keller, T., 2007. Soil stress as affected by wheel load and tyre inflation pressure. Soil Till. Res. 96, 284–291.

Arvidsson, J., Westlin, H., Keller, T., Gilbertsson, M., 2011. Rubber track systems for conventional tractors - Effects on soil compaction and traction. Soil and Tillage Research, 117, 103-109,

Blackwell, J., Horn, R., Jayawardane, N.S., White, R., Blackwell, P.S., 1989. Vertical stress distribution under tractor wheeling in a partially deep loosened typic paleustalf. Soil Till. Res. 13, 1–12.

Boussinesq, J. 1885. Application des potentielles a'l'etude de l'equilibre et du movement des soildes elastique. Paris: Gauthier Villais

Brown, H.J., Cruse, R.M., Erbach, D.C., Melvin, S.W., 1992. Tractive device effects on soil physical properties. Soil Till. Res. 22, 41–53.

Cottrell, A. 1964. The Mechanical Properties of Matter. Willey Series of Science and technology of Materials, first ed., New York.

Cui, K., Défossez, P., Richard, G., 2007. A new approach for modelling vertical stress distribution at the soil/tyre interface to predict the compaction of cultivated soils by using the PLAXIS code. Soil Till. Res. 95, 277–287.

Dąbrowski, J., 2001. Effect of rubber pads on tractive performance of tracked vehicles. PhD diss. Mil. Inst. Auto. Arm. Techn. Sulejówek.

Frochlich, O.K. 1934. *Druckverteilung im Baugrunde*. Wien: Julius Springer Verlag.

Golyandina, Nina., Nekrutkin, V. V. Vladimir V., & Zhigliāvskiĭ, A. A. Anatoliĭ A. (2001). *Analysis of time series structure : SSA and related techniques*. Chapman & Hall/CRC.

Harris, H.D., Bakker, D.M., 1994. A soil stress transducer for measuring in situ soil stresses. Soil Till. Res. 29(1), 35 - 48

Hetherington, J.G., Littleton, I., 1987. The role of mean maximum pressure in specifying cross-country mobility for armored fighting vehicle design. J. Terramech. 24, 263-280

Hetherington, J.G., White, J.N., 2002. An investigation of pressure under wheeled vehicles. J. Terramech. 39, 85-93

Horn, R., Blackwell, P.S., White, R., 1989. The effect of speed of wheeling on soil stresses, rut depth and soil physical properties in an ameliorated transitional red-brown earth. Soil Till. Res. 13, 353-364

Horn, R., Johnson, C., Semmel, H., Schafer, R., Lebert, M., 1992. Raumliche Spannungsmessungen mit dem Stress State Transducer (SST) in ungesattigten aggregierten Boden – theoretische Betrachtungen und erste Ergebnisse. Z. Pflanzenernähr. Bodenk. 155, 269–274.

Horn, R., Lebert, M., 1994. Soil Compactability and Compressibility. Chapter 3 in Developments in Agricultural Engineering, Editor(s): B.D. Soane, C. van Ouwerkerk, Elsevier, Vol. 11, 45-69

Horn, R., 2003. Stress-strain effects in structured unsaturated soils on coupled mechanical and hydraulic processes. Geoderma, 116, 77-88

Keller, T., Trautner, A., Arvidsson, J., 2002. Stress distribution and soil displacement under a rubber-tracked and a wheeled tractor during ploughing, both on-land and within furrows. Soil Till. Res. 68, 39–47.

Keller, T., 2005. A model to predict the contact area and the distribution of vertical stress below agricultural tyres from readily-available tyre parameters. Biosyst. Eng. 92, 85–96.

Keller, T., Arvidsson, J., 2004. Technical solutions to reduce the risk of subsoil compaction. Soil Till. Res. 79, 171–205.

Keller, T., Défossez, P., Weisskopf, P., Arvidsson, J., Richard, G., 2007. SoilFlex: a model for prediction of soil stresses and soil compaction due to agricultural field traffic including a synthesis of analytical approaches. Soil Till. Res. 93, 391–411.

Kinney, G.R., Erbach, D.C., Bern, C.J., 1992. Soil strain under three tractor configurations. Trans. ASAE 35 (4), 1135–1139.

Kirby, J.M., 1999a. Soil stress measurements: part I: transducer in a uniform stress field. J. Agric. Eng. Res. 72, 151–160.

Kirby, J.M., 1999b. Soil stress measurements: part II: transducer under a plate. J. Agric. Eng. Res. 72, 151–160.

Lamandé, M., Schjønning, P., Tøgersen, F.A., 2007. Mechanical behaviour of an undisturbed soil subjected to loadings: effects of load and contact area. Soil Till. Res. 97, 91–106.

Lamandé, M., Schjønning, P., 2011a. Transmission of vertical stress in a real soil profile. Part I. Site description, evaluation of the Séhne model, and the effect of topsoil tillage. Soil Till. Res. 114, 57–70.

Lamandé, M., Schjønning, P., 2011b. Transmission of vertical stress in a real soil profile. Part II. Effect of tyre size, inflation pressure and wheel load. Soil Till. Res. 114, 71–77.

Lamandé, M., Schjønning, P., 2011c. Transmission of vertical stress in a real soil profile. Part III. Effect of soil water content. Soil Till. Res. 114, 78–85.

Lamandé, M., Keller, T., Berisso F., Stettler M., Schjønning P., 2015. Accuracy of soil stress measurements as affected by transducer dimensions and shape. Soil Till. Res. 145, 72–77

Li, Q., Ayers, P.D., Anderson, A.B., 2005. Modeling of terrain impact caused by tracked vehicles. J. Terramech. 44, 395–410

Mirsayapov, I.T., Koroleva I.V., 2020. Softening and hardening clay soil under loading. IOP Conf. Series: Materials Science and Engineering, 890 (2020), 012070

Nichols, T.A., Bailey, A.C., Johnson, C.E., and Grisso, R.D., 1987. A stress state transducer for soil. Trans. ASAE 30(5),1237-1241

Palazzo, A.J., Jensen, K.B., Waldron, B.L., Cary, T.J., 2005. Effects of tank tracking on range grasses. J. Terramech. 42, 177–191

Pytka, J., Dąbrowski, J., 2001. Determination of stress-strain relationship for sandy soil in field experiments. J. Terramech. 38(4), 185-200

Pytka, J., Konstankiewicz, K., 2002. A new optical method for soil stress and deformation determination. Soil Till. Res. 65, 243-251

Pytka, J., Tarkowski, P., Dąbrowski, J., Bartler, S., Kalinowski, M., Konstankiewicz, K., 2004. Soil stress and deformation determination under a landing airplane on an unsurfaced airfield. J. Terramech. 40, 255-269

Pytka, J., 2005. Effects of repeated rolling of agricultural tractors on soil stress and deformation state in sand and loess, Soil Till. Res. 82, 77 - 88

Pytka, J., Dąbrowski, J., Zając, M., Tarkowski, P., 2006. Effects of reduced inflation pressure and vehicle loading on off-road traction and soil stress and deformation state. J. Terramech. 43, 469-485

Pytka, J., 2009. Determining and analysing the stress state under wheeled vehicles loads. Proc. Institution of Mechanical Engineers, Part D: Journal of Automobile Engineering, Vol. 223 No. D2, 233-253

Pytka, J.A., 2009. Design considerations and calibration of stress transducers for soil. J. Terramech. 46, 241–249.

Pytka, J., 2010. Determination of snow stresses under vehicle loads. Cold Regions Science and Technology 60, 137-145

Pytka, J.A., 2013. Dynamics of Wheel-Soil Systems: A Soil Stress and Deformation- Based Approach. Ground Vehicle Engineering Series. CRC Press, Taylor and Francis group, Boca Raton, FL, USA.

Raper, R.L., 2005. Agricultural traffic impacts on soil. J. Terramech. 42, 259–280.

Rowland, D., 1989. Tracked vehicle ground pressure and its effect on soft ground performance. Proc. 4th Int. Conf. ISTVS, 353-384.

Rybansky, M., 2020. Determination the ability of military vehicles to override vegetation, J. Terramech., 91, 129-138.

Shoop, S.A., Affleck R., Collins C., Larsen G., Barna L., Sullivan P., 2005. Maneuver analysis methodology to predict vehicle impacts on training lands. J. Terramech. 42(3-4): 281-303.

Schjønning, P., Lamandé, M., 2010. A note on the vertical stresses near the soil–tyre interface. Soil Till. Res. 108, 77–82.

Wiermann, C., Way, T.R., Horn, R., Bailey, A.C., Burt, E.C., 1999 Effect of various dynamic loads on stress and strain behavior of a Norfolk sandy loam Soil Till. Res. 50, 127-135
Wiermann, C., Werner, D., Horn, R., Rostek, J., Werner, B. 2000 Stress/strain processes in a structured unsaturated silty loam Luvisol under different tillage treatments in Germany Soil Till. Res. 53, 117-128
Zink, A., Fleige, H., Horn, R., 2011. Verification of harmful subsoil compaction in loess soils. Soil Till. Res. 114, 127-134.

2437| Improving predictive control methods on off-road vehicles with realistic steering preview inputs

Andries Peenze [a], P. Schalk Els [a, *]

[a] Department of Mechanical and Aeronautical Engineering, University of Pretoria, Pretoria, South Africa

* Corresponding author: schalk.els@up.ac.za

ABSTRACT

This paper investigates improvements to predictive control methods for off-road vehicles with the addition of realistic steering preview. The objective of this study is to improve the performance and efficacy of predictive controllers by accounting for significant time delays in active and semi-active systems on vehicles. The zero-order and first-order hold methods for steer preview are compared to a more realistic steer preview obtained from the curvature of the path, vehicle velocity, and tyre lateral force properties. A new semi-active suspension system, rear wheel steering, and individual brake control are used as the actuators on this off-road vehicle. The results show that the addition of a realistic steering preview improves the handling performance of the vehicle in a severe manoeuvre without affecting normal driving on smooth and rough roads. The controller can pre-empt and consider the effect of the actuator time delays, and the preview states from the predictive controller are more representative over the prediction horizon. In conclusion, the findings suggest that the addition of a realistic steering preview improves the performance of a predictive controller on vehicles, and further investigation of other disturbances and their preview effects on the system should be conducted to find further improvements for predictive control strategies on vehicles.

Keywords
Steering preview
Predictive control
Off-road vehicle dynamics

1. Introduction

Predictive control can be improved by using a realistic preview of the disturbances the system will experience over it's predictive time horizon. In vehicles there are many disturbances which the driver manipulates to change the vehicle's response. Steering wheel input, throttle and brake pedal input and gear changes are amongst some methods the driver can directly control the vehicle. Particularly the steering input will affect the vehicle's lateral response, which influences the handling, stability and safety of the driver and occupants. If the lateral dynamics of a vehicle want's needs to be influenced through some automated control algorithm it would be advantageous to have a good preview of the steering disturbance during the prediction time.

In literature the zero-order hold (ZOH) or first order hold (FOH) methods are used to provide a steer preview to the controller. Although these methods are easy to use, they come with some drawbacks which limits the reasonable preview time that can be used. If the control actuator delays are longer than what these preview times can provide, then the system might become unresponsive to fast changes to control set points or unstable.

In this paper we will be looking at these different steer preview methods to compare them to a proposed steer preview method that gives more realistic and smooth steer previews.

2. Vehicle model

The vehicle simulation model is based on a Land Rover Defender 110 which has undergone many modifications to become a test vehicle used to research off-road vehicle dynamics. The vehicle has been previously modelled in MSC Adams to incorporate all the non-linear elements such as tyres, suspension, bushes and suspension kinematics. In a previous iteration of the vehicle the simulation model has been validated with test data. With all the recent modifications made to the vehicle a re-validation is needed and therefore, for the purpose of this paper, the simulation model is considered unvalidated.

The vehicle simulation model is controlled through co-simulation with MSC Adams and Matlab Simulink. The suspension model, steering controller, driving torques and other actuators are modelled and controlled through Simulink, with the vehicle kinematics and kinetics as well as the tyre forces carried out in MSC Adams.

The tyre model used in this paper makes use of a Pacejka 89 (Pacejka, 1992) tyre model based on data from a validated FTire model of an off-road Michelin LTX SUV tyre (Bosch et al., 2016).

3. Vehicle actuator models and delays

The vehicle's suspension has been recently tested and characterised for different spring and damper characteristics. A model has been developed to capture the characteristics of the real suspension system. The suspension system is a hydro pneumatic system with four discrete spring characteristics and a continuously variable damper characteristic. The suspension also has ride height adjustment available. The system is considered to be a semi-active suspension system with slow-active ride height adjustments possible.

With the pneumatic system a total of 3 spring characteristics (hard, intermediate, and soft) can be achieved with an additional lock-out functionality if all the oil to the pneumatic accumulators is blocked off. The discrete springs characteristics are selected by

https://doi.org/10.56884/UZAR6470
ISBN 978-1-942112-55-6 (pdf)

opening or closing select hydraulic valves to manage the flow of oil to the two pneumatic accumulators. The oil from the suspension strut is forced through a damper pack, whereby a continuously variable bypass valve can vary the damping curve set between the extremes of lowest damping and highest damping, and any option in-between. This forms the continuously variable damper (CVD). With these different options of spring and damper characteristics the suspension response and reactions can be changed on the go with an adequate control algorithm.

The three available spring characteristics are shown if in Fig. 1a. Figure 1b shows the continuously variable damper characteristic for 5 values of the variable control current.

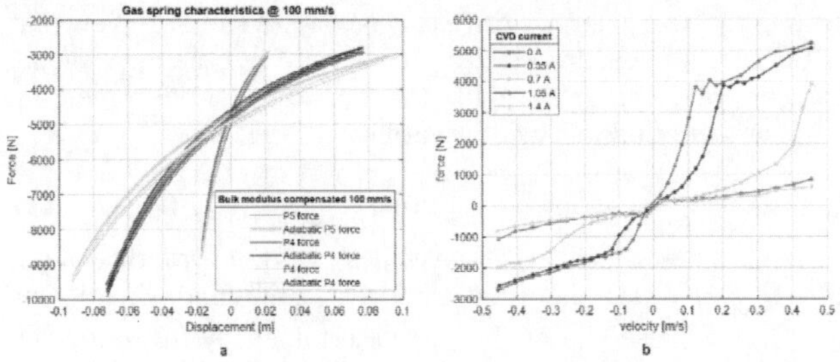

Fig. 1. Hydro pneumatic suspension system spring (a) and damper (b) characteristics

The vehicle has been fitted with a steerable rear axle. The rear wheel steering actuator makes use of an electric motor and ball screw linear actuator to adjust the angle of the rear wheels and vary the vehicle's dynamic response actively. Figure 2 shows the rear wheel steering actuator installed on a test trailer where the initial calibration and control development was performed.

Fig. 2. Rear wheel steering actuator mounted on steerable axle.

Finally, the vehicle has also been retrofitted with an ABS modulator with full control over the modulator's operation. This ABS modulator is used, together with a brake robot consisting of a linear actuator on the brake pedal, to achieve differential braking. By having the ability to block the valves in the modulator for some of the callipers on the vehicle the brake robot can vary and control the brake pressure going to the open valves' brake callipers and thereby achieve differential braking. This differential braking is often used to achieve direct yaw control of a vehicle.

With these systems the vehicle's roll, pitch, yaw, heave, and lateral dynamics can be controlled to improve the driver's comfort and achieve better handling and stability when needed.

Unfortunately, we live in a constrained world where these systems do not respond immediately, and actuator delays have to be taken into consideration. The delays have been experimentally measured. The delays were found to have a dead time and a first order response, therefore a first order plus dead time (FOPDT) delay is used to model the actuator and valve responses. These The damper, spring, rear wheel steering (RWS) and ABS responses are shown in Fig. 3.

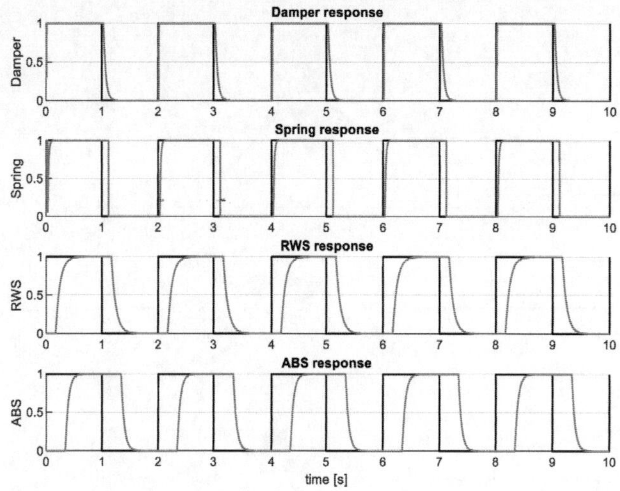

Fig. 3. First order plus dead time model of actuators

With a predictive control method these actuator delays can be managed so that they have less of a negative effect on the control of the vehicle.

4. Mathematical model of vehicle

A mathematical model of the vehicle is used to predict the future states given the current states and some preview of the disturbances on the vehicle. For this paper the mathematical model is developed to account for the non-linearities of the suspension system and the tyre forces. This model is deemed to be adequate to capture the necessary dynamics the simulation model will experience given the same disturbances applied to both. Figure 4 shows the output from the simulation model and the mathematical model for the vehicle performing a double lane change manoeuvre. The most important states we need to consider for this paper, namely side slip angle, yaw rate, lateral acceleration, roll angle and steer angle, are indicated.

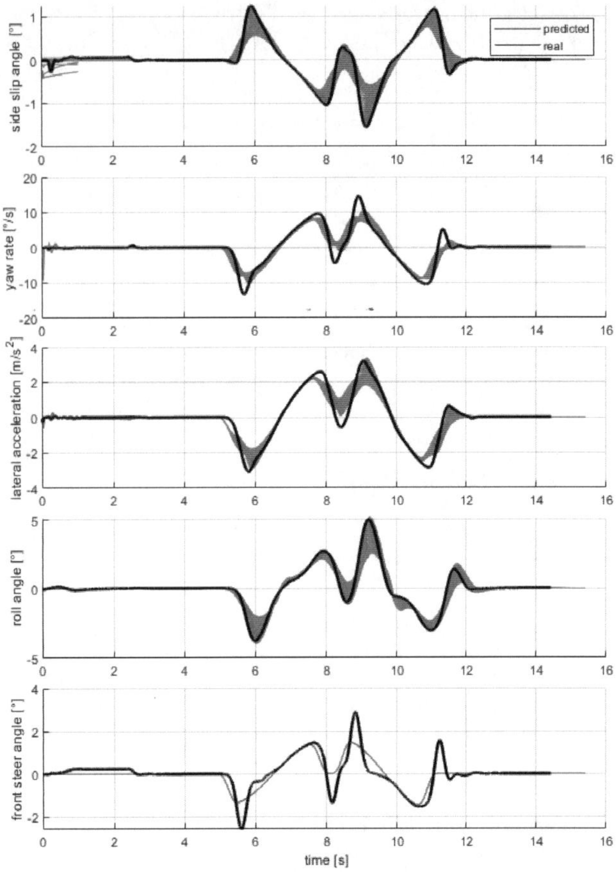

Fig. 4. Mathematical model response compared to simulation results.

Using a steering preview and the mathematical model we can predict the states into the future so that we can make informed choices of the control actuators to overcome their delays.

5. Steering preview

The first steering preview method considered is the zero-order hold (ZOH) method. This method is the simplest steering preview method where the assumption is made that the current steering angle the driver is making is kept constant throughout the entire preview time. This method can give good results if the preview time is relatively short compared to the dynamic response of the system. This approach might work well for systems where the delay of the control actuators is short, and the dynamic response of the system is slow.

The second steering preview method considered is the first-order hold (FOH) and saturation method. This method, similar to the ZOH method, makes use of the current steering angle and steering velocity to predict what the steering angle will be if it linearly extrapolates from the current steer angle. The saturation is added to limit how far the steer angle can extrapolate to achieve reasonable maximum and minimum steer angles. This method can excite more transient dynamics of the model as compared to the ZOH method when there is a significant steer velocity being applied to the vehicle. One downside of the FOH method is that it can very easily over- estimate what the steering angle will look like if the preview time is long and therefore make the mathematical model unstable. Due to the typical numerical differentiation needed of the steer angle to get the steer velocity, noise or high frequency inputs will also be amplified and render incorrect steer previews.

Finally, a steer preview method that makes use of the path information is considered. If there is some form of path planning happening on the vehicle or trajectory planning to avoid obstacles these paths can be used to extract the curvature of the road ahead. If the vehicle velocity is known and the curvature of the road ahead is known a steer preview can be made which is more realistic of what a driver would give as input to follow the curvature of the road (Nitin R. Kapania et al., 2015) has shown that a feedforward and feedback steering controller can be used to safely steer a vehicle at the limits of handling. If we consider the feedforward aspect of this controller, we can easily develop a steer preview that uses the current vehicle velocity and curvature of the road ahead. This steer preview is referred to as the curvature preview method. The equation to find this feedforward steer angle is:

$$\delta_{ffw} = L\kappa + \alpha_f - \alpha_r \tag{1}$$

Here L is the wheelbase of the vehicle, κ is the curvature of the road, α_f and α_r is the front and rear combined slip angles respectively. These slip angles are determined by inversing the magic tyre formula. To perform this inversion, we consider the combined lateral force desired at the front and rear axle due to the mass of the vehicle, its weight distribution and the lateral acceleration needed to keep the vehicle on the curve of the road. The combined lateral force on the front and rear axles are:

$$F_{y-front} = \frac{m \times b}{L} \times V^2 \times \kappa \tag{2}$$

$$F_{y-rear} = \frac{m \times a}{L} \times V^2 \times \kappa \tag{3}$$

with a and b the distance from the centre of gravity to the front and rear axles respectively. V is the longitudinal velcocity of the vehicle and m the mass of the vehicle.

α_f and α_r is then determined by inverting the magic tyre formula with parameters specific to the tyres used on the vehicle:

$$\alpha_f = \text{inv}MTF(F_{y-front}) \tag{4}$$

$$\alpha_r = \text{inv}MTF(F_{y-rear}) \tag{5}$$

With this method a realistic steer preview can be determined for the available road curvature preview. Naturally this steer preview cannot compensate for any lateral deviations from the intended path and the preview will deviate from what the vehicle will have to perform when it reaches that location down the road. Nevertheless, this steer preview can give good predictions of what a driver will do given a curved path in front.

6. Control method

The control of the semi-active suspension, rear wheel steering and differential braking is treated separately with different control algorithms for each. All of these control algorithms are reaction-based control. The controller delays are added to the simulation model for each of the control actuators such that a reaction-based control algorithm will not immediately affect the dynamics of the vehicle. With the help of the different steer previews and the mathematical model we can predict what the states of the vehicle will be in the future, just after the delay time, and apply control now so that when the delay has passed the controller is already at the correct control effort.

6.1 *Semi-active suspension control algorithm*

The semi-active suspension system is controlled by an algorithm that compares the running root mean square (RRMS) value of a moving window of both the lateral acceleration and vertical acceleration of the centre of gravity of the vehicle. The principal behind the control algorithm is that if the vertical RRMS value is higher than the lateral RRMS value, the vehicle is most likely driving over a rough road or driving in a straight line. Therefore, the suspension system should be in the soft spring characteristic and low damping to improve ride comfort. If the lateral RRMS value is higher than the vertical RRMS value, then the vehicle is experiencing a handling manoeuvre and the suspension should switch to the hard spring characteristic and high damping to improve handling. This control method addresses the ride-comfort and handling compromise passive suspension systems need to make by changing the spring and damper characteristics on the semi-active suspension system with a simple yet effective heuristic rule (Els, 2006).

Based on the semi-active suspension valve FOPDT model delay times the preview time for the semi-active suspension control need to be at least 150 ms.

6.2 *Rear wheel steering control algorithm*

For the rear wheel steering an off the shelf controller is used. Proposed by (Abe 2015) to drive the side slip angle of a vehicle to zero the following controller is used:

$$\delta_r = k_\delta \times \delta_f + k_{\dot{\psi}} \times \dot{\psi} \tag{6}$$

with:

$$k_\delta = -C_f / C_r \tag{7}$$

$$k_{\dot{\psi}} = \frac{m \times V^2 + 2(a \times C_f - b \times C_r)}{2 \times C_r \times V} \tag{8}$$

Here δ_f and δ_r is the front and rear steer angles respectively. $\dot{\psi}$ is the yaw rate of the vehicle. C_f and C_r is the combined front and rear cornering stiffness of the tyres.

Based on the front steering angle demanded by the driver and the experienced yaw rate as well as the current longitudinal velocity, the rear wheel steering angle is determined by multiplying these states to the respective gains.

From the rear wheel steering actuator FOPDT model we can see that we need at least 400 ms preview time to overcome the delay.

6.3 *Differential braking control algorithm*

The proposed direct yaw controller used by (Abe, 2015) is used for the differential braking control. The direct yaw controller finds a moment:

$$M = k_{\delta-DYC} \times \delta_f \tag{9}$$

with

$$k_{\delta-DYC} = \frac{4 \times L \times b \times C_f \times C_r \left(1 - \frac{m \times a}{2 \times L \times b \times C_r} V^2\right)}{m \times V^2 + 2(a \times C_f - b \times C_r)} \tag{10}$$

This is the direct yaw control moment that will lead to the side slip angle to be driven to zero. This equation is only dependent on the front steering angle and the longitudinal velocity of the vehicle. However, to optimally distribute this moment to the brakes the load on each tyre is needed as well as the lateral force experienced at that time to ensure the longitudinal braking force falls within the frictions circle.

The brake force distribution is done by solving this optimal problem for each direct yaw control moment determined:

$$\min J = \left(-M - \left(F_{xfl} \times \frac{df}{2} + F_{xrl} \times \frac{dr}{2} - \left(F_{xfr} \times \frac{df}{2} + F_{xrr} \times \frac{dr}{2}\right)\right)\right)^2 + 0.01\left(\sum_{i=fl}^{rr} F_{xi}^2\right) \tag{11}$$

such that:

$$-\sqrt{F_{zi}^2 - F_{yi}^2} \leq F_{xi} \leq 0 \text{ for } i = fl, fr, rl, rr \tag{12}$$

Here F_x is the longitudinal brake force that can be applied at each corner. F_z and F_y are the vertical and lateral forces at the instance considered with df and dr the front and rear track widths.

With this optimal problem solved the differential braking forces can be applied to the vehicle's wheels.

For the differential braking the FOPDT model a 500 ms preview time is required to overcome the delay.

7. **Testing methods**

The ISO 3888-1 double lane change (DLC) (ISO 3888-1:2018) is used as the manoeuvre to test the steer preview and controls on this vehicle for all three controllers. The test is done at 30 km/h and 80 km/h to simulate a normal safe driving scenario and a severe handling manoeuvre. The roads will be an ISO 8608 class A (ISO 8608:2016) random road as well as a Belgian paving road (ARMSCOR). These roads and speeds are chosen to represent a smoother road for on-road driving and a rough road for off-road driving. For the rear wheel steering and differential braking tests the suspension system of the vehicle is kept in the soft setting.

8. **Results and discussion**

8.1 *Semi-active suspension control results*

Figure 5 shows the results for the test on the DLC performed at 80 km/h on the class A road.

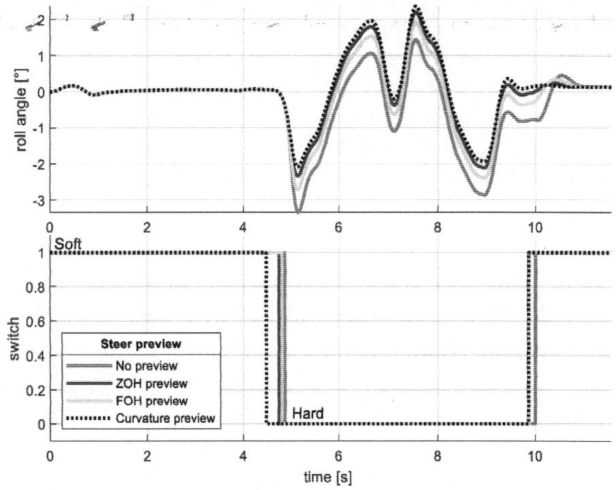

Fig. 5. Double lane change at 80 km/h on class A road with semi-active suspension control

From the results we can see that the curvature preview method gave the best results in keeping the roll angle to a minimum. This is due to the fact that the suspension was switched to a handling mode before any of the other methods. The curvature preview gave the controller warning that the lateral acceleration is going to change in the future and therefore the suspension system could switch to the handling mode before the other methods. The other preview methods also reacted faster than the no preview method but may not have successfully overcome the switching delays in the system.

When performing the same manoeuvre on the Belgian paving road at 30 km/h the vertical acceleration running root mean square value is higher than the lateral acceleration and therefore the suspension stays in ride comfort mode. Figure 6 shows the results for the DLC on the Belgian Paving at 30 km/h. Since all the controllers stayed in ride comfort mode the response on the RRMS acceleration graph is indistinguishable from each other.

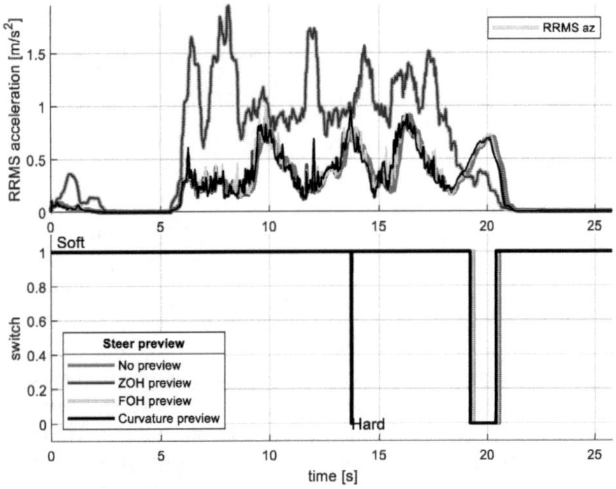

Fig. 6. Double lane change at 30 km/h on Belgian paving road with semi-active suspension control

Interestingly at the end of the manoeuvre the Belgian paving road ran out and the vehicle was back on smooth road. There the lateral RRMS was higher than the vertical RRMS and we can see close to 20 seconds the suspension went into a handling mode. From this figure we can see the curvature preview method performed best when switching to handling mode for a handling manoeuvre but gave a false positive during the ride comfort part of the path over the Belgian paving.

Overall, the curvature method performed well on smooth roads as compared to the other preview methods. There is a distinct advantage to have any steering preview, whether it is the ZOH, FOH or the curvature preview method, when changing suspension characteristics to overcome the delays.

8.2 *Rear wheel steering control results*

Figure 7 shows the results for the rear wheel steering control applied to the simulation model for a DLC on a class A road at 80 km/h.

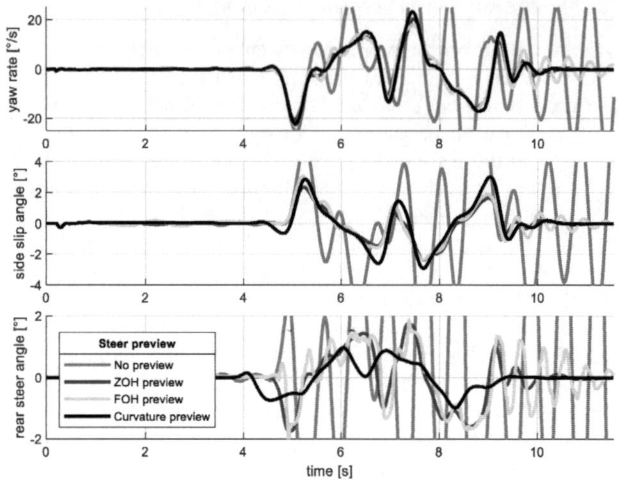

Fig. 7. Double lane change at 80 km/h on class A road with rear wheel steering control

From the figure we can see the "no steer preview" was unstable and did not perform well. For the other steer preview methods, the results for the yaw rate and side slip angle are similar, with the FOH method struggling to dampen out after the manoeuvre. The ZOH and FOH methods reduced the side slip angle the most by applying more rear wheel steering angle. Overall, the curvature preview method gave the best results with the least number of oscillations.

In Fig. 8 we can see that on rough roads like the Belgian paving at 30 km/h the ZOH, FOH and no steer preview methods gave erratic rear steer angles. The curvature preview method gave much smoother rear steering commands.

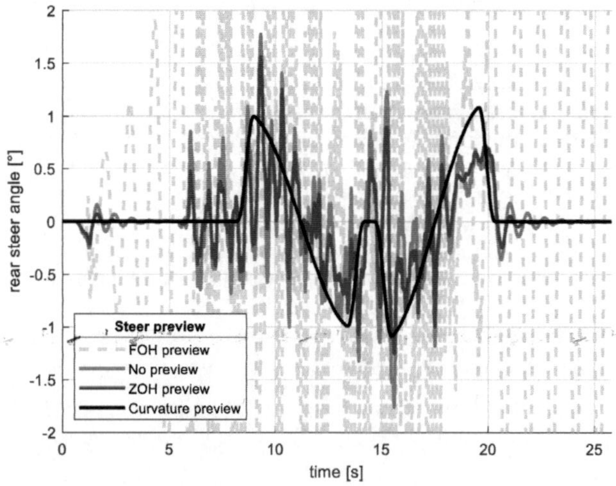

Fig. 8. Double lane change at 30 km/h on Belgian paving road with rear wheel steering control

The FOH method is particularly negatively affected by the rough road conditions since the front steering angle is constantly changing when driving on rough roads to keep the vehicle's heading correct.

8.3 *Differential braking control results*

For the differential braking control, the figures are presented with the independent variable being distance travelled in the x direction. Due to the different tests demanding different braking forces and times of activation of the braking the independent variable can't be time. The steer angle used for the ZOH and FOH methods are also different from the previous control examples. The steer angle is derived from the internal feedforward steer controller in the simulation model which also uses the feedforward feedback steering controller presented by (Nitin R. Kapania et al., 2015).

Figure 9 shows the results of DLCs on the class A road at 80 km/h. Here we can see the curvature preview method outperforming the other methods by having the lowest side slip angle. Probably the largest contributing factor to the decrease in the side slip angle is not necessarily the yaw moment being applied to the vehicle but rather due to the scrubbing of the speed of the vehicle due to the applied brakes.

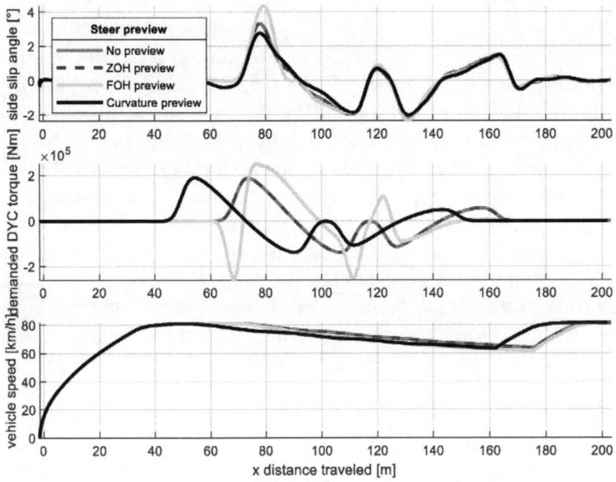

Fig. 9. Double lane change at 80 km/h on class A road with differential braking control

Performing the DLC on a Belgian paving road at 80 km/h the results are very similar to the class A road as can be seen in Fig. 10.

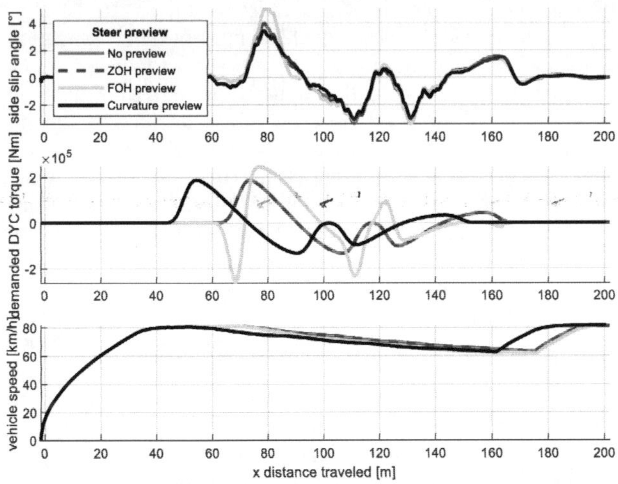

Fig. 10. Double lane change at 80 km/h on Belgian paving road with differential braking control

In general, all the steer preview methods worked well for the differential braking control with the curvature preview performing the best.

9. Conclusion

From the results we can see the advantage of adding steering preview to predictive control to overcome the actuator delays. In particular the curvature preview method performed the best out of all the steer preview methods proposed. The curvature preview method gave smoother and stable responses since the states that influence the controller is determined with a steer angle that depicts true driving closer. On average the curvature preview method also gave the fastest response due to the accurate prediction of the states. From this work it is evident that adding realistic steer preview to a predictive controller is advantageous, especially if there are actuator delays present in the system. It is recommended to investigate if there are other disturbances on the vehicle that can be modelled and previewed in a sensible way to further improve predictive control performance. Additionally, the preview can be extended further than just the actuator delays so that further states can be predicted and pre-emptive control applied. Also, in integrated chassis control schemes these different control actuators need to be combined in a single model to optimally distribute the control actuation considering each actuators constraints and delays.

10. References

Abe M. (2015) Vehicle Handling Dynamics, Theory and Application. Science Direct.

ARMSCOR Armaments Corporation of South Africa SOC Ltd. Gerotek Test Facilities [Online]. Available: https://www.armscor.co.za/?page_id=3967 [Accessed 31 May 2023]

Els, P. S., & University of Pretoria. Department of Mechanical and Aeronautical Engineering. (2006). The ride comfort vs. handling compromise for off-road vehicles (dissertation). University of Pretoria.

Pacejka H.B. & Bakker E. (1992) THE MAGIC FORMULA TYRE MODEL, Vehicle System Dynamics, 21:sup001, 1-18

Bosch H.R.B. et al. (2016) Parameterisation, validation and implementation of an all-terrain SUV FTire tyre model, Journal of Terramechanics, Volume 67, 11-23

International Organization for Standardization. (2018). Passenger cars — Test track for a severe lane-change manoeuvre — Part 1: Double lane-change (ISO 3888-1:2018)

International Organization for Standardization. (2016). Mechanical vibration — Road surface profiles — Reporting of measured data (ISO 8608:2016)

MathWorks Matlab and Simulink [Computer software]. 2020b

MSC Adams [Computer software]. 2019.2

Kapania N.R. & Gerdes J.C. (2015) Design of a feedback-feedforward steering controller for accurate path tracking and stability at the limits of handling, Vehicle System Dynamics, 53:12, 1687-1704

2765 | Obstacle detection vision system enabling autonomous mounding on clearcuts

Songyu Li [a,*], Håkan Lideskog [a], Magnus Karlberg [a], Ruben van Westendorp [b]

[a] *Department of Engineering Sciences and Mathematics, Luleå, Sweden*

[b] *Bracke Forest AB, Bräcke, Sweden*

** Corresponding author: songyu.li@ltu.se*

ABSTRACT

After final felling, a clearcut needs to be prepared as soon as possible for subsequent planting with the purpose of restoring forest reserves. In Sweden, the multi-row continuously advancing mounder is a mature and widely used solution for site preparation. It is driven by a base machine, where rotating mattock wheels make upside-down mounds on the area to prepare for subsequent planting. During operation, a plurality of obstacles such as stumps, residual trunks, single trees, large stones, and the like, remain after felling and are scattered at the site, forcing the operator to continuously watch out for them. These obstacles can to some extent be avoided through control of each mattock wheel arm, which causes inconvenience to the operator and reduces the efficiency of the mounding procedure. Thus, by adding automatic obstacle avoidance functionality to the mounder, the efficiency and working environment can be significantly improved.

In this paper, we present a feasible solution for real-time detection of obstacles in front of the mattock wheels of a mounder using a realistic simulated forest environment combined with a vision-based deep learning detection algorithm. The developed system can effectively identify most target obstacles and provide a reference for the operation of the mounder, either manually or by machine control. The successful implementation of this system also provides a starting point to further improve the automation level of mechanized site preparation, while further improving the mounders' working efficiency.

Keywords
Forestry
Forest rejuvenation
Reforestation
Automation
Driving simulation
Object detection
Site preparation
Obstacle avoidance

1. Introduction

In Nordic countries, forest owners are legally obligated to reforest after logging (Skogsstyrelsen, 2023; Finlex, 2023). Planting tree seedlings after site preparation is a common form of forest regeneration (Eriksson, 2013) to quickly restore tree numbers, accelerate forest renewal, restore forest stock, and establish stable forest ecosystems. Forest site preparation is a key step in reforestation to secure successful establishment of healthy seedlings who largely depends on creating optimal growth conditions. Threats to newly planted seedlings are common, as both biological and non-biological factors can reduce forest regeneration success (Hjelm et al., 2019). Site preparation is a method that effectively improves site conditions and increases the survival and growth of planted seedlings during the early stages (Saksa et al., 2005).

Mounding site preparation is a prevalent intermittent site preparation method in Sweden and Finland, typically conducted using mechanized approaches (Luke, 2016; Krekula et al., 2018). Mounding involves removing and flipping the forest ground and underlying mineral soil onto adjacent undisturbed forest ground (Sutton, 1993). This procedure forms mounds consisting of variable amounts of mineral soil, approximately 5-20 centimeters in height (Luoranen et al., 2007), positioned above a double layer of organic material. A well-formed mound, with the presence of a top layer of mineral soil, effectively prevents damage by pine weevils (Örlander and Nilsson, 1999; Nordlander et al., 2011). Moreover, the unique structure of mineral soil covering the double layer of organic material ensures stable soil moisture content, good insulation (Örlander et al., 1990; Von der Gönna, 1992; Prévost, 1992), and suppresses the growth of competitive vegetation (Prévost, 1992; Johansson et al., 2013), providing favorable conditions for seedling growth.

Mechanized mounders, often equipped with a transport unit like a forwarder as base machine, are commonly used for mound formation. Mounders typically have two to four independently adjustable hydraulic arms, each with a hydraulic-powered three-point or four-point mattock wheel at the end (Löf et al., 2015; McCarthy et al., 2017). The base machine drives the forward movement of the mounding machine, accompanied by the rotation of the Mattock wheels. This enables intermittent excavation, folding, stacking, and compaction of the soil to create mounds suitable for planting. During the mounding process, the adjustable arms need to be adjusted to avoid obstacles in the terrain, preventing direct impact on the arms or mattock wheels by hard obstacles such as large stones and remaining tree stumps, which could damage the mounding machine. In addition, every other mound attempt fails because of obstacle occurrence (Larsson, 2011) out of which surface boulders and stumps are a major cause. Currently, the control of the adjustable arms is entirely manual, requiring constant attention to the current status of each arm to control various displacement, lifting, and placing actions. These operations must be performed by the operator while driving the base machine, placing a significant demand on the operator's experience, increasing the difficulty of operation, and causing mental stress. Consequently, forestry machinery companies have raised new needs for improving the automation capabilities of mounders.

Taking the above considerations into account, a collaboration between Bracke Forest and the Machine Design research group at LTU (Luleå University of Technology) was initiated. Based on the Bracke M46.b four-row mounder, the Machine Design research group conducted fundamental research on automatic obstacle avoidance for the adjustable arms using a simulation approach.

https://doi.org/10.56884/VQZG2856
ISBN 978-1-942112-55-6 (pdf)

This work provided a preliminary but comprehensive solution to achieve our research objectives. It includes environment modeling, vehicle platform modeling, motion simulation, implementation of vision-based object detection, action determination for obstacle avoidance algorithm, and integration of these system functions. Through the construction of all these sub-functions, the preliminary foundation of a framework for automatic obstacle avoidance for mounders. has been established. Based on the virtual scenes created, the recognition of ground obstacles and corresponding determination of needed arm adjustments have been preliminarily achieved. The purpose of this article is to present these current research achievements and the implementation of the simulated environment, the obstacle detection function and the output used for obstacle avoidance.

2. Method

The method for implementing the entire project can be divided into three parts: Firstly, the construction of a simulated scene based on the Unity engine. Secondly, the establishment of a dataset and implementation of object detection. Lastly, building upon the simulated scene and the trained dataset, we develop the fundamental obstacle avoidance algorithm for the mounder's adjustable arms.

2.1. *Simulation environment*

The limitations of objective conditions make it difficult to achieve actual interactive operations between the physical mounder and the base machine in the short term. Therefore, constructing a simulation environment that closely resembles the actual operating environment has become the direction of our current research. The implementation of a complete simulation scene can be considered as consisting of two parts: the simulation of the combined mounder with base machine, and the simulation of the forest landscape as the working area.

2.1.1 *Mounder and base machine*

The modeling of the mounder is based on the Bracke M46.b four-row mounder from Bracke Forest AB, which features four articulated adjustable arms, hydraulic drive, a maximum working width of 8.4 meters, and weight of 7200 kg (see Fig. 1 left). The modeling of the base machine is based on the unmanned forest machine experimental platform developed by the Machine Design research group at LTU. This platform is a 10-tonne unmanned forest machine with articulated joint steering with a four luffing pendulum wheel-legs system (see Fig. 1 right).

Fig. 1. Mounder and base machine demonstration. Left: Bracke M46.b mounder. Right: LTU Unmanned Forest machine experimental platform.

The original modeling of the mounder and the base machine was done using Siemens NX, and then imported into Blender, a 3D creation suite, for assembly and constraints of each component after simplification. In Blender, the joints and links of each subcomponent were defined and constrained using the Phobos component (von Szadkowski et al., 2020) in the URDF (Unified Robot Description Format) format. This allowed the mounder and the base machine to be integrated into a cohesive unit, with the mounder body securely connected to the rear part of the base machine, resulting in the "BaseMachine-Mounder" combination, shown in Fig. 2.

To simplify operation during the motion simulation of the "BaseMachine-Mounder", we only enabled forward motion, reverse motion, four wheel braking, and articulate joint rotation of the base machine, allowing for forward/backward motion and steering control. As for the four pendulum arms, they were set to a fixed angle and not adjustable. For the mounder part, the two outer adjustable arms were set to a 35° opening angle and the two inner adjustable arms to a 10° opening angle to mimic an ideal position in actual operation, while allowing free rotation of the four mattock wheels. All other movable joints were set to a fixed state.

Fig. 2. The "BaseMachine-Mounder" combination in Blender.

Additionally, simplified collision volumes have been added to the front and rear sections of the base machine, as well as to the main body and mattock wheels of the mounder.

2.1.2 *Forest environment*

In order to create a forest scene that closely resembles a clear-cut area in a natural state, we utilized the Unity engine, a real-time 3D development engine launched by Unity Technologies, and incorporated scene assets and materials from the Unity Asset Store, an online store where you can download assets for the Unity engine, to efficiently generate terrain landscapes, texture terrains, and object models, allowing us to construct a realistically simulated forest scene (see Fig. 3).

During the scene development process, we placed two types of forest obstacles: stumps and stones (similar to 'surface boulders' used in Larsson, 2011), which serve as subsequent objects to-be detected (and avoided). In our scene, we arranged nine different styles of stone variants and eight different variants types of stumps. The placement of each object was based on derived rules that considered the terrain height, slope, and object density within a given area. These rules included specific placement densities, random variations in width within predefined ranges, and random orientations.

The simulation environment also includes other common objects and conditions found in a forest, such as:

- Vegetation: Multiple species of pine, cypress, grass, dry grass, bushes, shrubs, heather, etc.
- Environment: Simulated sunlight conditions, gentle breeze conditions, added terrain features such as gullies and small water bodies.

Fig. 3. Partial showcase of the simulated forest environment created in the Unity engine.

2.2 Dataset preparation and training

In this research, we employed vision-based deep learning methods to achieve object detection functionality in the simulated environment. To implement object detection, it was necessary to prepare a dataset and train a model based on it, resulting in a function capable of object detection on color images from the camera signal. We will introduce this process in two parts: dataset preparation and the selection of specific detection methods with training.

2.2.1 Dataset

The availability of annotated datasets specifically designed for forest applications is crucial for successful deployment of deep learning methods. Considering that there seems to be no openly available annotated dataset that effectively meets the training requirements of our research, we decided to construct our own training dataset.

The final raw dataset consists of 4,150 color images with a resolution of 1280*720 pixels, composed of a mixture of real photos and virtual images. Among them, there are 3,740 real photos taken in a forest area located in Bräcke, Jämtland County, Sweden, in September, taken in the time interval from 1:00 PM to 6:00 PM. These photos were randomly captured in an open area that had undergone clearing and was being prepared for reforestation. A stereocamera called Zed2 from StereoLabs was used for capturing RGB images. Additionally, there are 410 virtual images captured from our virtual forest scene, with a field of view of 110° horizontally and 70° vertically.

These 4,150 images were divided into a training set and a validation set. The training set contains 3,650 images, consisting of 3,341 real photos and 309 virtual photos. The validation set includes 500 images, with 399 real photos and 101 virtual photos. All the tree stumps and stones that need to be detected in the images were manually annotated, marking the location using bounding boxes. The training set has a total of 14,111 labeled target obstacles, including 6,895 stones and 7,216 tree stumps. The validation set has a total of 1,996 labeled target obstacles, including 967 stones and 1,029 tree stumps.

To further increase the quantity of data available for training and improve the performance of the trained detector, we performed data augmentation on the existing raw training dataset. The augmentation techniques used and the parameters of the augmented training set are shown in Table 1.

Table 1
Augmentation of training dataset

Augmentations	Parameters
Shear	±4° Horizontal, ±4° Vertical
Saturation	Between -15% and +15%
Exposure	Between -5% and +5%
Blur	Up to 0.25px
Noise	Up to 1% of pixels
Outputs per training example: 4	
Resize: Fit (black edges) in 640x640	

By applying these augmentation techniques with the specified parameters, we were able to generate additional training samples and expanded the diversity of the dataset, which usually leads to improved performance of the trained detector.
After dataset augmentation, the total number of images in the augmented training set reached 14,600. This set includes a total of 56,444 labeled obstacles, with 27,589 labeled stones and 28,864 labeled stumps.

2.2.2 Detector training

In the system design carried out in this work, one important functional component is the detection of target objects in images to be able to output their locations in the form of bounding boxes. This requires obtaining an object detector based on the prepared dataset. Visual-based deep learning has been widely applied in R&D, and we chose to use the popular YOLO (You Only Look Once) algorithm (Redmon et al., 2016) family to achieve the required object detection functionality in this study. Specifically, we chose the YOLOv7 (Wang et al., 2022) algorithm, which is the latest algorithm in the YOLO family at the time of this project. It is widely recognized for its excellent real-time performance and ease of deployment while maintaining high detection accuracy.

The choice of the YOLOv7 algorithm is further convenient for us to integrate into a ROS (Robot Operating System) environment. Ultimately, we deployed the object detection model based on YOLOv7-E6E on an Nvidia Jetson AGX embedded computer in the ROS Noetic environment used in our unmanned forest machine experimental platform. YOLOv7-E6E is a more complex model that combines compound scaling techniques and Extended ELAN (E-ELAN). Compared to the conventional version, this model improves detection accuracy and stability while sacrificing a slight decrease in detection speed.

For training, epochs were set to 300 and the mini-batch size was set to 8. All other parameters remained default. We used the provided yolov7-e6e_training.pt pre-trained weights for transfer learning.

After training with the training dataset, a detector corresponding to the labeled obstacles can be obtained. In this work, we define stumps and stones as the target obstacles to be detected. The detection of these two types of obstacles is performed continuously during the operation of the base machine. For each frame of the image, the bounding box information of all detected obstacles in the image is sent to the adjustable arm obstacle avoidance function for further computation.

2.3 Adjustable arm obstacle avoidance function

To utilize obstacle detection to allow the mounding arms to conduct obstacle avoidance, a new function is created whose main objective is to monitor whether obstacles are within the area in which the mattock wheels are deemed to move across in the image. The function is written in Python and implemented in ROS 1, where input data is bounding box information as a

Detection2DArray message topic (defined in vision_msgs), and the output comprises a custom message called IntList where four positions are defined in a list (e.g. [1,0,1,1]). Each position corresponds to each unique mattock wheel in which a '1' or a '0' at a certain position in the list corresponds to an obstacle being situated in the mattock wheel course areas or not.

These areas are defined as polygons in accordance with the Zed2 camera parameters and its spatial and angular position on the machine, see Fig.4.

Fig. 4. Defined polygon areas for where the mattock wheels are expected to move, exemplified with an image created in Unity with camera parameters corresponding to the Zed2.

At this point, the polygons are used to evaluate whether obstacles detected are found within it (1) or not (0). A detected obstacle is defined in the image using a rectangular shape that entirely enfolds the object, i.e., a bounding box. To be deemed as interfering with the mattock wheels' course, the lower left and right corners of a bounding box is utilized for evaluation. A line is defined from these two points (for each bounding box in the image) and the line position is compared to the defined polygon areas in three ways: (1) the line lies entirely within one of the polygons, (2) the line intersects one of the polygons, or (3) the line falls entirely outside the polygons. In cases (1) and (2) a '1' is added to the corresponding position in the custom message and immediately published to the ROS network for subsequent use for mattock movement control. The average time spent for this evaluation during a 30 second simulation fell below 3*ms* on a computer with laptop performance.

3. Results

The presentation of all the results obtained in this study is divided into three separate sections according to the three aspects of the method. These sections include the driving simulation of the "BaseMachine-Mounder" combination in the simulated forest environment, the performance evaluation of the stump and stone detector, and the implementation demonstration of the obstacle avoidance detection algorithm for the mounder's adjustable arms.

3.1. Mounder Driving Simulation

Once the "BaseMachine-Mounder" combination model is placed in the simulated forest environment, the operation of the mounder can be simulated. For the base machine part, we set the 'w' and 's' keys on the keyboard to control the forward and backward rotation of the four wheels, and the 'space' key is set to stop the wheels, enabling actions such as forward movement, backward movement, and braking of the base machine. The 'a' and 'd' keys on the keyboard control the left and right movement of the articulate joint, to achieve the steering of the base machine. Additionally, simplified collision volumes are set for the main body of the base machine, the adjustable arm of the mounder, and the mattock wheels.

In the simulated forest environment, stumps, stones, and trees have collision volumes, while vegetation such as grass and bushes do not have collision volumes. Objects with collision volumes will interact with the collision volumes of the "BaseMachine-Mounder" combination model. This setup better simulates the impact of different ground objects on the forestry machine during its movement, such as the jolts caused by driving over larger stones.

The visualized scenes obtained during the actual driving simulation process are shown in Fig.5.

Fig. 5. Scene demonstration of driving the "BaseMachine-Mounder" combination in the virtual forest environment.

A virtual camera, positioned at the rear of the base machine to capture the ground conditions in the area where the mattock wheels are about to reach, has been set up. The camera's field of view parameters are set to match those of the Zed2 camera, aiming to achieve similar capture performance. An example of a captured image is shown in Fig.6.

Fig. 6. Visualization of the imaging of the camera mounted at the rear end of the BaseMachine during the simulated driving process.

By manually adding Unity's official package extensions to the Unity engine and ROS, it is possible to create endpoints to accept ROS messages' transfer between the Unity scene and ROS. With the Unity-ROS communication established, the video stream captured by the virtual camera is converted in real-time to the sensor_msgs/Image format message within the ROS environment. This converted video stream is then published to a customized pre-defined topic in the ROS environment.

3.2 Stumps and stones detector

The trained YOLOv7 detector is used to detect stumps and stones in the video stream that has been published to ROS. The detection results, in the form of bounding box information, are then published to the ROS environment. These results are made available for subscription by the obstacle avoidance detection algorithm for the mounder's adjustable arms.

After 300 epochs of training, the obtained results are shown in Fig.7.

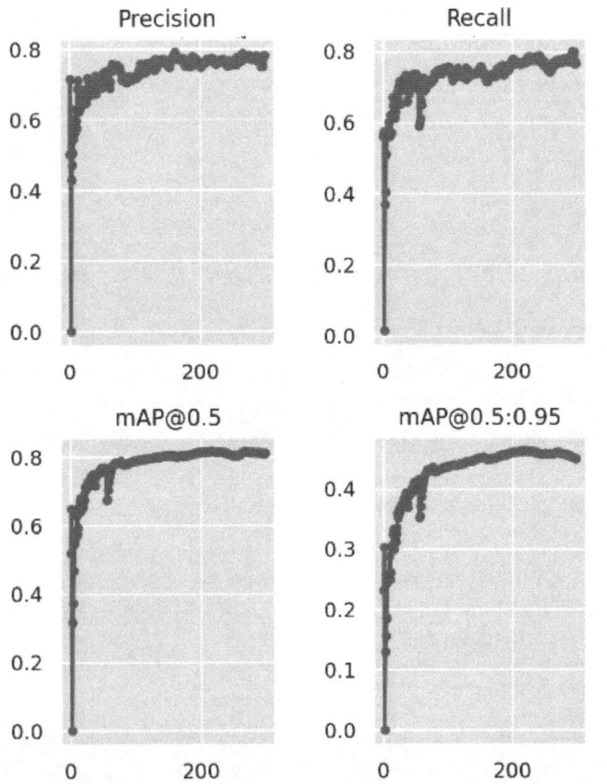

Fig. 7. Detection performance after 300 epochs of training. The x-axis represents the number of epochs, and the y-axis represents the values of corresponding performance metrics.

The precision and recall of the detector are both above 0.8, with mAP@0.5 of 0.81 and mAP@0.5:0.95 of 0.45. When considering the two types of target obstacles separately, the detector achieves mAP@0.5 of 0.771 and mAP@0.5:0.95 of 0.435 for stones, while for stumps, the mAP@0.5 is 0.849 and the mAP@0.5:0.95 is 0.465.

3.3. Adjustable arm obstacle avoidance

Along with the detector publishing the detection results in the form of bounding box information to ROS, this function will subscribe to the published bounding box information, calculate and provide whether the obstacles will affect the four polygonal regions. This result is presented in the form of a list containing four numbers, so it is not visualized during the actual operation of the project. However, an example image is shown in the Fig.8.

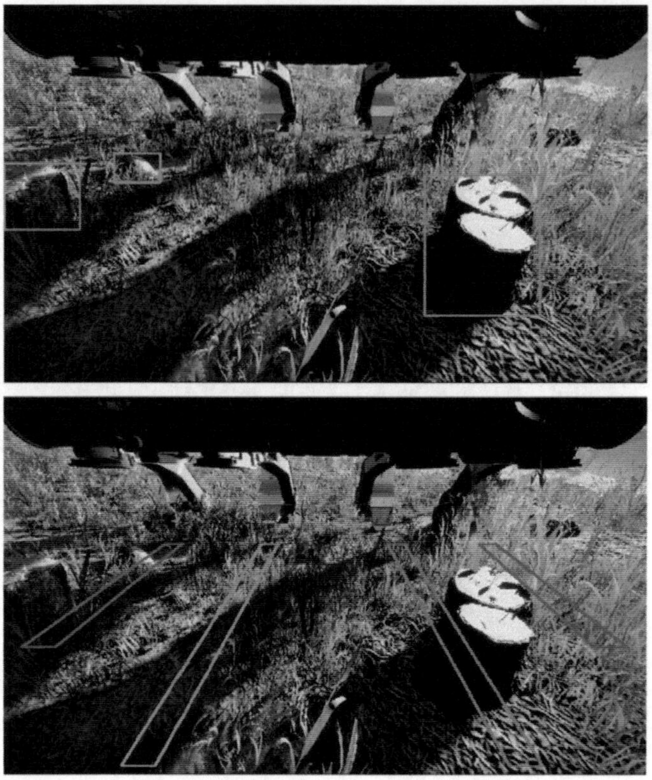

Fig. 8. Visualization of object detection and obstacle analysis for the same frame of the image. (*Up*) Display of detection results for the target obstacles in the captured scene, represented by bounding boxes. (*Down*) Four polygonal regions corresponding to the four articulated arms in the captured scene.

The bounding boxes shown in the Fig.8 represent the actual detection results obtained by the detector for this frame of the image. In this case, the obstacle avoidance algorithm provides a result of [1,0,1,0], which means that in the current state, the first and third arms, when counted from left to right, will be affected by obstacles and require appropriate action.

In this paper, we have presented a series of functionalities aimed at achieving the goal of automatic obstacle avoidance for the adjustable arms of a four-row mounder. These functionalities include:

- Construction of a simulated forest environment after final felling.
- Construction of a combined "BaseMachine-Mounder" simulation model.
- Creation of a mixed forest image dataset consisting of virtual images captured in the simulation environment and real photos.
- Training of a YOLOv7-based detector for detecting stumps and stones with significant detection accuracy.
- Development of a real-time detection feature to determine whether the adjustable arms of the mounder would collide with obstacles.

In this paper, we have established the connections between these functionalities, resulting in a solution that simulates the operation of a mounder in a simulated environment and provides real-time obstacle avoidance detection for the adjustable arms. Each functionality has been verified in simulation and proven to be feasible and effective.

The current state of the achievements meets our intermediate goals for this work, but there are still areas that require further improvements and additional features. Two main directions for future work include further refinement of the vehicle model and improvement of the adjustable arm obstacle detection algorithm.

Regarding the "BaseMachine-Mounder" combined model, many of the movable joint parts are currently set in fixed positions and angles to simplify the control of movement. Further improvements are needed in controlling the four pendulum arms for the wheels of the base machine and the angles of the joints on the adjustable arms of the mounder in future work.

For now, because of the fixed positions of the wheels and the fixed angle of the adjustable arms, the four independent obstacle detection areas corresponding to the four mattock wheels are also fixed. This needs improvement after refining the motion control of the model. The position of each independent detection area needs to be dynamically adjusted based on two factors. One is the current posture of the base machine, mainly considering the effects of pitch and roll on the offset of the detection area. This is expected to be further studied by adding a virtual IMU (Inertial Measurement Unit) to the base machine model. The other factor is the opening angle of each adjustable arm. After implementing independent control for each adjustable arm, the position of each detection area will be linked to the actual opening angle of each adjustable arm to ensure accurate detection.

Apart from these two main areas, there are also several other aspects that can be further improved, such as constructing a larger and more comprehensive dataset and upgrading the object detection algorithm used.

4. Data availability

The programs and code involved in this work will be made publicly available on community platforms such as Github and Gitee after all the work is completed. Additionally, our training dataset and training results will be released to the community as part of a broader effort to develop deep learning methods for forestry applications.

5. Acknowledgements

This research was partially funded by VINNOVA, [Grant No 2020-04202], by Interreg Aurora [Reference nr. 20357984], and by the EU [Grant Agreement No 869580].

6. Declaration of competing interest

The authors declare that they have no known competing financial interests or personal relationships that could have appeared to influence the work reported in this paper.

7. References

Eriksson, T., 2013. Silviculture and Environmental Consideration. In: Skogsstatistisk Årsbok [Swedish Statistical Yearbook of Forestry] Skogsstyrelsen [Swedish Forest Agency], p. 112-148.

FINLEX, 2023. Metsälaki 12.12.1996/1093. https://finlex.fi/fi/laki/ajantasa/1996/19961093/ (accessed 28 May 2023).

Von der Gönna, M., 1992. Fundamentals of mechanical sitepreparation. FRDA Report, 178.

Hjelm, K., Nilsson, U., Johansson, U., Nordin, P., 2019. Effects of mechanical site preparation and slash removal on long-term productivity of conifer plantations in Sweden. Canadian Journal of Forest Research, 49(10), 1311-1319.

Johansson, K., Ring, E., Högbom, L. 2013. Effects of pre-harvest fertilization and subsequent soil scarification on the growth of planted Pinus sylvestris seedlings and ground vegetation after clear-felling. Silva Fennica. 47(4). https://doi.org/10.14214/sf.1016.

Krekula, B., Bergqvist, J., Fries, C., Gällerspång, J., Reisek, J., Ringagård, J., Sollander, E., Svensson, L. & Wågström K., 2018. Föreskrifter för anläggning av skog -Regeringsuppdrag. Skogsstyrelsen. Rapport 2018/13.

Larsson A., 2011. Val av markbehandlingsmetod inom Sveaskogs innehav i norra Sverige. Second cycle, A1E. Umeå: SLU, Dept. of Forest Ecology and Management.

Löf M., Ersson B.T., Hjältén J., Nordfjell T., Oliet J.A., Willoughby I., 2015. Chapter 5: Site preparation techniques for forest restoration. In: Stanturf J.A. (ed.). Restoration of boreal and temperate forests. 2nd edition. CRC Press. pp. 85–102.

Luke, 2016. Silvicultural and forest improvement work 2015 (provisional). https://www.luke.fi/en/statistics/silvicultural-and-forest-improvement-work/silvicultural-and-forest-improvement-work-2015-provisional (accessed 28 May 2023).

Luoranen J., Saksa T., Finér L., Tamminen P., 2007. Metsämaan muokkausopas. [Soil scarification manual for forest lands]. Finnish Forest Research Institute, Vantaa. 75 p. http://urn.fi/URN:ISBN:978-951-40-2059-9. [In Finnish].

McCarthy R., Johansson F., Bergkvist I., 2017. Mounding with three- and four-pointed mattock wheels. Skogforsk, Arbetsrapport nr 925. 24 p. [In Swedish with English summary].

Nordlander, G., Hellqvist, C., Johansson, K., Nordenhem, H., 2011. Regeneration of European boreal forests: Effectiveness of measures against seedling mortality caused by the pine weevil Hylobius abietis. Forest Ecology and Management. 262, 2354-2363.

Örlander, G., Gemmel, P., Hunt, J., 1990. Site preparation: A Swedish overview. FRDA Report, 105.

Örlander G., Nilsson U., 1999. Effect of reforestation methods on pine weevil (Hylobius abietis) damage and seedling survival. Scandinavian Journal of Forest Research. 14(4), 341–354. https://doi.org/10.1080/02827589950152665.

Prévost, M., 1992. Effets du scarifiage sur les propriétés du sol, la croissance des semis et la compétition: revue des connaissances actuelles et perspectives de recherches au Québec. Annales des sciences forestières. 49(3), 277–296.

Redmon, J., Divvala, S., Girshick, R., Farhadi, A., 2016. You only look once: Unified, real-time object detection. IEEE Conference on Computer Vision and Pattern Recognition (CVPR). Las Vegas, NV, USA, pp. 779-788. https://10.1109/CVPR.2016.91.

Saksa, T., Heiskanen, J., Miina, J., Tuomola, J., Kolström, T., 2005. Multilevel modelling of height growth in young Norway spruce plantations in southern Finland. Silva Fennica. https://doi.org/10.14214/sf.403.

Skogsstyrelsen, 2023. The Forestry Act. https://www.skogsstyrelsen.se/en/laws-and regulations/skogsvardslagen/ (accessed 28 May 2023).

von Szadkowski, K., Reichel, S., 2020. Phobos: A tool for creating complex robot models. Journal of Open-Source Software, 5(45), 1326, https://doi.org/10.21105/joss.01326.

Sutton, R. F., 1993. Mounding site preparation: a review of European and North American experience. New forests. 7, 151-192. https://doi.org/10.1007/BF00034198.

Wang, C. Y., Bochkovskiy, A., Liao, H. Y. M., 2023. YOLOv7: Trainable bag-of-freebies sets new state-of-the-art for real-time object detectors. Proceedings of the IEEE/CVF Conference on Computer Vision and Pattern Recognition (CVPR). pp. 7464-7475.

2801 | In situ soil property estimation for autonomous earthmoving using physics-infused neural networks

W. Jacob Wagner [a,b,*], Ahmet Soylemezoglu [b], Dustin Nottage [b], Katherine Driggs-Campbell [a]

[a] *Department of Electrical and Computer Engineering, University of Illinois at Urbana-Champaign, Urbana, IL, United States*

[b] *Construction Engineering Research Laboratory, Engineer Research and Development Center, United States Army Corps of Engineers, Champaign, IL, United States*

* *Corresponding author: William.J.Wagner@erdc.dren.mil*

ABSTRACT

A novel, learning-based method for in situ estimation of soil properties using a physics-infused neural network (PINN) is presented. The network is trained to produce estimates of soil cohesion, angle of internal friction, soil-tool friction, soil failure angle, and residual depth of cut which are then passed through an earthmoving model based on the fundamental equation of earthmoving (FEE) to produce an estimated force. The network ingests a short history of kinematic observations along with past control commands and predicts interaction forces accurately with average error of less than 2kN, 13% of the measured force. To validate the approach, an earthmoving simulation of a bladed vehicle is developed using Vortex Studio, enabling comparison of the estimated parameters to pseudo-ground-truth values which is challenging in real-world experiments. The proposed approach is shown to enable accurate estimation of interaction forces and produces meaningful parameter estimates even when the model and the environmental physics deviate substantially.

Keywords
Soil Property Estimation
Fundamental Equation of Earthmoving
Physics-Infused Neural Networks
Autonomous Earthmoving

1. Introduction

Autonomous earthmoving is important for enhancing the safety, productivity and efficiency of construction and mining operations, especially in hazardous and remote environments where human operators face significant risks. One big challenge in designing these systems is enabling the equipment to operate efficiently and robustly across a wide variety of soils. Soils are complex heterogeneous granular materials whose properties vary significantly, not just between disparate locations, but also locally. As soil strength properties vary, the force required to shear the soil changes as well, which has a large effect on the earthmoving process. For example, in low strength soils a bulldozer can fill its blade in a short distance by making a deep cut without the vehicle stalling or tracks slipping. In contrast, cut depth is limited for higher strength soils because force limits of the machine are exceeded by higher required shear forces that increase with depth.

1.1 Autonomous Earthmoving

Early approaches to autonomous excavation are posed as trajectory control problems where a desired trajectory is generated, typically based on some heuristic such as ensuring the swept volume is equal to the capacity of the bucket, and tracked using a position controller (Bradley et al., 1998). However, these kinematic trajectory control approaches fail when the generated trajectory is infeasible due to force limits of the machine. The simplest approach to address this problem is to assume the soil is very high-strength and to conservatively select trajectories that ensure that the machine does not stall and tracks do not slip during execution of the motion. Unfortunately, this is very inefficient in lower-strength soils and limits the practicality of autonomous earthmoving.

To address these issues, alternative control strategies have been proposed that enable adaptation to changing soil parameters. In bulldozers, blade control technologies that adjust blade position to ensure the machine is exerting a force near the optimal load without producing track slip are commercially available (Hayashi et al., 2013; Jackson, 2017). Others have proposed impedance control (Ha et al., 2000) or power maximization (Sotiropoulos et al., 2019) as a way to balance the position control objective with force limitations. Instead of following a kinematic trajectory, a prototypical force-torque trajectory can be tracked which produces kinematic trajectories that vary depending on soil composition (Jud et al., 2017). More recently, reinforcement learning methods have been used to train control policies that achieve greater fill-factors in excavation and scooping tasks (Azulay et al., 2021; Backman et al., 2021; Egli et al., 2022).

However, local adaptation to soil conditions does not ensure efficient completion of a global terrain shaping task, implying that adaptation should be considered at the planning level as well. For an excavation task, iterative learning control has been used to adjust the desired trajectory between dig cycles to generate more feasible paths that result in increased bucket fill (Maeda et al., 2012). Another approach is to estimate soil properties in situ and use these estimated properties to inform planning and control (Singh, 1995a). At the planning level, these estimated properties can be used in combination with a model of earthmoving to ensure generated trajectories are feasible and efficient (Singh, 1995b). To improve tracking of the trajectory, the interaction forces can be predicted and used with impedance control (Tan, 2005).

https://doi.org/10.56884/JDXP2382
ISBN 978-1-942112-55-6 (pdf)

1.2 The Fundamental Equation of Earthmoving

The first step in estimating soil properties is to assume a model, as the properties are tautologically tied to a soil model. In most soil property identification for autonomous earthmoving work, the Mohr-Coulomb model of soil shear strength is assumed which relates the shear strength of the soil, τ, to the applied normal force, $\sigma_n$via

$$\tau = c + \sigma_n \tan(\phi) \tag{1}$$

where the parameters c and ϕ are the soil cohesion and angle of internal friction respectively.

Autonomous earthmoving requires a model of how the soil properties affect this process. Using the method of trial wedges where soil is assumed to fail along a plane producing a wedge, (Mckyes, 1989; Reece, 1964), an equation can be derived that describes the applied force F required to shear soil with a flat blade or bucket moving horizontally through the soil, as depicted in Fig. 1. The remaining forces acting on this soil wedge include the force of the loose soil accumulated on the blade which is referred to as surcharge or Q, the weight of the soil wedge W, the frictional component of the soil shear force combined with the normal force R, the soil cohesive force, and the soil-tool adhesive force. The forces are assumed to be in equilibrium, i.e., neither the soil nor the blade is accelerating, and are summed up along $\vec{x}$ and $\vec{z}$ directions to arrive at the fundamental equation of earthmoving (FEE):

$$F = f_{FEE}(\Theta) = \gamma d^2 w N_\gamma + c d w N_c + Q N_Q + c_a w N_a \tag{2}$$

where the coefficients are given by

$$N_\gamma = \frac{[\cot(\rho)+\cot(\beta)]\sin(\alpha+\phi+\beta)}{2\sin(\eta)}, \quad N_c = \frac{\cos(\phi)}{\sin(\beta)\sin(\eta)} \tag{3-6}$$

$$N_Q = \frac{\sin(\alpha+\phi+\beta)}{\sin(\eta)}, \quad N_a = \frac{-\cos(\rho+\phi+\beta)}{\sin(\rho)\sin(\eta)}$$

where $\eta = \delta + \rho + \phi + \beta$ is defined for notational convenience. This particular formulation of the FEE was developed by Holz et al., 2013 to enable consideration of inclined surfaces. Initially, the equation has three unknown variables (F, R, β), but applying the equilibrium conditions reduces this down to the single unknown soil failure angle β. Typically, it is assumed that the optimal soil failure angle should be chosen such that

$$\beta^* = \arg\min_\beta N_\gamma \tag{7}$$

In some cases, it is possible to derive a closed form expression for β^*, but often numerical methods are relied upon. The interaction force can be broken into its $\vec{x}\vec{z}$ subcomponents by

$$\boldsymbol{F} = f_{FEE}(\Theta) = [F\cos(90° - \rho - \delta + \alpha),\ F\sin(90° - \rho - \delta + \alpha)] \tag{8}$$

During bulldozing, the blade may move vertically in addition to horizontally as the desired cut depth is adjusted. This causes a change in the relative movement between the soil wedge and the blade. In the case that the wedge is moving down with respect to the blade, this results in the soil-tool adhesive and frictional forces changing direction. In order to account for these effects, Holz et al., 2013 make the following modification

$$\delta' = \tanh(-C_1(\vec{\iota_b} \cdot \vec{v}))\,\delta,\ c_a' = \tanh(-C_1(\vec{\iota_b} \cdot \vec{v}))\,c_a' \tag{9-10}$$

where C_1 is a positive constant, $\vec{\iota_b}$ is the unit vector pointing upwards along the blade, and $\vec{v}$ is the velocity vector of the blade represented in the $\vec{x}\vec{z}$ coordinate system. The values δ' and c_a' are used in place of δ and c in equations 2-6.

1.3 Soil Property Estimation

There has been considerable attention on developing methods for estimating and predicting vehicle traversability for the purposes of autonomous off-road navigation (Borges et al., 2022). Of particular interest for this work is the characterization of soil properties from on-board vehicle sensing including vision and proprioception, sometimes referred to as inverse terramechanics (IT) estimation. In-situ soil property estimation using these methods has been studied within the field of terramechanics and is often motivated by the need for autonomous navigation of planetary rovers (Lopez-Arreguin et. al. 2021). This problem is formulated as the estimation of parameters of a soil-wheel interaction model and is often solved using optimization-based approaches (Iagnemma et al., 2004), although filtering methods have been developed (Dallas et al., 2020) and learning based methods have been used (Lopez-Arreguin et al., 2021).

For autonomous earthmoving tasks, a similar IT approach can be taken. However, as the vehicle is more forcefully interacting with the terrain, different models must be assumed. Typically, a subset of the model parameters is assumed to be known, and the remaining parameters are found by fitting the model to observed interaction forces (Singh, 1995a; Luengo et al., 1998; Tan et al., 2005; Althoefer et al., 2009). It is often difficult to determine the accuracy of the estimated parameters directly, but Tan et al., 2005 report predictions of the soil failure force to within 10%-30%.

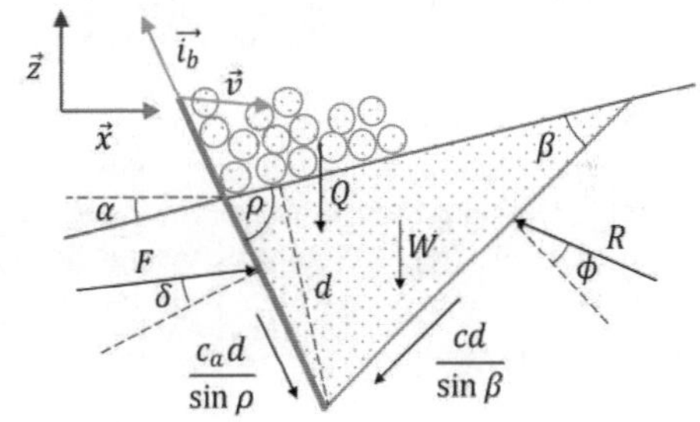

Fig. 1. FEE soil wedge geometry and force diagram. This is a cross-sectional view depicting a blade, shown in gray, moving in the $\vec{v}$ direction through the soil at a depth of cut d and failing the soil along the purple line at an angle β with respect to the surface. The soil surface, shown in green, is inclined at the angle α from the horizon and the flat blade is at an angle ρ with respect to the soil surface. The passive failure case is assumed where the soil wedge moves up with respect to the blade and forward with respect to the un-sheared soil. The cohesive forces resist this movement and are drawn accordingly. (Holz et. al. 2013)

1.4 Contributions

The estimation of soil properties is a key challenge for autonomous earthmoving operations, such as bulldozing. Existing IT methods have mainly focused on the soil-tire interaction for vehicle navigation applications and, while motivating, these methods do not directly apply to earthmoving tasks due to different underlying physics. The relatively limited prior work in IT methods for construction tasks leverage optimization approaches to find parameters of the earthmoving model given observations of the interaction force. In this work, we propose a novel IT method based on physics-infused neural networks

(PINNs) that can estimate soil properties in situ from kinematic and control observations of a bulldozer blade. The main contributions of this work are:

- Development of a PINN-based in situ soil-property estimation method that incorporates physical laws and constraints into the learning process without requiring knowledge of true soil properties
- Evaluation of the accuracy and robustness of the proposed method using simulation data with ground-truth soil properties
- Introduction of a physical inconsistency loss function that enables enforcement of minimization or maximization of certain physical quantities
- Demonstration of a novel technique to estimate the blade-soil interaction forces directly from kinematic and control data

2. Methods

The creation of a simulation to study the in-situ soil-property estimation problem is first discussed, and the PINN approach to soil property estimation is then developed.

2.1 Simulation

In this work, the proposed soil property estimation method is evaluated to determine how capable the system is at extracting characteristics of the soil. Real world evaluation a challenging task, as evaluation of the full system would require significant instrumentation of a bulldozer, time consuming real-world experimentation across various soil types and levels of compaction, and careful setup (e.g., compaction, moisture content) and characterization of these soils to establish a soil property ground-truth to enable validation of the approach. Instead, using a simulation as an analog for experiments on a real bulldozer is proposed. To perform these experiments, a simulator that exhibits similar behavior to a real system is required, but faithful representation of the dynamics of a specific piece of earthmoving equipment for a specific soil is not strictly necessary.

2.1.1 Vortex Studio

Vortex Studio[1] is a simulation software which enables modelling of multi-body dynamics and supports accurate real-time soil-tool interaction physics. In Vortex, the soil is modeled using a hybrid heightfield-particle representation. An FEE based model is used to determine the reaction force of the soil prior to shearing. As the relative density I_d of the soil increases, ϕ and c are varied to increase the resistance of the soil to shearing due to stronger particle interlocking, and γ is varied to adjust the weight of the soil wedge and surcharge (Holz, 2009).

When the force exerted by the tool on the soil reaches this threshold, the portion of the heightfield in contact with the blade is converted into particles whose behavior is governed by a discrete element model (DEM) simulation which more accurately characterizes the behavior of the disturbed soil. As particles come to rest, they are then reintegrated back into the heightfield representation. The initial relative density is defined as a fixed value but may vary throughout the simulation as the soil is compacted by the tool or as sheared particles are merged back into the heightfield (Holz, 2009; Holz et al., 2013). The remainder of the soil properties remain constant for a given soil type and do not vary within a simulation episode.

Haeri et al., 2020 showed that Vortex Studio is capable of predicting the soil-tool interaction forces for a lunar simulant material within 20%-30% of the measured force obtained from real-world experimentation. They introduce some modifications to the original Vortex model, including a term that reduces the contribution of the surcharge by a factor of 10 in equation 2. Additionally, they find that selecting β such that the top surface of the soil wedge matches extent of the surcharge pile, better matches the experimental data as compared to minimization of N_γ as in Equation 7.

One benefit of using a simulation to study soil-property estimation is that it can provide ground-truth knowledge of the soil parameters for validation of a given approach. However, in Vortex, this is not straightforward as the simulation is not a purely physics-based model, but rather a combination of physics and heuristics that have been added to improve the realism, stability, and performance of the system. In addition to the modifications discussed by Haeri et al., 2020, these heuristics include various adjustments to the FEE force calculation, such as limiting the force based on the submerged surface area, the addition of penetration forces, scaling of the FEE force, inclusion of additional frictional, elastic, and damping contact forces between the heightmap and tool, etc. These heuristics are not directly derived from the soil properties, but rather must be tuned to match experimental data or may be modified to incorporate operator feedback, improving the realism of the simulation.

Vortex provides four default soil configurations (Clay, Loam, Sand, and Gravel), but further tuning is recommended for specific applications. However, it is not obvious how to configure these parameters to achieve realistic performance for different soil types and blade geometries. Despite these limitations, Vortex meets the needs for a system that exhibits realistic earthmoving phenomenon and is sufficiently complex to act as an analog for experiments on a real machine.

Fig. 2. Rendering of simple bladed vehicle simulation. The chassis is shown in green and the flat blade collision geometry is shown in blue. The soil particles generated due to the blade failing the soil constitute the surcharge.

2.1.2 Simple Bladed Vehicle Model

A simplified simulation of a bladed earthmoving vehicle is developed, see Figure 2, and consists of two dynamic

[1] https://www.cm-labs.com/en/vortex-studio/

components: a chassis and a blade. The chassis is modeled as a rectangular prism (3m, 2m, 2m) with a mass of 5,000kg. The blade is flat 3.164m wide, 0.660m tall, 0.001m thick[2] with a mass of 400kg.

Vortex constraints, which enable restricting the motion between two or more parts, are used to control the motion of the chassis and the blade. Motorized constraints enable control of a part's velocity and exert a force proportional to the velocity error. The inverse of the loss parameter specifies this constant. Locked constraints enable control of a part's position and exert a viscoelastic force on the part equivalent to a spring-damper system (CM Labs Simulations, 2016). A motorized constraint is used to control the forward velocity of the chassis and a locked constraint is used control its vertical position. The lateral position and orientation of the chassis are locked, limiting the motion of the vehicle to the $\vec{x}\vec{z}$ plane. The blade is attached to the chassis body using a locked constraint enabling control of the vertical position relative to the chassis.

To emulate limits on a real machine, the relative position of the blade with respect to the chassis is limited to enable the bottom edge of the blade to reach 0.3m. below and 1.0m above the surface when the chassis is on level ground. The soil height is measured at four locations underneath the chassis and the average of these height is used to command the chassis vertical position to emulate a simple vehicle suspension. Additionally, the vertical and horizontal constraint forces are limited to 30kN and 20kN respectively to emulate the tractive and penetrative force limits of a bladed earthmoving machine.

On a bulldozer, hydraulic pressures in the vehicle hydrostatic drive circuit and in the cylinders controlling the blade position can be measured to derive estimates of the tractive force given knowledge of the machine's dynamics (Yamamoto et al., 1997). The simulation developed in this effort has a simpler drive system, therefore, the simulator chassis-blade constraint forces are used to measure the interaction force instead. Significant effort was given to tune the loss, stiffness, and damping coefficients of the constraints to yield a stable simulation with minimal oscillations in the constraint forces and part positions. However, oscillations on the order of 5kN and 10kN in the x and z forces respectively are commonly observed in the experiments, see Fig. 3. These oscillations arise from the interaction of the constraints used to control the vehicle and the constraints imposed by the soil-tool interaction model. In particular, when a portion of the heightfield is converted into particles, the blade momentarily loses contact with the heightfield resulting in a significant disturbance in the force applied to the blade. This amounts to a step input to the control constraints and can lead to transient violations of the force limits, as Vortex constraints are essentially dynamical systems.

During simulation, at a rate of 60Hz, observations are collected of the blade x and z positions and chassis z positions, $\boldsymbol{p} = [p_x^b, p_z^b, p_z^c]$, the blade z velocity and chassis x and z velocities, $\boldsymbol{v} = [v_z^b, v_x^c, v_z^c]$, and the x and z cutting forces, $\boldsymbol{F} = [f_x^b, f_z^b]$. Additionally, the commanded blade relative and absolute z position, and the chassis x velocity $\boldsymbol{u} = [u_{z_r}^b, u_{z_a}^b, u_x^c]$ are collected. At each timestep t, the observations and actions are concatenated into the observation vector $\boldsymbol{o_t} = [\boldsymbol{p_t}, \boldsymbol{v_t}, \boldsymbol{u_t}]$ which is consumed by the PINN. Pseudo-ground truth soil-properties are collected from the simulation as $\boldsymbol{\Theta} = [\phi, c, \delta, c_a, \gamma, \rho, \alpha, w, d, Q, \vec{\boldsymbol{v}}]$, where Q is obtained using the Vortex soil mass sensor plugin. The simulation does not provide access to the soil failure angle β, but for notational simplicity it will be included in $\boldsymbol{\Theta}$ for the remainder of the text.

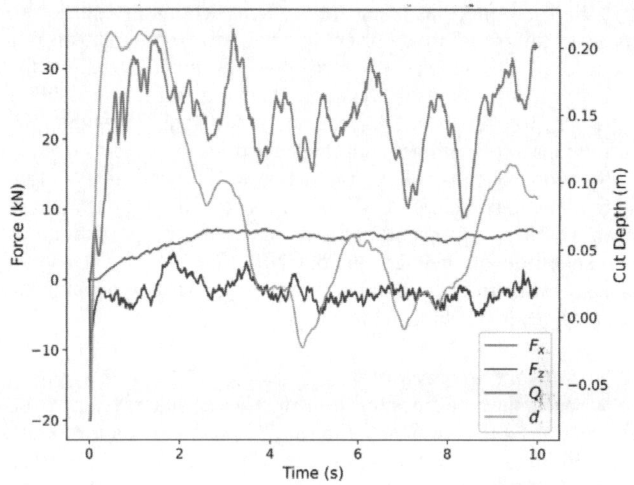

Fig 3. Measured forces and cut depth for a typical episode within the D_{default} dataset. This episode was recorded for a loam soil with $I_d = 60\%$ and illustrates the large amplitude noise in force measurements observed particularly for denser soils. Also, note how the surcharge grows with time indicating the accumulation of soil on the blade from the cutting operation.

2.1.3 Data Collection Controller

To ensure clean data is collected to train an effective model, a simple data collection controller was designed. In the case that the force required to shear the soil is larger than the machine limits, the blade becomes stalled, and the soil is not sheared. Therefore, an anti-stall blade controller is developed which enables automatic data collection.

For a given desired forward velocity $\bar{v}_x$ and target absolute cut depth $\bar{d}_z$, the controller adjusts the blade relative position command u_{z_r} to ensure that the velocity tracking error e_{v_x} does not drop below a target velocity dependent threshold $e_{v_{min}} = \bar{v}_x c_{v_x}$. This is accomplished using a gain-scheduled integral controller which reduces the desired depth of cut by a depth offset $\Delta\bar{d}_z^k$ at each timestep k; this quantity is limited to enable raising of the blade only a small amount above the surface. The controller is defined as follows

$$\Delta\bar{d}_z^k = K_{\Delta p}(e_{v_x}^k - e_{v_{min}}) + \Delta\bar{d}_z^{k-1} \quad \in [0, \bar{d}_z + \Delta\bar{d}_{z_{max}}] \tag{11}$$

$$u_{z_r} = -(\bar{d}_z - \Delta\bar{d}_z^k) \tag{12}$$

The scheduled depth offset gain term, $K_{\Delta p} = K_v / \bar{v}_x$, modulates the depth offset integral gain K_v to enable improved forward velocity tracking for low desired velocities $\bar{v}_x$. Overall, the effect of this controller is that vehicle maintains forward motion with some low-frequency oscillation in the depth of cut and forward velocity around the targets. As the soil becomes harder to shear and vehicle force limits are reached, the desired

[2] Originally, the blade was made to have a more reasonable thickness, but initial experimentation revealed problems with the vehicle being able to penetrate the soil surface even with large forces for denser soil configurations. The authors believe that this is an artifact of the way contact detection is used to compute the soil blade angle ρ internal to Vortex. The hypothesis is that the system computes the angle using the bottom surface of the blade instead of the leading edge causing the simulator to produce unrealistically large reaction forces. Reducing the blade width alleviated most of these problems.

depth of cut becomes harder to track, but stalling is prevented ensuring the soil failure condition assumed by the FEE is met.

2.2 Physics-Infused Neural Network (PINN)

Broadly, the approach taken in this effort is to train a neural network to produce estimates of the soil parameters which are then fed through the FEE model to produce an estimated force. The network ingests a T step history of observations and control commands $\boldsymbol{o}_{[t:t+T]}$ along with known FEE parameters $\boldsymbol{\Theta_k} = [c_a, \gamma, \rho, \alpha, w, d, Q, \vec{\boldsymbol{v}}]$, and outputs an estimate of the unknown parameters $\widehat{\boldsymbol{\Theta}}_{\boldsymbol{u}} = [\hat{\phi}, \hat{c}, \hat{\delta}, \hat{\beta}, \Delta d]$ and the estimated interaction force $\widehat{\boldsymbol{F}}$. It is important to include control actions in addition to kinematic information, as tracking error can be thought of as a measure of the interaction force. For example, Bradley and Seward (1998) use the position tracking error to achieve "software force feedback" that can be used to determine when the bucket comes into contact with the ground.

By properly constraining the network, the hypothesis is that the model will converge to estimating reasonable soil parameters because this representation is the most parsimonious representation of the data. In other words, the model should learn to estimate meaningful soil properties as they explain the observations in the simplest form.

2.2.1 Network Architecture

While it may be possible to estimate FEE parameters from observations and commands at a single time-step, this is difficult given the noise present. Therefore, observations and actions across $T = 60$ timesteps (1 sec.) are incorporated, and a temporal convolution is leveraged to compress the $T \times 9$ dimensional observation $\boldsymbol{o}_{[t:t+T]}$ into a $l \times N$ vector, where l and N are model hyperparameters. This vector is passed through a stacked transformer style encoder network with multi-headed attention, modeled after the network presented by Zerveas et al., 2021, producing another $l \times N$ vector. This vector is then combined with the known FEE parameters $\boldsymbol{\Theta_k}$ and passed through a dense network with 2 hidden layers of 20 neurons each that is referred to as the integration network. The role of this network is to incorporate the knowledge of the existing soil parameters with the latent state produced by the transformer from the observation history. This enables estimation of the unknown soil parameters. Without this network, the system cannot learn to account for variation in $\boldsymbol{\Theta_k}$ by changing $\widehat{\boldsymbol{\Theta}}_{\boldsymbol{u}}$. A linear output layer produces the estimated unknown FEE parameters $\widehat{\boldsymbol{\Theta}}_{\boldsymbol{u}}$ and the residual end to end force $\boldsymbol{F}_r$. The FEE parameters are combined $\widehat{\boldsymbol{\Theta}} = [\boldsymbol{\Theta_k}, \widehat{\boldsymbol{\Theta}}_{\boldsymbol{u}}]$, clipped to remain within the limits defined in Table 1, and passed to the FEE network to produce the FEE reaction force $\widehat{\boldsymbol{F}}^{\boldsymbol{FEE}} = \boldsymbol{f_{FEE}}(\widehat{\boldsymbol{\Theta}})$. The residual and FEE reaction forces are then summed to produce the estimated interaction force $\widehat{\boldsymbol{F}} = \boldsymbol{F}^{\boldsymbol{FEE}} + \boldsymbol{F}^r$. See Fig. 4 for a visual depiction of this architecture.

The residual depth of cut which augments the depth of cut by $d' = d + \Delta d$, was added to the model after initial experimentation revealed significant sensitivity of the FEE model to the depth of cut; major improvements in reconstructing the observed force entirely from the ground-truth parameters was shown to be possible with hand tuning of Δd. Similarly, the residual interaction force $\boldsymbol{F}^r$ was added to the network to avoid non-FEE components of the simulation (e.g., noise in the measured force and non-zero measured force when $d < 0$) corrupting the parameter estimates.

2.2.2 Network Training

To allow for stable training, observations $\boldsymbol{o}$ are normalized to have unit mean and variance where the normalizing mean and variances are computed from the training set observations. Values for the residual force are normalized $\boldsymbol{F}^r$ to 10% of the value of $\boldsymbol{F}$. Both known and unknown FEE parameters are scaled using min-max normalization with ranges provided in Table 1. The ranges for ϕ, c, δ, and c_a are obtained from multiple sources listing geotechnical parameters (Geotechdata.info 2023; MATHalino, 2023; and Fine Software, 2023) while the remaining parameters are scaled using knowledge of machine limits and geometry of the FEE. During tuning of the model, some liberty was taken to adjust parameter ranges for β and c from the initially assumed values in order to avoid the model converging to a poor local optimum.

Table 1
FEE parameter normalization ranges and limits.

Parameter	Norm. Range	Limits $[l_l, l_u]$	Units
ϕ	[17, 45]	[0, 90]	°
c	[0, 10e3]	[0, -]	Pa
δ	[11, 35]	[0, 90]	deg.
c_a	[0, 10e3]	[0, -]	Pa
γ	[14e3, 22e3]	[0, -]	N/m³
ρ	[2, 178]	[2, 178]	°
α	[10, 10]	[-30, 30]	°
w	[0, 3.164]	[0, 3.164]	m
d'	[0, 0.3]	[0, 0.660]	m
Δd	[-5e-2, 5e-2]	[-5e-2, 5e-2]	m
Q	[0, 10e3]	[0, -]	N
v_x	[0,1]	[-2, 2]	m/s
v_z	[-1, 1]	[-2, 2]	m/s
β	[11.5, 34.5]	[11.5, 34.5]	°
$\partial N_\gamma / \partial \beta$	[-10, 10]	[-, -]	unitless
η	[2, 178]	[2, 178]	°
$\zeta = \phi - \delta$	[11, 35]	[0, -]	°

The loss function for the network is comprised of five types of losses: a mean average error force prediction loss ($\mathcal{L}_{MAE}^{F}$), a MAE regularization of the residual force ($\mathcal{L}_{MAE}^{F_r}$), ReLU-based regularization of FEE parameters ($\mathcal{L}_{\text{ReLU}}^{\chi}$), a mean square error (MSE) based regularization of the residual depth of cut ($\mathcal{L}_{MSE}^{\Delta d}$), and a MAE regularization of the gradient $\partial N_\gamma / \partial \beta$ ($\mathcal{L}_{MAE}^{\partial N_\gamma/\partial \beta}$). It is defined as follows

$$\mathcal{L}(\boldsymbol{F}, \widehat{\boldsymbol{\Theta}}, \widehat{\boldsymbol{F}}, \boldsymbol{F}^r) = \lambda_{\boldsymbol{F}} \mathcal{L}_{MAE}^{\boldsymbol{F}}(\boldsymbol{F}, \widehat{\boldsymbol{F}}, \mathbf{w}_{xz}) + \lambda_{\boldsymbol{F}_r} \mathcal{L}_{MAE}^{\boldsymbol{F}_r}(\boldsymbol{F}^r, \boldsymbol{0}, \mathbf{w}_{xz}) + \sum_{\chi \in X}[\lambda_\chi \mathcal{L}_{\text{ReLU}}^{\chi}(\chi, \boldsymbol{l}^\chi)] + \lambda_{\Delta d_{MSE}} \mathcal{L}_{MSE}^{\Delta d}(\Delta \bar{d}, \Delta d) + \lambda_{\partial N_\gamma/\partial \beta} \mathcal{L}_{MAE}^{\partial N_\gamma/\partial \beta}\left(\frac{\partial N_\gamma}{\partial \beta}, \boldsymbol{0}, \mathbf{w}_{\partial N_\gamma/\partial \beta}\right) \quad (13)$$

$$\mathcal{L}_{MAE}^{\boldsymbol{y}}(\bar{\boldsymbol{y}}, \boldsymbol{y}, \mathbf{w}) = \sum_{i=0}^{|\boldsymbol{y}|} \mathbf{w}_i |\bar{\boldsymbol{y}}_i - \boldsymbol{y}_i| \quad (14)$$

$$\mathcal{L}_{MSE}^{\Delta d}(\Delta \bar{d}, \Delta d) = (\Delta \bar{d} - \Delta d)^2 \quad (15)$$

$$\mathcal{L}_{\text{ReLU}}^{\chi}(\chi, \boldsymbol{l}^\chi) = max(0, l_l^\chi - \chi) + max(0, \chi - l_u^\chi) \quad (16)$$

The FEE parameters regularized to remain within the limits listed in Table 1 are $\boldsymbol{X} = [\phi, c, \delta, \beta, \eta, \zeta, \Delta d, d']$ and the weights $\lambda_\chi = 1$ for all of these parameters except for $\lambda_\eta, \lambda_\beta = 10$. These values are chosen to be higher because when violated, they lead to singularities in Eq. 2 causing $\hat{\mathbf{F}}$ to become unstable, derailing training of the model. The other loss weights are tuned to achieve low force estimation error and low residual contributions. Their values are: $\lambda_{\boldsymbol{F}} = 0.5$, $\lambda_{\boldsymbol{F}_r} = 3e-2$, $\lambda_{\Delta dMSE} = 5e-3$, $\lambda_{\partial N_\gamma/\partial \beta} = 0.2$.

Although not explicitly expressed in Eq. 13, averaging is performed over all samples in the batch for $\mathcal{L}^{\chi}_{\text{ReLU}}$, $\mathcal{L}^{\partial N_\gamma/\partial\beta}_{MAE}$, and $\mathcal{L}^{\Delta\delta}_{MSE}$. The $\mathcal{L}^{F}_{MAE}$ loss is only averaged over samples for which the limit constraints for η and β are not violated to avoid the gradient step causing large changes to the network parameters when near a singularity of the FEE model. Similarly, $\mathcal{L}^{F_r}_{MAE}$ is only averaged over samples for which the residual augmented depth of cut $d' > 0$ to avoid penalizing the residual force from compensating for when the FEE model produces a zero-output force. All of the losses are computed using the normalized values, which helps to expedite hyperparameter tuning.

To encourage the model to estimate β so that it is consistent with the assumptions of the FEE model, it is limited to lie within an interval for which the minimum is expected to exist using ReLU-based physical inconsistency regularization, Equation 16. Additionally, $\partial N_\gamma/\partial\beta$ is regularized to be small. It is important to note that only gradients of this loss $\mathcal{L}^{\partial N_\gamma/\partial\beta}_{MAE}$ with respect to β are allowed to flow, i.e., gradients with respect to the other model parameters of $\partial N_\gamma/\partial\beta$ are zero. Combining these losses encourages the network to converge to the critical point β^* that is the global optimum of Eq. 7 in a *soft* fashion, allowing for deviations from this optimum when significant force prediction accuracy improvements can be achieved. This approach is novel and expands the type of physical inconsistency loss functions that can be accounted for when developing a PINN beyond equality and inequality constraints, as outlined by Karpatne et al., 2022.

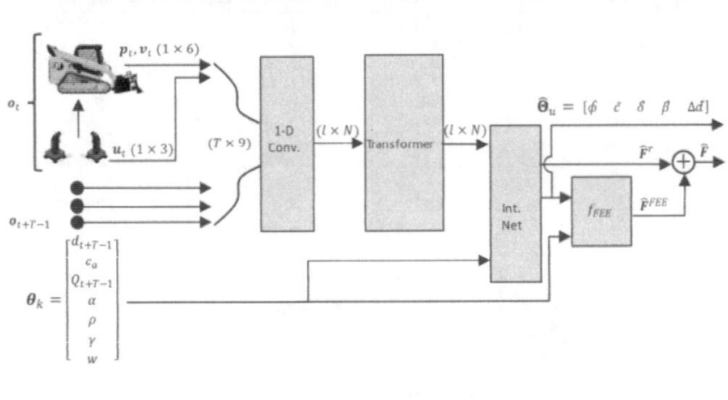

Fig. 4. FEE parameter estimation PINN diagram.

3. Results & Discussion

To evaluate the approach proposed in this effort, the PINN derived parameters are compared with the ground truth parameters used in the simulation.

3.1 Experimental Setup

Two sets of experiments using two different datasets collected from different configurations of the Vortex soil model are performed. The first dataset, D_{FEE}, is collected by configuring the soil model where the heuristics discussed in Section 2.1.1 are disabled such that only the reaction forces generated by the FEE model are exerted on the blade. Additionally, particle generation is disabled, i.e., $Q = 0$, to ensure that the forces generated by particles do not confound the parameter estimation. The goal of making these changes is to collect a dataset for which accurate ground-truth knowledge of the soil parameters is available. This enables evaluation of the accuracy of the parameter estimation approach in an ideal scenario, where the physics that governs the system matches the model that is assumed.

The second dataset, D_{default}, uses the default configuration for the soils, leaving the generation of particles and all of the heuristics enabled. In this case, there exists a significant mismatch between the physics governing the system and the assumed model. As discussed previously, while the simulation does not necessarily reflect the real-world physics to some quantifiable accuracy, the system exhibits similar phenomena as observed in real-world earthmoving operations. The D_{default} dataset can therefore be treated as an analog to experimentation on a real-world system.

For all experiments, the ground is assumed to be flat and level, meaning that $\alpha = 0$ and $d = p^b_z$. The mass of the particles in front of the blade is obtained using a soil mass sensor yielding the surcharge force Q. A simple windowed average is used to filter the velocity measurement $\vec{v}_t = \text{avg}([v^c_x, v^b_z]_{t-10:t})$ to reduce oscillations in the velocity-based scaling of the soil-tool parameters. The blade angle is fixed for all experiments at $\rho = 80°$ as is the tool width at $w = 3.164$m. The initial values of ϕ, c, and γ are obtained at the beginning of an episode based on the initial relative density I_d, as it is not supported to obtain the changing parameter values while running the simulation. This is not problematic in practice as the system only moves forward and does not interact with any previously disturbed terrain. The adhesion is fixed for all soil types, c_a=200 Pa, as is the soil-tool friction angle, $\delta = 10°$. Since the blade angle is fixed at $\rho = 80°$, the FEE predicts zero force in the z direction for larger forward velocities. To enable observation of some non-zero forces in the z direction for D_{FEE}, the default soil-tool angle is overridden to $\delta = 15°$. All of these values are collected to form the pseudo-ground-truth FEE parameters $\boldsymbol{\Theta}$.

The D_{FEE} dataset consists of 4,320 sequences collected from 432 separate 10-second-long episodes using the default simulation update frequency of 60 Hz. Data is collected across all 4 default soil types and the relative density of the soil is varied across episodes, $I_d \in [0,100]$, as are target velocity and cut depth commands, $\bar{v}_x \in [0.3,1.0]$m/s, $\bar{d}_z \in [0.05,0.3]$m. The D_{default} dataset contains 5,760 sequences from 576 episodes and is collected similarly. The PINN is implemented in PyTorch, training performed on a laptop equipped with a GPU, and the network loss in Eq. 13 is minimized using the Adam optimizer.

3.2 Results

The model does well predicting the interaction force, achieving an average magnitude error $|\boldsymbol{F} - \widehat{\boldsymbol{F}}|$ of approximately 730N, or 11% of the measured force, on D_{FEE} and 1940N, 13% of measured, on D_{default}. To put this in perspective, normalizing the error by machine limits reveals 2% error on D_{FEE} and 5% error on D_{default}. This is compelling and shows that the network is doing a good job of learning the dynamics of both the simplified and complex system. To evaluate the ability of the model to identify soil parameters, parameter estimates $\widehat{\boldsymbol{\Theta}}$ and estimated forces $\widehat{\boldsymbol{F}}^{FEE}$ are compared to their pseudo-ground-truth counterparts as a function of relative density. Parameter estimates from each episode within the dataset where relative density I_d and soil type are the same are aggregated.

$$\overline{\boldsymbol{\Theta}}^j_i = \left[\widehat{\boldsymbol{\Theta}}^j \in \boldsymbol{D} \text{ s.t. } I_d = i\right] \text{ for } j \in \widehat{\boldsymbol{\Theta}} \quad (17)$$

The mean and variance of the aggregated parameters are then computed at each I_d for each soil type and visually depicted in the upper half of Figs. 5 and 6. This enables direct comparison between individual parameter estimates and their corresponding pseudo-ground-truth.

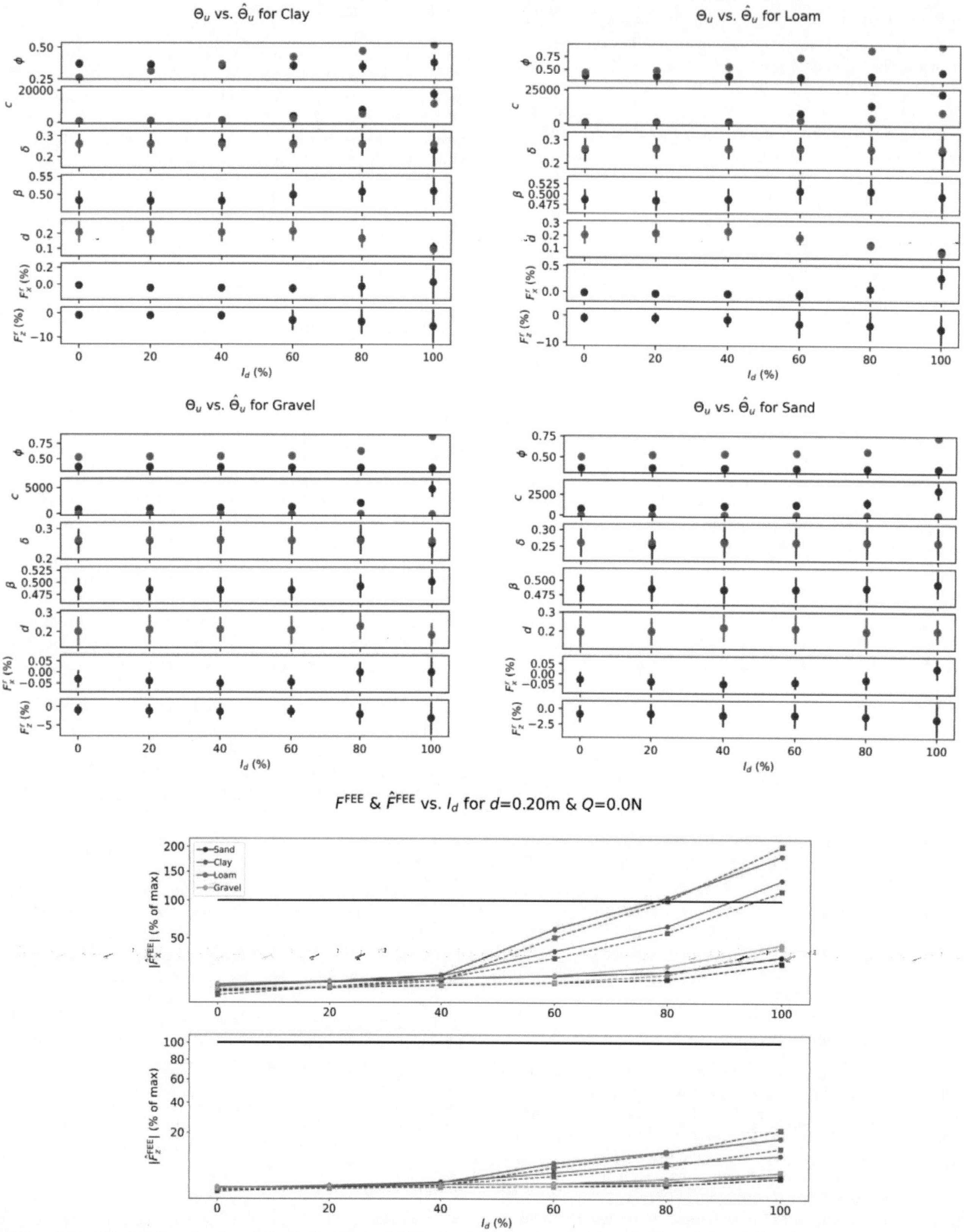

Fig.. 5 D_{FEE} dataset (top) FEE parameters estimated by the model (blue) and the pseudo-ground-truth values obtained from Vortex. (green), (bottom) Soil failure force predicted by the FEE component of the model. Units for all angles are in radians.

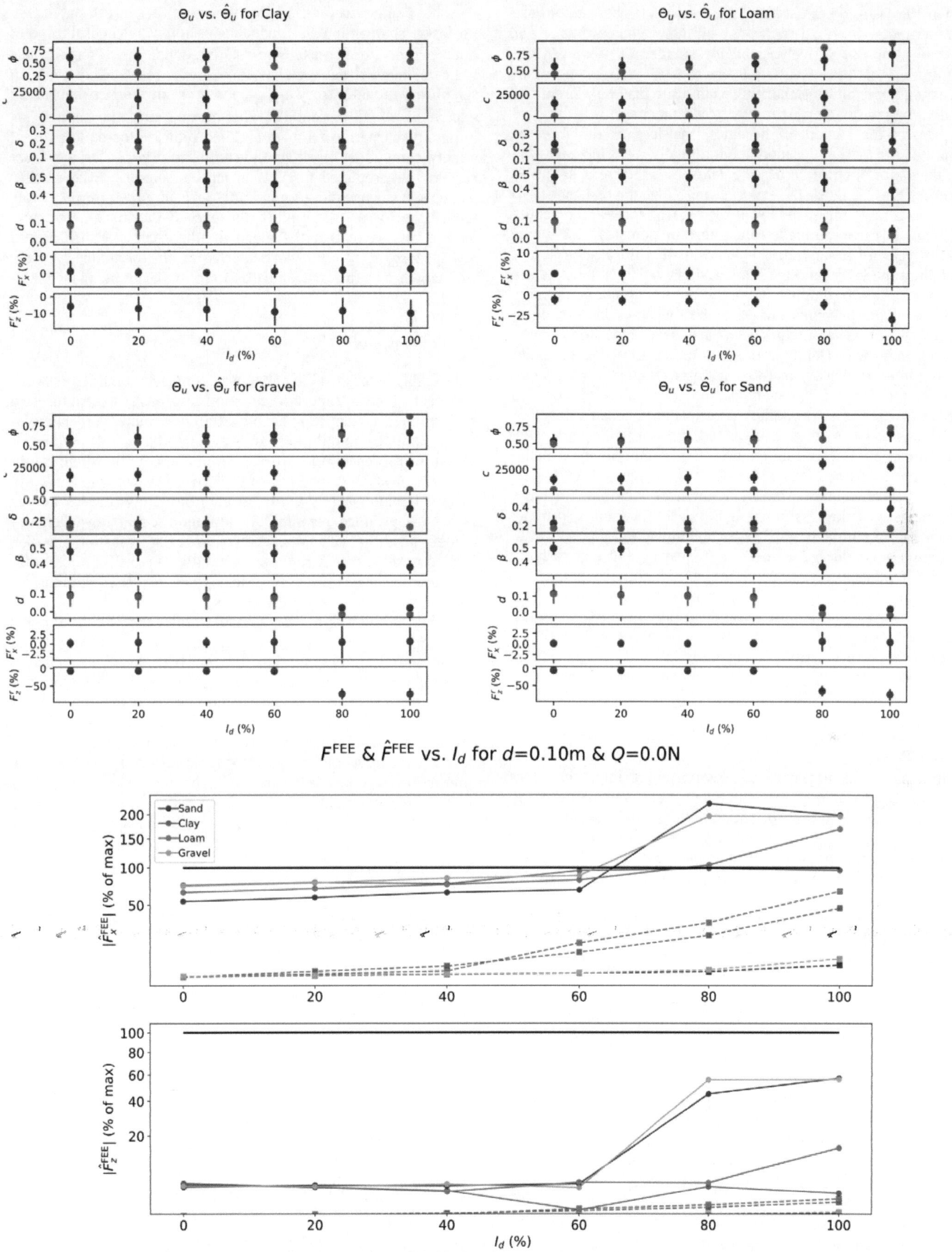

Fig. 6 D_{default} dataset (top) FEE parameters estimated by the model (blue) and the pseudo-ground-truth values obtained from Vortex. (green), (bottom) Soil failure force predicted by the FEE component of the model. Units for all angles are in radians.

For the D_{FEE}, we can see that the network latches on to these true parameter values. It struggles a bit in correctly estimating ϕ, but the error is relatively small for smaller I_d. However, the magnitude of the residual force $\boldsymbol{F}^r$ also increases correspondingly, hinting that this magnitude may be a predictor of degrading parameter estimation performance. This is intuitive because as the larger $\boldsymbol{F}^r$ becomes, the less of the force is accounted for by the FEE model. Additionally, as I_d increases the average depth of cut d across the dataset also decreases. This is because the vehicle is experiencing higher resistance to movement and increased stalling, which forces the blade controller to raise the blade to account for poor tracking of the forward velocity command. This indicates that the assumption of soil failure made by the FEE is increasingly being violated as the soil becomes more compact. When this assumption does not hold, then the model becomes invalid, and it becomes impossible to correctly estimate parameters from the observations alone.

This analysis is useful but does not inform the effect errors in the parameters have on the estimated failure force. Small parameter error is a sufficient but not necessary condition to achieve low error of the failure force. This is because the FEE model, Eqs. 2-10, is under constrained, meaning that there are a multitude of parameter combinations that will produce the same failure force. This presents a significant challenge, particularly as the number of unknown model parameters increases. In large part, this is why the loss function, Eqs. 13-16, contains so many regularization components. These components all constrain the estimation to encourage the network to learn the true underlying parameter values.

To complement the direct parameter error analysis and provide insight regarding the effect these parameters have on the force, the mean of the parameters at each relative density are used to produce predictions of a hypothetical soil failure force for which the depth of cut is fixed, and the surcharge is zero. This means of visualizing the results provides an alternative for evaluating the quality of the parameter estimation and illustrates the effect of the ambiguity present in the FEE model (see Figs. 5 and 6).

In general, this approach enables reasoning about the effect the estimated parameters have on the expected soil failure force, which is the relevant information autonomous soil-property aware earthmoving planners require. For the D_{FEE} dataset, the predicted force obtained from the averaged estimated parameters $\hat{\boldsymbol{F}}^{\boldsymbol{FEE}} = f_{FEE}(\overline{\boldsymbol{\Theta}}|\boldsymbol{Q} = \mathbf{0}, \boldsymbol{d} = \mathbf{0.1})$ agrees well with the predicted force obtained from the ground-truth parameters $\boldsymbol{F}^{\boldsymbol{FEE}} = f_{FEE}(\boldsymbol{\Theta}|\boldsymbol{Q} = \mathbf{0}, \boldsymbol{d} = \mathbf{0.1})$. This indicates that while the parameter estimates do not match exactly the ground-truth, the resulting predicted force is still accurate. This is important from an autonomy perspective, because if the parameter estimates are consistently biased, meaning that the system estimates the same parameter values for the same soil conditions, and the system is able to accurately predict the cutting force, then the model can be used to map soil conditions during earthmoving and utilize this map to plan efficiently.

Moving to the real-world analog dataset D_{default}, the analysis becomes a bit more challenging. The parameter estimates are noisier than in the simplified dataset, and largely only $\hat{\delta}$ matches with the pseudo-ground-truth values. This is expected, due to the increased complexity of the full Vortex model that results in interaction forces only partially derived from the FEE and the pseudo-ground-truth parameters. Broadly, the forces observed in this dataset are larger than those in D_{FEE}, in particular $\boldsymbol{F_z}$ is much larger.

That being said, for the non-cohesive soils, ϕ is estimated well for lower compaction. For the cohesive soils, the general trend of both $\hat{\phi}$ and $\hat{c}$ to increase with relative density is observed. This indicates that the network has uncovered a relationship between the observational history $\boldsymbol{o}_{[t:t+T]}$ and these properties that can be explained by I_d. Essentially, this means that the model has implicitly learned to estimate I_d. The predicted force plots reveal a more compelling result. As the compaction increases, a gradual rise in the predicted force is observed, mirroring the smaller changes of the pseudo-ground-truth derived forces, until $I_d > 60$, where the force rises rapidly and exceed the machine tractive force limits. Note that the mean depth of cut d also drops to nearly zero at $I_d = 80$, particularly for the non-cohesive soil, again indicating that soil failure assumption is not being met.

4. Conclusions

In this work, a novel physics-infused neural network approach to soil property estimation approach is introduced, a simulation is developed to enable attaining ground truth values of soil parameters, and the system is evaluated across two different datasets representing two environments with significantly different physics. The approach is shown to work well when the assumed model aligns with the environment's physics and is shown to produce meaningful parameter estimates when the model and the environmental physics deviate substantially. For both datasets, the parameter estimates enable prediction of interaction forces that are informative for the planning of autonomous earthmoving operations.

While this work provides a novel framework for soil property estimation, a number of future improvements are being considered including: estimation of parameter uncertainty, development of a soil-property mapping system, extending the model to account for changes in the surface profile by varying α, estimation of additional parameters of the FEE model (γ, Q, and c_a), and modifications to the network to enable supervision of the model with only knowledge of kinematic history and without interaction force measurements to enable more rapid integration on existing equipment.

5. Nomenclature

c	Soil cohesion	[Pa]
c_a	Soil-tool adhesion	[Pa]
d	Depth of cut	[m]
F	Cutting force	[N]
N_a	FEE Soil-tool adhesion coeff.	[unitless]
N_c	FEE Soil cohesion coeff.	[unitless]
N_γ	FEE Soil wedge weight coeff.	[unitless]
N_Q	FEE Surcharge coeff.	[unitless]
Q	Surcharge force	[N]
R	Soil friction and normal force	[N]
$\vec{x}$	Longitudinal direction	[unit vector]
W	Soil wedge weight	[N]
w	Tool width	[m]
$\vec{z}$	Vertical direction	[unit vector]
α	Soil surface inclination	[rad]
β	Soil failure angle	[rad]
γ	Soil moist unit weight	[N/m^3]
δ	Soil-tool friction angle	[rad]
ρ	Soil-tool angle	[rad]
σ_n	Normal force	[N]
τ	Shear force	[N]
Φ	Soil internal friction angle	[rad]

6. Acknowledgements

The authors are grateful to Xavier Trudeau-Morin, Marek Teichmann, and Laszlo Kovacs from CM Labs for their support on the Vortex soil simulation and for Xavier's assistance in obtaining force measurements in the simulator.

7. References

Althoefer, K., Tan, C. P., Zweiri, Y. H., and Seneviratne, L. D., 2009. Hybrid soil parameter measurement and estimation scheme for excavation automation. IEEE Transactions on Instrumentation and Measurement, 58(10), 3633-3641.

Azulay, O. and Shapiro, A., 2021. Wheel loader scooping controller using deep reinforcement learning. IEEE Access, 9, 24145-24154.

Backman, S., Lindmark, D., Bodin, K., Servin, M., Mörk, J., and Löfgren, H., 2021. Continuous control of an underground loader using deep reinforcement learning. Machines, 9(10), 216.

Borges, P. V. K., Peynot, T., Liang, S., Arain, B., Wildie, M., Minareci, M. G., Lichman, S., Samvedi, G., Sa, I., Hudson, N., Milford, M., Moghadam, P., and Corke, P. 2022. A survey on terrain traversability analysis for autonomous ground vehicles: methods, sensors, and challenges. Field Robotics, 2(1), 1567-1627.

Bradley, D.A. and Seward, D.W., 1998. The development, control and operation of an autonomous robotic excavator. J. Intelligent and Robotic Sys.: Theory and Applications, 21, 73-97.

CM Labs Simulations, 2016. Theory guide: Vortex software's multibody dynamics engine. https://vortexstudio.atlassian.net/wiki/spaces/VSD21A/pages/2823143940/Vortex+Theory+Guide+Document (accessed 1 Mar. 2023)

Dallas, J., Jain, K., Dong, Z., Sapronov, L., Cole, M. P., Jayakumar, P., and Ersal, T., 2020. Online terrain estimation for autonomous vehicles on deformable terrains. Journal of Terramechanics, 91, 11-22.

Egli, P., Gaschen, D., Kerscher, S., Jud, D., and Hutter, M., 2022. Soil-adaptive excavation using reinforcement learning. IEEE Robotics and Automation Letters, 7(4), 9778-9785.

Fine Software, GEO5 software: table of ultimate friction factors for dissimilar materials. https://www.finesoftware.eu/help/geo5/en/table-of-ultimate-friction-factors-for-dissimilar-materials-01/. (accessed 1 Mar. 2023)

Geotechdata.info, Geotechnical Parameters, https://www.geotechdata.info/parameter, (accessed 1 Jan. 2023).

Ha, Q. P., Nguyen, Q. H., Rye, D. C., and Durrant-Whyte, H. F., 2000, Impedance control of a hydraulically actuated robotic excavator. Automation in Construction, 9, 421-435.

Haeri, A., Tremblay, D., Skonieczny, K., Holz, D., and Teichmann, M., 2020. Efficient numerical methods for accurate modelling of soil cutting operations. Proceeds of the Int. Symposium on Automation and Robotics in Construction, 37, 608-615.

Hayashi, K., Shimada, K., Ishibashi, E., Okamoto, K., and Yonezawa, Y., 2013. Development of D61EXi/PXi-23 Bulldozer with automatic control system of work equipment. Komatsu Tech. Report, 59(166).

Holz, D., Azimi, A., and Teichmann, M., 2013. Real-time simulation of mining and earthmoving operations: a level set-based model for tool-induced terrain deformations. Proceedings of the Int. Symposium on Automation and Robotics in Construction, 30.

Holz, D. 2009. Deformable Terrain for Real-Time Simulation. Thesis, RWTH Aachen University.

Iagnemma, K., Kang, S., Shibly, H., and Dubowsky S., 2004. Online terrain parameter estimation for wheeled mobile robots with application to planetary rovers. IEEE Transactions on Robotics, 20(5), 921-927.

Jackson, T., 2017. Site-by-side: comparing Cat, Deere, Komatsu, and Topcon mast-less dozer GPS systems. Equipment World, www.equipmentworld.com/side-by-side-comparing-cat-deere-komatsu-and-topcon-mast-less-dozer-gps-systems, (accessed: 1 March 2022).

Jud, D., Hottiger, G., Leemann, P., and Hutter, M., 2017. Planning and control for autonomous excavation. IEEE Robotics and Automation Letters, 2(4), 2151-2158.

Karpatne, A., Watkins, W., Read, J., and Kumar, V., 2022. Physics-guided neural networks (PGNN): an application in lake temperature modeling, in: Knowledge Guided Machine Learning, edited by Karpatne, A., Kannan, R., and Kumar, V., Chapman and Hall/CRC, New York.

Lopez-Arreguin, A. J. R., Montenegro, S., and Dilger, E. 2021. Towards in-situ characterization of regolith strength by inverse terramechanics and machine learning: A survey and applications to planetary rovers. Planetary and Space Science, 204, 105271.

Lopez-Arreguin, A. J. R. and Montenegro S., 2021. Machine learning in planetary rovers: A survey of learning versus classical estimation methods in terramechanics for in situ exploration. Journal of Terramechanics, 97, 1-17.

Luengo, O., Singh, S., and Cannon, H., 1998. Modelling and identification of soil-tool interaction in automated excavation. IEEE International Conference on Intelligent Robotic Systems, 1900-906.

Maeda, G.J. and Rye, D.C., 2012. Learning Disturbances in Autonomous Excavation. IEEE Conf. on Intelligent Robots and Autonomous Systems, 2599-2605.

MATHalino: Engineering Mathematics, Geotechnical engineering: unit weights and densities of soils. https://mathalino.com/reviewer/geotechnical-engineering/unit-weights-and-densities-soil, (accessed 1 Mar. 2023)

Mckyes, E., 1989. Lateral earth pressures, in: Developments in Agricultural Engineering, Elsevier Science Publishing Company Inc., New York, pp. 137-171.

Reece, A.R., 1964. The fundamental equation of earth-moving mechanics. Proceedings of the Institution of Mechanical Engineers, 179(6), 16-22.

Singh, S., 1995a. Learning to predict resistive forces during robotic excavation. IEEE Int. Conf. on Robotics and Automation, 2, 2102-2107.

Singh, S. 1995b. Synthesis of tactical plans for robotic excavation. Doctoral thesis, The Robotics Institute, Carnegie Mellon University.

Sotiropoulos, F. E. and Asada, H. H., 2019. A model-free extremum-seeking approach to autonomous excavator control based on output power maximization. IEEE Robotics and Automation Letters, 4(2), 1005-1012.

Tan, C.P., Zweiri, Y.H., Althoefer, K., and Seneviratne, L.D., 2005. Online soil parameter estimation scheme based on Newton-Raphson method for autonomous excavation. IEEE/ASME Tran. on Mechatronics, 10(2), 221-229.

Yamamoto, S., Nagase, H., Itogawa, H., and Kamikawa, N. 1997. Dozing system for a bulldozer. U.S. Patent No. 5950141A.

Zerveas, G., Jayaraman, S., Patel, D., Bhamidipaty, A., and Eickhoff, C., 2021. A transformer-based framework for multivariate time series representation learning. Proceedings of the ACM SIGKDD Conference on Knowledge Discovery & Data Mining, 27, 2114–2124.

2936 | Moisture content impacts on soil load-bearing capacity and its spectral behaviour

Ahmed Elawad Eltayeb Ahmed [a,*], György Pillinger [b], Péter Kiss [b]

[a] *Mechanical Engineering Doctoral School - Hungarian University of Agriculture and Life Sciences (MATE), 2100 Gödöllő, Páter Károly u. 1., Hungary.*

[b] *Institute of Technology, Department of Vehicle Technology - Hungarian University of Agriculture and Life Sciences (MATE), 2100 Gödöllő, Páter Károly u. 1., Hungary.*

* *Corresponding author: wadelawad2004@gmail.com*

ABSTRACT

Soil moisture content plays a crucial role in determining the soil's load-bearing capacity, which is an important mechanical property in off-road engineering since it shows the soil's ability to withstand a load without excessive deformation or failure. The term soil spectral behaviour refers to how soil interacts with the light of different wavelengths, either by absorbing or reflecting it. One of the factors that can affect the soil's spectral behaviour is its moisture content. From an autonomous vehicle's perspective, understanding soil moisture content's impact on spectral reflectance and bearing capacity is critical. This study investigates the influence of soil moisture content on the soil load-bearing capacity (pressure-sinkage relationship) and its spectral behaviour. The Bevameter technique was utilized to determine the soil-bearing capacity, using the moisture analyzer for quantifying the soil moisture content. In addition, a spectrophotometer was employed to analyze the soil colour (spectral behaviour). The results show that the soil moisture content significantly affects both the load-bearing capacity and the spectral reflectance of the tested soil. The findings have significant implications for off-road vehicles since the soil's load-bearing capacity is a critical factor affecting the vehicle's performance over soil terrain. Understanding the effect of moisture content on the load-bearing capacity and the spectral behaviour of soil can help in optimizing a vehicle's design and its operation on various soil types under different moisture conditions.

Keywords
Soil moisture content
Load-bearing capacity
Soil spectral behaviour
Spectral reflectance
Pressure-sinkage relationship
Bevameter technique
Soil colour
Spectrophotometer

1. Introduction

Soil moisture content is a key factor affecting soil load-bearing capacity and spectral behaviour in the off-road engineering field. It is well known that soil moisture content significantly impacts soil mechanical properties. Therefore, it is important to understand the relationship between soil moisture content and both load-bearing capacity and soil spectral behaviour. In this article, we will discuss the effects of soil moisture content on load-bearing capacity and soil spectral behaviour in off-road engineering.

Load-bearing capacity refers to the capability of the soil to support loads without undergoing excessive deformation or failure. The load-bearing capacity of soil is often influenced by way of factors such as soil type, compaction, and moisture content. Many scientists have investigated the impact of moisture content on soil load-bearing capacity. For instance, a study by Aziz et al. (2019) found that the load-bearing capacity of soil decreased significantly as moisture content increased. Similarly, a study by Sharma et al. (2020) concluded that soil strength decreased as moisture content increased, resulting in reduced load-bearing capacity. Also, in a study by Pillinger et al. (2018) The soil's physical composition, wetness (water content), density, and initial compression condition affect the soil deformation caused by the vehicle's passage. The soil becomes more compact as the moisture increases. Furthermore, studies by Ahmed et al. (2021) The change in the moisture content of the soil changes its mechanical properties, thus influencing its load-bearing capacity, Ahmed et al. (2023) The moisture content in the soil influences its bearing capacity, thus influencing its vertical deformation resulting from an excitation load.

Moreover, soil spectral behaviour is another critical aspect that is influenced by soil moisture content. Soil spectral reflectance is a function of its physical and chemical properties, including soil moisture content. According to Singh and Chahal (2020), the spectral reflectance of the soil increases with the increase in the moisture content due to the increase in soil water content. This finding suggests that soil moisture content can influence the spectral behaviour of soils, which has important implications for precision agriculture.

Furthermore, soil moisture content has a significant impact on the mechanical behaviour of soil. The relationship between soil moisture content and mechanical behaviour is complex and depends on various factors such as soil type and compaction. Hu et al. (2018) investigated the effects of soil moisture content on loess's mechanical behaviour and spectral reflectance. They found that the soil's compressive strength decreased with increasing moisture content, while the spectral reflectance increased. This finding suggests that the mechanical behaviour and spectral reflectance of soil are interdependent and influenced by soil moisture content.

In addition to its impact on load-bearing capacity, moisture content also affects the spectral behaviour of soils. Soil spectral behaviour refers to the way soil interacts with electromagnetic radiation across different wavelengths. Moisture content affects soil reflectance, absorbance, and transmittance properties, which can be measured using different spectral techniques.

Several studies have investigated the spectral behaviour of soils under different moisture conditions. For example, a study by Yegizaw et al. (2018) found that the spectral reflectance of soil decreased as moisture content increased. Similarly, a study by Rittenour et al. (2017) concluded that moisture content significantly impacted the spectral behaviour of soils, particularly in the near-infrared (NIR) region. Also, a study by Ahmed et al.

https://doi.org/10.56884/DTUX6735
ISBN 978-1-942112-55-6 (pdf)

(2023) and El Hariri et al. (2023) illustrates that the soil spectral behaviour at different moisture contents increases the moisture content with the sent wavelength fixed, and the reflected wavelength decreases. Pillinger et al. (2023) conducted an investigation to observe the alteration in colour exhibited by three discrete sandy soil types in response to variations in their moisture content. The soils under examination were characterized by comparable grain sizes. The study mentioned that the soil becomes darker with the increase in the moisture content.

In conclusion, the soil moisture content is a crucial factor that affects the load-bearing capacity and soil spectral behaviour in off-road engineering. Understanding the effect of moisture content on the load-bearing capacity and the spectral behaviour of soil can help in optimizing a vehicle's design and its operation on various soil types under different moisture conditions. Knowing the effect of soil's moisture content on both load-bearing capacity and its colour (spectral reflectance) and having it programmed in a remote sensor is a brilliant idea to be implemented in off-road engineering vehicles, especially in off-road autonomous vehicles moving on soil terrains. The sensor will detect the soil load-bearing capacity regarding its spectral reflectance based on the moisture content; thus, based on its value, the driver or the autonomous vehicle computer will decide to go over the soil zone or avoid it (in case of high sinkage).

2. Soil Type and the Measuring Techniques

2.1 Moisture Content Measurement (Moisture Analyzer)

The moisture analyzer (HE53 230 V) was used to measure the soil's moisture content. It works based on the principle of drying. An amount of soil is placed on the tray inside the moisture analyzer and the heating plates inside the moisture analyzer start to dry the sample until reaching unchangeable soil mass (dry mass).

The analyzer calculates the moisture percentage by subtracting the remaining mass from the initial mass and the answer is divided by the initial mass (the moisture content obtained is the wet-based result). The moisture analyzer is shown in Fig.1.

Fig. 1. The moisture analyzer (HE53 230 V) shows the percentage of moisture content.

2.2 Load-bearing Capacity Measurement (Bevameter Test)

The Bevameter measurement (*see* Fig.2.) was made at Hungarian University of Agriculture and Life Sciences (Salman et al. 2020) using the silty loam soil type with a mechanical composition of soil: sand (11.63%), silt (62.18%), and clay (26.18%). Since the soil infield is inhomogeneous and anisotropic, it is complicated to study the soil (Máthé, 2014 & Pillinger, 2016). For simplifying this engineering case (studying the soil's mechanical properties), the soil was sieved to study it as a homogeneous material. Having the soil initially at ambient conditions, holding moisture due to the surrounding humidity, the soil was sieved and prepared in the bin of the Bevameter, with height H_1 = 12 cm and an inner diameter D_1 = 22 cm.

Having the soil filled to the height of H_2 =10 cm was centered under the pressing plate diameter D_2 = 20 cm of the Bevameter as shown in Fig.2. Having the soil filled in the bin with $H_2/D_2 \leq 0.5$ showed that the soil behaved in a bin similarly to when tested in infinite space (Salman et al, 2022).

At the top of the Bevameter plate, there are two sensors; one sensor is used for measuring the vertical displacement (mm) and the other is used for measuring the applied load (Force; Newton is the unit) (Salman et al, 2020). The results were transferred to a computer using a data logger (Spider 8) and read through the CATMAN 4.5 software (force–displacement graph as a function of time) see Fig.3.

The first test was carried out on the soil at ambient conditions; the test was repeated on the same soil sample but with an increase in moisture content. After every measurement, the soil was removed from the bin and placed in a bucket, and water was added to increase the moisture content. The new moisture content was measured by the moisture analyzer. The soil was tested at 6.55%, 13.06%, 16.35%, and 18.93% moisture content. These moisture content values were chosen randomly since it is difficult to set precise moisture values in the soil. What is important in the moisture value is its effect on the physical status of the soil; thus, upon adding water to the soil when moisturing it and recognizing a change in the soil behaviour (even if a slight change), the moisture was measured by the analyzer and the measurement was carried out on the soil sample (the repeated measurements were on the soil with close moisture values).

The mass of the soil sample (at every moisture value) and the final height (after being compressed by the plate; H_3) were measured, knowing that the soil was initially at height H_2. These values were recorded for calculating the soil density before and after applying the load.

Fig. 2. The main structure of the Bevameter equipment

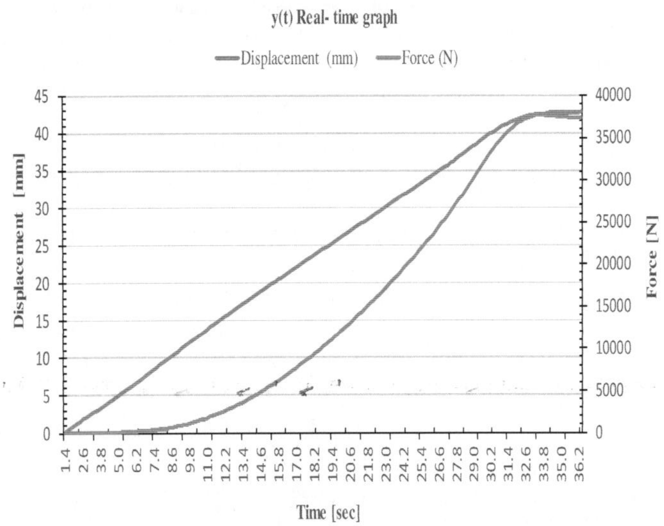

Fig. 3. The temporal profiles of force and displacement were obtained from a sample dataset recorded using the data logger (Spider 8) in conjunction with the CATMAN 4.5 software. Subsequently, these datasets were analyzed and evaluated using Microsoft Excel to visualize their variations over time

2.3 Spectral Behaviour Measurements (Colour Test)

The spectrophotometer CM-700d as in Fig.4. was used to measure the soil's colour (soil spectral behaviour). The spectral behaviour of the silty loam soil was determined by the spectrophotometer at each moisture content (6.55%, 13.06%, 16.35%, and 18.93%) before the Bevameter test.

The spectrophotometer sends electromagnetic waves in the visible range (400–700 nm; light visible band), and the colour of the soil was determined based on the reflected wavelength (spectral reflectance %). Three measurements were taken by the spectrophotometer for the soil at each moisture content, and the average of these was considered as the result.

Fig. 4. The spectrophotometer (CM-700d).

3. Result

The properties and the measurement parameters of the four tested soil samples are shown in Table 1. Each sample had a different moisture content (MC%). The dimensions of the soil container were height H_1 = 12 cm and an inner diameter D_1 = 22 cm. The diameter size of the Bevameter pressure plate, D_2, was 20 cm, and the height of the soil before and after the measurement was H_2 = 10 cm and H_3, respectively. The soil was initially filled to the height of H_2. The soil mass (m) was measured to calculate the density of the soil at different moisture contents.

The initial density of the soil ρ_1 and the density of the soil ρ_2 after being compressed by the Bevameter were calculated using Equation 1. Keeping the H_2/D_2 value fixed (0.5) at different moisture contents, the additional amounts of the moistured soil (remaining after filling the bin to the fixed H_2/D_2) were removed (after stirring the water and the soil since the water occupies space in the soil). The mass of the soil at different moisture contents used for filling the bin to the required height (keeping the H_2/D_2 value equal to 0.5) was recorded.

The density was calculated using the mass and the height.

$$\rho_{1,2} = \frac{m}{\pi . r^2 . H_{1,2}} \quad (1)$$

Table 1

Four samples of silty loam soil were measured at different moisture contents.

Sample No.	H_2/D_2	MC (%)	Mass (kg)	Height (cm)	Density (g/cm3)	Density Ratio (-)
1	0.5	6.55	3.37	H_2= 10, H_3= 7.9	ρ_1 = 1.47, ρ_2 = 1.75	0.27
2	0.5	13.06	3.32	H_2= 10, H_3= 6.5	ρ_1 = 1.44, ρ_2 = 2.09	0.54
3	0.5	16.35	3.11	H_2= 10, H_3= 5.4	ρ_1 = 1.35, ρ_2 = 2.28	0.82
4	0.5	18.93	3.42	H_2= 10, H_3= 5.2	ρ_1 = 1.44, ρ_2 = 2.43	0.92

3.1 Bevameter Results (Pressure–Sinkage Relationship)

The pressure–sinkage curves of the sandy loam soil resulting from the Bevameter test were read through the CATMAN 4.5 software (see Table 1). The results are shown in Fig.5.

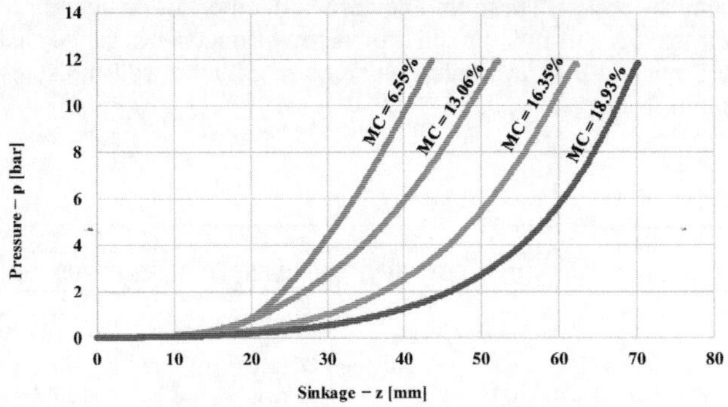

Fig. 5. Silty loam pressure–sinkage relationship at different moisture contents

At each moisture content, the sinkage increases with the applied pressure (applied force divided by the plate area).

At different moisture contents, with the same applied pressure, the resulting sinkage increases with the increase in moisture content. The pressure–sinkage curves follow the same trajectory (hyperbolic) at different moisture contents. At the maximum applied pressure (around 12 bar), it is recognized that the highest sinkage is at the highest moisture content and that the sinkage increases with the increase in moisture at the fixed pressure.

3.2 Spectrophotometer Results (Spectral Reflectance)

The soil spectral behaviour results obtained using the spectrophotometer are shown in Fig.7. The spectral reflectance of the soil at each light wavelength can be obtained from the curves. The sent light wavelengths for measuring the reflectance of the soil at each moisture content are in the range of 400–700 nm. Starting with the lowest moisture content, with the increase in the value of the sent wavelength, the reflected wavelength increases. This applies to the soil at all tested moisture contents.

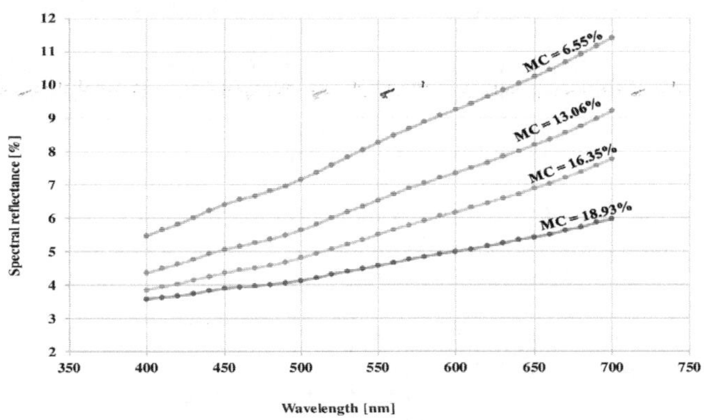

Fig. 6. Silty loam Soil spectral reflectance (spectral behaviour) to the wavelength sent from the spectrophotometer at different moisture contents.

When comparing the soil spectral behaviour at different moisture contents, the obtained results show that with the increase in the moisture content with the sent wavelength fixed, the reflected wavelength decreases. For the tested soil it's obvious that with an increase in the moisture content, the soil becomes darker which means the soil is absorbing more light and reflecting less (see Fig. 7.)

Fig.7. The tested soil with different moisture content.

4. Discussion

4.1 The Influence of Moisture Content on the Soil's Density

The density of the moistured soil after compaction under the same load shows an increase in its value. With the increase in the moisture content, this soil becomes more compressible, hence ending with a lower soil volume (higher sinkage after compaction at higher moisture contents). The same mass divided by the decreasing volume results in an increase in the density.

The compressibility of the soil can be determined by plotting the density ratio as a function of moisture content (see Fig. 8). The density ratio Δρ is calculated using Equation 2. With the increase in the moisture content, the density ratio increases.

$$\Delta\rho = \frac{\rho_2 - \rho_1}{\rho_1} \tag{2}$$

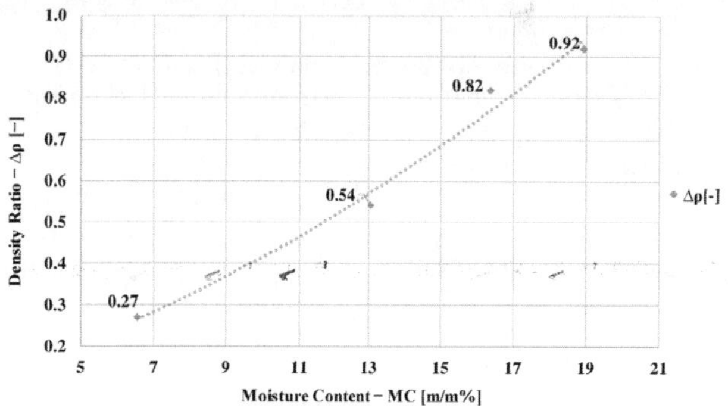

Fig. 8. The change in the density ratio (before and after the Bevameter test) of silty loam soil with moisture.

4.2 The Effect of Moisture Content on the Soil's Maximum Sinkage

Figure 9. shows the sinkage at the maximum applied pressure (12 bar) as a function of moisture content. The results show that with the increase in the moisture content, the sinkage increases, which emphasizes that with the increase in the moisture content, the soil load-bearing capacity decreases under the same compressing load.

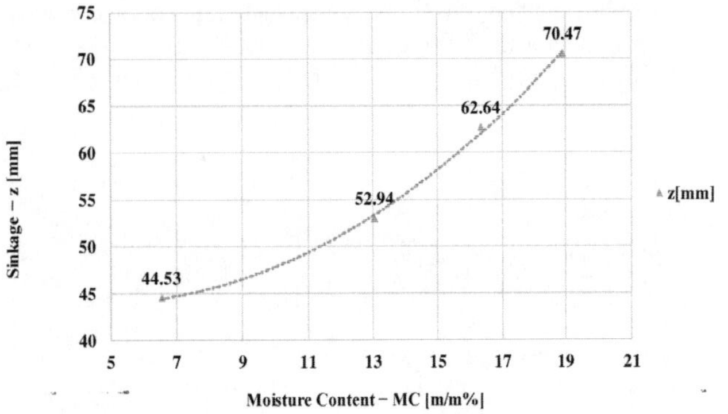

Fig. 9. The change in the sinkage of silty loam soil with moisture content under the maximum applied load (12 bar).

4.3 Spectral Reflectance Affected by Moisture Contents

For confirming the results shown in Fig.6. that the increase in the moisture leads to a decrease in the soil reflectance behaviour, average values of the reflectance results of 400–500nm and 400–700nm were calculated. Equations 3 and 4 were used for calculating the mean of the reflected wavelength at different moisture contents.

$$R_m^{4/5} = \sum_{400}^{500} R_i \quad (3)$$

$$R_m^{4/7} = \sum_{400}^{700} R_i \quad (4)$$

where:

R_i: The reflected wavelength at a specific spectrum, 400–700nm.

$R_m^{4/5}$: The mean for wavelengths in the 400–500nm range.

$R_m^{4/7}$: The mean for wavelengths in the 400–700nm range.

When comparing the Z (10°/D65) value, a spectrophotometer colour-measuring tristimulus parameter (in the spectrophotometer system), the results show that with an increase in the moisture, the Z parameter decreases, thus complying with the reflectance behaviour.

Moreover, in comparing the reflected wavelength at 700 nm at the maximum sent wavelength (spectrum, 700 nm), the results show that there is a decrease in the reflected value (see Table 2) with an increase in moisture.

Figure 8. shows the average (R_m; Equations 3 and 4), Z (10°/D65), and reflected wavelengths at 700 nm (Ref. at 700 nm) values as a function of moisture content.

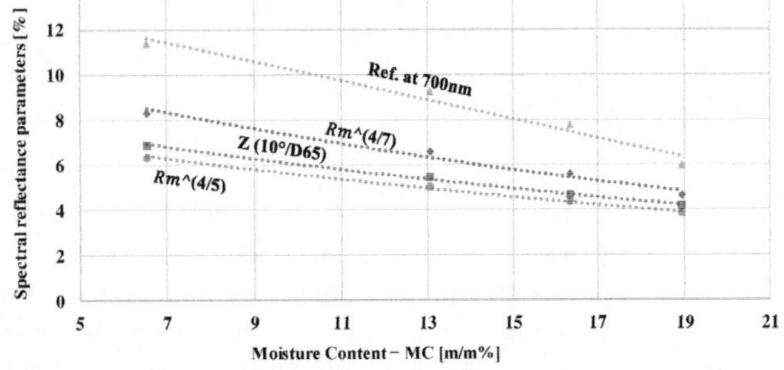

Fig. 10. Spectral reflectance parameter–moisture content relationship.

Table 2 presents the averaged results obtained from the measurements at 400–500 and 400–700 nm, also including the tristimulus values X(10°/D65), Y(10°/D65), and Z(10°/D65), as well as the reflectance at the maximum visible spectrum. The data demonstrate that all the measured values decrease with increasing moisture content. These findings provide important insights into the impact of soil moisture on the spectral properties of soil, which can be useful for developing methods to estimate soil moisture content from spectral data.

Table 2
Important spectral reflectance parameters of silty loam soil.

Sample No.	MC (%)	$R_m^{(4/5)}$	$R_m^{(4/7)}$	X (10°/D65)	Y (10°/D65)	Z (10°/D65)	Ref. at 700 nm
1	6.55	6.34	8.29	11.56	11.71	11.71	16.25
2	13.06	5.01	6.59	9.68	9.77	9.77	13.92
3	16.35	4.33	5.59	7.40	7.41	7.41	10.92
4	18.93	3.86	4.63	5.94	5.95	5.95	8.67

5. Conclusions

The moisture content is a critical factor that impacts the load-bearing capacity of soils and their spectral behaviour. Understanding the relationship between moisture content and soil properties behaviour is important for various applications, such as off-road engineering, geotechnical engineering, agriculture, and environmental monitoring. The findings in this article are highly relevant to the fields of geotechnical engineering, soil science, and civil engineering. By accounting for the impact of moisture content on soil behaviour, we can develop more accurate and effective models for predicting the performance of off-road vehicles in different environmental conditions.

The moisture content in the soil influences its bearing capacity, thus influencing its vertical deformation resulting from an excitation load. At the same moisture content, an increase in the applied pressure leads to an increase in the sinkage of the soil. With the same consolidation load applied to the soil and an increase in the moisture content, the sinkage increases. Furthermore, the influence of the moisture content on the soil-bearing capacity is clear in Fig. 5, 8, and 9; thus, with the increase in the moisture, the load-bearing capacity of the soil decreases.

Additionally, the colour results obtained from the measurements show that with the increase in the moisture content, the wavelength reflectance decreases. Moreover, the influence of the moisture content on the soil spectral reflectance is clear in Fig. 7, 10, and Table 2.

Knowing the load applied from the tractive element on the soil, in addition to having the pressure–sinkage curves of the soil at different moisture contents, the resulting sinkage of the tractive element can be deduced from a curve at a moisture value. The moisture value will be predicted from the colour measured using the spectrophotometer; thus, the response of the terrain (regarding the sinkage) is identified from the soil colour and pressure–sinkage curves (from laboratory work).

Finally, using the results of the moisture content on the soil spectral reflectance and its load-bearing capacity at different moisture contents, the load-bearing capacity and the reflectance can be linked at a moisture value. This idea is beneficial in off-road engineering projects for studying the performance of vehicles on soil terrains, especially in autonomous vehicles. The proposed method could help us predict the load-bearing capacity of the soil, serving in choosing the best design (design of the tractive element;

dimensions of the wheel and the track) needed to operate on a specific soil terrain. The findings in this article provide a foundation for future research in this area, with significant implications for a range of practical applications.

6. Acknowledgements

Project no. 2022-2.1.1-NL-2022-00012 has been implemented with the support provided by the Ministry of Culture and Innovation of Hungary from the National Research, Development, and Innovation Fund, financed under the 2022-2.1.1-NL funding scheme.

7. Declaration of competing interest

The authors declare that they have no known competing financial interests or personal relationships that could have appeared to influence the work reported in this paper.

8. References

Ahmed, A. E. E.; El Hariri, A.; Kiss, P., 2021. Soil strength and load bearing capacity measurement techniques. HAE, 40, pp. 16-27. ISSN 0864-7410.https://doi.org/10.17676/HAE.2021.40.16.

Ahmed, A. E. E.; El Hariri, A.; Kiss, P., 2023. Soil Spectral Behavior Related to Its Load-Bearing Capacity Based on Moisture Content. Applied Sciences 13, no. 6: 3498. https://doi.org/10.3390/app13063498

Aziz, M. A., Lashari, S. K., Khattak, N. S., & Rahman, K. U., 2019. Effect of moisture content on load-bearing capacity of soil. Journal of Engineering, Design and Technology, 17(6), 1237-1252.

Hu, C., Li, Y., Zhang, J., Liu, Y., and Wang, Y., 2018. Effects of soil moisture content on mechanical behaviour and spectral reflectance of loess, Journal of Mountain Science, vol. 15, pp. 11. https://doi.org/2336-2347. 10.1007/s11629-018-5062-9.

El Hariri, A.; Ahmed, A.E.E.; Kiss, P., 2023. Sandy Loam Soil Shear Strength Parameters and Its Colour. Appl. Sci., 13, 3847. https://doi.org/10.3390/app13063847

Máthé, L. Analysis of the Motion of Vehicles Running onto Terrain. Ph.D. Dissertation, Szent Istvan University, Gödöllő, Hungary, 2014.

Pillinger, G.; Kiss, P.; Géczy, A., 2018. Hudoba, Z. Determination of soil density by cone index data. J. Terramech. Vol. 77, Pp.69–74.https://doi.org/10.1016/j.jterra.2018.03.003.

Pillinger, G. Deformation and Damping of Soil under Tire. Ph.D. Dissertation, Szent Istvan University, Gödöllő, Hungary, 2016.

Pillinger, G.; Ahmed, E.E.A, Kornél B., Kiss, P., 2023. Correlations between moisture content and colour spectrum of sandy soils, Journal of Terramechanics, Vol.108, Pp.39-45, ISSN0022-4898, https://doi.org/10.1016/j.jterra.2023.05.002.

Rittenour, T. M., Wemple, B. C., & Shroder Jr, J. F., 2017. The impact of soil moisture on spectral properties of arid region soils. Remote Sensing of Environment, 191, 1-9.

Salman, N.D.; Pillinger, G.; Hanon, M.M.; Kiss, P., 2020. Design and performance evaluation of Bevameter equipment. J. Adv. Mech. Des. Syst. Manuf. 14, AMDSM0084. https://doi.org/10.1299/jamdsm.2020jamdsm0084.

Salman, N.D.; Pillinger, G.; Kiss, P., 2022. Soil behaviour of shallow homogenous upper layer soil. J. Appl. Sci. Eng. 25, 159–164. https://doi.org/10.6180/jase.202202_25(1).0016.

Salman, N.D.; Pillinger, G.; Sitkei, G.; Kiss, P., 2023. Load bearing capacity of finite half space agricultural homogeneous soil, Journal of Terramechanics, Vol.107, Pp. 35-46,ISSN0022-4898.https://doi.org/10.1016/j.jterra.2023.03.001.

Sharma, A., Singh, G., & Kumar, A., 2020. Effect of moisture content on strength characteristics of fine-grained soils. Geomechanics and Geoengineering, 15(4), 269-277.

Singh, B., and Chahal, N., 2020. "Impact of moisture content on soil spectral reflectance and its implications in precision agriculture," Current Science, vol. 118, no. 5, pp. 717-722.

Yegizaw, Y. A., Elaalem, M., Dari, B., & Al-Quraishi, A. M., 2018. Effect of soil moisture content on the spectral reflectance of soils in the visible to near-infrared region. Journal of Applied Remote Sensing, 12(2), 026016.

3280 | Comparison of tweels and pneumatic tires on ltatv military vehicle

Michael Parker [a,*], Clifford Witte [a], Allan Wheeler [a], Sally Shoop [a], Nathan Kamprath [b]

[a] *US Army Corps of Engineers, Engineer Research and Development Center (ERDC), Cold Regions Research and Engineering Laboratory (CRREL), Hanover, New Hampshire, United States*

[b] *US Army CCDC Ground Vehicle Systems Center (GVSC), Warren, MI, United States*

* *Corresponding author: Michael.w.parker@erdc.dren.mil*

ABSTRACT

Tire technology and design can have a significant impact on a vehicle's performance across different surfaces, seasons, and operating environments. Two examples of this are winter tires and airless tires. Winter tires are typically used in the United States in areas where snow and ice are present for a significant portion of the year like in the northern states and Alaska. Similarly, airless tires have recently started to immerge as viable options for skid steers and light utility vehicles like side-by-side vehicles. The military operates in cold regions and austere regions where off-road mobility is paramount and winter tire and airless tire technology could provide a tactical advantage. Airless tires are less susceptible to puncture from difficult terrain conditions, shrap metal, bullets and could be a viable option to run flat tire designs. This research focuses on the performance of a lightweight tactical all-terrain vehicle military vehicle on snow and ice with the standard issue mud and snow nematic tires as compared to studded and standard airless tires.

Keywords
Off-road
Snow
Vehicle
Tires
Airless tires
Military

1. Introduction

The US Army has recently expressed an interest in winter tires for tactical vehicles and assigned a National Stock Number (NSN) number for a High Mobility Multipurpose Wheeled Vehicle (HMMWV) winter tire (Shoop, et al., 2021). Follow-on studies have targeted the Joint Light Tactical Vehicle (JLTV) (Witte, et al., 2023) (continuing in 2024), and will include a new start winter tire research program for the Heavy Tactical Vehicle fleet (HVT) in 2024. This shows a clear commitment to winter tire research but does not address any of the Lightweight Tactical All-Terrain Tactical Vehicle (LTATV) sized vehicles. This research will focus on evaluating airless tire performance against stock pneumatic tires on lightweight tactical vehicles.

1.1 Background

The concept of airless tires has been around since 1938 when J.V. Martin (Fig. 1) invented an airless safety tire (Martin, 1938) followed by designs from the University of Wisconsin-Madison (Fig. 2) (Rutherford, 2008) and Michelin (Golson, 2019). Throughout this development there have been many limitations including inadequate spoke stiffness, spokes packing full of debris, and durability issues. Currently, some designs limit the maximum speed of the airless tires and therefore limits the overall market. Despite these limitations, airless tires, such as Tweels, offer a great advantage over pneumatic tires in combat environments since they are not susceptible to punctures. While a puncture or flat on a military tire is initially mitigated with bead lock and runflat technology they have a limited top speed and duration while operating on the runflat. An airless tire has a great advantage in this respect since it does not go flat because of puncture.

Fig. 1. 1938 airless tire design by J.V. Martin.

Fig. 2. University of Wisconsin-Madison airless tire design.

https://doi.org/10.56884/DPER8822
ISBN 978-1-942112-55-6 (pdf)

Airless tires have gained interest from the US military because of a recent variant specifically designed for the Polaris military razor (MRZR) Alpha. The airless tire design has multiple benefits; it is a tire and a run-flat all in one tire and is resilient to small arms fire with continued use without dependence on repair to keep the vehicle moving at operational speeds. With this technology, there are some restrictions such as the rugged airless tires are currently only designed for specific vehicles (i.e. they are not universal). They are also heavier than a run-flat pneumatic tire of the same size, and recently tested designs successfully clear the spokes to the point that they spray the material (snow, for this testing) onto the occupants.

An airless tire option called a Tweel (Michelin) shows promise and was evaluated for winter performance on a Polaris MRZR Alpha LTAVT during February of 2023. This rugged Tweel was designed to remain operable even if three adjacent rungs or spokes were damaged. Both the standard version and a studded version were tested alongside the standard pneumatic tire with a runflat for motion resistance, drawbar pull, and dynamic maneuvers on packed snow, ice, and fresh snow surfaces. Results of the testing are discussed within this report.

2. Experimental Testing

Winter testing of the three tire configurations was completed in February of 2023 at the Keweenaw Research Center in the Upper Peninsula of Michigan. This is a location well known for highly consistent winter test conditions, something that is critical when evaluating multiple tire configurations.

2.1 Vehicle and Instrumentation

A Polaris MRZR Alpha (Fig. 3) was used for this work because this is the latest in the Polaris MRZR family of vehicles and it is being adopted and, in many cases, replacing the MRZR D4 for the LTATV. It is a purpose-built military vehicle, unlike the D4, that is both longer and wider than the D4 providing more room for the occupants in full tactical gear as well as additional cargo space. Since units are switching to this platform some in the tire industry have taken this as an opportunity to develop an airless tire option for the platform. This was a critical component for this study since airless tires are being evaluated against standard pneumatic tires.

Fig. 3. Polaris MRZR Alpha during winter tire testing, February 2023, Upper Peninsula of Michigan.

The Alpha was outfitted with an Aim Evo 5 data logger capable of logging multiple channels of different types of data all within a compact and waterproof package. Data channels recorded include individual wheel speeds, GPS for position and vehicle speed, inline load cell, brake pedal force, throttle position, steering wheel angle, and six degrees of freedom from the IMU in the Aim (surge, sway, heave, roll, pitch, yaw). Individual wheel speeds were measured using hall effect sensors on the cross drilled holes in the stock rotors and the brake pedal force was measured with a puck style load cell mounted directly to the brake pedal. Zero to 100 percent throttle pedal position and lock to lock steering were measured using string potentiometers directly attached to the throttle pedal and steering shaft respectively. Key parameters for the Polaris Alpha are listed in Table 1.

2.2 Tire Configurations

Three separate tire configurations were evaluated during this winter test event. Figure 4, 5 and 6 show the standard Alpha pneumatic tire (IPT Ultra Cross R Spec 32X10.00R15), a non-studded X Tweel UTV A/T2, and a studded X Tweel UTV A/T2 (both by Michelin) respectively. The tread depth and non-directional tread pattern on the studded and non-studded Tweels were identical with the only difference being the studs. All three sets of tires were run 25 miles on a plowed asphalt and gravel road surface prior to testing.

Table 1
Vehicle Parameters

Parameter	Polaris Alpha
Vehicle Weight (empty, standard tire) (lbs)	3450
Vehicle Weight (as tested, standard tire, driver and passenger) (lbs)	3737
Center of Gravity Height (in)	26.2
Vehicle Width (in)	59.3
Vehicle Length (in)	173.5
Track Length (in)	133.5
Tread Width (both front and rear, Center to Center (in)	51
Standard tire weight with rim and runflat (lbs)	71.25
Non-studded Tweel (lbs)	107.5
Studded Tweel (lbs)	113.75

Each set of tires was weighed after testing to identify the differences in weight between the three options. The non-studded and studded Tweels were 40.5% and 46% heavier than the standard tires respectively. Air pressure for the standard tires was set each day prior to testing with each front set at 15 PSI and each rear set at 20 PSI and all tires were acclimated to the ambient temperature prior to being tested.

Fig. 4. Standard Pneumatic Polaris MRZR Alpha Tire with Runflat.

Fig. 5. Non-studded Tweel on Polaris Alpha.

Fig. 6. Studded Tweel on Polaris Alpha.

2.3 Surface Characterization

Testing was conducted primarily on three different winter surfaces (ice, packed snow, and fresh snow) with hard surface rolling resistance being conducted on asphalt. Each winter surface was characterized throughout the day during each test event to ensure the tire testing was conducted on similar surface conditions. A subset of these tests includes surface friction (surface slipperiness), CTI (snow compaction), Clegg Impact Hammer (surface strength), Yamaha Drop Cone (soft surface strength), and density. Shoop, et al., 2019 describes in more detail how each of these sensors are used and what they specifically measure.

Typically, the CTI was measured on packed snow and Ice, Clegg on packed snow, Yamaha Drop Cone on fresh snow, density on fresh and packed snow, and surface friction on packed snow and ice.

- Fresh snow testing was conducted February 1st on an average of 5 inches of very light and fluffy snow with an average density of 0.06 g/cm3. This was on top of a hard packed snow surface approximately 5 inches thick where the density of the top inch was 0.68 g/cm3.
- Ice testing was conducted February 1st on a large flat ice rink with an average friction coefficient of 0.077. Density is not measured on ice.
- Packed snow testing on February 3rd was conducted on a uniformly hard packed snow surface approximately 5 inches thick where the density of the top inch was 0.73g/cm3. The average friction coefficient was 0.253.

2.4 Test Methods

Four different test methods were conducted for this evaluation. Hard surface rolling resistance (HSRR), Motion Resistance (MR), Draw-bar Pull (DBP), and Circle break out (CBO). Parker, et al., 2019 describes in detail the procedure for the first three tests. CBO is conducted by slowly accelerating the test vehicle around a constant radius circle (fixed steering angle) until the vehicle can no longer maintain the radius and breaks out. This is a measure of the maximum lateral acceleration achievable by the vehicle/tire/surface combination.

3. Results

Figure 7 shows the HSRR for all three tire configurations. As expected, the stock tire has the least resistance because it has the smallest contact patch, followed closely by the non-studded Tweel. The studded Tweel has a 36% higher rolling resistance caused largely by the stud to asphalt interaction.

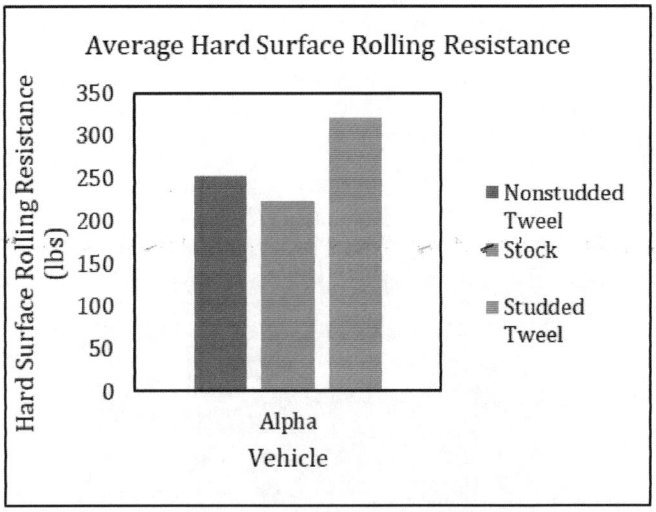

Fig. 7. Average Hard Surface Rolling Resistance for all three tires.

Motion resistance was measured on each surface prior to the draw-bar pull test. Figure 8 shows the results of the motion resistance testing for each tire on the three surfaces. Similar to the HSRR test, the non-studded Tweel appears to have a slightly higher motion resistance value than the stock tire on each surface. The studded Tweel also shows a significantly higher motion resistance value caused by the stud/terrain interaction. This is seen on the fresh snow test as well because its low density allowed the fresh snow to be compacted enough to allow the studs to interact with the packed snow surface below.

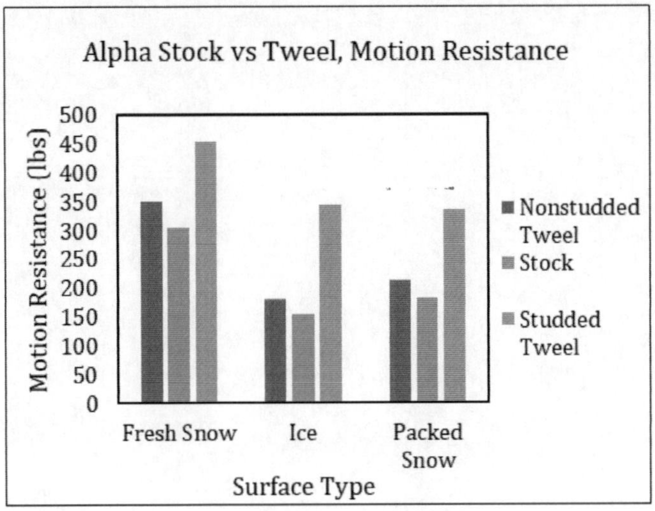

Fig. 8. Motion resistance for each tire on all three surfaces.

Gross traction, calculated by adding motion resistance to draw-bar values and subtracting the hard surface rolling resistance, is shown in Fig. 9. The difference between the standard tire and the Tweel is minimal on all three surfaces, but the studded Tweels provided a noticeable traction advantage ranging from 38% on fresh snow to 137% on ice.

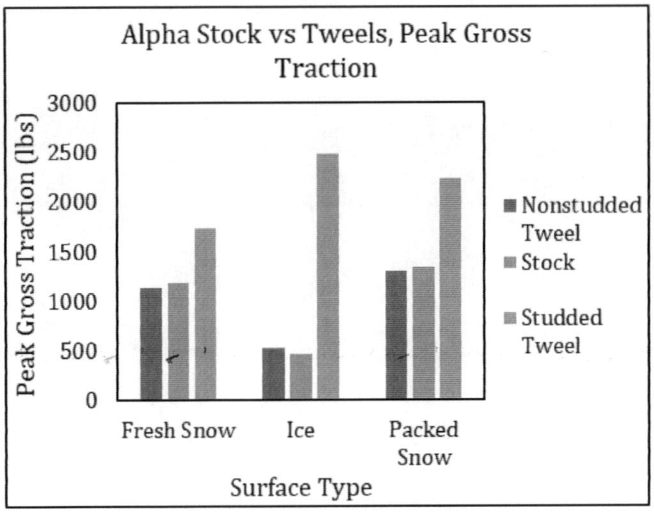

Fig. 9. Gross traction for each tire on the three test surfaces.

Lateral testing was conducted for each tire using the circle breakout test to achieve the peak traction. The vehicle was driven around a 25m diameter circle increasing speed until the vehicle could no longer maintain this constant radius and broke out. Peak lateral traction values for each tire are shown in Fig. 10. The peak lateral traction values show a similar trend as the peak gross traction results where the studded Tweel outperforms the stock tire by 52% on fresh snow and 142% on Ice.

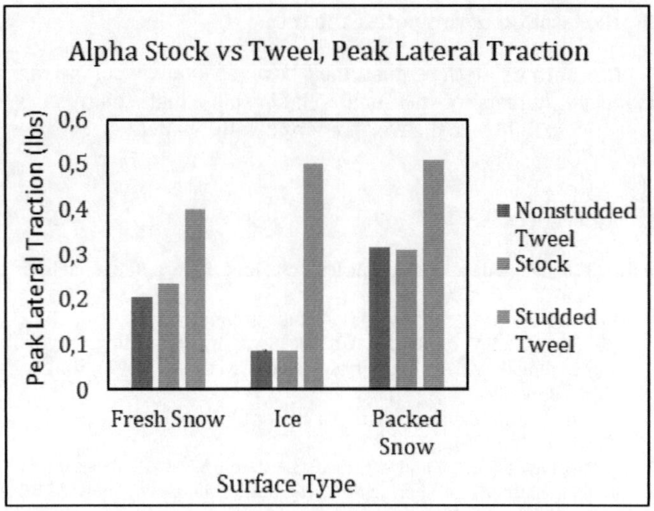

Fig. 10. Peak lateral traction for each tire type on the three different surfaces.

4. Conclusion

Results presented in this report show no real significant improvements for non-studded Tweels over standard tires, whereas the studded variant showed the opposite with noteworthy improvements in all reported results.

The Tweels are significantly heavier than the standard tire with runflat technology, but are able to run at operational speeds even after accepting punctures to the tread.

This is the initial Tweel test on the Alpha and additional improvements may be achievable by altering the tread design and possibly opening up the tread spacing similar to the standard tire tread configuration.

Tweels do a great job of keeping the spokes free of debris to the point where the evacuated material sprays up covering the occupants. Additional testing may need to be don't to minimize this to maintain visibility in a tactical environment.

5. Nomenclature

CBO – Circle Break Out
DBP – Draw-bar pull
GPS – Global Positioning System
HSRR – Hard Surface Rolling resistance
IMU – Inertial Measurement unit
in – Inches
LTATV – Lightweight Tactical All-Terrain Vehicle
MR – Motion Resistance
MRZR – Military Razor
pcf – pounds per cubic foot
PSI – pounds per square inch
UVT – Utility Vehicle

6. Acknowledgements

Funding for this effort was provided by the Cold Regions Methodologies for Autonomy and Terramechanics project under the 6.2 BG202 Task: Autonomous Vehicle/Terrain Interactions work package under Modeling and Simulation for Manned Unmanned Teaming.

7. Declaration of competing interest

The authors declare that they have no known competing financial interests or personal relationships that could have appeared to influence the work reported in this paper.

8. References

Golson, J., Michelin develops new airless tyre. June 6 2019. Retrieved June 28, 2023.

Parker, M.W., Stott, A.R., Bodie, M.O., Frankenstein, S., Shoop, S.A., 2019. Vehicle Mobility on Highly Organic Soils, In Proceedings of the International Society for Terrain-Vehicle Systems (ISTVS), Prague, Czech Republic.

Rutherford, M., 2008. New honeycomb tire is 'bulletproof'". CNET. Retrieved June 28, 2023.

Shoop, S.A., Weider, W.L., Elder, B.C., Beal, S.A., Deeb, E.J., 2019. Assessment of Field methods for Measuring Mechanical Properties of Snow, ERDC Technical Report, TR-19-17.

Shoop, S.A., Witte, C.F, Karwaczynski, S.K., Ellis, C.S., Matthews, E.M, Bishel, S.G., Bomier, B., McCullough, R.T., Parker, M.W., Martin, S.C., Kamprath, N.A., Welling, O.Z., Elder, B.C., 2021. Improving Winter Traction for Vehicles in Northern Operations, In Proceedings of the Ground Vehicle Systems Engineering and Technology Symposium (GVSETS), NDIA, Novi, MI

Witte, C.F., et al., 2023. Modifying Military Tires for Improved Winter Traction", In Proceedings of the Ground Vehicle Systems Engineering and Technology Symposium (GVSETS), NDIA, Novi, MI.

Martin, J.V., 1938. Rubber Spokes Give Bounce to Airless Safety Tires". Popular Science. Retrieved June 28, 2023.

3493 | Numerical modeling of a tire on undrained saturated clay using FEM, ALE, and SPH

Varsha S. Swamy [a,*], Rashna Pandit [b], Alba Yerro [b], Corina Sandu [a], Denise M. Rizzo [c], Katherine Sebeck [c], David Gorsich [c]

[a] *Terramechanics, Multibody, and Vehicle Systems (TMVS) Laboratory Department of Mechanical Engineering, Virginia Tech, Blacksburg, Virginia, United States*

[b] *Department of Civil and Environmental Engineering, Virginia Tech, Blacksburg, Virginia, United States*

[a] *U.S. Army DEVCOM Ground Vehicle Systems Center, Warren, MI 48397*

* *Corresponding author: varshasswamy@vt.edu*

ABSTRACT

Modeling and performance prediction of tires on wet, plastic, cohesive soils is challenging. Unlike dry soils, in wet soils, the undrained shear strength is lesser. The worst scenario is when the soil is fully saturated. This work aims to model highly deformable saturated clay to predict the performance of pneumatic tires on such terrain. The short-term effect on the soil due to single tire pass is of interest. First, the modeling and validation of the soil material model are discussed. Dealing with saturated soils is important to understand the effect of water in the pores. The external loads acting on the soil (i.e., total stresses) can be carried by the soil skeleton (i.e., effective stresses) and/or water (i.e., pore water pressure). Fundamentally, it is the effective stresses that determine soil failure. Material models can be defined using two frameworks: (a) total stress and (b) effective stress. In the total stress analysis commonly found in the literature to address rapid loading, soil and water are modeled as one medium; this is the simplest approach, but it requires different calibrations for different initial soil conditions. In the effective stress analysis, the pore pressure evolution can be tracked by means of hydromechanical formulations with the advantage that a unique calibration can deal with a larger range of initial stress states. The performance of the tire using different numerical techniques, including the Finite Element Method (FEM), Arbitrary Lagrangian Eulerian (ALE), and Smoothed Particle Hydrodynamics (SPH) are compared using total stress and effective stress material models. Finally, it is concluded that the effective stress model captures an accumulation of excess pore water pressure after one tire pass, which may potentially affect the tire performance of later passes. In addition, the FEM model fails to give acceptable solutions for higher normal loads and slip ratios due to excessive deformation. ALE and SPH give more stable solutions for large deformations.

Keywords
Tire soil interaction
Tire clay
Wet soil
Soil modeling and calibration
Pore water pressure

1. Introduction

The ability of ground vehicles to traverse challenging and unprepared terrains has presented an enduring challenge over the years. The application of terramechanics-based modeling has considerably contributed to understanding this complex behavior, facilitating accurate predictions of off-road vehicle mobility and performance. The development of robust terrain models and virtual proving grounds plays a crucial role in these predictions. Although many state-of-the-art works have been done on dry soils, wet soils are still challenging to model using physics-based techniques. It is known that soils comprise three phases; the solid skeleton (solid particles), the pore water, and the pore air. The significant impact of pore water on soil shear strength and its consequential implication on mobility has been well documented experimentally. Generally, the worst-case scenario in terms of soil strength and mobility is for soils in the fully-saturated form (i.e., soils where all the voids are mainly filled with water). In particular, since fine-grained soils can retain substantial water content, and excess pore water pressures take a large time to dissipate because of the small hydraulic conductivity, the effect of the water cannot be ignored. Due to this, with regards to fully-saturated clays, the total vehicle loads (pressure and shear) are endured by the soil skeleton (pressure and shear) and the pore water (pressure only) together. Large water contents lead to non-linear history-dependent material behavior, large deformations, and strain-rate effects, further causing challenges associated with the characterization and modeling of the soil. The capabilities of existing approaches to accurately capture such complex scenario are the research focus in the terramechanics community.

Different physics-based numerical techniques including Finite Element Method (FEM), Coupled Eulerian Lagrangian (CEL), Arbitrary Lagrangian Eulerian (ALE) and Smoothed Particle Hydrodynamics (SPH) have been employed over the years in terramechanics. A short comprehensive review of these techniques and studies on clay/wet clay modeling is discussed in Swamy et al., 2023 and is summarized. Hambleton and Drescher, 2008 modeled clay using Finite Element Method (FEM) and the Drucker-Prager material model to understand the influence of tire width, normal loads, and soil layering technique on mobility. To allow for larger deformations, Jafari, 2015 used CEL to model clayey loam soil. This led to the more accurate prediction of soil deformation along with the study of the effect of normal load, inflation pressure, and velocity of the tire on the clay. With the advancement of mesh-less techniques like Smoothed Particle Hydrodynamics (SPH) in soil modeling, Dhillon, 2013 modeled clayey soil and heavy clay. It was noted from the study that FEM soil gave stiffer solutions and unable to model the shear and penetration effect. FEM soil showed considerably lower rolling resistance than the SPH soil. Further, Sharifi, 2020 used SPH to model clayey loam soil and studied the steering characteristics along with the rolling resistance.

Further, a few studies focus on modeling the effect of moisture on mobility. In FEM, Farhadi et al., 2020, 2019 modeled clay loam soil with water contents of 11.20%, 14.86% and 18.68% using the Drucker-Prager material model with associative flow. The CD triaxial test and direct shear test were conducted to find the total properties of the soil and water mixture and used for the modeling. Interestingly, Maria et al., 2021 modeled firm, soft and wet clayey loams in FEM to identify soil damage for agricultural purposes. The cone penetrometer test (CPT) was employed to obtain material parameters for the Drucker-Prager cap model with empirical correlations. El-Sayegh,

https://doi.org/10.56884/TWIL1196 — DISTRIBUTION STATEMENT A. Approved for public release; distribution is unlimited. OPSEC #7710
ISBN 978-1-942112-55-6 (pdf)

2020 extended the modeling of loamy clay at different moisture contents using SPH. The soil and water are modeled separately and integrated using Darcy's Law. Based on the moisture content (0, 10, 17, and 24%), water particles are pressurized into the soil and calibrated.

As seen, in the terramechanics literature, rapid loading on saturated fine-grained soils has been only approached considering a total strength framework such as the Tresca model. This framework is relatively simple and accounts for the strength of the soil (solid particles and water combined behavior). However, different strengths might be required for different initial stress conditions, and the evolution of pore water pressure is not considered. Fundamentally, from soil mechanics, the effective stress acting on the soil particles leads to failure, and different material models with different complexities and capabilities have been developed in this framework. In this approach, the response of the solid skeleton is evaluated, and the pore pressure evolution is tracked by means of hydro-mechanical formulations. These models are typically more complex than the traditional total stress approach but have the advantage that a unique framework is valid for a range of initial stress states and drainage conditions, which can be crucial to evaluate multi-pass and long-term conditions. Ideally, the same material model can be used to model both dry soil, fully-saturated soil, and short and long-term conditions.

In this work, the traction study of a pneumatic, lugged tire on a fully-saturated low plasticity clayey terrain (CL) is conducted. Two main aspects are discussed. The first focus is on the assessment of the relative abilities of different numerical techniques (FEM, ALE, and SPH) for modeling tire-clay interaction using the traditional total stress material framework. In view of capturing the significant deformation induced by the plastic soil at high slip ratios, it is necessary to employ the most suitable modeling technique to accurately predict the rut profile. The second aspect offers a new perspective and extends the literature on modeling wet clays using effective stress approach. A comparison of capabilities of total stress and effective stress material frameworks in FEM is presented since the effective stress framework is only implementable with FEM. Studies on valid, low deformation based slip ratios are conducted and compared. A novel contribution to the state of the art is the visualization of the pore water pressure for tire-soil interactions in the effective stress framework.

The paper is organized as follows: first, an introduction to the soil mechanics background and terminologies is provided in section 2. Next, the material modeling of the saturated soil is illustrated with descriptions of material calibration for both total stress and effective stress approaches. Later, the tire model and tire-clay model are presented. Results using FEM, ALE, and SPH are compared followed by a discussion on the importance of considering the pore water pressure is highlighted. Finally, the conclusions are summarized.

2. Soil mechanics background and terminologies

This section introduces the three fundamental theories in soil mechanics that are required to understand the numerical modeling of fully saturated soils: (a) the total and effective stress principle, (b) the drainage condition of the saturated soil (i.e., undrained, drained, and fully hydromechanically coupled), and (c) loading stress histories of soil and resultant classification (i.e., normally consolidated and overconsolidated), (d) total stress and effective stress analyses.

The total and effective stress principle proposed by Terzaghi, 1936 is the first theory. It states that the total external load applied on a saturated porous medium (e.g., soil) is taken by the solid skeleton (i.e., effective stress) and pore water (i.e., pore water pressure). The second concept is on the drainage condition of the saturated soil (Das, 2021). Upon loading, as the soil skeleton is compressible, but water is not, the mean stress is first taken by the water in the porous, so-called excess pore water pressure. The excess pore water pressure accumulated due to, e.g., rapid loading, dissipates (or drains) with time. Depending on the permeability of the soil and the loading rate, this process, known as consolidation, is faster or slower. Regarding the case of fine-grained soils like clays, the permeability is very low; hence, when rapid, short-term effects are of concern, it is reasonable to presume that no pore water dissipates in that time frame. This condition is termed as undrained, and the volume of the soil remains constant if water is assumed incompressible. Instead, a drained condition is relevant when a long-term analysis is of concern, in which the excess pore water pressure is completely dissipated (zero). In reality, however, the hydro-mechanical behavior of the material depends on the soil permeability and the loading rate. A fully coupled hydro-mechanical analysis (i.e., both excess pore pressure buildup and dissipation can both happen simultaneously at different rates) is preferred if the process of consolidation is of interest in which the total volume of the soil decreases because the water escapes, and as a result, particles rearrange and densify. This work concentrates on the short-term effect of one tire passing (rapid load); therefore, an undrained analysis is implemented. Note that coupled hydromechanical analysis would be needed to account for multi-pass effects.

The next concept involves comprehending that clays are history-dependent materials and that the previous loading history affects the present behavior. Based on this, clays are classified into normally consolidated (NC) and over consolidated (OC) (Das, 2021). The highest effective pressure experienced by the soil is called the preconsolidation pressure. When the current effective pressure on the soil is the highest pressure the soil has experienced (i.e., equal to the preconsolidation pressure), the soil is NC. Instead, soils are OC if the current stress is lower than the preconsolidation pressure, indicating that the soil has experienced unloading. NC and OC clays pressurized to the same initial effective pressure (loading condition) behave differently. While NC will be more compressible and tend to contract when sheared, OC soils are less compressible and tend to dilate. As a result, for the same initial effective stress, the undrained shear strength of OC is higher than for NC (Mitchell and Soga, 2005).

The last concept is the discussion on total stress and effective stress analysis frameworks. The total stress framework models the soil as the unique single medium of soil particles mixed with water. The total soil strength can be easily obtained and calibrated. Although practical for most applications, these models do not have any information regarding the evolution of pore water pressure in soil. Prospectively, in long term scenarios with the case of multipass, this history along with the soil effective strength may be crucial to track in order to predict the traction for each following tire pass. The effective stress framework considers the pore water and effective stress evolution. When implemented with the most representative drainage condition, a wide range of soil demonstrative of a range of initial stresses and drainage conditions can be modeled using the same effective strength based material model.

3. Clay material model

Implementations using the continuum mechanics framework require a material model to describe the stress-strain behavior of the material. Material models based on the total stress (3.1.1) and effective stress (3.1.2) framework is considered. The material models inputs require different geotechnical laboratory testing to be conducted, as described in section 3.2. Based on the testing data, material parameters are calibrated and validated (3.3)

3.1 Material model

3.1.1 Total stress approach

The total stress material model used in this analysis is the "material soil and foam" on LS DYNA, a simple and robust linearly elastic perfectly plastic model proposed by Krieg (1972) based on isotropic plasticity. The shear modulus and the bulk modulus are required as input. The yield surface can be pressure dependent although in this framework is assumed constant. The coefficients of the quadratic function are derived from the UU triaxial test. The three-dimensional representation of the yield surface can be seen in Fig. 1. A great advantage of this model is that it can be implemented for Lagrangian FEM, ALE, and SPH elements (LSTC, 2017). However, limitations of the model include that it is a simple linear elastic perfectly plastic model, strain hardening or softening cannot be modeled, it does not include post-failure modeling or strain rate effects (Harris, 2006).

3.1.2 Effective stress approach

The material model described based on effective strengths contains properties of the soil skeleton only. CU test results provide the effective strength parameters of the soil and are used to parameterize the "geological cap material model" available in LS DYNA proposed by Sandler and Rubin, 1979 and Simo et al., 1988. It is a history-dependent, isotropic two-invariant cap material model. The shear strength surface in the $\sqrt{J_2}$ - I_1 space is made of three surfaces (Fig. 2): (a) a fixed failure envelope, (b) an elliptical cap, and (c) a tension cutoff. The failure envelope is an exponential equation; however, it can be represented as a linear equation with inputs from the CU test. The elliptical cap is defined using the cap axis ratio (R). The shear and volumetric deformations are coupled on the cap surface. The cap can harden and move through the simulation based on the hardening parameter (κ). The intersection of the cap with the I_1 axis, illustrated as X represents three times the preconsolidation pressure. The initial preconsolidation pressure can be input through the X_0 parameter. Additionally, L corresponds to the value of I_1 where the cap intersects the failure envelope.

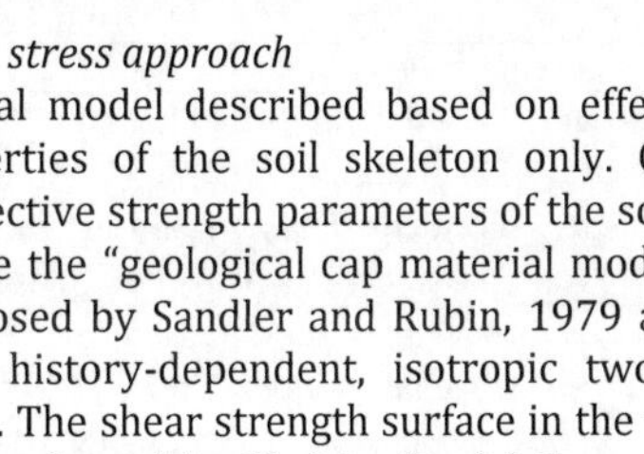

Fig. 1. Yield surface of the soil and foam model (Shin and Kim, 2016)

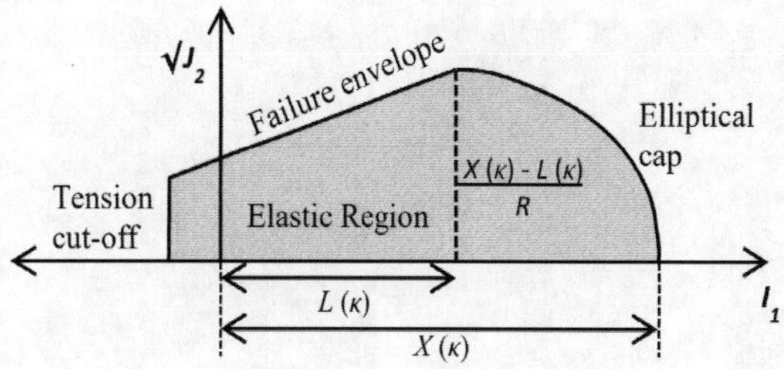

Fig. 2. Yield surface of the geological cap model (modified from LSTC, 2017)

The Koiter's associative flow rule is present over all three surfaces. In order to control unrealistic excess dilatancy, the cap shrinks to the current stress point once the stress point meets the failure envelope. The hardening law, which is a relationship between the hardening parameter (κ) and the plastic volume change, is defined and requires two inputs, including the void fraction of the uncompressed sample (W) and the hardening law exponent (D). The hardening law exponent governs the initial hydrostatic response and is not obtained directly from testing; it requires calibration.

The geological cap model has a few drawbacks. First, it is only available in LS DYNA when combined with a FEM and SPH framework (Wright, 2012). Also, advanced features like strain softening, rate dependency, post-failure modeling cannot be modeled. However, it is a more advanced physics-based model compared to the soil and foam material due to the presence of the cap, flow rule, and hardening. The material model in combination with a hydro-mechanical framework can capture the evolution of effective stresses.

3.2 Geotechnical lab testing

A range of testing in accordance with the ASTM standards is conducted to characterize the clay material. The clay is fully saturated with a natural water content of 24% and has a plastic consistency. It classifies as low plasticity inorganic clay (CL). It has a limit liquid of 41.11% and a plastic limit of 19.57%, both calculated with the Atterberg limits. The specific gravity is 2.66, which is used to find the total density of the soil. Additionally, the dry density of the soil is 1630 kg/m^3. The shear modulus is obtained from the direct simple shear (DSS) test. The DSS is used to determine the stress-strain behavior of cohesive soils under undrained conditions. Unlike the commonly done direct shear test, where the specimen is forced to shear in the middle, the DSS test is uniformly deformed, allowing it to fail at the weakest point. Further, triaxial tests are conducted to determine shear strength parameters. In a triaxial test, a soil specimen is axially loaded till it fails while maintaining the same confinement pressure. It consists of two phases, namely the hydrostatic (confinement) phase and the shearing phase. Based on whether soil consolidation is allowed in the hydrostatic phase and the drainage condition (drained or undrained) of the soil in the shear phase, there are different types of triaxial tests. In particular, for the characterization of saturated fine-grained soils, there are: (a) Unconsolidated Undrained (UU) test to obtain the total strength parameters and (b) Consolidated Undrained (CU) test to obtain the effective strength parameters. The loading rates are in par with ASTM recommendations. The pore water pressure is measured in the case of the CU test, making it possible to obtain effective strength parameters. Each test was performed at four different confinement pressures (34.47 kPa, 68.95 kPa,

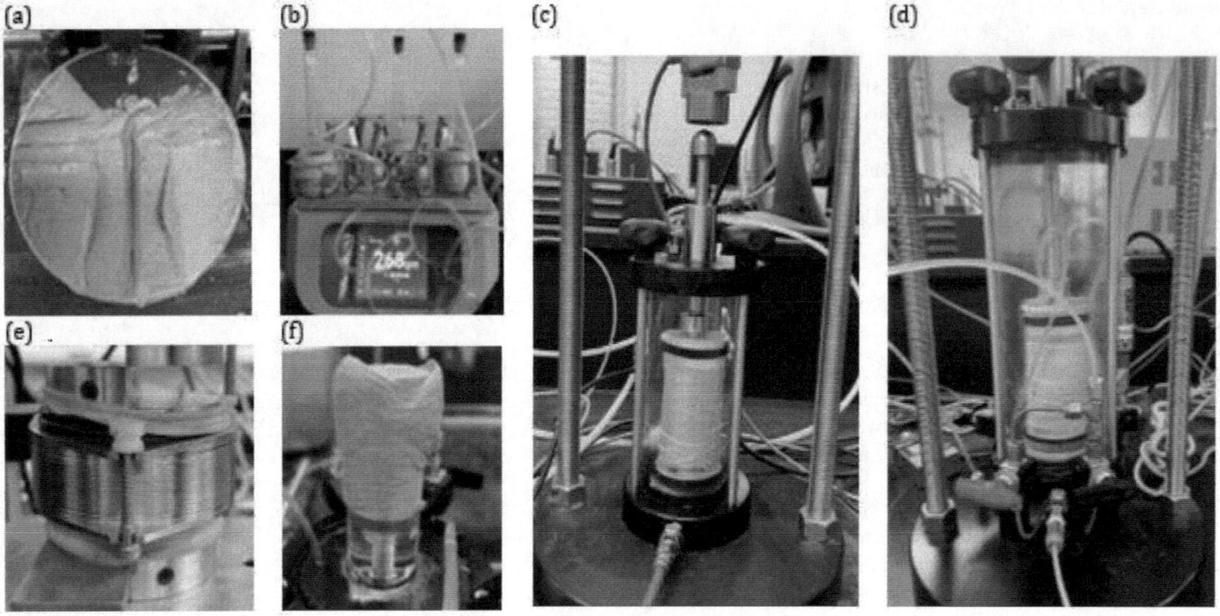

Fig. 3. Geotechnical laboratory experiments: (a) Atterberg limit test, (b) Specific Gravity test, (c) UU triaxial test, (d) CU triaxial test, (e) Direct simple shear test, (f) Clay specimen after failure in UU test

Table 1
Summary of geotechnical test results for saturated clays

Sl. No.	Test	Description	Outputs
1.	Atterberg limit test (ASTM D4318)	Conducted to experimentally determine soil classification and the plasticity characteristics of the soil. It defines the moisture content when the soil changes its state from plastic to liquid.	Liquid Limit - 41.11% Plastic Limit -19.57%, Plasticity index = 21.54 % Classification - CL - Low plasticity inorganic clay
2.	Specific gravity test (ASTM D854 – Method A)	Conducted to experimentally determine the specific gravity of the soil	Specific gravity = 2.66 Dry density = 1630 kg/m3 Natural density = 2167 kg/m3
3.	Direct simple shear test (ASTM D6528) 5%/hr	Conducted to experimentally determine the consolidated undrained shear strength of the soil.	Shear Modulus = G_{50} = 1.39 MPa at 5 psi normal loading
4.	Triaxial Unconsolidated Undrained (UU) test (ASTM D2850) at 1%/min	Conducted to experimentally determine the total yield surface of the soil. Consolidation is not allowed in the hydrostatic phase, and shear load is applied rapidly without volume change permitted.	Undrained shear strength = Su = $\frac{\sigma_1 - \sigma_3}{2}$ = 26.68 kPa
5.	Triaxial Consolidated Undrained (CU) test (ASTM D4767) at 0.1%/hr	Conducted to experimentally determine the effective yield surface of the soil. Consolidation is allowed in the hydrostatic phase, and shear load is applied rapidly without volume change permitted.	Cohesion, c' = 13.78 kPa Friction angle, φ' = 13.87°

137.895 kPa, and 241.32 kPa), with the lowest confinement available in the apparatus being 34.47 kPa. The results of the UU test shows the soil strength to be independent of the confining pressure (φ = 0) with an undrained shear strength of 26.68 kPa. The effective strength obtained from the CU triaxial test however does have a friction angle of 13.87 degrees and cohesion of 13.78 kPa. Figure 3 shows a representation of the laboratory tests performed at the geotechnical lab at Virginia Tech, and Table 1 summarizes the tests conducted and the parameters obtained.

3.3 Material model calibration and validation

3.3.1 Total stress approach

All parameters are obtained from the geotechnical testing as indicated in Table 2. To validate the material model, a strain-based UU triaxial test using FEM is simulated and compared to experimental data. The UU test is modeled in two phases. In the first hydrostatic phase, hydrostatic pressure or confining pressure (same as what is applied in the laboratory) is loaded, and the sample is only allowed to deform axially. Next, the top nodes are axially displaced down to model the shear phase. The principal stresses ($\sigma_1 > \sigma_2 > \sigma_3$) resulting from this displacement boundary condition are recorded for the center-most element. In a triaxial test, $\sigma_2 = \sigma_3$, and hence the deviatoric stress, q, is defined to be ($\sigma_1 - \sigma_3$). The deviatoric stress is plotted as a function of the axial strain for different confining pressures and compared to the laboratory test in Fig. 4. It is seen that the material model is able to capture the peak behavior of the clay. However, as expected the softening behavior at higher strains is not captured.

Table 2
Material parameter inputs for total strength based "material soil and foam" model

Parameter	Description	Value	Details
RO (kg/m³)	Density	2660	Obtained from Specific Gravity test
BULK (MPa)	Bulk Modulus	3.014	Obtained from direct simple shear test (G50)
G (MPa)	Shear Modulus	1.39	Poisson's ratio
A0 (MPa)	Yield surface parameters in J2 – P space	9.43e-04	Obtained from the UU test
A1		0	
A2		0	
PC (MPa)	Tension Cut Off	0.03	Assumed based on literature

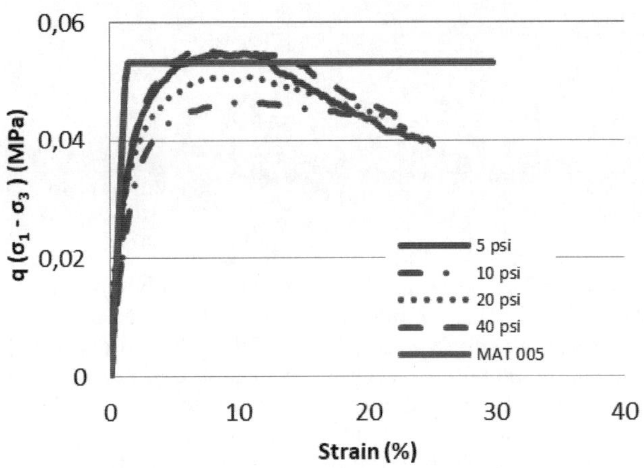

Fig. 4. Comparison of experimental and simulation results for UU test using total stress based material model

3.3.2 Effective stress approach

The input parameters and their values are given in Table 3. A sensitivity analysis is conducted to understand the effect of different input parameters. It was seen that when the soil is normally consolidated, the hardening law exponent (D) had the most impact on the result by controlling the amount of volumetric deformation. Further, the cap ratio, R, also had an influence on the result in terms of how fast the peak stress is achieved. Both these parameters need to be calibrated to obtain a good model performance. Similar observations were noted by Majzoobi and Jannesari, 2015.

The soil parameters are calibrated using the CU test data. Two sets of experimental data, one of OC soil (34.47 kPa confinement) and one of NC soil (241.32 kPa confinement) were considered for reference. Such tests were modeled in FEM, and the numerical results were compared with the lab data in terms of total stress and effective stress paths (p vs. q and p vs. q) for the center most element in the model (Fig. 5). Note that the pore water drainage given to the FEM model was consistent with the conditions during the CU testing. In the hydrostatic phase, the model is 1-D consolidated by a drained analysis (drainage allowed at the top surface of the soil sample). Only axial deformations are allowed in. During the shear phase, the model is

Table 3
Material parameter inputs for effective strength based "geological cap" model

Parameter	Description	Value	Details
RO (kg/m³)	Density	1630	Obtained from dry density test
BULK (MPa)	Bulk Modulus	3.01	Obtained from direct simple shear test (G50) and Poisson's ratio
G (MPa)	Shear Modulus	1.39	
ALPHA (MPa)	Yield surface parameters	0.0145	Obtained from the peak shear strengths of the CU test
THETA		0.1078	
R	Cap surface axis ratio	8	Calibration required
D	Hardening law exponent	0.15	Calibration required
W	Hardening law coefficient or Void ratio	0.65	Obtained from consolidation phase of CU test
X0 (MPa)	Hardening law exponent	0.12	Obtained from CU test
TOFF (MPa)	Tension Cut Off	0.03	Assumed based on literature

switched to the undrained regime, and all the soil boundaries are made impervious to water flow. The upper nodes are displaced down axially, similar to the UU test. After the calibration was performed, the model was validated by running the other set of tests with different initial confining pressures (68.95 kPa and 137.9 kPa). These validation results are also presented in Fig. 5 with the experimental data and calibration dataset. The strain-softening effects seen by the soil at lower confinements cannot be captured as expected. Also, negative excess pore water pressure (when the effective stress path curves to the right), seen in the OC clay case at 34.47 kPa confinement, cannot be replicated. Despite these differences, the numerical model captures the overall behavior reasonably well.

Further, as another round of validation, the UU test is numerically recreated using the effective stress model. Consolidation is not allowed in the first hydrostatic stage, and the boundary is impervious throughout the calculation. The results are provided in Fig. 6, which show that the "geological cap material model" calibrated using CU tests predicts the undrained shear strength predicted by the UU laboratory tests pretty well. The soil strength is slightly under predicted, and hence conservative. An important note is that in the case of the total stress model, the undrained shear strength is given as an input to the model. In the effective stress model, the undrained strength is predicted by the model. It is hence more physics-based.

4. FEM Tire model

The modeling of a FEM tire is challenging due to the complex structure and non-linear rubber material. The FEM tire model used for this study is a simplified tire model. The tire to be modeled is a lugged, bias-ply tire named BKT TR 319 29X12.50-15. The outer layers are modeled with Mooney-Rivlin material model with data obtained from El-Sayegh, 2020, for the tread compound.

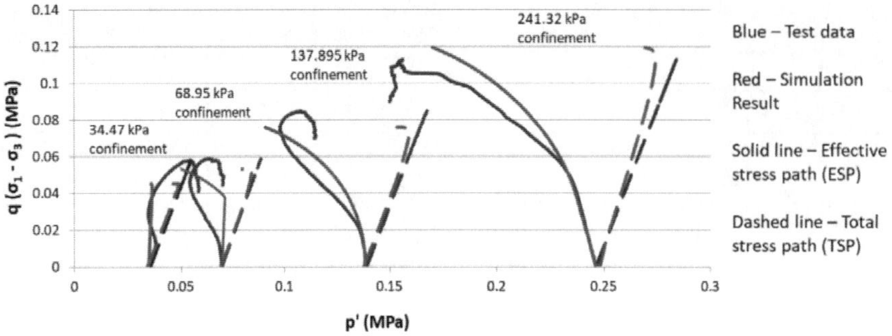

Fig. 5. Comparison of experimental and simulation results for CU test using effective stress based material model.

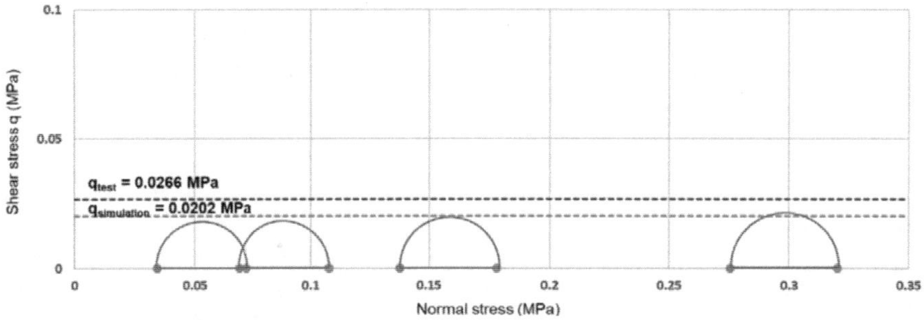

Fig. 6. Comparison of experimental and simulation results for UU test using effective stress-based material model

Fig. 7. FEM tire model

The constant stress element is employed to ensure stability at higher deformations. Hourglass energies are monitored to be within 10% of the internal energy as recommended by the LS DYNA manual to ensure over-stiff solutions are not obtained. Hourglass control 5 in LS DYNA, the Flanagan-Belytschko stiffness form with exact volume integration for solid elements, is seen as the most effective for the tire. The rim and carcass are made rigid as an initial attempt. Figure 7 represents the tire model in FEM used for this study.

5. Saturated Clay – Tire model

The FEM tire is made to roll on the validated soil model. The first focus is on the comparison of capabilities of different numerical techniques i.e. FEM, ALE, and SPH for representing the saturated clay. The total stress framework is employed with the 'soil and foam' material model. The tire rolling process is done in three phases: (a) gravity loading on the soil, (b) tire placement and normal loading (5000 N for this study) on the soil, (c) tire rotation to achieve a predefined constant slip ratio (10-70% in steps of 20). For all cases, the interaction between the FEM tire and soil is established using penalty-based contact with a friction coefficient of 0.4, as recommended by J. Hui and Zhou Min, 2014. To compare the relative performances of the different numerical techniques for mobility purposes, a similar mesh size (25 mm width) is maintained for this study. The stresses in the soil and the plastic rut deformation of the soil are then investigated.

The second focus deals with the comparison of the results employing the total and effective stress framework for the material model. The effective stress based framework is only available for FEM. The tire is rotated to a low slip ratio of 10% where the deformations in FEM are still valid. The simulation is drained during the gravity initialization and tire placement, but switched to undrained when the tire is rotated.

Fine-tuning the numerical setup for good results is necessary for any simulation and discussed in this paragraph. The FEM of soil is modeled with a constant stress element since higher-order elements showed stability issues at large deformations. As the mesh is sensitive to the hourglass energy, regulation is done with the implementation of a slightly rectangular mesh. Additionally, hourglass control 6, the Belytschko-Bindeman assumed strain co-rotational stiffness form, is found to be the most successful. For the ALE soil, the Multi-Material ALE (MM-ALE) formulation is opted with a layer of air defined above the soil. The FEM tire is interacted with the ALE soil using the constrained Lagrange in solid function. In the results, it is ensured that no leakage of ALE material into the FEM. For the SPH soil, a uniform mesh is recommended due to stability in the characteristic length. SPH "form 1" with renormalization approximation, the default cubic spline kernel function, and the Monaghan artificial viscosity formulation is used. The coefficients of artificial viscosity are set to ensure stability of the simulation. Higher slip ratios require higher coefficients. Lastly, since the explicit dynamics scheme is used for all three numerical types, it is very important to ensure the soil stresses settle at the end of each phase.

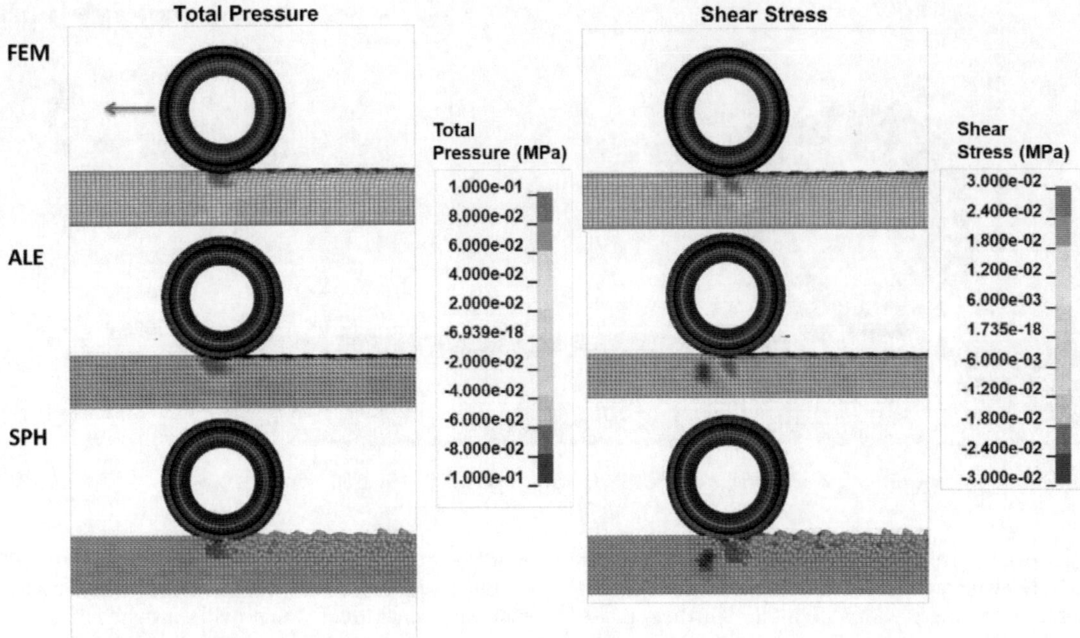

Fig. 8. Total pressure and shear stress distribution of an FEM tire on FEM (total), ALE, and SPH Saturated clay (Normal Load = 5000 N and Slip ratio = 10%)

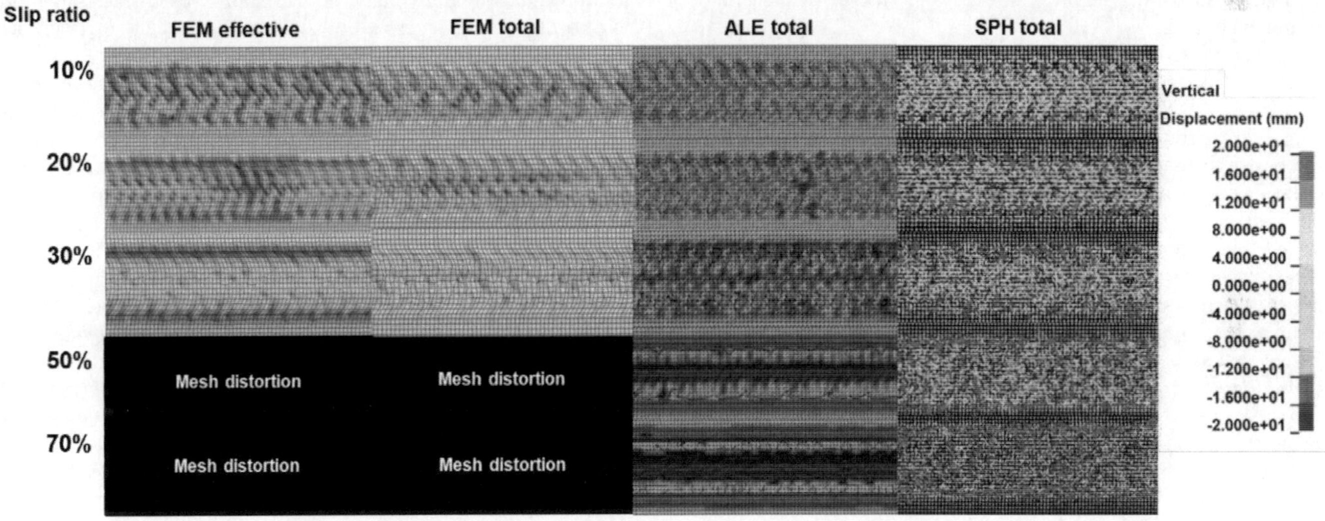

Fig. 9. Rut profile: FEM tire on FEM effective, FEM total, ALE total, and SPH total saturated clay at different slip ratios (Normal Load = 5000 N)

6. Results

The simulation results are discussed in this section in terms of (a) Total stresses and shear stresses in FEM, ALE, and SPH, (b) Displacements and rut profiles for different slip ratios FEM, ALE, and SPH, (c) Pore water pressure using the effective stress framework in FEM.

The total stresses and shear acting on the mid-plane of the saturated soil for all three cases are represented in Fig. 8. Similar total stress bulbs are obtained from all three techniques. The shear stress contours have two zones with opposite signs. In theory, since the tire is driven, the front negative shear stress is negligible and the positive shear stress in the rear of the tire contributes to the traction. In the simulation too, it is seen that the tire is majorly in contact with the negative shear stress region. In general, the resultant forces on the tire are smooth in FEM and SPH. ALE on the other hand displays noisy results. This is due to the constant re-meshing scheme in the algorithm of ALE. Also, at the same mesh density, SPH, due to the particle nature, has a few particles with concentrated stress loads. A more uniform distribution is seen in FEM. Finally, it was seen that in terms of computational time, ALE required the most time and FEM required the least.

Next, modeling different slip ratios is approached using different numerical techniques as shown in Fig. 9. It was seen that at the load of 5000 N, FEM (total) ceases to give reliable results

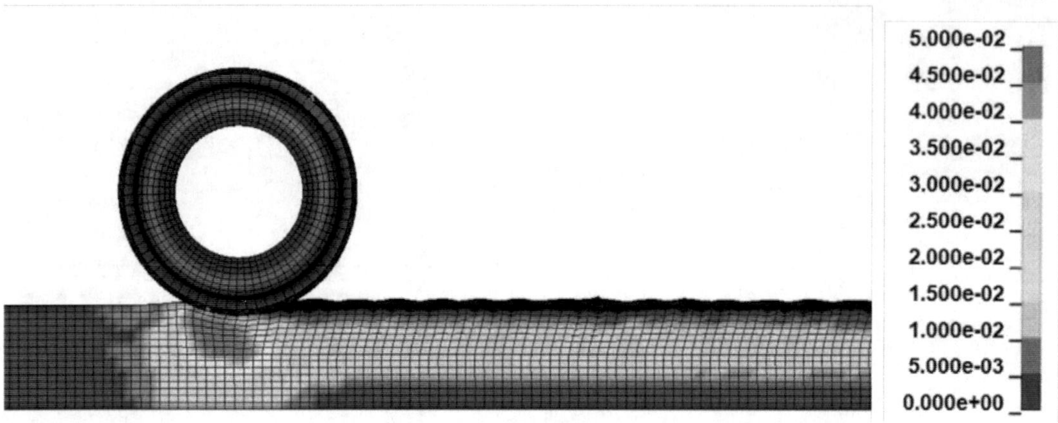

Fig. 10. Excess pore water pressure in the case of saturated clay in FEM (Normal Load = 5000 N and Slip ratio = 10%)

at 30% slip; the results are stiff. When the normal load was increased to 8000 N, even lower slips resulted in mesh distortion errors. This problem is not seen with ALE or SPH. Further, unlike FEM (total), the rut profile due to the tire lugs can be clearly seen in ALE and SPH. At lower slips, a much more pronounced rut profile is left by the tire. High displacement and sinkage are seen at higher slips. In general, SPH shows more displacement compared to ALE.

In the case of the effective stress based model in FEM, the resultant displacements are more than the total stress model due to a more realistic representation of the soil stress-strain behavior. Moreover, the pore water pressure and excess pore water pressure can also be evaluated. The excess pore water pressure is the difference between the pore water pressure and the hydrostatic pore water pressure. When the soil is gravity loaded, hydrostatic pressure exists in the soil. The excess pore water pressure is zero. On the rolling of the tire, the pore pressure increases immediately to bear the pressure loads. The shear on the other hand is only taken by the soil skeleton. It is known that soils develop shear strength only under compression loads. Since the soil skeleton does not take the compressive pressure loads, the shear strength is considerably less. As a result, the top layer of soil subjected to high shearing due to the tire rotation at different slip ratio fails causing high deformations. Since the soil is not elastic, on unloading, there are still volumetric deformations due to the shearing. This volumetric deformation of the soil leads to excess pore water pressure to get accumulated after one pass and represented in Fig. 10. Since clays have high permeability and the dissipation of this pressure takes immense time, the second tire following it will see this terrain which has already failed and has excess pore pressures. As suggested from the result, this pore water pressure may affect the stresses in the soil and hence response in long term multi-pass conditions.

7. Conclusions

This research provides a comprehensive analysis of modeling of physics-based saturated clayey terrains for mobility purposes. A number of tests including the Atterberg limit test, specific gravity test, direct simple shear test, UU triaxial test and CU triaxial test are conducted to replicate the soil behavior. Two material models, one based on total stress and another on effective stresses are parameterized and validated. Traditionally in the literature, total stress based approaches have been explored, which does not provide insight into the fundamental behavior of the soil skeleton and the water. The effective stress based modeling is proposed to be a good solution to model soil for a large range of drainage and initial conditions. For example, the undrained shear strength, which is an input directly given in the case of the total stress model, is predicted as an output in the effective stress model.

A driven tire at different slip ratios is allowed to traverse on both these validated soil materials. The performance of different numerical techniques for the soil including FEM, ALE, and SPH are investigated and compared. Some key finding was that FEM was unable to perform under high shear deformations reliably, causing erroneous and stiff results. ALE was seen to give very noisy results even after leaving ample time to settle. FEM and SPH gave smooth results. Further mesh refinement in SPH may be necessary to obtain more uniform results. Other future investigations to improve the simulations include incorporating the strain rate effect in the soil material model and advancing the tire material model with additional testing. The current soil model is modeled with slower loading rate leading to lower undrained shear strengths of saturated clays, and hence representing the lower bound of measurable strength values. In the future, additional tests performed at higher strain rates, including cyclic tests might be conducted to further complete the characterization of the clay.

Importantly, from the results obtained from this work, the effective stress based material model in FEM demonstrates the importance of considering effective strengths and pore water pressure. More information and conclusions are obtained using the history-dependent effective stress material model. Importantly, soil testing is not required after one pass to retrieve the soil properties again since the material model records the history and reacts based on this and the current load. A broader impact of this work in terramechanics including the modeling of multi-pass for wet, cohesive, plastic soils is to be investigated in the future using this modeling framework.

8. Acknowledgements

This study has been supported by the Automotive Research Center (ARC) under the Cooperative Agreement W56HZV-19-2-

0001 U.S. Army DEVCOM Ground Vehicle Systems Center, as well as by the TMVS Lab and Geotechnical Engineering Lab at Virginia Tech.

Declaration of competing interest

The authors declare that they have no known competing financial interests or personal relationships that could have appeared to influence the work reported in this paper.

9. References

ASTM Committee D-18 on Soil and Rock, 2011. Standard test method for consolidated undrained triaxial compression test for cohesive soils. ASTM International.

ASTM, D.4.0., 2010. Standard test methods for liquid limit, plastic limit, and plasticity index of soils. D4318-10.

ASTM D6528-17, 2017. Standard test method for consolidated undrained direct simple shear testing of fine grain soils.

Das, B. M. (2021). Principles of geotechnical engineering. Cengage learning.

Dhillon, R.S., 2013. Development of truck tire-terrain finite element analysis models. https://doi.org/10.1190/segam2013-0137.1

Farhadi, P., Golmohammadi, A., Malvajerdi, A.S., Shahgholi, G., 2020. Tire and soil effects on power loss: Measurement and comparison with finite element model results. J. Terramechanics 92, 13–22.

Farhadi, P., Golmohammadia, A., Malvajerdib, A.S., Shahgholi, G., 2019. Finite element modeling of the interaction of a treaded tire with clay-loam soil. Comput. Electron. Agric. 162, 793–806.

El-sayegh, Z., 2020. Modeling and Analysis of Truck Tire - Terrain Interaction.

Hambleton, J.P., Drescher, A., 2008. On modeling a rolling wheel in the presence of plastic deformation as a three- or two-dimensional process. J. Terramechanics 51, 846–855. https://doi.org/10.1016/j.ijmecsci.2009.09.024

Harris, D., 2006. Computer Material Models for Soil, Rock, and Concrete using FLAC and DYNA. Dam Safety Technology Development Program. DSO-06-01

Jafari, H., 2015. 3D finite element model to simulate the rolling resistance of a radial ply tire validated by experiments in soil bin testing facility. J. Adv. Veh. Eng. 1, 37–45.

Krieg, R.D., 1972. A simple constitutive description for cellular concrete. Sandia National Laboratories.

LSTC, R., 2017. LS-DYNA Keyword User's Manual Volume II—Material Models.

Majzoobi, G.H. and Jannesari, S., 2015. Determination of the constants of cap model for compaction of three metal powders. Advanced Powder Technology, 26(3), pp.928-934.

Maria, E. di, Reina, G., Ishii, K., Giannoccaro, N.I., 2021. Rolling resistance and sinkage analysis by comparing FEM and experimental data for a grape transporting vehicle. J. Terramechanics 97, 59–70.

Mitchell, J. K., and Soga, K. (2005). Fundamentals of soil behavior. Vol. 3, p. USA. New York: John Wiley & Sons.

Sharifi, M., 2020. Modelling and Validation of Agricultural Tire-Terrain Interaction.

Shin, H. and Kim, J.B., 2016. A refinement of the yield surface of a pressure-dependent and elastic-perfectly plastic constitutive model for a particulate compact by considering specimen barreling in triaxial testing. Powder Technology, 301, pp.1275-1283.

Simo, J.C., Ju, J.W., Pister, K.S. and Taylor, R.L., 1988. Assessment of cap model: consistent return algorithms and rate-dependent extension. Journal of Engineering Mechanics, 114, 191-218.

Standard, A.S.T.M., 2007. Standard Test Method for Unconsolidated-Undrained Triaxial Compression Test on Cohesive Soils.

Standard, A.S.T.M., 2016. D854-14; Standard Test Methods for Specific Gravity of Soil Solids by Water Pycnometer. ASTM International: West Conshohocken, PA, USA.

Sandler, I.S. and Rubin, D., 1979. An algorithm and a modular subroutine for the cap model. International Journal for Numerical and Analytical Methods in Geomechanics, 3, 173-186.

Swamy, V.S., Pandit, R., Yerro-Colom, A., Sandu, C., Rizzo,D., Sebeck, K., Gorsich, D., 2023. Review of modeling and validation techniques for Tire-Deformable soil interactions. J. Terramechanics 109, 73-92.

Terzaghi, K.V., 1936. The shearing resistance of saturated soils and the angle between the planes of shear. In First international conference on soil Mechanics, 1, 54-59.

Wright, A., 2012. Tire / soil interaction modelling within a virtual proving ground environment.

3507 | Determination of the tractive forces of a tractor based on the reconstruction of tire – soil contact surface obtained by means of photogrammetry

Tomasz Śliczniak [a], Jarosław Pytka [b,*], Leszek Gardyński [b], Aleksander Czajka [b], Sławomir Tarkowski [b]

[a] *Akademia Zamojska, Zamość, Poland*

[b] *Faculty of Mechanical Engineering, Lublin University of Technology, Lublin, Poland*

* *Corresponding author: j.pytka@pollub.pl*

ABSTRACT

The aim of the work was to develop a method for determining tractive forces based on a three-dimensional reconstruction of the tire - soil contact surface. The photogrammetry method was used to reconstruct the contact surface between the tractor tire and the soil, further the traction forces were determined using the Bekker model. The value of the contact pressure was determined on the basis of the three-dimensional model of the contact surface. The tow test method was used to measure traction forces as part of verification tests. Field measurements were carried out for different wheel load values, on two soils, sandy and loamy, at different moisture. It was found that the obtained values of the drawbar pull force, determined analytically using photogrammetry, show small discrepancies in relation to the values measured in field tests using a dynamometer. In the case of loamy soil, it was noticed that with increasing moisture, the sinkage of the wheel (rut depth) as well as the horizontal displacement of the soil was smaller. By analyzing the obtained 3D tire - soil interfaces, cracks on the soil surface were identified, resulting from the mechanism of soil tension under vertical load. It was discovered that these cracks in sandy soil were bigger (of greater width) than in clay soil. Their width has been correlated with net traction force generated during test rides. It was found that in the case of sandy soil, the obtained relationships were increasing (the larger the crack, the higher the net traction value), while for clay soil, they were decreasing. Further research may include development of a tire-soil interaction model, taking into account the mechanism of cracks formation as a result of soil tension.

Keywords
Agricultural tractor
Off-road traction
Tire-soil contact area
Tractive forces
Photogrammetry
Soil tensile strength
Field measurement

1. Introduction

Determining the tire footprint on soil has been the main task of terramechanics for years. The methods used can be generally divided into two groups, namely analytical, in which the contact area is calculated on the basis of the assumed shape of the footprint, and experimental, consisting in mapping the actual footprint on the soil using various techniques. In the past, simple methods were used to make a footprint on paper, foil or other flexible material spread on the soil. The result was the reflection of the contact surface, which was photographed or scanned and then planimetered.

Improvements in image processing techniques have enabled more accurate assessment of tire footprint. Gysi et al. (2001) applied lime powder to the soil surface before loading and took pictures of the soil surface after making the tire footprint on the soil surface. The images were analyzed using an image processing technique to measure the contact area. Diserens et al. (2011) obtained a complete image of the tire contact area by means of analyzing tire footprint images using graphical software.

Photogrammetry is a method of reconstructing the three-dimensional surface of an object based on photographic images. Digital image capture and photogrammetric processing includes several strictly defined stages, thanks to which the final product can be obtained as a digital 2D or 3D model. Image coordinates define the position of the object's points in the photograph. By means of the known distance of two points, scale lines or known fixed points, connections are made to basic measurement units.

Stereo photogrammetry is a method of photogrammetry consisting in estimating three-dimensional coordinates of points on the tested object on the basis of measurements from two or more photographic images taken from different perspectives. Then, common points are found in each image. A line of sight can be constructed from the position of the camera to a point on the test object. The intersection of these lines determines the three-dimensional location of the point. It is an accurate non-contact measurement technique that can be used for dynamic characterization and shape reconstruction (Muro, 1992).

Photogrammetry has been frequently used in terramechanics. Botha and Els (2015a, 2015b, 2015c) used digital image correlation techniques to determine the longitudinal and lateral slippage of a tire on the soil. Digital Image Correlation (DIC) is a method by which optical methods are used to track changes in the image. This method can be used to obtain various measurements from particle displacements and velocities to deformation of the object in the image. The basic methodology is that it is to match regions in one image to regions in another image by tracing those regions throughout the image. The cited papers presented three algorithms capable of determining the longitudinal slip and side-slip angle from overlapping images. The first is a simple fast planar method. The second is a more complex algorithm that can extract not only the side-slip angle, but also its rotation velocities and scaled translational velocities. The last uses a calibrated stereo-rig to get all rotations and translational motion in world coordinates. The last two methods are more geared towards off-road terrain applications where the terrain induces motion components other than the typical dominant motion in the yaw plane.

Kenarsari et al. (2017) used short-range photogrammetry to assess the depth, area, and volume of a tractor tire footprint using Agisoft® graphical software to construct 3D models of a tractor tire footprint created in the laboratory conditions. The technique involved creating photogrammetric 3D models of the soil surface before and after the tire was run over. The two 3D models are then subtracted in MATLAB® to measure the soil deformation. Two

https://doi.org/10.56884/XSXK4952
ISBN 978-1-942112-55-6 (pdf)

types of tire ruts were analyzed. The first tire rut was created by lowering a full-size tractor tire onto the soil surface and applying increasing loads. The second tire rut was made by manually rolling the tire over a layer of sand on a laboratory concrete floor, creating a continuous rut. The volume, surface area and depth of the ruts were successfully determined using a close-range photogrammetry technique.

Botha et al (2019) investigated the feasibility of using cameras to measure 3D terrain profiles from which a single rut depth measurement is obtained. Tests were carried out on different vehicles, for different vehicle dynamic maneuvers, on different terrains, including sand, mud, grassland, snow and ice. The results were verified using the traditional ruler method. Real-time rut depth measurements described in this work can also be used to determine vehicle mobility in off-road conditions that can change rapidly under environmental conditions such as rain or snow.

In this paper, the authors set themselves the goal, using the classic photogrammetry method, to obtain a 3D footprint of a tractor tire on two arable soils at different soil moisture contents and wheel loads. On the basis of the obtained tire footprint, traction forces were not only determined using the Beekker method, but also an attempt was made to verify the hypothesis based on the work of Sohne (1952), in which the traction force was correlated with cracks on the soil surface, resulting from the tension of the soil, which may be a mechanism of interaction between soil and a tractor tire with high lugs.

2. Materials and methods

2.1 Soils

The field experiments were carried out on two soils: sandy (medium sand) and clay (dusty clay), at different moisture levels. In order to carry out the tests at different soil moistures, the field tests were divided into two stages. The first stage was completed in autumn 2019. It included tests on sandy and clayey soils with a moisture content of 24%. The second stage of the research was carried out in the spring of 2020 in extremely different conditions - the moisture of the same soils was 4%, which made it possible to compare the measured traction forces. The soils were selected due to different physical and mechanical parameters, which are significantly different for loose and cohesive soils. The measurement sites were located in south-eastern Poland. The study area belongs to the West Roztocze and is located in the southern part of the Lublin Upland. The tests on clay soil were carried out within the area of Hosznia Abramowska in the commune of Goraj, and the tests on sandy soil were carried out within the area of Gorajec Zagroble in the commune of Radecznica.

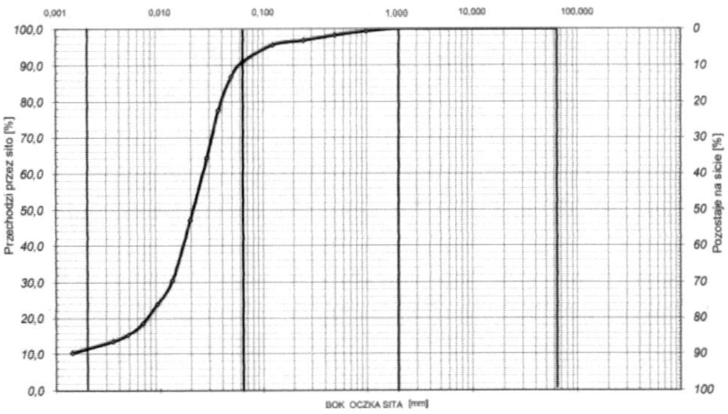

Fig. 1. Granulation of clay soil

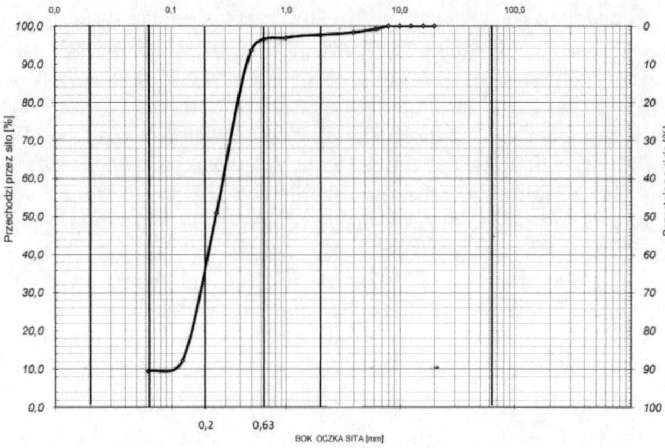

Fig. 2. Granulation of medium sand soil

Figures 1 and 2 show grain size distribution of both tested soils.

2.2 Photogrammetry

In this work, the photogrammetric method was used to obtain a 3D model of the tractor tire footprint, useful for determining the horizontal displacement of the soil, rut depth, length and width of tire-soil contact area, and to reconstruct the tire-soil interface using computer software based on 3D stereo-photogrammetry algorithms. Determination of the horizontal soil deformation was carried out using the AutoCad software. The reconstruction of the rut was carried out in the ReCap Photo software, and the Gom Inspect software was used to analyze the tire sinkage.

Pictures were taken with a standard digital camera due to its easy accessibility and sufficient quality of photographs (Fig. 3). Images of ruts imprinted in the soil by tractor wheels were taken from several different viewing directions (perspectives). Dozens of photos were taken for each test variant.

Fig. 3. Photographing a rut in the soil

In order to reconstruct the tire-soil interface for a tractor, an area was selected where the track left by the tractor tire was complete. The track left by the driven wheels was studied. Then, a section of the trace with a length of approx. 1 m was selected, in which markers were placed on. Markers in the form of beads (balls) with a diameter of about 15 mm served as reference points to facilitate scaling of the photos. A rule was that there should be at least 7 markers, due to the analysis of a complex three-dimensional surface with irregular shapes. In the next stage, photos were taken from different heights and from different directions of view. The photos of the interface were taken using the markers described above and a ruler or square with a millimeter scale, as shown in Figure 4.

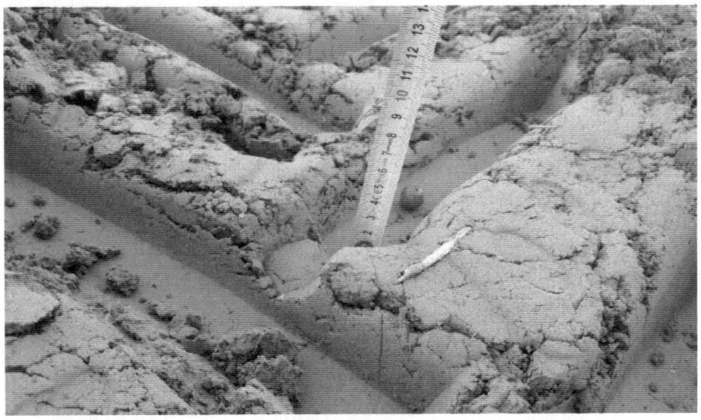

Fig. 4. A typical photograph of the rut with markers and a ruler

2.3 Analysis of images

A set of photographs for each experiment variant was used to obtain a 3D tire-soil interface reconstruction. For this purpose, computer software was used, based on 3D stereo-photogrammetry algorithms and enabling the determination of 3D point maps from many photographs of a given object. In order to obtain a 3D model, several dozen photos of each combination of passes on different soils were imported. Scaling can be done using any markers with known dimensions.

An example of the obtained 3D model views of such photos for clayey soil with a moisture content of 6% is shown in Fig. 5.

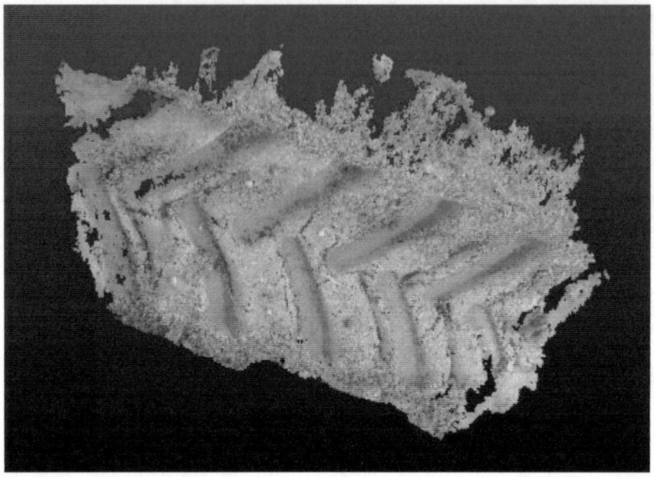

Fig. 5. Example 3D reconstruction of the rut, obtained in the ReCap Photo software

Individual 3D models of the tire imprint in the soil were then imported into a coordinate measuring method class software for scaling and depth analysis. This software allows to create and analyze a triangle mesh. Thanks to inspection tools, measurements of the rut depth (Fig. 6) and the contact area can be performed with a high level of precision.

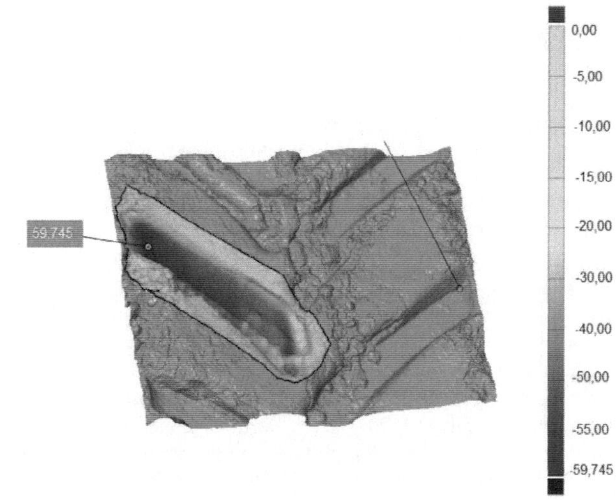

Fig. 6. Determining the rut depth using the Gom Inspect software

Next, the tire – soil contact surface was determined. To do this, it was first necessary to limit the area corresponding to the tire footprint on the ground when the wheel is stationary (for an infinitesimal driving time). It was assumed that the shape of the area is rectangular and the dimensions of the rectangle, length and width, were determined as follows:

- the length of the tire-soil contact area depends on the tire sinkage and can be determined knowing the rut depth and the tire diameter;
- the width of the contact area was determined for individual variants by analyzing the 3D image, the side edges of the area were determined in the place where the tyre's visual impact on the soil ends.

2.4 *Experimental determination of drawbar pull force*

Traction force was determined for 2 agricultural tractors: CASE Farmall 95A and JOHN DEERE 5075E, each with four-wheel drive. An agricultural trailer with a curb weight of 2400 kg was used in the tests. In order to obtain the pull force, the trailer was loaded with loose material (agricultural fertilizer) in bags weighing 600 kg each. Figure 7 shows a typical set up of two vehicles, ready for test runs.

The tests were carried out with two different trailer loads of 2400 kg and 4800 kg, because no horizontal displacement of the ground could be observed during the drive without those loads.

During the measurement, the tractor operator maintained a constant engine speed in the range of 2000 rpm. During a test ride, the traction force was measured using a strain gauge force sensor connected with a digital data recorder.

Fig. 7. A test vehicle set: a tractor with a trailer, linked with a load cell

3. Results

Exemplary results of the analysis of tire-soil interface: contact surface area and rut depth are presented in Tables 1 and 2. They also contain the results of calculations carried out using the Bekker model, their results are the values of traction forces. Then, the results of the calculations were compared with the results of the measurements of the pull force and the rolling resistance force (details of the methodological measurements are included in section 4.1).

Table 1
The results of the tire footprint analysis together with traction forces, CASE Farmall 95A tractor, W = 39.26kN

Soil	*MC* [%]	*Q* [kN]	F_{DP}^{calc} [kN]	F_{DP}^{exp} [kN]
Sand	6	23.54	10.84	11.10
	24	47.08	13.80	14.30
	6	23.54	11.35	11.70
	24	47.08	15.37	15.58
Clay	6	23.54	10.89	10.55
	24	47.08	15.53	14.88
	6	23.54	11.49	11.25
	24	47.08	17.35	16.75

Table 2
The results of the tire footprint analysis together with traction forces, John Deere 5075E tractor, W=32.57kN

Soil	*MC* [%]	*Q* [kN]	F_{DP}^{calc} [kN]	F_{DP}^{exp} [kN]
Sand	6	23.54	10.40	10.49
	24	47.08	13.60	14.05
	6	23.54	9.98	9.95
	24	47.08	14.40	15.08
Clay	6	23.54	9.95	9.70
	24	47.08	14.45	14.94
	6	23.54	11.43	11.07
	24	47.08	14.24	14.76

Complete results are included in the doctoral thesis of the first author (Śliczniak, 2023). The results of both the photogrammetric analysis of the ruts and the results of the traction forces, the traction force and the rolling resistance force, determined on the basis of the contact surface parameters, are included there. It was found, among others, that the results of the traction force determined from the calculations using photogrammetry show a very good correlation with the values of the traction force determined in field tests using a dynamometer measurement. The largest difference of 1.14 kN between the value of the traction force determined in the field tests and the value calculated was found for driving the CASE tractor on a sandy ground with a moisture content of 24% and trailer's load of Q=47.08 kN. The discrepancy between these values was 9.98%.

4. Tensile cracks in soil under wheel loading

One of the possible ways of using the 3D tire footprint on soil is the analysis of the contact surface, including the identification of cracks. This section describes the study of an alternative mechanism of tractor tire – soil interaction using the obtained 3D reconstructions of the tire footprint.

In the case when the slip of the driving wheel is relative small (less than 10%), the mechanism of generating the driving force in the area of tire-soil interaction consists in friction between the front surface of the tread and the soil. In addition, the soil is compressed and tensed between the vertical lugs of the tire tread, which is especially visible on tractor tires. Sohne (1952) had presented this mechanism graphically, noting the horizontal deformation of the soil in the rut (Fig. 7). Supposedly, the magnitude of the horizontal deformation of the soil under the action of the drive wheel protectors can be correlated with the traction force. The deformation of the soil takes place in its half-space, while the visible evidence are the cracks visible on the soil surface in the tire footprint. Therefore, research was undertaken to: 1) identify these cracks, 2) estimate the size of soil deformation based on cracks width, 3) correlate it with the experimentally measured value of the tractor's pull force.

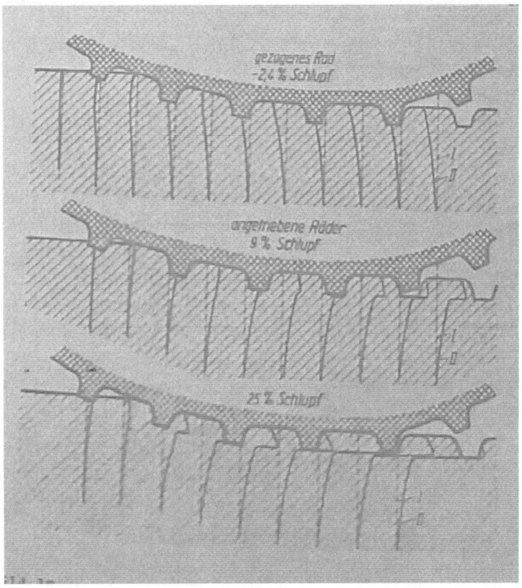

Fig. 7. Soil deformation as an effect of tractor tire loading, during tractors' pass at various wheel slip, after Sohne, 1952

The described phenomenon may take place in the case of relatively low values of slip, of the order of 3-7%, which results with a small horizontal displacement of the soil (almost imperceptible), while the straining of the soil under the horizontal pressure of the tire lugs may cause cracks visible on the surface of the tire footprint, especially in the places between the lugs (in the tread cavities, which are placed on the "hills" of the tire footprint).

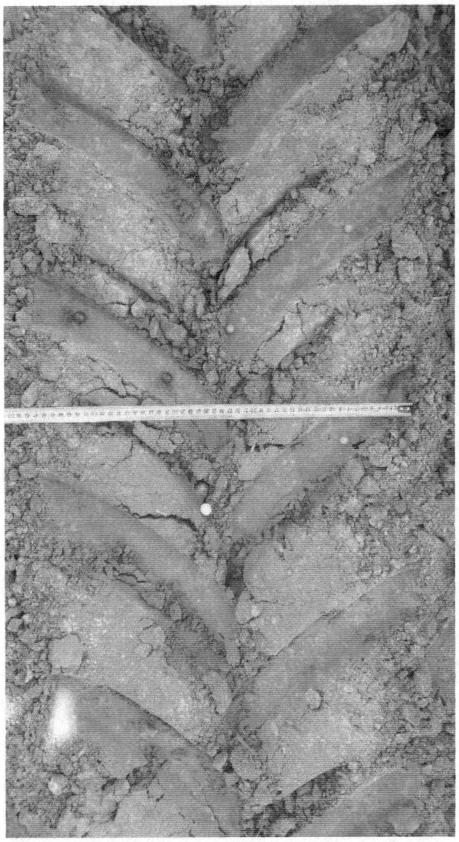

Fig. 8. Sample tire footprint with soil cracks, clay soil, MC=6%

Fig. 9. Sample tire footprint with soil cracks: upper photo, sand, MC=24%

Selected photos of tire footprints were analyzed in a CAD software, the photos were scaled and the cracks in the soil were measured (Figs. 8 and 9). Selected values of horizontal deformation measurements for all variants are listed in Table 3. They were used to determine the traction force using the analytical method.

Table 3
A comparison of drawbar pull force, measured and calculated from soil crack

Soil	*MC* [%]	Q_W [kN]	Crack [mm]	F_{DP}^{calc} [kN]	F_{DP}^{exp}[kN]
Sand	6	5.88	1.88	10.84	11.10
	24	11.77	2.41	13.80	14.30
Clay	6	5.88	3.32	11.35	11.70
	24	11.77	4.03	15.37	15.58

4.1 Relationship between drawbar pull and soil cracks

The obtained results: the measured pull force and the calculated pull force based on the cracks' width from the Janosi-Hanamoto model were compiled for individual variants of soil type and moisture, and tractor load. Results are included in Table 3 and in Figs. 10 and 11.

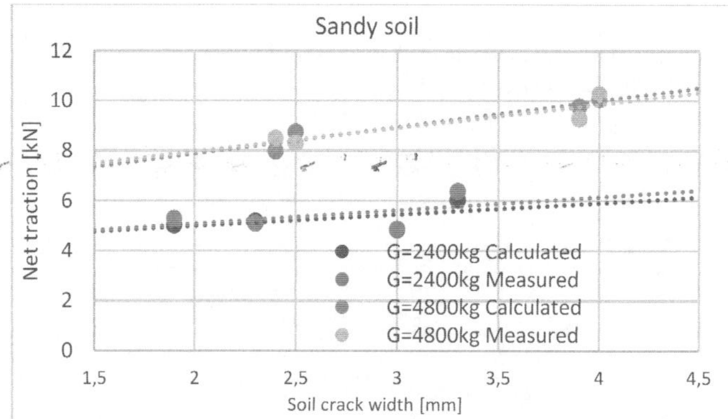

Fig. 10. Relationships between soil cracks width and calculated as well as measured net traction for sandy soil

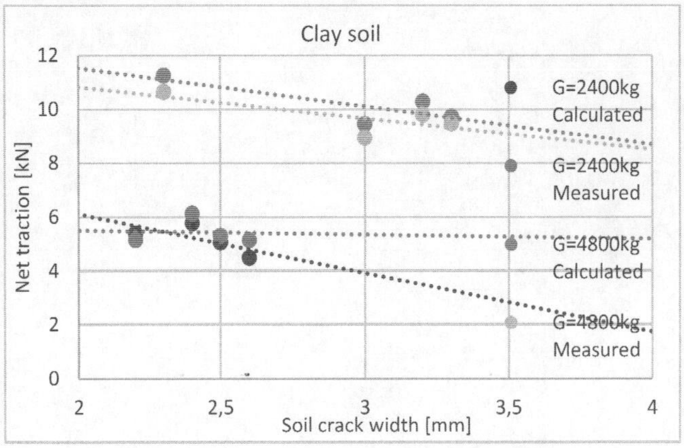

Fig. 11. Relationships between soil cracks width and calculated as well as measured net traction for clay soil

It was found that in the case of sandy soil, the obtained relationships were increasing (the larger the crack, the higher the net traction value), while for clay soil, they were decreasing. This is inrtiguing considering the intuitively expected rising character for both soils, similar to the Janosi-Hanamoto model where the driving force is proportional to the horizontal deformation of the soil. In the case under consideration, however, we are dealing with cracks probably caused by compressive and tensile stress, not shear. Therefore, an explanation of the obtained results should be sought on the basis of the mechanisms of soil deformation due to traffic.

Ten Damme et al (2021a) reported that an increase in drawbar pull caused a redistribution of the tractor's load from the front to the rear axle. The effect of the increased wheel load on mean ground pressure and in the contact area was compensated by an increase in the tyre-soil contact area. Moreover, an increase in drawbar pull evened out the distribution of vertical contact stresses beneath the tractor rear wheel, which cancelled out the effect of the higher wheel load on maximum vertical stress in the contact area. Ten Damme et al (2021b) also pointed out that the propagation of horizontal stress through the soil profile still is poorly understood. They obtained different results between soil response at 0.16 m depth (effect of traction) and stress measurements at 0.33 m depth (no effect of traction). From those studies and on the basis of other references (Koolen et al 1992, Kirby 1989, De Pue et al 2020) it follows that the main mechanism for transferring loads from the vehicle to the soil is the vertical wheel load.

The search for an explanation of the obtained results on the basis of soil mechanics and geotechnical engineering does not give satisfactory answers, because the methods used refer to the static load of the soil and long-term processes. Yin and Vanapalli (2018) distinguished two different mechanisms that contribute to the capillary force and consequently tensile strength of unsaturated cohesionless soils: matric suction and surface tension. In the capillary regime, matric suction predominantly contributes to the tensile strength, while surface tension is significant in the pendular regime. In the funicular regime, those two mechanisms jointly contribute to the overall tensile strength, which is consistent with "the superimposed theory" proposed by Schubert (1973, 1975). Yin and Vanapalli (2018) developed a model for calculation of soil tensile strength, in which the total uniaxial tensile strength of unsaturated cohesionless soils is the superimposition of the tensile strength component induced by the matric suction within the saturated pores, and the tensile strength component induced by the surface tension associated with water bridges.

However, in the case of highly dynamic processes, we are dealing with instantaneous deformation and the above models will not provide a proper theoretical basis for explaining the obtained dependencies.

An initial explanation for the decreasing nature of the relationship between soil crack width and net traction is the effect of water in clay soil, where soil pores and particles are several orders of magnitude smaller than in the case of sandy soil. Capillary effects in clay soil may contribute to limiting the development of cracks, the stronger the greater the load resulting from the applied drive torque on the road wheel. This is of course a hypothesis, not fully confirmed.

Concluding, the obtained dependencies are difficult to assess on the basis of the current state of knowledge and references, therefore it seems necessary to conduct further research in order to describe the mechanism of soil cracking in the track of the tire on the soil.

5. Conclusions and future work

Field tests were carried out using agricultural tractors, consisting in determining the tire footprint on soil and measuring the traction force and rolling resistance force. On the basis of photographs of selected areas of the tire track on the soil, a 3D reconstruction of the tire - soil interface was obtained, and then the values of the traction force and rolling resistance were calculated using the Bekker model. The calculated traction forces were compared with the results of the measurements of the traction force and the rolling resistance force. Good or very good match was obtained.

From a practical point of view, it was found that using the above method in the case of traction analysis of the whole vehicle (not a single wheel) is troublesome and may give erroneous results if the rear wheels differ in size from the front wheels (similarly, the load on the front and rear wheels). This is due to the fact that the analyzed tire footprint on the soil is imprinted by the rear wheel.

Further, by analyzing the obtained 3D tire - soil interfaces, cracks on the soil surface were identified, resulting from the mechanism of soil tension under vertical load. It was discovered that these cracks in sandy soil were bigger (of greater width) than in clay soil. Their width has been correlated with net traction force generated during test rides. It was found that in the case of sandy soil, the obtained relationships were increasing (the larger the crack, the higher the net traction value), while for clay soil, they were decreasing.

As it is proper practice in scientific research to strive to develop a generalizing model of the studied phenomenon, the next step in the undertaken research by the authors may be to develop a tire-soil interaction model, taking into account the mechanism of cracks formation as a result of soil tension.

6. Nomenclature

A	Tire-soil contact area	[m^2]
F_{DP}	Drawbar pull	[kN]
F_{DP}^{calc}	Drawbar pull calculated	[kN]
F_{DP}^{meas}	Drawbar pull measured	[kN]
MC	Soil moisture content	[%]
Q	Trailer load	[kN]
Q_W	Wheel load	[kN]
W	Tractor's weight	[kN]

7. Acknowledgements

This work was prepared within the project KONF/SP/0330/2023/1 entitled "16th European-African Regional Conference of the ISTVS", financed by the Ministry of Education and Science (Poland) as a part of the Doskonała Nauka II Programme.

8. Declaration of competing interest

The authors declare that they have no known competing financial interests or personal relationships that could have appeared to influence the work reported in this paper.

9. References

Botha T.R. and Els P.S., 2015a. Digital image correlation techniques for measuring tyre-road interface parameters: Part 1-Side-slip angle measurement on rough terrain, J. Terramech., 61, pp. 87-100

Botha T.R. and Els P.S., 2015b. Digital image correlation techniques for measuring tyre-road interface parameters: Part 2-Longitudinal tyre slip ratio measurement, J. Terramech., 61, pp. 101-112

Botha T.R. and Els P.S., 2015c. Rough terrain profiling using digital image correlation, J. Terramech., 59 (2015), pp. 1-17

Botha T.R., Johnson D., Els P.S., Shoop S.A., 2019. Real time rut profile measurement in varying terrain types using digital image correlation. Journal of Terramechanics, Volume 82, April 2019, Pages 53-61.

De Pue, J., Lamande, M., Cornelis, W., 2020. DEM simulation of stress transmission under agricultural traffic Part 2: shear stress at the tyre-soil interface. Soil Tillage Res. 203 https://doi.org/10.1016/j.still.2020.104660.

Diserens E., Défossez P., Duboisset A., Alaoui A., 2011. Prediction of the contact area of agricultural traction tyres on firm soil Biosys. Eng., 110 (2) (2011), pp. 73-82

Gysi M., Maeder V., Weisskopf P., 2001. Pressure distribution underneath tires of agricultural vehicles. Transact. ASAE, 44 (6) (2001), pp. 1385-1389.

Farhadi P., Golmohammadi A., Sharifi A., Shahgholi G., 2018. Potential of three-dimensional footprint mold in investigating the effect of tractor tire contact volume changes on rolling resistance, Journal of Terramechanics, Volume 78, 2018, Pages 63-72.

Kirby, J.M., 1989. Shear damage beneath agricultural tyres: a theoretical study. J. Agric. Eng. Res. 44, 217–230. https://doi.org/10.1016/S0021-8634(89)80083-6.

Koolen, A.J., Lerink, P., Kurstjens, D.A.G., van den Akker, J.J.H., Arts, W.B.M., 1992. Prediction of aspects of soil-wheel systems. Soil Tillage Res. 24, 381–396. https:// doi.org/10.1016/0167-1987(92)90120-Z.

Schubert, H. 1973. Kapillardruck und Zugfestigkeit von feuchten Haufwerken aus kornigen Stoffen. Chemie Ingenieur Technik, 45(6): 396–401. [In German.] doi:10.1002/cite.330450613.

Schubert, H. 1975. Tensile strength of agglomerates. Powder Technology, 11(2): 107–119. doi:10.1016/0032-5910(75)80036-2.

Söhne W., 1952. Die Kraftubertragung zwischen Schlepperreifen und Ackerboden. Grundlagen der Landtechnik, Heft 3/1952

ten Damme, L., Schjønning, P., Munkholm, L.J., Green, O., Nielsen, S.K., Lamand´e, M., 2021a. Traction and repeated wheeling – effects on contact area characteristics and stresses in the upper subsoil. Soil Tillage Res. 211 https://doi.org/10.1016/j. still.2021.105020.

ten Damme, L., Schjønning, P., Munkholm, L.J., Green, O., Nielsen, S.K., Lamand´e, M., 2021b. Soil structure response to field traffic: Effects of traction and repeated wheeling Soil Tillage Res. 213 https://doi.org/10.1016/j.still.2021.105128

Yin P. and Vanapalli S.K., 2018. Model for predicting tensile strength of unsaturated cohesionless soil. Canadian Geotechnical Journal, Jan. 2018 https://doi.org/10.1139/cgj-2017-0376

3796 | The issue of assessing the suitability of own simulation models for testing off-road vehicles

Przemysław Simiński [a,*], Krzysztof Kosiuczenko [a], Marek Nowakowski [a], Alexander Kravcov [b]

[a] *Military Institute of Armour and Automotive Technology, Sulejówek, Poland*

[b] *Czech Technical University. Prague, Czech*

* *Corresponding author: przemyslaw.siminski@witpis.eu*

ABSTRACT

This paper presents a comprehensive study on the terrain-tire interaction considering the dynamics and mobility of military trucks on various terrains. The theoretical basis of the developed model is presented, and simulation results are discussed, focusing on the driving force's dependence on tire types, characteristics, and ground properties. Authors introduced a mathematical model of a military truck built using MBS software DADS which was validated on deformable ground, considering well-known tire characteristics. The obtained results are analyzed in terms of military truck dynamics and mobility across diverse terrain and tire conditions. Potential applications and future research directions of developed model are indicated.

Keywords
Dynamic test
FEM model
MBS model
Off-road vehicle
Static test

1. Introduction

The development of military trucks capable of effective maneuverability across diverse terrains poses numerous challenges (Mazal *et al.* 2020). Innovative solutions, such as evacuation features, are being incorporated into specialized designs (Janos *et al.* 2017) but traditional trucks with the capacity to carry heavy loads remain at the core of military operations. The conditions under which these vehicles are operated, including driving characteristics and tire selection, play a crucial role in their performance (El-Sayegh *et al.* 2020, Hassan *et al.* 2020). These factors impact stability, traction, and the vehicle's breaking properties (Borawski *et al.* 2020). To address these challenges, there are advanced modeling tools available that offer a wide range of possibilities for simulating and analyzing military truck dynamics and mobility (Bezin *et al.* 2021, Łukaszewicz *et al.* 2023). A multidisciplinary approach, integrating engineering, materials science, simulations, and military operations, is necessary to optimize the design of military vehicles and enhance their overall performance in various operational environments.

High mobility is a critical requirement for heavy military trucks, as it directly relates to their ability to maneuver effectively across different terrains. The selection of tires plays a crucial role in determining the performance of the military vehicles. The operational environment demonstrate volatile mechanical properties and surface irregularities, which significantly impact the vehicle's behavior due to the presence of random input functions. The complexity of vehicle behaviors makes it challenging to precisely describe the dynamic changes in terrain that influence mechanical properties and accurately characterize the input functions affecting the vehicle. As a result, simplified methods have become more prevalent in both vehicle motion theory and design-research practices as a practical approach to address these intricate complexities.

2. Phenomena description

The ability of a vehicle to traverse diverse terrains relies on the bearing properties of the soil utilization. Analyzing the balance conditions of forces influencing the vehicle it should be highlighted that successful motion requires the driving force, related to the driving torque and terrain type, that overcome the vehicle's rolling resistance. The determination of the driving force value relies on the vehicle's towing force.

The changing properties of the soil significantly impact the actual towing force. Precisely determining soil parameters and their mutual statistical dependence can present challenges; however, the towing force remains a valuable and indispensable parameter for assessing the vehicle's off-road capabilities. It is a key indicator that determines the vehicle's mobility on deformable ground surface. The rolling resistance is directly related to the deformation of the ground caused by the tire. This deformation generates resistance, which opposes the vehicle's forward motion, making it more challenging for the tires to roll smoothly without additional effort. Additionally, the rolling resistance is influenced by internal friction within the tire, especially when the tire's air pressure is at small values. This internal friction contributes to the energy losses and further increases the overall rolling resistance of the vehicle. The rolling resistance ratio, resulting from ground surface deformations, is calculated by dividing the work lost due to ground deformation by the travelled distance and vertical load on the wheel.

https://doi.org/10.56884/ ELGV3830
ISBN 978-1-942112-55-6 (pdf)

Detailed definition of the vehicle mobility on the deformable surface ground requires conducting extensive experimental research on various types of terrains, which involves a considerable investment in measuring and securing equipment generating significant costs. The verified simulation model provides a powerful tool for researchers to conduct comprehensive and cost-effective investigations, leading to enhanced insights into military truck dynamics and improved decision-making for operational planning.

3. Vehicle MBS model

The selection of appropriate and accurate assumptions directly determines the amount of experimental and literature data necessary for the successful implementation of the simulation. In this study, the cross-country heavy truck JELCZ 442.32 – Fig. 1 was used as base for experimental and simulation models. This vehicle features a 4×4 power transmission system, which was utilized during data collection and measurements.

Fig. 1. Cross-country heavy truck JELCZ 442.32 [source WITPIS]

The simulation model is created using DADS software, classified as a multibody simulation (MBS) program. MBS employs numerical simulation to model multibody systems with rigid or elastic bodies. Connections between the bodies can be modelled with kinematic constraints (such as joints) or force elements (such as spring dampers). Unilateral constraints and Coulomb-friction can also be used to model frictional contacts between bodies. DADS program has been used to generate the heavy truck structure and the model of interaction between a tire wheel and the deformable ground. The DADS package automatically generates motion equations in a numerical way. In the first step, upon entering, the object structure is being developed and its variables are being entered (coordinates, velocities, external forces, simulation conditions).The Newton-Euler formalism for multibody systems is subsequently applied to generate and solve the equations. Finally, procedures are employed to visualize and animate the motion based on the obtained solutions. A useful form of description in numerical applications is a matrix description, presented by Zboiński and Dusza (2006), where a general form of Newton-Euler's equations that are applied in numerical MBS type of programs has been as the following formula (1):

$$I\dot{x}_{II} = Q + \Lambda + \Phi^{T}(x_{I},t)\lambda \tag{1}$$

The above equation is supplemented with equations of 1st grade kinematical constraints having the form of (2):

$$\Phi(x_{I},t)x_{II} + \gamma(x_{I},t) = 0 \tag{2}$$

The solution of the system of equations is possible upon its supplementation with kinematical relations (3):

$$F_{x}\dot{x} = K(x_{I})x_{II} = m\frac{dv}{dt} \tag{3}$$

where:

I - symmetrical matrix of inertia (6n x 6n), covering masses and components of tensors of the solid's inertia moment,

Q - vector (6n x 1) of inertial forces moments,

Λ - vector (6n x 1) of external moments and forces,

γ - vector (r x 1),

x_{I} - vector of coordinates (6n x 1) describing linear and angular motions of solids, defined for adopted bases of unit vectors,

x_{II} - velocity vector (6n x 1), establishing along with Cartesian coordinates the so-called status coordinates,

n - number of solids of the multibody system,

r - number of 1st grade kinematical constraints.

4. MBS model of a tire wheel and a description of its cooperation with the surface

During simulation tests of vehicle motion within the driver-vehicle-road system, a detailed understanding of the model related to the wheel rolling phenomena is crucial. This model is described by the tire-terrain interaction dependence, which incorporates mechanical properties of the tire framing and tread, as well as frictional characteristics describing tire-ground interface.

The comprehensive description of tire-ground cooperation includes the force components acting in three directions and the three components of the moment. The ground-tire interaction is function of the axial force, which relies on the tire deformation coefficient. In the tire-wheel model within the DADS package, the component relationships of contact lateral and longitudinal forces (complying with the "adhesion ellipse") are associated with the axial force. Additionally, kinematic relations in the steering system, such as the wheel turning angle and inclination angle, impact the wheel torque and wheel slip during rotary motion around its axle.

The direction of the axial force vector on deformable surface ground is defined by the road profile. The standard surface for the axial force vector is determined by the middle of the non-deformed driving wheel(non-deformed), however, the effects of sharp terrain protrusions are accounted for in the calculations. To facilitate the calculations, the wheel is divided into slices, and the locations of two extreme points (determined by the intersection of the wheel outline and the road profile) are compared to the height of the road profile. These points determine an arch chord, which defines the terrain fragment considered in calculating the trapezoid Ai surface area. If these points are below the road profile, the plain trapezoid (Fig.2) is taken into consideration.

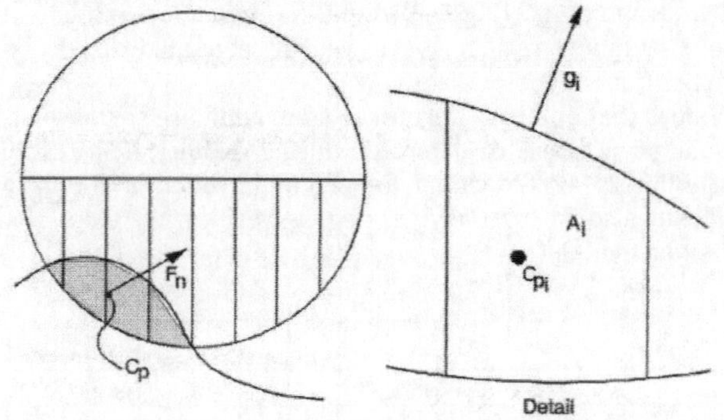

Fig. 2. The way for determining application of the axial force and the road profile [10]

The authors built a tire model based on DADS calculations and taking into account other computational models.

Calculations presented below are made by DADS program to determine values of contact forces in cooperation of tire wheel with a deformable surface ground (DADS Manual).

The point of the axial force application is determined out from the following relationship (4):

$$C_p = \frac{\sum A_i C_{p_i}}{\sum A_i} \tag{4}$$

where:

A_i - trapezoid surface area,

C_{pi} - location of the axial resultant force for individual slices of the wheel.

The model has considered the tire deformation effect. It is described by relationships (5, 6)

$$\dot{D} = V \tag{5}$$

$$\dot{V} = \frac{[F_{patch} - (KD + CV)]}{m_{carcass}} \tag{6}$$

where:

D - matrix describing deformation of the tire fragment in normal, lateral and longitudinal direction,

V - matrix of velocity deformation in selected directions,

F_{patch} - matrix of the force value in contact of the wheel with the surface in selected directions,

C - matrix of tire damping,

K - matrix of tire rigidity,

$M_{carcass}$ - mass matrix of the tire fragment being deformed.

Deformation is calculated separately for each of the directions. Also, the tire resultant deformation is calculated.

Calculation of the contact longitudinal force in a contact with the ground is described by the correlation (7).

$$F_{rr} = -C_{rr} F_N \operatorname{sign}((V_c)_{long}) \tag{7}$$

where:

C_{rr} - rolling resistance coefficient,

F_N - axial force,

$(V_c)_{long}$ - velocity in progressive motion of the middle of the driving wheel.

For the deformable surface, the C_{rr} coefficient is replaced by coefficient C'_{rr} (8).

$$C'_{rr} = C_{rr} + \frac{1}{B_n} + \left|\frac{0{,}5S}{\sqrt{B_n}}\right| \tag{8}$$

where:

S - longitudinal relative slip,

B_n - mobility level.

The mobility level is calculated from the following relationship:

$$C'_{rr} = C_{rr} + \frac{1}{B_n} + \left|\frac{0{,}5S}{\sqrt{B_n}}\right| \tag{9}$$

where:

δ - deformation of a tire in a normal direction,

b - tire breadth,

d - tire diameter,

h - height of tire intersection,

CI - Cone Index coefficient.

For the deformable surface, the longitudinal adhesion coefficient is calculated from the relationship (10):

$$\mu_{long} = \mu(1 - e^{-0{,}1B_n})(1 - e^{75S}) + C_{rr} \tag{10}$$

Contact lateral force is determined from the Magic Formula model described by Pacejka and Bakker (1993).

5. Verification tests

The simulation model, developed in DADS computer program (Fig. 3), was verified by comparing to the real object.

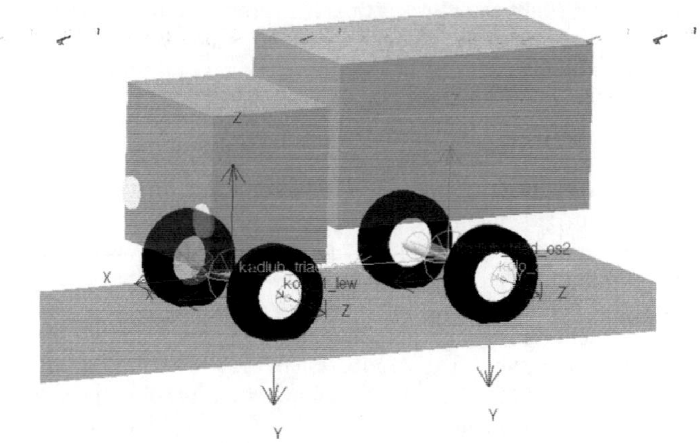

Fig. 3. MBS model of the vehicle

To conduct verification tests and assess the model's agreement with the real object, military field measurements were taken.

These measurements recorded the force on the tow hook, linear velocity, and angular velocity of the vehicle wheels while it was being towed by another vehicle on a deformable ground (Fig. 4).

Fig. 4. JELCZ 442.32 during verification tests (source: MIAAT Sulejówek)

In this way, towing resistance on a deformable surface ground was determined for a given type of ground and applied tires.

Out of a wide range of generated results, those time runs have been selected that are considered as the most repeatable and representative. The graphs (Fig. 6-9) present results that obtained during military field tests.

The obtained towing force was used as input value to the simulation model. The hereto elaborated model has also considered resistances effecting from internal resistances while rolling driving wheels. The value of the vehicle linear velocity, obtained during experimental tests and that from the model, has been adopted for verification. The chart below (Fig. 6) presents values of obtained velocities of the vehicle and the computer model of the cross-country heavy truck.

The conducted verification tests have shown that the model accurately reflects the characteristics of the real object with a high degree of accuracy. Therefore, it can be confidently used for further simulation tests to determine the traction capacities of military cross-country heavy vehicles.

6. Simulation MBS tests

Simulation tests of the cross-country heavy vehicle have been conducted for two different types of tires whose respective characteristics have been determined during static and dynamic tests (Technical Report, WITPIS Sulejówek, 2003), and then implemented to the model describing a cooperation of the tire with the ground. The conic index value has been assumed at CI = 1050 kPa. The first version of tires is tires with cross-country tread T-43 being part of the WP equipment, while the second version of tires is tires with universal tread U-34. Both versions of tires have been subjects for traction tests at MIAAT Sulejówek.

A decision has been made to define through simulation tests mobility of the vehicle via its capacity to develop a maximum towing force. For that purpose, an elastic connection between the vehicle point (towing hook) and a ground point (anchor) has been modeled. The vehicle model's motion has been forced by application of such a torque value to the wheels in order to obtain relocation of the vehicle at a defined speed. The force value in the elastic element has been growing along with progressive relocation of the vehicle until adhesion of the wheels has been lost (wheel slip = 100%) and at a given point of time has been the value of maximum towing force.

Below are shown the obtained simulation results in a form of time runs of the towing force and the slip of driving wheels as well as the towing force value in the wheels slip function.

The graph below (Fig. 9) presents a comparison of the towing force simulation results for a vehicle equipped in the two aforementioned types of tires.

Values that obtained during the simulation are convergent by its quality with results generated during military field tests for vehicle JELCZ 442.32 that is equipped in universal tires U34 (Fig. 5). It can also be partially treated as verification tests of the developed model.

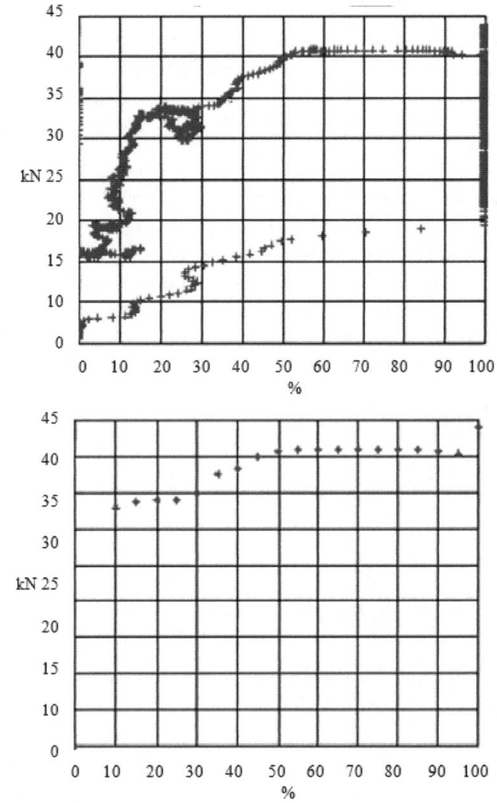

Fig. 5. Results of experimental tests of the vehicle with U34 tires

Comparisons of the velocity obtained from the experiment and that calculated from the model are shown in Fig. 6.

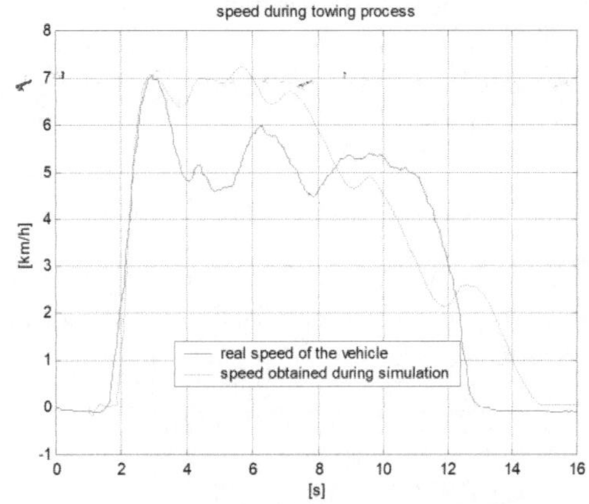

Fig. 6. Verification of the velocity values from simulation and experiment

Towing force in a slip function and the wheel slide in a time function for a vehicle with T43 tires and the illustration of the simulation model are shown in Fig. 7, 8 and 9.

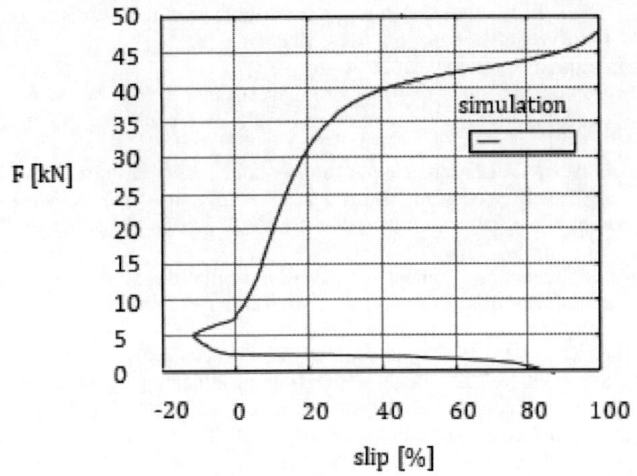

Fig. 7. Tow force as a function of wheel slip

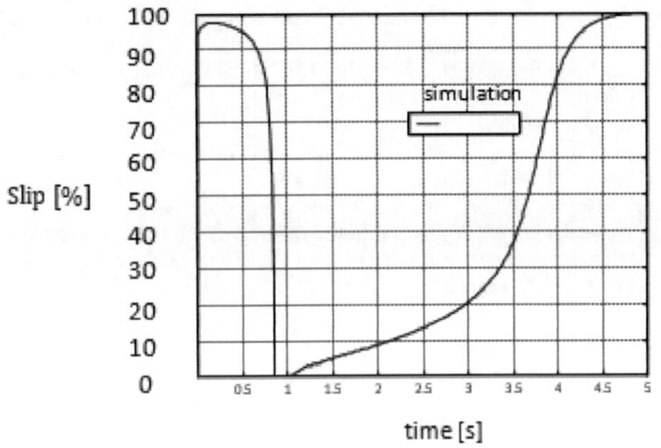

Fig. 8. Wheel slip as a function of time

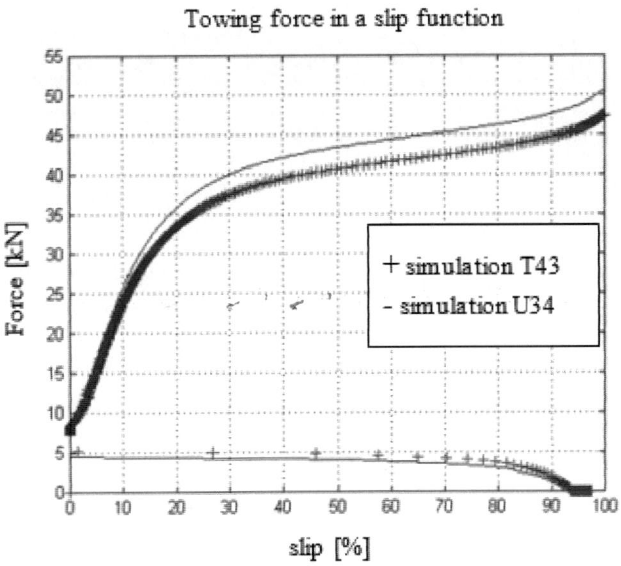

Fig. 9. Comparison of towing forces values in a slip function for two types of tires

7. Vehicle FEM model

The "multibody" simulation conducted in this study expanded the scope of vehicle analysis by incorporating strength analysis. Utilizing information about the current state of the vehicle model, constructed from rigid bodies, enabled the collection of data on external forces impacting the vehicle. This data was utilized to develop a flexible structure model for conducting strength analysis.

To perform the strength analysis, a numerical model was created using the FEA finite element method (Manujło *et al.* 2018). The Simcenter Femap program was employed for this purpose. The model's design involved describing the frame's geometry using a mesh of SHELL elements. Mass elements representing the vehicle's heavy components, such as the engine, cargo box, and cab, were then connected to the frame via RBE3 rigid interpolation elements.

Moreover, special attention was given to modeling the springs and tires in detail. SPRING spring-damping elements were utilized for this purpose, with properties assigned to replicate those of real springs (damping and stiffness). The resulting FEM model, shown in Fig. 10, incorporates these features and facilitates in-depth strength analysis of the vehicle.

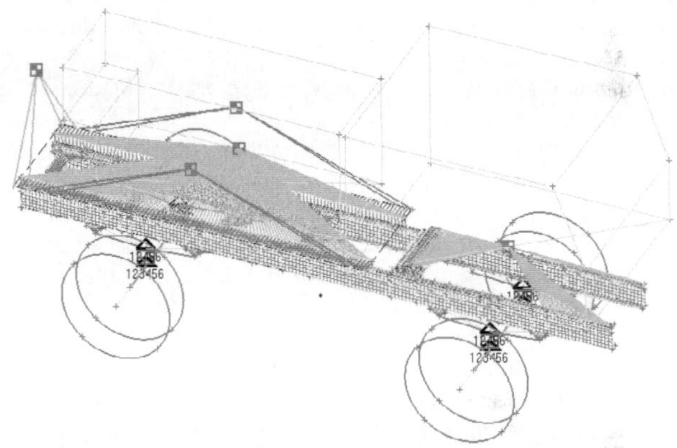

Fig. 10. FEM model of the analyzed JELCZ 442.32 vehicle

The frame was subjected to bending by loading due to gravity on elements with non-zero mass. In addition, the frame was subjected to torsion by raising the wheels asymmetrically to a height corresponding to terrain irregularities (several calculation variants were performed). As a result of the calculations, reduced stress maps were obtained in the stringers and cross members of the frame (Fig. 11).

In all design variants, the reduced stresses (108÷126 MPa) were lower than the allowable stresses (about 355 MPa). No additional dangerous stress concentrations appeared in any of the analyzed variants.

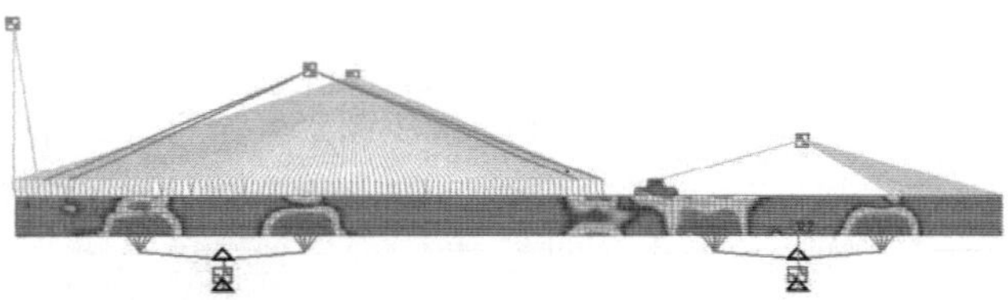

Fig. 11. Results of numerical calculations for the model (Huber-Mises-Hencky's stress)

8. Conclusion

The interaction of a vehicle's wheel with deformable ground is a complex and widely studied phenomenon, with different research centers worldwide working to describe it with varying degrees of success. In this study, an experimental review of a heavy truck model, developed using the Multi Body System (MBS) type of program, demonstrated its substantial compliance with the real object. The accuracy of the achieved results was notably influenced by the detailed characteristics of the tires and vehicle parameters considered by the authors.

The positive evaluation of the model enabled tests of the towing force on deformable ground, specifically exploring two different types of tires, solely through simulation. Notably, the utilization of U34 tires during vehicle load testing resulted in higher towing forces (Fig.7), a finding confirmed in subsequent military exercises conducted in the field. This highlights the practical aspects of the developed model, as it allows to carry out comprehensive testing and evaluation of various tire types and surfaces. It should be stress that despite of MBS simulations, a Finite Element Method (FEM) numerical model of the vehicle was developed, providing valuable information on the structural strength of the superstructure that cannot be obtained through MBS simulations alone By employing both FEM and MBS computational methods, a complementary numerical model of the vehicle was obtained, offering insights into both the vehicle's on-road behavior and the strain experienced by its load-bearing structure.

Based on the received results, the developed model allows for comprehensive assessments of the traction properties of cross-country heavy military vehicles and their potential for fording various types of terrain, contingent on the specific tires applied. This approach offers a multitude of benefits, not only in terms of reducing research time and costs but also in mitigating potential risks that experimental studies may pose to the research team. Furthermore, simulations can be performed using plenty of variants as well as diverse weather conditions. The integration of MBS and FEM methods significantly enhances the analysis and evaluation of military vehicle performance, making contributions to the advancement of military vehicle design and operational planning.

9. Acknowledgments

The article has utilized partially research data gathered through the implementation of the research project No. 4T12C06028..

10. Declaration of competing interest

The authors declare that they have no known competing financial interests or personal relationships that could have appeared to influence the work reported in this paper.

11. References

Bezin, Y., Pålsson B.A., Kik, W., Schreiber, P., Clarke J., 2021. Multibody simulation benchmark for dynamic vehicle-track interaction in switches and crossings: results and method statements. Vehicle System Dynamics, 1-38.

Borawski, A., Szpica, D., Mieczkowski, G., Borawska, E., Awad, M.M., Shalaby, R.M., Sallah, M., 2020. Theoretical Analysis of the Motorcycle Front Brake Heating Process during High Initial Speed Emergency Braking. J App Comp Mech, 6, pp. 1431-1437.

DADS Manual, Revision 9.6. CADSI

El-Sayegh, Z., El-Gindy, M., 2020. Rolling resistance prediction of off-road tire using advanced simulation and analytical techniques. SN Appl. Sci. 2, 1620.

Hassan, M.A., Abdelkareem, M.A.A., Moheyeldein, M.M., Elagouz, A., Tan, G., 2020. Advanced study of tire characteristics and their influence on vehicle lateral stability and untripped rollover threshold, A Eng Journal, 59, 3,1613-1628.

Janos, R., Sukop, M., Semjon, J., 2017. Conceptual design of a leg-wheel chassis for rescue operations. International Journal of Advanced Robotic Systems, 14(6).

Łukaszewicz, A., Skorulski, G., Melnyk, M., Kernytskyy, A., Zdobytskyi, A., Kolesnyk, K., 2023. Engineering drawing education using CAD tools. In: Banaś M, Pytel K, Lobur M, Zdobytskyi A, (eds) CAD in Machinery Design Implementation and Educational Issues. XXIX International Polish-Ukrainian Conference, 95–102.

Manujło, A., Gotowcki, P., Simiński, P., Szudrowicz, M., Kosiuczenko, K., 2018. Application of fiber Bragg gratings for stress analysis of high mobility vehicle frame, Proc of SPIE, Vol. 10808.

Mazal, J., Rybanský, M., Bruzzone, A.G., Kutěj, L., Scurek, R., Foltin, P., Zlatník, D., 2020. Modelling of the microrelief impact to the cross country movement. Proceedings of the 22nd International Conference on Harbor, Maritime and Multimodal Logistic Modeling & Simulation(HMS 2020), pp. 66-70.

Pacejka, H.B, Bakker, E., 1993. The Magic Formula tire model, Supplement Vehicle System Dynamics vol.21, 1-18.

Technical Report WITPiS nr 56/ZPS/2003, Sulejówek 2003

Zboiński K., Dusza M., 2006. Computer method of modeling in simulation of railway vehicles motion. Warsaw University of Technology Series: TRANSPORT, Issue 56.

3797 | Disturbed and naturally recovered soil surface as a ground for subsequent vehicle mobility operations

Kersti Vennik [a,*]

[a] Estonian Military Academy, Tartu, Estonia

* Corresponding author: kersti.vennik@mil.ee

ABSTRACT

The passages for off-road military vehicles will leave behind disturbed soil conditions because vehicles compact and displace the layers of soil. One of the first visual signs of disturbance are ruts. Thus, former active combat areas can have a serious effect on subsequent operations and the mobility of vehicles. More precisely, an uneven soil surface will remarkably increase the vibration of travelling vehicles and over-compacted soil layers affect soil water dynamics that, in turn, have a long-lasting effect on soil strength. Previous studies have revealed that disturbed conditions can affect the physical parameters of soil for decades. Nevertheless, soil can naturally recover from being disturbed thanks to freezing-thawing cycles, wetting-drying cycles, and biological activity. However, the rate of recovery depends on the physical parameters of soil. The objective of this study was to monitor and summarize the natural recovery rate of soils with different texture and organic material content. As an experiment, we carried out a single pass and repeated passes with a military truck (total weight of 70 kN). The ruts that formed were monitored and soil samples were collected up to 2 years after wheeling. Measurements in the field included rut depth determination, penetrometer and bulk density as well as field-saturated hydraulic conductivity measurements. The results about the rate and severity of natural recovery on different soils are presented and discussed herein, including the overall conclusions about mobility conditions for follow-on military operations.

Keywords
Ruts
Surface roughness
Compaction
Bulk density
Cone index

1. Introduction

In military missions, different kinds of military vehicles have to travel on natural soil surfaces. For example, a company-sized unit can consist of around 7 to 14 wheeled vehicles like Stryker or military trucks (ATP 3-21.11, 2020; Jalaväekompanii lahingutegevus, 2012) which could move dispersedly within a specified distance of each other or in a column, following the same track (Ayers, Rice, 2017). If a military conflict lasts longer, in some cases for several years, there is a high probability that military maneuvers are conducted over a previously trafficked area, i.e. combat and non-combat vehicles need to travel over a soil surface whose mechanical conditions are disturbed from previous travels. In addition to battlefields, this type of soil conditions are typical for military training areas that have to sustain constant traffic and training over decades (Sullivan, Anderson, 2000). Thus, there is the certain necessity to understand the conditions of disturbed soil strength and its effect on trafficability.

For a successful passage of vehicles, the strength of soil must be sufficient to support the vehicles' weight and to provide the thrust required for propulsion (Bekker, 1960). Thus, soil strength of up to 60 cm of depth is of interest for mobility and, more precisely, the soil strength status of the top surface layer is particularly important for thrust generation. Besides strength, surface roughness influences comfort, on-road handling of, and the speed of vehicles (Durst *et al.* 2011). In the worst case, an uneven soil surface that includes furrows, or the bumps with dimensions higher than the vehicle clearance, will cause immobilization.

When a vehicle travels over a terrain, it causes compaction and displacement of soil (Wong, 2008). Often, the traces of disturbances caused by moving vehicles are revealed in the form of rut (Raper, 2005). Rut includes groove with different depths and, for some cases, also the piles around the groove. The depth of rut as well as the height of the pile is determined in relation to the adjacent undisturbed soil surface. Previous studies in the field of trafficability reveal that the maneuvering of military vehicles can cause ruts of up to 0.6 m of depth (Affleck, 2005), and repeated passage of vehicles forms deeper ruts (Affleck, 2005; Liu *et al.* 2011). In summary, trafficking can form a very uneven surface with a high value for surface roughness.

The compaction of soil can be dependent on soil and vehicle parameters appear on different levels of depth in the soil profile (Arvidsson, Keller, 2007; Hamza, Anderson, 2005). Soil compaction if often determined by bulk density or cone index, whereas cone index is measured with a penetrometer (Raper, 2005). On the one hand, compacted layers are denser and, thus, possess higher strength, but on the other hand, dense soil layers will seriously impact water infiltration. A compacted soil layer on the top of the soil surface entirely prevents water infiltration into the soil. Thus, a relatively thin layer (up to few centimeters) covered and saturated with water can form on top of the soil, which will influence thrust generation. Compaction in deeper layers impedes the downwards movement of water, and wet soil layer with different thickness can form on top of it.

Former studies have shown that soil can naturally recover from disturbance caused by military activities (Prose, 1985). More precisely, recovery means that the cross-sectional profiles of rut can smooth up, and compacted layers and infiltration rates can alleviate. Natural recovery takes place because of the freeze-thaw cycles and wetting-drying cycles of cold periods and due to the biological activity in soil (Webb, 2002). Nevertheless, recovery is slow – it can take decades or even centuries to alleviate seriously compacted soil layers (Prose, 1985; Trumbull *et al.* 1994; Webb, 2002).

The aim of this study is to investigate the disturbance of soil caused by military vehicle traffic and the natural recovery of soil during a 2-year period.

https://doi.org/10.56884/ELGV3830
ISBN 978-1-942112-55-6 (pdf)

2. Methods

Soil disturbance analyzed herein was caused by a 2-axle military truck with a total weight of 7 tons (wheel load 18 kN, inflation pressure 650 kPa). Trafficking experiments were carried out in 2013 and 2014 (Table 1). The test vehicle conducted one pass in the first track and ten passes in the adjacent track. The test area was at least 50 m long and the vehicle travelled with a speed of 5 km h^{-1}. Altogether, there were four test sites in Estonia, named after the texture (by FAO) or organic matter content (i.e. 18% of SOC served as boundary value between organic soil and peat): peat, organic silt, silty sand, clay loam. The physical properties of tested soil are presented in Table 2. The sites of peat and clay loam were covered with approximately 1 m high and dense grasses, whereas the other two sites (organic silt and silty sand) were arable land, i.e. without permanent grass cover.

Table 1
Measurements outline

Soil	Measurements dates and outline		
	Traffiking	After 0.5 to 1 year	After 2 years
Peat	*2013-09-18*: RUT; BD; CI	*2014-05-23*: RUT; BD; CI; K_{fs}	*2015-11-06*: RUT; BD; CI; K_{fs}
Organic silt	*2013-11-08*: RUT; BD; CI	*2014-11-16*: RUT; BD; CI; K_{fs}	*2015-11-05*: RUT; BD; CI; K_{fs}
Silty sand	*2014-01-07*: RUT; BD; CI	*2014-10-07*: RUT; BD; CI; K_{fs}	*2015-09-26*: RUT; BD; K_{fs} *2015-12-12*: CI
Clay loam	*2014-05-30*: RUT; CI	*2015-06-12*: RUT; BD; CI; K_{fs}	*2016-05-14*: RUT; BD; CI; K_{fs}

RUT - groove depth and pile height measurements; BD – bulk density measurements; CI - cone index measurements; K_{fs} - field-saturated hydraulic conductivity measurements.

Table 2
Tested soils physical properties

Soil	Depth, cm	Sand, %	Silt, %	Clay, %	SOC, % by weight
Peat	0-20	-	-	-	25.6
	20-40	-	-	-	18.3
Organic silt	0-35	31.8	67.8	0.4	11.3
	35-50	27.1	53.0	19.9	0.1
Silty sand	0-35	61.5	30.7	7.8	0.6
	35-50	74.9	21.4	3.7	1.2
Clay loam	0-30	32.1	37.5	30.4	1.7
	30-50	34.7	39.0	26.3	0.3

Sand (2-0.063 mm); silt (0.063-0.002mm); clay (<0.002 mm)

After passes, rut depth and the height of soil piles in relation to the adjacent undisturbed soil surface was measured with a measuring stick in both tracks. In addition, the cone index was measured with a handheld cone penetrometer (30° conical tip, 3.33 cm^2 cone base area, provided by Geotest Instrument Corp). Readings were taken from an undisturbed surface and from both tracks in ten repetitions. The penetrometer was inserted into the ground at a steady speed of about 2 cm/s. For bulk densities and moisture determination, undisturbed soil samples were collected with steel cylinders (5.3 cm of diameter, 4 cm of height) from undisturbed soil and from both tracks. Cylinders were extracted at 0, 5, 10, 15, 20, 30 and 40-cm of depth (below the corresponding surface, Fig. 1). Additionally, soil was collected from topsoil and subsoil for organic matter and texture analysis. The soil organic carbon content (SOC) was determined with the Tjurin method (Vorobyova, 1998) and texture was analyzed with the pipette method (Robinson, 1922).

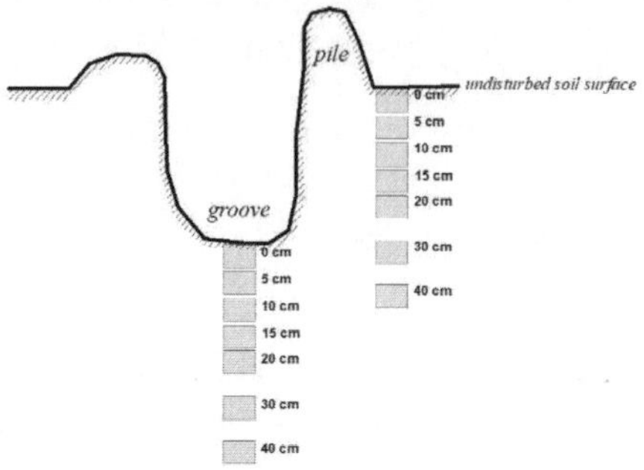

Fig. 1. Cross-sectional profile of rut and collection plan of cylinders.

All the abovementioned measuring was repeated after 0.5 to 1 years and after 2 years of trafficking. Detailed measurements dates are presented in Table 1. Monthly average temperature and precipitation data for study period are presented in Appendix A, Table A.1. In addition, field-saturated hydraulic conductivity (K_{fs}) was measured on the undisturbed soil surface area and in both tracks with five replications. For this, a simplified falling-head technique (SFH), as presented by Bagarello *et al.* (2004), was used. The SFH technique means that the known volume of water if relatively quickly added to the confined ring which is inserted into the soil surface. Here, a cylinder with a diameter of 10 cm and a height of 20 cm was inserted 10 cm into the soil. The volume of water was 0.3 liters. The period from the application of water to the total infiltration of water into the soil was measured. For calculations, the following formula was used:

$$K_{fs} = \frac{\Delta\theta}{(1-\Delta\theta)\,t_a}\left[\frac{D}{\Delta\theta} - \frac{D+\left(\frac{1}{\alpha*}\right)}{1-\Delta\theta}\ln\left(1 + \frac{(1-\Delta\theta)D}{\Delta\theta\left(D+\left(\frac{1}{\alpha*}\right)\right)}\right)\right] \quad (1)$$

Δθ stands for the difference between the field's saturated and initial volumetric soil water content; D stands for the depth of water in the ring after the application of water; t_a is the elapsed time for total water infiltration; α* expresses soil texture or structure parameter. For this research, the value of α* was estimated according to a study by Elrick and Reynolds (1992), and set to 12 m^{-1}. In order to determine Δθ, soil samples were collected from the ring (after the infiltration of water) and right next to the ring with steel cylinders (height 4 cm, diameter 5.3 cm).

3. Results

Rut depth, including groove depth and pile height, measured immediately after trafficking, in 0.5 to 1 year, and in 2 years after trafficking, are presented in Fig. 2.

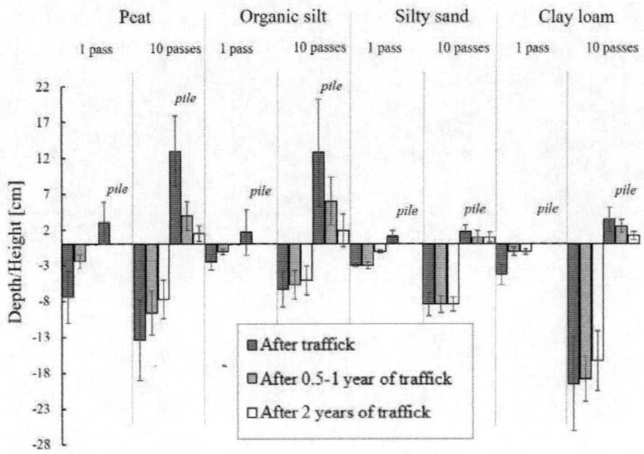

Fig. 2. Measured rut depths

After trafficking and the first pass, rut depth was rather low: in the range of 3 to 7.3 cm on average. After ten passes, remarkably deeper ruts were formed: 6.3–19.5 cm on average, whereas the deepest grooves evolved in soils with the highest clay content. In addition, in repeatedly passed tracks in organic soils, the pile height was remarkable: in peat soil, maximum height reached up to 18 cm, and in organic silt, it reached up to 29.3 cm. Overall, the relative height difference between groove and pile after ten passes was from 3.8 to 46 cm.

After a 0.5 to 1-year period, the once-passed ruts were hard to distinguish from an undisturbed soil surface but the ten times passed tracks had undergone significant changes in the cross-sectional profiles of rut (Fig. 2). For mineral soils (silty sand soil and clay loam soil), the rut depth values were mostly in the same range as the previous year, but for organic soils (peat and organic silt soil), the piles had significantly lowered, i.e. the average pile decrease was 7–9 cm. After a 2-year period, the organic soil piles of ten times passed tracks were further lowered and their grooves were smoother, whereas for mineral soils, no remarkable changes in the external shape of rut had taken place. One important aspect to underline here is that the rut material of organic soil was soft and easy to push around, whereas for mineral soils, all the rut material was stiff. Overall, after two years of trafficking, repeatedly passed tracks will still significantly influence the speed of vehicles maneuvering in the areas. According to Durst *et al.* (2011), surface roughness of 7.6–10.2 cm reduces the speed of wheeled and tracked vehicles down to 16–21 km/h^{-1}. The effect of the ruts' cross-sectional profile on speed is higher for mineral soils compared to organic soils.

Bulk densities of undisturbed areas, and of once-passed and ten times passed ruts right after trafficking as well as 0.5 and 2 years after trafficking for peat are presented in Fig. 3. As for bulk densities, there was no compaction after one pass in the whole soil profile, but after ten passes, there were slight signs for compaction in the soil layer of 15–25 cm of depth. Repeated measurements after 0.5 years confirmed the existence of a compacted layer at this depth. Two years after the experiments, compaction still existed around the depth of 22–25 cm.

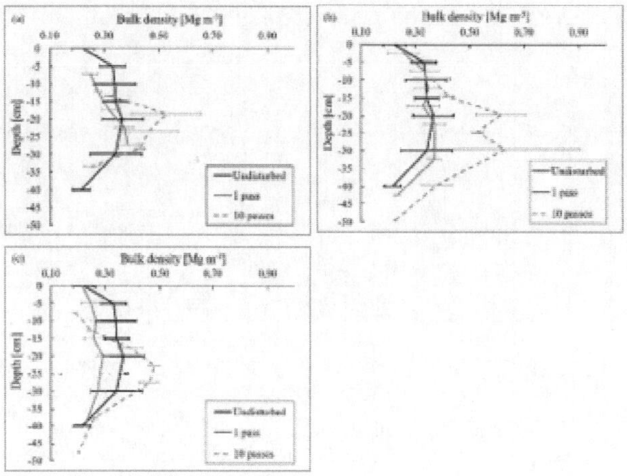

Fig. 3. Measured bulk density as a function of depth: (a) after trafficking, (b) after 0.5 years of trafficking, (c) after 2 years of trafficking on peat soil. Horizontal bars indicate standard deviation

Table 3
Gravimetric water content (%) as a function of depth: after trafficking, after 0.5 years of trafficking, after 2 years of trafficking on undisturbed peat soil

Soil layer, cm	After trafficking	After 0.5 years of trafficking	After 2 years of trafficking
0-15	106.4	307.6	186.2
15-30	114.8	164.3	182.4
30-40	245.2	245.3	235.2

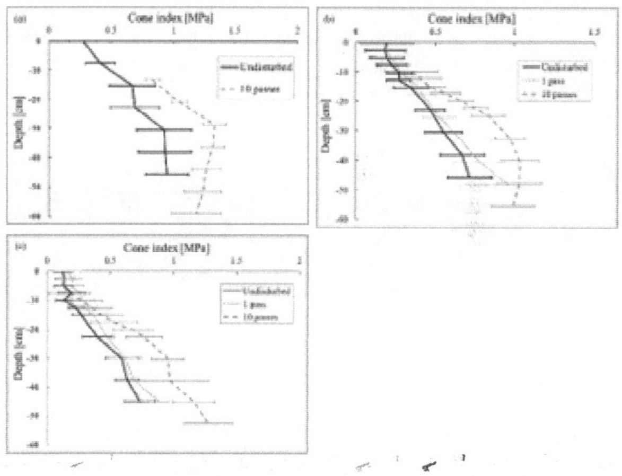

Fig. 4. Measured cone indexes as a function of depth: (a) after trafficking, (b) after 0.5 years of trafficking, (c) after 2 years of trafficking on peat soil. Horizontal bars indicate standard deviation

Cone index values presented in Fig. 4 support the conclusion about the existence of a compacted layer in deeper layers after trafficking and during the following years in a track passed ten times. Gravimetric water contents for studied period are illustrated on Table 3.

Field-saturated hydraulic conductivity (K_{fs}) values (Fig. 5) indicate that organic soils, including peat soil, in an undisturbed state have higher water infiltration rates than mineral soils. 0.5 years after trafficking, there was a significant difference in K_{fs} for ten times passed tracks, but not for a once passed rut. Two years after trafficking, infiltration in the surface of ten times passed tracks more or less recovered.

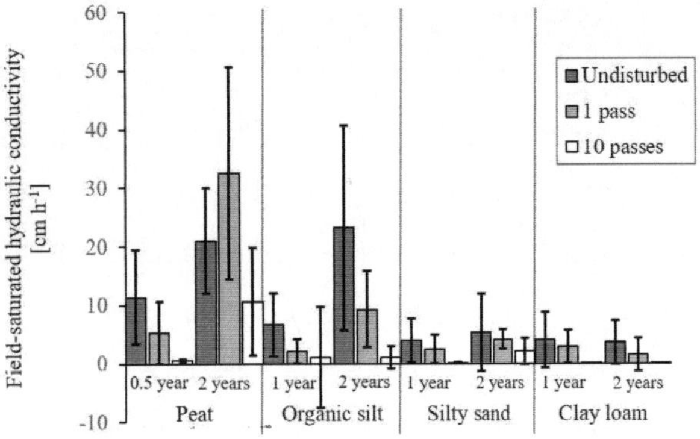

Fig. 5. Field-saturated hydraulic conductivity (K_{fs}) in tracks formed by one and ten passes after 0.5-1 year and after 2 years. Error bars indicate standard deviation.

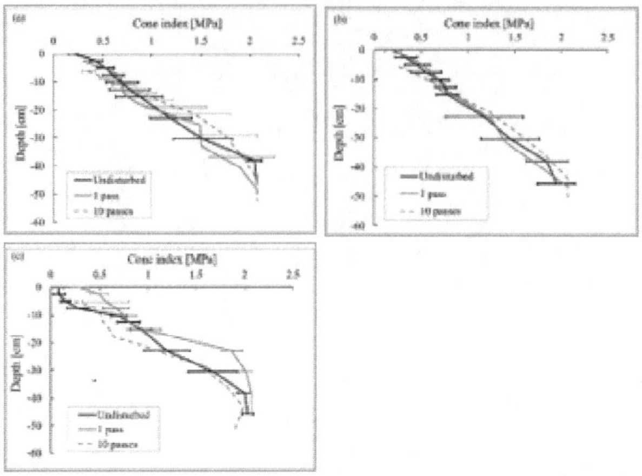

Fig. 7. Measured cone indexes as a function of depth: (a) after trafficking, (b) after 1 year of trafficking, (c) after 2 years of trafficking on organic silt soil. Horizontal bars indicate standard deviation.

The bulk density profile of undisturbed organic silt soil (Fig. 6) reveals that the layer with a higher organic matter content, i.e. 11.3% of SOC (Table 2) with a thickness of 30 cm, is laying on a denser mineral soil layer. With one pass, the test truck compacted the organic material laying on stronger mineral soil to the denser soil layer. Moreover, this compacted top layer does not disappear or alleviate in 2 years. During the succeeding passes, instead of compaction, the soft organic material of the top layer was displaced to the sides of the tracks, which was apparent from the high soil piles after ten passes (Fig. 2). On the other hand, cone indexes do not reveal any obvious compaction either in once passed nor ten times passed tracks (Fig. 7). K_{fs} values (Fig. 5) were lower for once passed tracks in 1 and 2 years after trafficking, compared to the undisturbed area. For ten times passed tracks, K_{fs} was significantly lower than the K_{fs} of adjacent undisturbed areas even after two years of experiments. Water contents for organic silt soil measurements are presented in Table 4.

Table 4
Gravimetric water content (%) as a function of depth: after trafficking, after 1 year of trafficking, after 2 years of trafficking on undisturbed organic silt soil.

Soil layer, cm	After trafficking	After 1 year of trafficking	After 2 years of trafficking
0-15	72.4	61.1	87.8
15-30	71.9	60.4	104.0
30-45	23.3	17.1	30.4

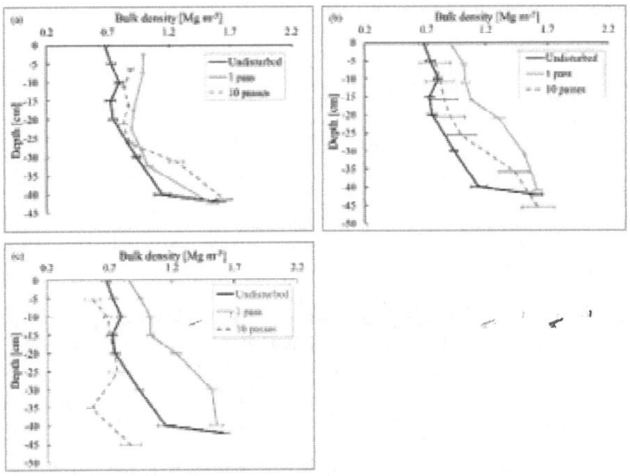

Fig. 6. Measured bulk densities as a function of depth: (a) after trafficking, (b) after 1 year of trafficking, (c) after 2 years of trafficking on organic silt soil. Horizontal bars indicate standard deviation.

According to the bulk densities, silty sand soil was compacted after one pass in the depth of 7–20 cm and with ten passes in the depth of 10–27 cm (Fig. 8). After one year, there still existed a compacted layer in the depth of approx. 17 cm in the once passed rut, but it alleviated after 2 years. For a ten times passed rut, according to the bulk densities, the compacted layer in the depth of 25–30 cm still exists after two years of experiments. Cone index values indicate no compaction immediately after the traffick (Fig. 9). This can be explained with a very high water content during the experiments (Table 5). According to Benghough *et al.* (2001), at high water contents, the cone index is practically insensitive to changes in bulk density. A year after the experiments, the soil was in a very dry state (Table 5) during the measurements and, thus, the maximum value of the penetrometer (2 MPa) was reached in the top layer of rut and it was not possible to determine any differences between the areas. Cone indexes measured two years after the experiments confirm the conclusions made from bulk densities, i.e. for ten times passed tracks, there still exist compacted layers in around 20–27 cm of depth. The significant difference in K_{fs} of undisturbed and once passed rut disappears in one year, and for ten times passed tracks, it disappears in two years (Fig. 5).

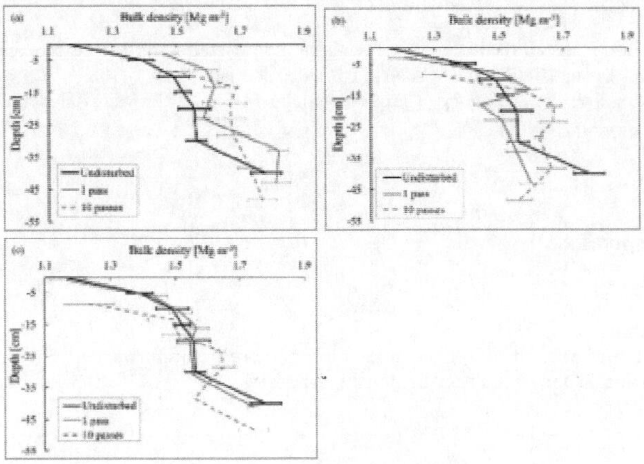

Fig. 8. Measured bulk densities as a function of depth: (a) after trafficking, (b) after 1 year of trafficking, (c) after 2 years of trafficking on silty sand soil. Horizontal bars indicate standard deviation.

Table 5
Gravimetric water content (%) as a function of depth: after trafficking, after 1 year of trafficking, after 2 years of trafficking on undisturbed silty sand soil.

Soil layer, cm	After trafficking	After 1 year of trafficking	After 2 years of trafficking
0-15	25.9	13.1	48.4
15-30	21.5	11.6	18.2
30-45	16.8	8.2	17.0

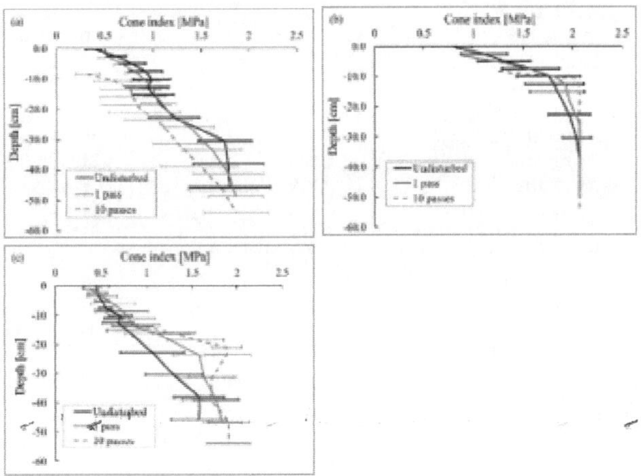

Fig. 9. Measured cone indexes as a function of depth: (a) after trafficking, (b) after 1 year of trafficking, (c) after 2 years of trafficking on silty sand soil. Horizontal bars indicate standard deviation.

In clay loam soil the soil particles were not compacted significantly neither with one nor ten passes. Obviously the deep grooves after ten passes (Fig. 2) were formed due to the soil displacement to the sides. This conclusion was supported by the measurement results of bulk density (Fig. 10) and cone indexes (Fig. 11). However, the K_{fs} indicates (Fig. 5) that water infiltration is decreased in both tracks in comparison to the area in natural state. Moreover, there are no significant changes in water infiltration during 2-years period. Thus, obviously the pore system of clay-rich soil will be seriously affected during the traffick and the recovery takes longer than 2 years. Gravimetric water contents for studied period are presented in Table 6.

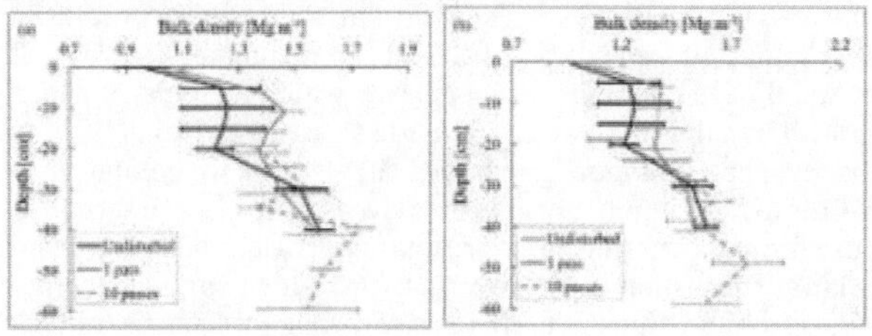

Fig. 10. Measured bulk densities as a function of depth: (a) after 1 year of trafficking, (b) after 2 years of trafficking on clay loam soil. Horizontal bars indicate standard deviation.

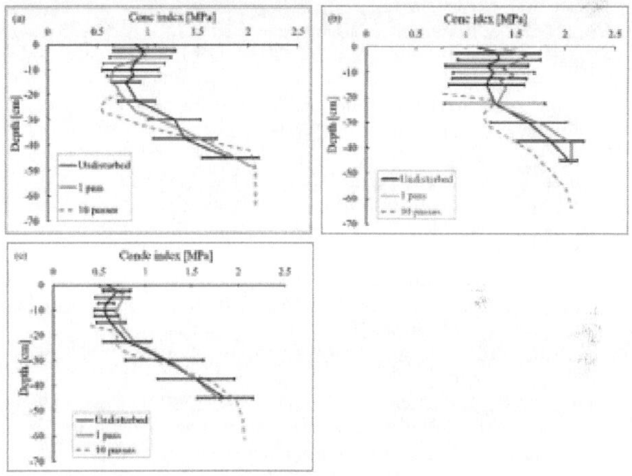

Fig. 11. Measured cone indexes as a function of depth: (a) after trafficking, (b) after 1 year of trafficking, (c) after 2 years of trafficking on clay loam soil. Horizontal bars indicate standard deviation.

Table 6
Gravimetric water content (%) as a function of depth: after trafficking, after 1 year of trafficking, after 2 years of trafficking on undisturbed clay loam soil.

Soil layer, cm	After trafficking	After 1 year of trafficking	After 2 years of trafficking
0-15	29.9	21.1	34.4
15-30	32.5	30.4	37.6
30-45	21.3	21.8	25.3

4. Conclusions

This study reveals that with high probability the largest mobility issues for follow-on battles at the previously trafficked area will be related to the high surface roughness, compacted soil layers in deeper depth and impeded water infiltration rates. The primary question determining the disturbance severity as well as the alleviation is the movement scheme used previously. In case dispersed maneuvering by the military vehicles was used, there could be dependent on soil texture and organic matter content compacted layers around 10–15 cm depth that will impede water infiltration and can affect negatively thrust generation. These soil properties will have an effect on trafficability app. one year after traffick. For column type movement the disturbance is more severe and its effects are reflected in first place in high surface roughness index but also in disturbed water infiltration processes. The compacted layer around 20–30 cm depth as well as the disrupted pore continuity will interfere water free movement in top layer and thus have an effect on soil strength state and bearing capacity. The suggestion from battlefield disturbance point of view are first, to keep number of passes as low as possible, and secondly, to use for maneuvering more organic matter rich soils. However, the latter suggestion is controversial due to the poor trafficability characteristics of peat, i.e. there is constant threat to get stuck because of low shear strength and high compressibility of peat.

5. References

Affleck, R.T., 2005. Disturbance Measurements from Off-Road Vehicles on Seasonal Terrain. ERDC/CRREL TR-05-12. US Army Corps of Engineers. Engineer Research and Development Center.

ATP 3-21.11, 2020. Stryker Brigade Combat Team Infantry Rifle Company, November 2020. Headquarters, Department of the Army.

Arvidsson, J., Keller, T., 2007. Soil stress as affected by wheel load and tyre inflation pressure. Soil Tillage Res. 96, 284–291.

Ayers, P., Rice, M., 2017. Analysis of vehicle platoon movement and speed-spacing relationships during military exercises. J. Terramech. 73, 37–47.

Bagarello, V., Iovino, M., Elrick, D., 2004. A simplified falling-head technique for rapid determination of field-saturated hydraulic conductivity. Soil Sci. Soc. America J. 68, 66–73.

Bekker, M.G., 1960. Off the road locomotion. Ann Arbor, Michigan: University of Michigan Press.

Benghough, A.G., Campbell, D.J., O'Sullivan, M.F., 2001. Penetrometer techniques in relation to soil compaction and root growth, in: Smith, K.A., Mullins, C.E. (Eds.), Soil and Environmental Analysis, 2nd ed. Marcel Dekker, New York, pp. 377–403.

Durst, P.J., Mason, G.L., McKinley, B., Baylot, A., 2011. Predicting RMS surface roughness using fractal dimension and PSD parameters. J. Terramech. 48, 105–111.

Elrick, D.E., Reynolds, W.D., 1992. Methods for analyzing constant-head well permeameter data. Soil Sci. Soc. America J. 56, 320–323.

Hamza, M.A., Anderson, W.K., 2005. Soil compaction in cropping systems. A review of the nature, causes and possible solutions. Soil Tillage Res. 82, 121–145.

Jalaväekompanii lahingutegevus, 2012. Kaitseväe Ühendatud Õppeasutused. Taktika õppetooli õppematerjal, 221 pp.

Liu, K., Ayers, P., Howard, H., Anderson, A., Kane, K., 2011. Multi-pass rutting study for turning wheeles and tracked vehicles. Trans. ASABE 54(1), 5–12.

Prose, D.V., 1985. Persistent effects of armored military maneuvers on some soils of the Mojave Desert. Environ. Geol. Water. Sci. 7 (3), 163–70.

Raper, R.L., 2005. Agricultural traffic impacts on soil. J. Terramec. 42, 259–280.

Robinson, G.W., 1922. A new method for the mechanical analysis of soil and other dispersions. J. Agric. Sci. 12, 306–321.

Sullivan, P.M., Anderson, A.B., 2000. A Methodology for Estimating Army Training and Testing Area Carrying Capacity (ATTACC) Vehicle Severity Factors and Local Condition Factors. ERDC TR-00-2. Engineer Research and Development Centre. US Army Corps of Engineers.

Trumbull, V.L., Dubois, P.C., Brozka, R.J., Guyette, R., 1994. Military camping impacts on vegetation and soils of the Ozark Plateau. J. Environ. Manage. 40, 329–339.

Vorobyova, L.A., 1998. Chemical Analysis of Soils. Moscow University Press, Moscow.

Webb, R.H., 2002. Recovery of severely compacted soils in the Mojave Desert, California, USA. Arid Land Res. Manag. 16, 291–305.

Wong, J.Y., 2008. Theory of ground vehicles (4th ed.). NJ: John Wiley and Sons, Inc.

Appendix

Table A.1
Estonian average monthly temperatures and precipitation for the study period. (Composed based on ilmateenistus.ee)

Month, year	Average temperature, C°	Average precipitation, mm
September, 2013	12.2	51
October, 2013	7.6	53
November, 2013	4.8	75
December, 2013	2.5	58
January, 2014	-6.4	40
February, 2014	-0.3	30
March, 2014	2.1	33
April, 2014	5.8	16
May, 2014	11.1	62
June, 2014	13.2	88
July, 2014	19.6	38
August, 2014	17.5	124
September, 2014	12.9	30
October, 2014	6.5	42
November, 2014	2.3	25
December, 2014	-0.2	68
January, 2015	-0.8	55
February, 2015	-0.1	27
March, 2015	2.5	32
April, 2015	5.2	57
May, 2015	9.8	40
June, 2015	13.9	45
July, 2015	16.1	81
August, 2015	17.1	37
September, 2015	13.3	61
October, 2015	5.9	11
November, 2015	4.9	66
December, 2015	3.3	54
January, 2016	-7.7	61
February, 2016	0.7	70
March, 2016	0.2	20
April, 2016	5.3	48
May, 2016	13.1	17

3835 | Effect of steer angle rate upon tire lateral force generation on two different soils

Dawid Tatarynow [a,*], Paweł Tomiło [b], Jarosław Pytka [a], Piotr Budzyńki [a], Michał Kuszneruk [a]

[a] *Faculty of Mechanical Engineering, Lublin University of Technology, Lublin, Poland*

[b] *Faculty of Management, Lublin University of Technology, Lublin, Poland*

* *Corresponding author: d.tatarynow@pollub.pl*

ABSTRACT

The mechanism of formation and the characteristics of the tire lateral force play an important role in terms of driving dynamics, safety and comfort. In the case of an off-road vehicle, the generation of tire lateral force is, among others, dependent on the soil. The effects of the three-phase nature of the soil medium may cause the characteristics of the lateral force build-up to be degressive or progressive. Experimental research was undertaken to determine tire lateral force on the steered wheel of the Suzuki Vitara off-road vehicle driven on loess and sandy soil as well as on wet snow. The vehicle was equipped with rotating wheel dynamometers and a steering robot. The measurement of the lateral force was carried out for three values of the steer angle rate, 100, 500 and 1500 deg/sec. Based on the obtained results, $F_Y(\delta)$ characteristics were obtained and further analyzed in order to determine the mathematical function describing the investigated mechanism of tyre – soil interaction.

Keywords
Off-road vehicle dynamics
Tire dynamics
Tire lateral force
Non-linear modeling
Model identification
Field measurement

1. Introduction

Tire lateral force measurements are of interest to vehicle designers for steerability and stability, comfort as well as safety (Crolla and El-razaz, 1987, Mitschke and Wallentowitz, 2005). Experimental data of lateral forces are needed for verification and validation in vehicle dynamics modelling and simulation studies. In general, there are two methods for tire lateral force experimental determination: laboratory and field tests (Albinsson et al, 2019, Feldesi et al 2020). Laboratory experiments have the advantage of better control of soil conditions and, consequently, can obtain repeatable and accurate results (Armbruster, 1986). By comparison, the more realistic field surfaces in outdoor testing promise results closer to the reality (Ferhadbegovic et al, 2006).

Laboratory tests are usually performed in a controlled environment of a soil bin. While flat track and large drum tire test rigs are used for most on-road applications, conditions for off-road measurements require so-called wheel testers. Two examples of wheel testers are the Hohenheim University single wheel tester (Armbruster, 1986), in which trailing arms guide the test wheel and a hydraulic cylinder applies tire load, and the Harper Adams trailed single wheel tire tester (Schlotter, Keen, 2003), designed to represent a quarter vehicle.

Field testing uses either wheel testers pulled by a carrier vehicle or instrumented vehicles. One example of such instrumented vehicle is the U.S. Engineer Research and Development Center, Cold Regions Research and Engineering Laboratory (ERDC-CRREL) Instrumented Vehicle CIV (Couretmarsh and Shoop, 2009, Shoop et al, 1994).

In vehicle dynamics testing, two general modes of steering wheel adjustment are in practical use: sine wave excitation and ramp change or trapeze. Schlotter and Keen, (2003) reported on measured lateral forces on large agricultural tires with the Hohenheim Tyre Tester at a sinusoidal adjustment of the tire slip angle. This method is suitable for stationary tire dynamics analysis. Results are usually presented in a form of Bode diagram together with the cornering frequency parameter for the amplitude ration and the phase shift.

The ramp change of the steering wheel or tire slip angle provides another method of excitation. The relaxation length parameter can be obtained from such experiments,. By definition this parameter is driving speed independent. This procedure is used for non-stationary analysis of vehicle lateral dynamics.

Pytka et al, (2011) and Pytka, (2014) have developed an experimental setup consisting of instrumentation capable of measuring the tire lateral forces in the field, installed in a test vehicle with a steering robot and a computer system to record the data. This vehicle – based sensor was used for field measurements of vehicle dynamics related quantities with tire lateral force among them. These authors pointed that when driving on deformable terrain with low or medium inflation pressure, analysis of lateral forces should take into account the soil effect on the force generation mechanism. Unlike tires with a rigid structure, massive tread, low-pressure tires flex significantly, which is important for the characteristics of the lateral force.

Based on the above analysis of the literature and having access to database with experimental data of tire lateral force, the authors aimed at searching for a mathematical description of the mechanism of generating the side force of the tire when the car is driving in the turning mode on soft soil.

https://doi.org/10.56884/PHHZ8245
ISBN 978-1-942112-55-6 (pdf)

2. Experimental methods

Since it was assumed that the mathematical model would be reconstructed on the basis of experimental data, it was necessary to develop a measurement method and prepare and conduct an experiment that enabled the direct measurement of the lateral force on the steered wheel. In addition, it was assumed that the measurement would be carried out using a full-size vehicle while running in real conditions.

Details of the research methodology are presented below.

2.1 Wheel dynamometer

A rotating wheel dynamometer was designed to measure six elements: three forces and three moments acting on a road wheel of a vehicle. The wheel dynamometer system consists of a sensor with strain gage transducers. A modified wheel rim enables to install this device on a wheel hub, without any modifications to the vehicle. Another item is the data recording system, which was developed for collecting measurement data and to communicate with the dynamometer microprocessor unit wirelessly. The measurements can be performed with the test vehicle in either on-road or off-road conditions, at maximum longitudinal speed of 60km/h. For the present study, two wheel dynamometers were used and they were installed on the front (steered) wheels of the test vehicle. Figure 1 shows the dynamometer installed on the left wheel of the test vehicle. More details on the dynamometer can be found in Pytka et al (2011) and Pytka, (2014).

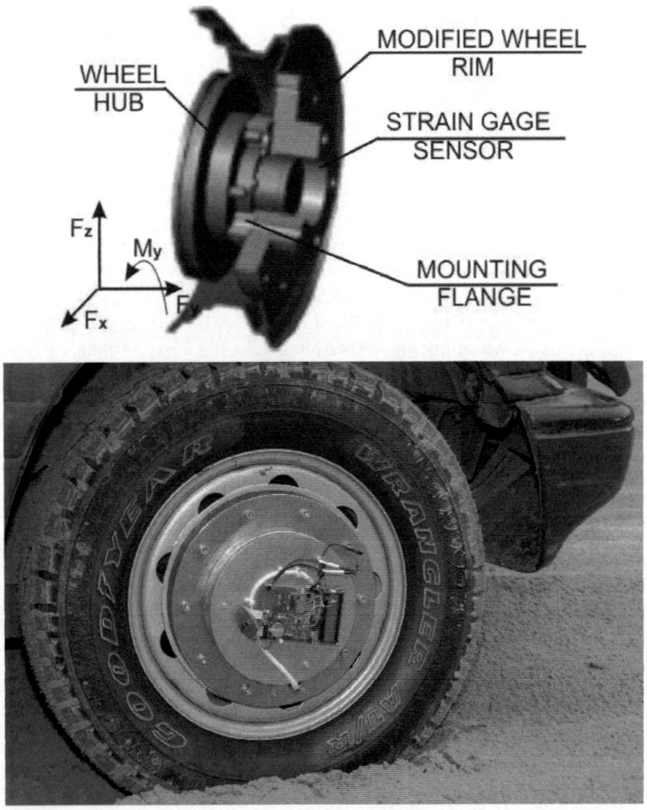

Fig. 1. Wheel dynamometer – schematic and the device installed on the test vehicle

2.2 Test vehicle as a sensor

The idea of the use of an instrumented vehicle as a sensor is a good alternative to laboratory tests but also for using of the so called wheel testers.

The test vehicle, based on the Suzuki Vitara 1.6, was instrumented by installing two dynamometers on the front wheels and the rest of the instrumentation. Measurements of tire forces acting while driving require that the slip angle of the vehicle is controlled depending on the excitation method (sine wave or ramp change/trapeze).

For this purpose, the method described by Coutermarsh and Shoop (2009) was used, namely, the test vehicle is towed by a heavy vehicle, which minimizes the feedback of the reaction force and changes in the required straight trajectory. The two vehicles were coupled by means of a custom made truss with three joints: one major spherical joint that allowed lateral movements between the two vehicles during pre-maneuvers, and two single-axis joints that allowed limited vertical movements of the test vehicle to compensate for surface deflections or unevenness.

Fig. 2 shows a diagram and Fig. 3 shows a photograph of the towing method.

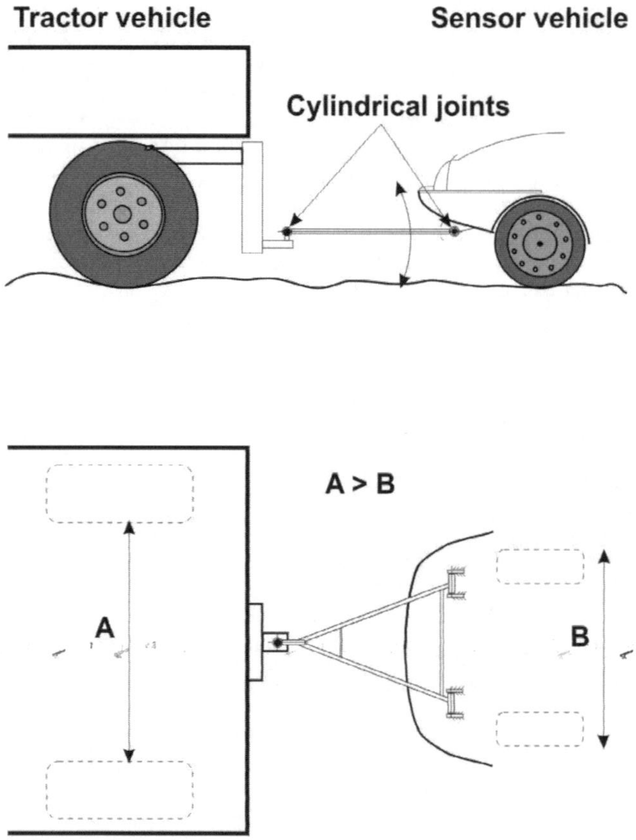

Fig. 2. A schematic of the towing truss

Fig. 3. The two vehicles during a test ride: the tow vehicle and the sensor-vehicle. A closeup view in the lower photo.

2.3 Steer angle control using a steering robot

In order to ensure precise and repeatable control of the steering wheel, the sensor-vehicle was equipped with a steering robot, which consisted of a ring motor mounted on the steering wheel of the vehicle, a reaction column and steering torque and steering angle sensors.

Fig. 4. The ring motor – a part of the steering robot, installed on the steering wheel of the test vehicle

Fig. 5. The reaction column and steering torque sensors

The photographs in Figures 4 and 5 show details of the steering robot in the test vehicle. Figure 6 illustrates the schematic of the instrumentation installed in the sensor – vehicle.

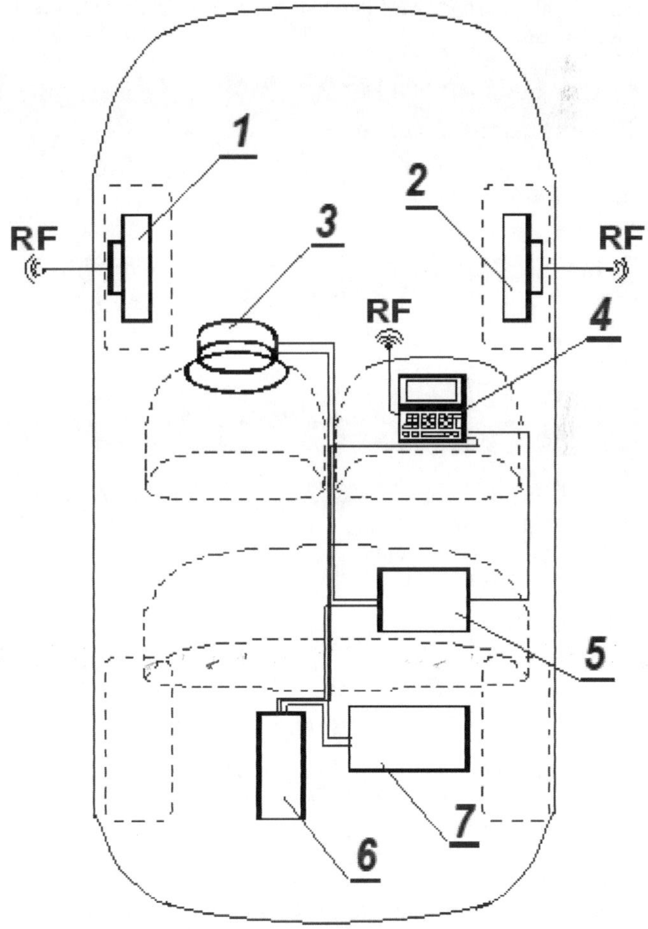

Fig. 6. The instrumentation of the sensor- vehicle. 1, 2 – rotating wheel dynamometers, 3 – ring motor of the steering robot, 4 – notebook computer, user interface for control of the robot and communication with dynamometers, 5 – steering robot computer, 6 – DGPS and inertial measurement unit, 7 – power supply, RF = radio frequency communication device

2.4 Field procedures and test program

Measurements were carried out on two different soils, sand and loess agricultural soil. A typical loess soil formed from periglacial or aeolian sediment, containing about 20% of clay with a balance of roughly equal parts sand and silt. Loess soils occur in the Lublin Upland, near Nałęczów and Kazimierz Dolny. They are beneficial to a variety of crops. The period of the measurements, November - December, ensured small fluctuations in soil moisture due to low insolation and weak wind. There was no precipitation in the week before and during the measurements.

In order to obtain representative results, the measurements were repeated at least 5 times.

Preparation of the site for the tests consisted in leveling and loosening the soil over a length of approx. 150 - 200 m, thanks to which it was possible to carry out the required number of repetitions of the experiment in one pass along the entire length of the track. Leveling and loosening were carried out using a tractor with agricultural equipment.

The test program included measurements of the lateral force on the front wheels of the sensor-vehicle on both soils, with selected steering parameters for sine wave and ramp change excitation. Force measurements were carried out at a constant longitudinal (towing) speed of the sensor-vehicle, which was about 5-6 km/h.

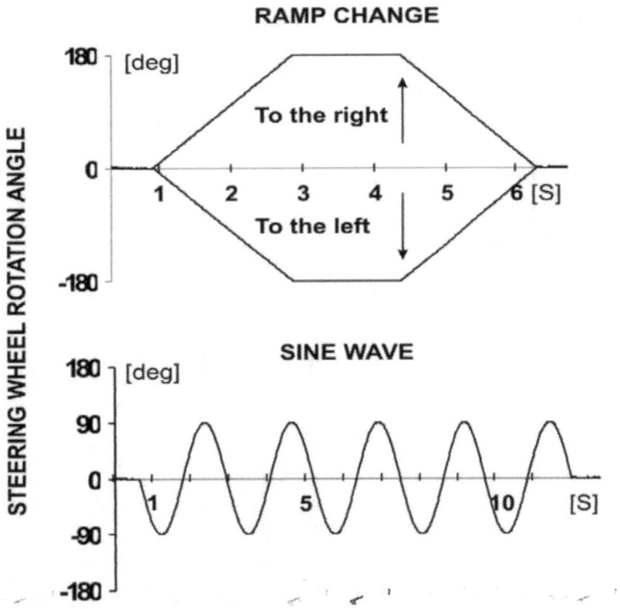

Fig. 7. A graph showing the two steering wheel angle excitation modes

Two modes of excitation were used in the tests: sine wave, with frequencies of 0.5, 1.0 and 2.5 Hz, amplitude of 180deg and ramp change or trapeze excitation of 180 deg steering angle range and 5 seconds duration. Steering angle ranges were selected taking into account the ratio of the steering reduction gear of the test vehicle, which is 18. This steering ratio is constant and almost linear for the Vitara vehicle. The test vehicle's steering system has been carefully overhauled before the tests in order to minimize effects of worn parts. In the case of sine wave excitation, in which the full amplitude of the steering wheel angle was 180 degrees, the values of the side slip angle were 5 degrees a side. For ramp change excitation, the full range of the steering angle was 180 degrees, so the side slip angle at maximum steering angle was 10 degrees. Three values of steering wheel angle rate have been set up for the ramp change mode: 100, 500 and 1500 deg/sec. The rate of 100 deg/sec corresponds to very slow steering, in practice it does not apply to maneuvering and only occurs when driving a curve. The rate of 500 deg/sec is a value typical for dynamic maneuvers, achievable for most drivers. The highest steering rate, 1500 deg/sec, is unattainable in practice, only in the case of outstanding rally drivers, and for much smaller ranges of steer angle. It was selected to be tested in extreme conditions, with rapid steering. Figure 7 illustrates time course of the two excitation modes with amplitude and time intervals.

3. Results of field experiment

Because a full presentation of the results of the described research was included in the reference by Pytka (2014), this paragraph presents a summary of these results. The final result consists of both time course of the tire lateral force as well as peak force coefficient and time. In each of the considered cases is the arithmetic mean calculated from the individual repetitions for a given case (steer angle rate, type of excitation – sine or trapeze, type of soil). In addition, the graphs presenting the individual results also included the course of the steer angle. Each of the graphs contains a set of two curves, which are the course of the side force on a given wheel (right or left), for turning left and right. Figure 8 includes sample curves of tire lateral force, one for the sine wave and another for the ramp change excitation.

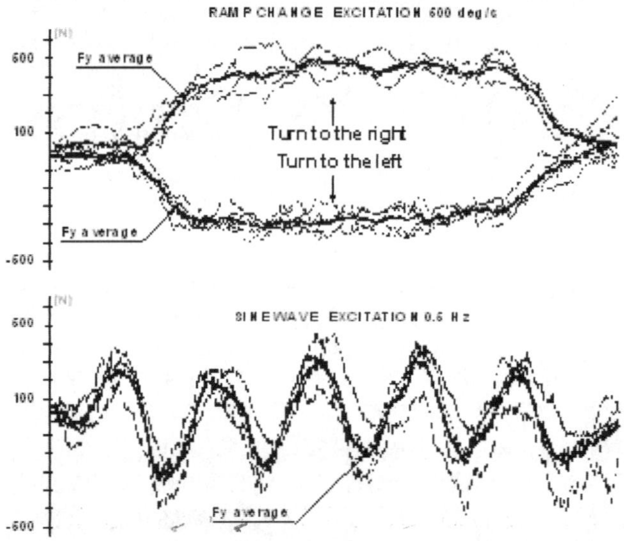

Fig. 8. Tire lateral force for sine wave excitation (upper graph) and ramp change excitation (lower graph)

Data reduction for ramp change excitation included searching for peak tire lateral force values within ramp change results. An average value for the F_y was calculated for each test variant (a moving average of 100 points taken from a saturation stage). Based on this data, a peak tire lateral force factor $F_{ypeak}/F_{yaverage}$ was determined. The peak force time, another measure that was derived from the experimental data, named analogously to the peak response time (Mitschke, Wallentowitz, 2005). This parameter describes a delay of the steering wheel – tyre – surface system. Figure 9 illustrates these measures. A complete set of results in included in the cited literature (Pytka, 2014).

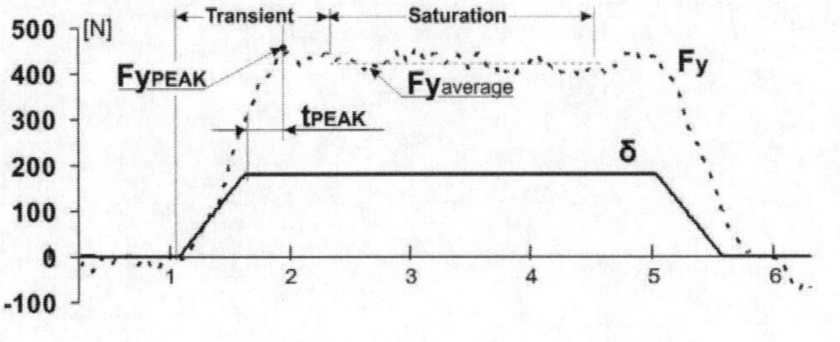

Fig. 9. Definition of peak force coefficient and peak force time

3.1 *Effect of steer angle rate and frequency*

Based on the results, no significant effect of ramp change rate on tire lateral force for the loess soil was detected. This effect is evident on sandy soil: the highest values of the F_y were indicated at 500deg/s. One explanation of this could be the effect of high steer angle rate, which weakened the tire lateral force generation due to soil inertia.

Peak force time values showed the effect of steer angle rate on time delay of the system. The trend was similar for the two soil surfaces: the greatest delay occurred at the slowest steer angle rate, 100deg/s. This trend was opposite to that on the hardpan, for which slowly turning the wheels results in less inertia effect and shorter time of delay. On the other hand, the values of the peak force time are greater for sandy soil than for loess, as an effect of higher deformability of sand.

At sine wave excitation, the trend in tire lateral force is similar to trapeze excitation mode results: the lateral force F_y is about 1.5 to 2 times greater for the loess soil surface than for the sand.

3.2 *Effect of soil surface, loess vs sand*

Values of tire lateral force captured on loess soil surface were about two times higher than on sandy soil surface. This could be an effect of high deformability of sandy soil that resulted in deeper rut. Typically, rut depth in sand was about 170-210mm, while in loess soil it was smaller, namely 70-130mm. High deformation of the soil surface results in low stresses and less intensive soil response. Here, soil deformability should be understood as multidirectional, not only in one, vertical direction, that causes rut formation.

Concluding, the following was pointed out. Tire lateral forces determined on loess surface were approx. 2 times greater than those obtained on the two remaining surfaces. The steer angle rate effect on the peak tire force coefficient was observed only for sandy soil. This effect was proportional, for higher rates, the coefficients were greater. For loess soil, the parameter showed no dependency on angle rate. The effect of angle rate was also significant for peak force time, which quantifies delay of the system; for both soil surfaces, this parameter decreased with increased angle rate. The effect of surface was evident for peak tire lateral force coefficient. This parameter is constant and equal to 1 for the loess soil, while it increased from about 0.9 to 1.3 for sandy soil. Another effect of surface was for peak force time, which was the greatest for sandy soil due to surface deformability.

4. Mathematical model of tire lateral force generating mechanism

4.1 *Introduction*

Further analysis of the obtained results may provide an answer to the question asked in the introduction, namely, what is the character of the lateral force courses for the two tested soils. The visual evaluation of the course of forces gives only preliminary information about trends, it is a qualitative result. Therefore, research was undertaken toward a quantitative analysis of the results obtained in field experiments.

Singular-Spectrum Analysis (SSA) was used to analyze the forces on the wheels when turning. SSA is a nonparametric spectral estimation method based on embedding a time series into lagged vectors - Trajectory Matrix. The algorithm decomposes the matrix by applying singular value decomposition (SVD) (Golyandina et al., 2001).

4.2 *Data reduction*

Each of the time series of the variability of the forces on the wheels was divided into 50 components. The first component for each time series was used as a smoothed version of it, the other components were added together. As a result, two time courses F_0 and F_{1-49} were obtained. An exemplary application of the SSA on the discussed time series is shown in Figure 10.

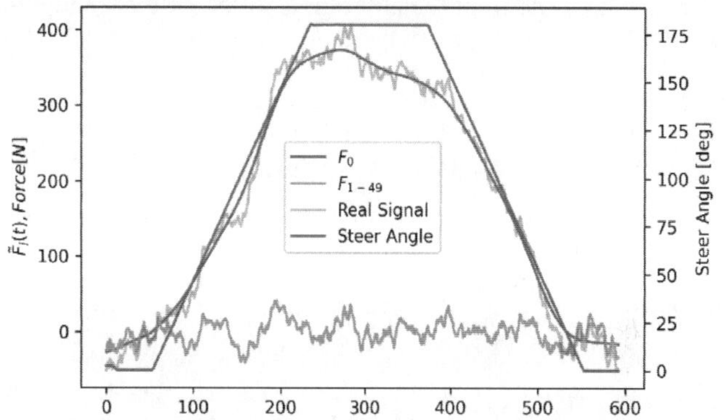

Fig. 10. An exemplary application of the SSA on the discussed time series

The resulting components F_0 and the sum of the components from 1 to 49 - F_{1-49} were analyzed statistically, for the components F_{1-49} the mean and standard deviation (std) were calculated, and for F_0 the maximum value. Table 1 presents the discussed values for the tested types of soil and rotation speed.

Table 1
Statistical analysis of F_0 and F_{1-49} components

Soil type	Steer angle rate ($\dot{\delta}$) [deg/s]	F_{1-49} Mean	F_{1-49} Std	F_0 Max
Loess	100	-0.2341	10.9851	458.9832
	500	-0.6887	35.8731	463.9799
	1500	0.2316	50.7891	466.1673
Sand	100	-0.7626	19.3253	178.2823
	500	-0.0519	19.0587	230.4148
	1500	0.1260	25.9091	168.2012

The value of F_0 was normalized by dividing by the maximum value, the same treatment was applied to steer angle rate $\dot{\delta}$. The resulting time series were subtracted from each other: $D_R = \dot{\delta} -$

F_0. Figures 11 thru 13 show the averaged D_R timeseries. The figures show an area that was created from the maximum and minimum values for each sample. Markings on the charts: P0 - the point where the wheel steering started, P1 - maintaining the wheel position at an side slip angle, P2 - steering the wheel in the opposite direction, P3 - stopping the steering action.

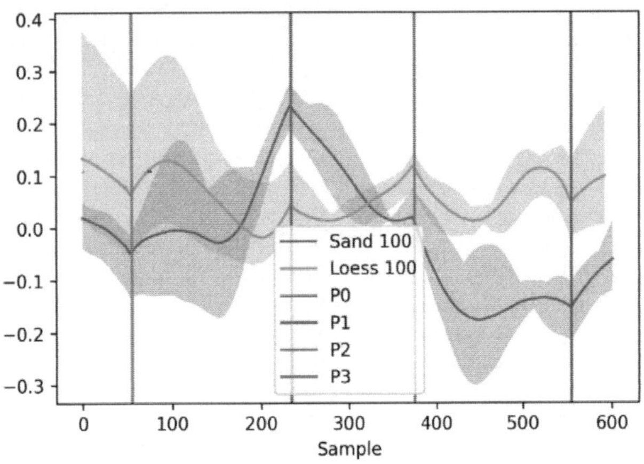

Fig. 11. Mean D_R for sand and loess at steer angle rate 100 deg/s

For the steer angle rate of 100 deg/s, the time series for loess remains within the trapezoidal function, in the case of sand, the values after point P2 go outside the area of the trapezoidal function. This means that during the return of the wheel to the initial position, the forces were higher than for the P0-P1. In the case of both averaged timeseries, there is a phenomenon of maintaining the value within the trapezoidal function P0-P3.

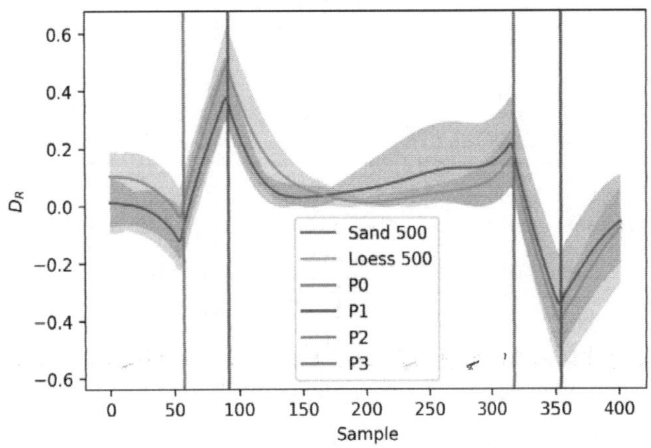

Fig. 12. Mean D_R for sand and loess at steer angle rate 500 deg/s

In the initial phase of the P1-P2 interval, for the steer angle rate of 500 deg/s, in the case of loess, there is quite a large variation between individual measurements, and for sand, the variation increases to the point P2.

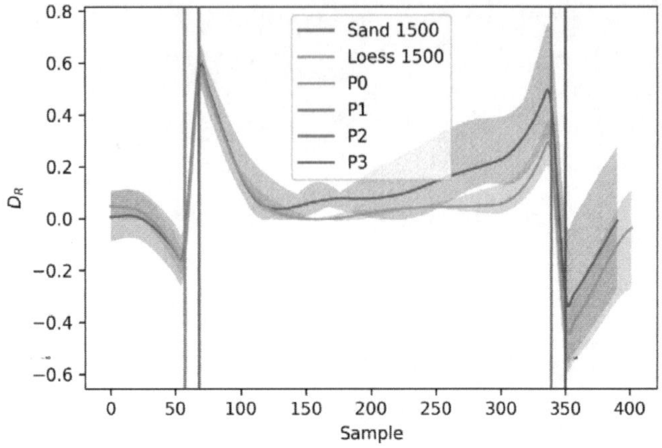

Fig. 13. Mean D_R for sand and loess at steer angle rate 1500 deg/s

For the steer angle rate of 1500 deg/s, the timeseries for loess in the P0-P3 were characterized by high repeatability. In the case of sand in the P1-P2 (maintaining wheel position), the difference between the time series increased until the P2.

In the case of the steer angle rate of 100 deg/s, loess soil is characterized by greater stability of values. In the case of both the steer angle rate of 500 deg/s and 1500 deg/s, for sand there is an increase in the difference between the values for individual measurements. For the same steer angle rate of the wheel, loess shows greater repeatability. Table 2 presents the discussed values for individual points.

Table 2
Comparison of values in points P0-P3

		Sand	Loess	Sand	Loess	Sand	Loess
$\dot{\delta}$ [deg/s]		100		500		1500	
P0	Mean	-0.0460	0.0668	-0.1141	0.0007	-0.0980	-0.1142
	d_y	0.1865	0.3095	0.1865	0.2877	0.1931	0.1076
P1	Mean	0.1240	0.0441	0.3595	0.4793	0.5757	0.5526
	d_y	0.2331	0.1165	0.2165	0.3320	0.1778	0.0912
P2	Mean	0.0876	0.1147	0.1838	0.1411	0.4350	0.2401
	d_y	0.0167	0.0710	0.3195	0.1885	0.4631	0.2056
P3	Mean	-0.5166	0.0499	-0.3462	-0.4017	0.2989	-0.4193
	d_y	0.1121	0.1518	0.3410	0.2583	0.5095	0.2419

In the case of loess soil, the maximum value of the tire lateral force on the wheel increases with the increase in the steer angle rate, similarly to the standard deviation. The sand is characterized by the achievement of the maximum tire lateral force value for the steer angle rate of 500deg/s.

Based on the dy values given in Table 2, the stability of the tire lateral force value can be assessed. The differences between the two tested soils are quite significant here. In the case of loess soil, in the range P1-P2, stability of measured values of Fy was found. However, the values of this force measured on sandy soil show significant dispersion.

5. Conclusions and future work

The article presented experimental studies, the aim of which was to determine the time courses of tire lateral force on two soils: loess and sand, at different values of steer angle rate, 100, 500 and 1500 deg/s.

The SVD method allowed for a quantitative analysis of tire lateral force courses obtained in measurements on two different soils. It was found that the deviation between the excitation and the first averaged component D_R from all measurements reaches high values in the case of sandy soil, especially at the low steer angle rate of 100 deg/s. For higher rates, the D_R parameter decreases, with the values for loess soil being generally higher compared to sand.

Taking into account the result of the analysis of tire lateral force measurements, it is proposed in the future to undertake research using machine learning methods, in particular the model-based method, of a decision-making nature. A system could be developed that uses a predictive model of the vehicle dynamics to ask questions of the form "what will happen if I do x?" to choose the best x1. In the alternative model-free approach, the modeling step is bypassed altogether in favor of learning a control policy directly. This can bring beneficial results in the systems of supporting the steering of an off-road vehicle, moving on terrain with various properties.

6. Nomenclature

D_R	Deviation between the excitation and the first averaged component for all measurements	[-]
dy	Difference between max and min at a point, normalized	[-]
F_o	The first component of decomposition	[N]
$F_{1\text{-}49}$	Other components	[N]
F_{Ypeak}	Peak tire lateral force	[N]
$F_{Yaverage}$	Average tire lateral force	[N]
t_{peak}	Tire lateral peak force time	[s]
$\dot{\delta}$	Steer angle rate	[deg/s]

7. Acknowledgements

This work was prepared within the project KONF/SP/0330/2023/1 entitled "16th European-African Regional Conference of the ISTVS", financed by the Ministry of Education and Science (Poland) as a part of the Doskonała Nauka II Programme.

8. Declaration of competing interest

The authors declare that they have no known competing financial interests or personal relationships that could have appeared to influence the work reported in this paper.

9. References

Albinsson A., Bruzelius F., Els P.S., Jacobson B., Bakker E., 2019. Tire Lateral Vibration Considerations in Vehicle-Based Tire Testing. Tire Sci Techn, 47 (3), 211–231.

Armbruster K., 1986. Development of a single wheel tester for measuremnts on driven angled wheels. Proc. 4th ISTVS European Conference, Wageningen, The Netherlands.

Coutermarsh B.A., Shoop S.A., 2009. Tire slip-angle force measurements on winter surfaces. J. Terramech. 46, 157-163.

Crolla D.A., El-Razaz A.S.A., 1987. A review of the combined lateral and longitudinal force generation of tyres on deformable surfaces. J.of Terramech. 24, 199-225.

Feldesi F., Botha T. R., Els P. S., 2020. Full-field strain measurements of the inside of an agricultural tyre using digital image correlation. J Terramech, 91, 309 - 318

Ferhadbegovic B., Brinkmann Ch., Kutzbach H.D., Bottinger S., 2006. Hohenheim tyre model – a transient model for driving dynamics simulation. Proc. of the ISTVS Conference, Budapest, Hungary.

Golyandina, N., Nekrutkin, V., Zhigliavski, A., 2001. Analysis of time series structure : SSA and related techniques, Chapman & Hall/CRC, Boca Raton.

Mitschke M, Wallentowitz H., 2005. Dynamik der Kraftfahrzeuge, Springer Verlag, Berlin-Heidelberg.

Schlotter V., Keen A., 2003. The dynamic characteristic of off-road tyres. Proc. of the 9th European Conference of the ISTVS, Harper Adams, UK.

Pytka, P. Tarkowski, S. Fijałkowski, P. Budzyński, J. Dąbrowski, W. Kupicz, P. Pytka, 2011. An Instrumented Vehicle for Off-Road Dynamics Testing, J Terramech, 48(5), 384 – 395.

Pytka J., 2014. Effects of steering dynamics upon tyre lateral forces on deformable surfaces. Int. Journal of Vehicle Design, 64 (2/3/4), 170-194.

Shoop, S., Young, B., Alger, R., Davis, J. 1994. Effect of test method on winter traction measurements, J Terramech 31(3), 153-161

4141 | Modeling and verification of a full-scale forestry vehicle real-time multi-physics digital twin

Mattias Lehto [a,*], Antti Tikanmäki [b], Torbjörn Lindbäck [a], Håkan Lideskog [a], Magnus Karlberg [a]

[a] *Luleå University of Technology, 97187 Luleå, Sweden*

[b] *University of Oulu, Pentti Kaiteran katu 1, 90570 Oulu, Finland*

* *Corresponding author: mattias.lehto@ltu.se*

ABSTRACT

At Luleå University of Technology, an offroad vehicle capable of performing autonomous forwarding of full-scale logs, has been developed and built. This vehicle serves as a modular research platform, where different hardware, software, and algorithms can be exchanged, tested, and validated. To further facilitate this research, a real-time multi-physics digital twin of the vehicle has been developed. The digital twin can be executed either in fully virtual or in Hardware-in-the-Loop mode. Virtual and hybrid testing and development are expected to increase the efficiency, safety, and sustainability of these activities. Future envisioned use-cases also include machine learning and generation of auto-annotated detector training data. The digital twin features multibody simulation, hydraulics, contacts and collisions, flexible tires, emulated sensors, and co-simulation capabilities. The test environment is a physical environment that has been recreated digitally, using data from a UAV-mounted camera and lidar, as well as handheld RTK GPS measurements. Therefore, accurate GPS positions can be obtained from the digital environment. The digital twin is equipped with an emulated GPS sensor and can thus follow the same GPS tracks as the physical vehicle. This paper presents the development and verification of the digital twin model. Calibration and testing of the digital twin's hydraulically actuated articulated joint is detailed, and the results are analyzed. The modeled hydraulics of the articulated joint show realistic dynamics. It is also shown that the level of detail of the wheel motors' hydraulic circuits, affects the calibration of the hydraulics model of the articulated joint, as they affect the resistance during turning.

Keywords
Multibody simulation
MBS
Offroad
Forwarder
Harvester
Hydraulics
Articulated vehicle
GPS
Simulation
Hardware-in-the-Loop

1. Introduction

The CTL (cut-to-length) tree harvesting practiced in the Nordic countries is at the technological forefront of current forestry methods. In recent years a lot of research and innovation has also been performed to increase the level of automation (LOA) of CTL forestry vehicles, such as forwarders and harvesters (Lindroos et al., 2017). At Luleå University of Technology, a 10-tonne modular offroad vehicle, capable of performing autonomous forwarding of full-scale logs, along with other forestry operations, has been developed and built. For more information about the vehicle, see Lideskog et al. (2015). This vehicle performed the world's first fully autonomous forwarding of full-scale logs in 2021, to the best knowledge of the authors. The mentioned vehicle serves as a modular research platform, where different hardware, software, and algorithms can be exchanged, tested, and validated. E.g., Li & Lideskog (2021) used the sensor hardware suite of the vehicle to perform successful real-time object detection and localization of stones and tree stumps on a clearcut area (harvested forest land). This was another important step towards autonomous forestry vehicles, since it enables autonomous object avoidance and targeting during operation.

To further facilitate this research, a real-time multi-physics digital twin of the vehicle has been developed. The digital twin can be executed either in fully virtual or in Hardware-in-the-Loop (HiL) mode. Virtual and hybrid testing and development are expected to increase the efficiency, safety, and sustainability of these activities. Future envisioned use-cases also include machine learning and generation of auto-annotated detector training data. All the aforementioned use-cases are supported by the opportunities brought up by Choi et al. (2021), except the environmental sustainability aspect which is not mentioned in their paper. Increased environmental sustainability of the testing is nonetheless seen by the authors as an effect of relocating more testing from physical to virtual environments. In addition to being less material intensive, virtual testing and machine learning activities also won't cause damage to the natural environment, such as soil damage or other potential disturbances of the flora and fauna.

To do HiL-simulation, where parts of the physical vehicle, or the entire physical vehicle, is embedded in the simulation, the simulation must be run in real-time (Fathy et al., 2006). Real-time simulation in agriculture and forestry has been documented in previous research to some extent, but mostly with heavy focus on one aspect and simplifications in other aspects. For example, HiL-testing of specific components, such as hydraulic actuators or ECU's, often with detailed vehicle or subsystem representations, but in very simplified or unspecified virtual environments (Raikwar et al., 2019; Engelmann et al., 2017; Lee et al., 2017; Kelber et. al., 2016). Other studies include simulators for operator training or sensor emulation, in richer virtual environments, but lacking the intention or vehicle fidelity, such as detailed hydraulics modeling, to be used for detailed vehicle analysis and development purposes (Tsolakis et al., 2019; Zheng et al., 2018; Dabbiru et al., 2020; Wolf et al., 2020; Baharudin et al., 2014). There have also been papers focusing on real-time simulation of offroad vehicles driving in soft soil, where the wheel-soil contact or the soil itself has been the focus, but other aspects of the virtual environment, as well as the internal vehicle dynamics, has not been a focus (Peiret et al., 2021, Dallas et al., 2021).

The simulation model closest to the model described in this paper is found in the paper by Baharudin et al., (2014). Their model is also implemented in an earlier version of the same software that has been used in this paper, Mevea (Mevea). While

https://doi.org/10.56884/TSOG1043
ISBN 978-1-942112-55-6 (pdf)

the purpose of their model is solely operator training, their vehicle model still has a higher level of detail than in other similar papers that have been found, such as a more detailed modeling of the hydraulics. The main differences between their model and the one described in this paper is that they are not aiming to use their model for HiL-simulations or vehicle development activities. They also haven't implemented any exteroceptive sensors (such as GPS) or automated driving into their model, nor are they driving in a scanned real-world environment.

The digital twin described in the following paper intends to fill the mentioned gaps: a detailed HiL-capable multi-physics simulation model of a forestry vehicle with high validity, that also enables driving in highly realistic environments while simultaneously emulating both proprioceptive (vehicle internal) and exteroceptive sensor signals. Describing the early modeling and verification of this digital twin is the aim of the following paper.

2. Method and materials

The overarching goal of this section is to describe how different parts of the vehicle and the environment were implemented into the real-time simulation model. Before the start of this work, the vehicle existed both physically (see Fig. 1) and as a detailed CAD-model (see Fig. 2).

Fig. 1. The physical vehicle

2.1 Data from CAD model

Since the physical vehicle was developed in-house, a detailed CAD-model of the vehicle was available. It can be seen in Fig. 2.

Fig. 2. A detailed CAD-model of the vehicle

Rigid bodies and their mass properties, such as mass, center of mass, and moments of inertia were extracted from the CAD-model. Joint positions were measured. Detailed visualization graphics and simplified collision graphics were also exported from the CAD-system, see Fig. 3. below.

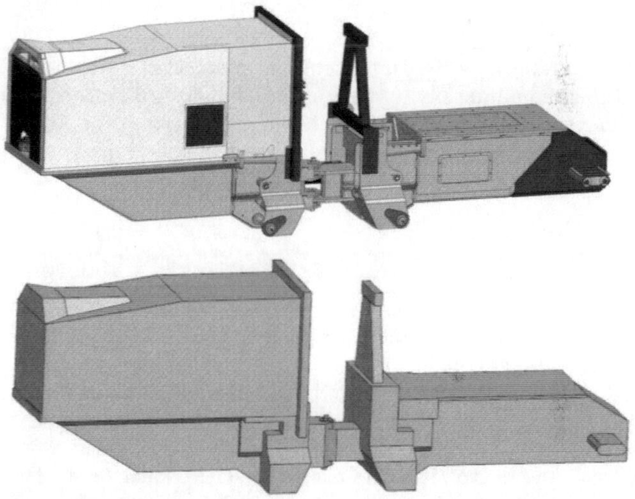

Fig. 3. Examples of detailed visualization graphics and the corresponding simplified collision graphics

2.2 Data from datasheets and drawings

To be able to accurately implement all the mechanical functionality of the vehicle, such as the hydraulic drivetrain and pendulum arms, there are a lot of complex components that need to be accurately described. Most of the needed information was found in the datasheets and drawings of the components. The obtained information includes basic information like mass, center of mass, and moments of inertia, as well as detailed information such as dimensions of hydraulic cylinders, characteristics of pumps, hydraulic valves, hydraulic motors, hydraulic hoses, diesel engine, etc. Although a large majority of the needed information was possible to acquire from drawings and datasheets, some of it was not available and had to be acquired through contacts with manufacturers, scientific papers, empirical testing, and approximations. The acquisition of these remaining unknown parameters was seen as the most challenging and time-consuming aspect of the modeling of the digital twin. But since the number of remaining unknown parameters was relatively low, the process was still manageable.

2.3 Hydraulics

The hydraulics simulation implemented in the used software Mevea is based on lumped fluid theory (Watton, 1989), where each hydraulic circuit is divided into discrete volumes with uniform internal pressures. Each volume is modeled as a hose or pipe with a diameter, length and Young's modulus, which are normally set to be as close as possible to the characteristics of the real connecting hose or pipe (multiple hoses and/or pipes can also be assigned to the same volume if needed). These volumes are separated by restricting valves of different types. The valves are modeled with a semi-empirical approach, where the valve type and its flow characteristics are used as input parameters. For example, the turbulent flow rate Q over a simple restricting valve, for positive Δp, is calculated as (Rahikainen et al., 2018):

$$Q = UC_v\sqrt{\Delta p} \tag{1}$$

where U is a parameter defining the relative valve opening, Δp is the pressure drop over the valve, and C_v is the semi-empirical flow rate coefficient of the valve defined as:

$$C_v = C_d A_t \sqrt{\frac{2}{\rho}} \tag{2}$$

where C_d is the flow discharge coefficient, A_t is the area of the valve and ρ is the oil density. In Mevea, C_v is assumed to be constant around the operating point of the nominal pressure drop Δp_{nom}, measured at the nominal flow rate of the fully open valve Q_{nom}, it then follows from Eq. 1 that C_v can be calculated as:

$$C_v = \frac{Q_{nom}}{\sqrt{\Delta p_{nom}}} \tag{3}$$

The flow characteristics Q_{nom} and Δp_{nom} can often be found in the datasheets of the individual components. The flow over a valve is considered to be laminar if the pressure difference is below 1 bar, otherwise it's considered to be turbulent, this is currently a fixed setting in Mevea. The oil bulk modulus is included in the calculations and can be adjusted based on the oil being used. Since hydraulic systems often have a relatively high frequency response, the timestep of the hydraulics calculation is possible to adjust to be the main simulation timestep divided by any chosen integer value. For more details on the software internal hydraulics equations, see Baharudin et al. (2014).

To increase the readability, simplified component names are occasionally used in this paper, e.g. "Pump 1", etc. These simplified names and their corresponding actual component names are listed in Table 1.

Table 1
Simplified and actual component names

Simplified component name	Component name
Articulated joint cylinder	Side System Hydagent 80/40-265
Crane	Cranab FC8 with CR250 gripper
Diesel engine	John Deere Powertech 4045
Differential valve	Poclain INT-VMA-H15
Pendulum arm cylinder	Side System Hydagent 80/40-370
Pump 1	Poclain P90R130
Pump 2	Bosch A10VO140
Valve	Parker L90LS
Wheel motor / Motor	Poclain MSE18

A simplified overview of the hydraulic system of the modeled vehicle is shown in Fig. 4 below.

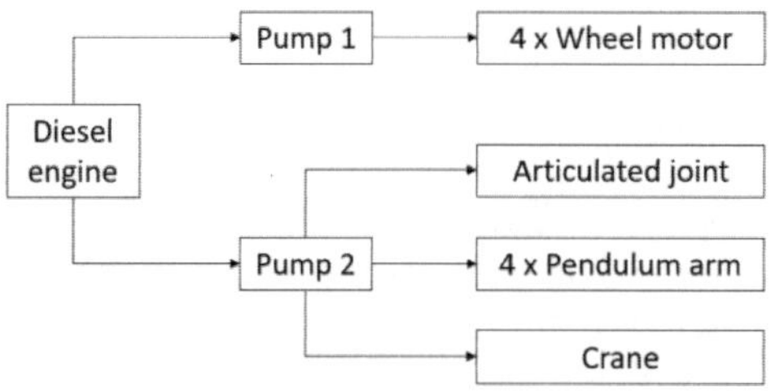

Fig. 4. Simplified overview of the vehicle's hydraulic system

The entire hydraulics system described in Fig. 4 was modeled in detail, except for the crane hydraulics. The crane hydraulics were left as future work since they were not relevant for the planned first stage of verification and validation testing.

For the first validation tests the most important part of the hydraulic system to model accurately was the articulated joint. An overview of the articulated joint hydraulics can be seen in Fig. 5 below. The articulated joint is actuated by two cylinders, where the A-side of one of the cylinders is connected to the B-side of the other cylinder, and vice versa. This configuration enables them to work in tandem.

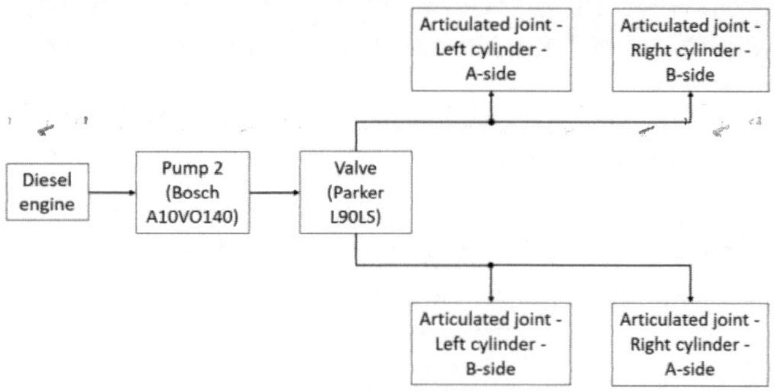

Fig. 5. Overview of the articulated joint hydraulics

The diesel engine was modeled by inserting its Speed-Torque curve, which was available online, into the virtual model as a spline curve. Thereafter, the virtual engine controller was set to keep the same constant angular speed as the physical diesel engine does. The Bosch pump, the L90LS valve, and the articulation cylinders were modeled by inserting the parameters found in the corresponding datasheets and drawings into the virtual model. A semi-empirical generic model of a L90LS valve was available in the simulation software and was configured to match the model used on the physical vehicle. This saved some time since the built in components, such as pressure compensator and shock valves, did not have to be modeled separately. For the pump and the cylinders, generic corresponding components were configured to match the ones on the physical vehicle. Thereafter, connections between the different components were established. The lengths of the connecting hoses were measured, and their diameters and rated pressures were checked. The Young's moduli of the hydraulic hoses were unknown. The best data found was from Hružík et al. (2017), where the Young's modulus of a hydraulic hose rated for 130 MPa was found to be 4.82×10^9 Pa. Higher rated pressure generally calls for the hose to have a correspondingly higher Young's modulus. Therefore, the Young's moduli of the hoses were assumed to increase linearly with increased rated pressure. Thus, as an example, a hose with a rated pressure of 420 MPa was assumed to have a Young's modulus of (420/130)× 4.82×10^9 = 15.57×10^9 Pa.

The hydraulic circuits for the wheel motors and the pendulum arm cylinders were modeled using the same methodology that was described for the articulated joint hydraulics.

2.4 Tyres and rolling resistance

The tyres are defined as multiple discs with radii and positions defined by a spline curve. The spline curve depicts the outer profile of the tyre's cross-section and was derived from a CAD-model of the tyre, received from the tyre manufacturer. The contact between the contact point of each disc and the ground is described by the penalty method, where the chosen spring stiffness and damping constant define the flexible behavior of the tyres. The friction between the discs and the ground is described by the lumped LuGre friction model, with separate longitudinal and lateral frictions, and one point of contact for each disc.

Choosing the most accurate values for the tyre parameters described above, would probably require physical testing of the tyres, since these parameters depend on multiple unknown factors. Such physical tests were outside the scope of this study. For this study the spring stiffness and damping parameters were defined by iteration, to visually resemble the behavior of the tyres of the physical vehicle. The LuGre friction parameters were taken from a wheel-loader tyre model provided by Mevea. Physical testing of the tyres and subsequent adjustment of the tyre parameters is planned as future work.

When checking the hydraulic motor pressure required for the vehicles to run at a constant speed of 0.9 m/s on flat ground, it was noted that the physical vehicle required 55 bar, while the digital twin required only 32 bar. This was somewhat expected since the rolling resistance of the tyres, as well as other losses in the driveline, were not yet fully modeled. These losses could be analyzed individually and modeled in detail. However, as a simplification at this early stage, disc brakes with a low but constant pressing force were added to the individual wheels of the model to make the resulting friction account for the missing losses. The constant pressing force of the disc brakes was adjusted so that the digital twin required the same hydraulic motor pressure as the physical vehicle, to travel at the same constant speed on flat ground.

2.5 Articulated joint hydraulics testing and calibration

All the important parameters for modeling the articulated joint hydraulics, with one exception, were found through datasheets, drawings, contacts with manufacturers, measurements, or through well founded approximations. The parameter not found through any of the aforementioned means, was the pressure drop over the L90LS valve at nominal flow, Δp_{nom}. Determining this parameter directly, through isolated testing of the L90LS valve in a fully controlled environment, was not possible within the scope of this study. Hence, the pressure drop was approximated indirectly, through testing of the full vehicle. Since the pressure drop was the last remaining significant parameter, it could thus be isolated.

The physical test was performed by measuring the time required to do a full turn, at full turning speed, on a plane asphalt surface. Thereafter the pressure drop parameter in the simulation model was adjusted so that the digital twin of the vehicle would do a full turn in the same amount of time as the physical vehicle, given the same input signal, and on a similar surface. The digital twin model used for this calibration was made as accurate and detailed as possible. High emphasis was put on the level of detail of the hydraulic motor circuit and the rolling resistance of the wheels. For more details and a discussion about the results of this calibration, see section 3.2.

2.6 Scanned virtual environment

Having a digital twin of the test facility provides a faster way of testing systems before deploying the actual machines to the site. This provides more ready products with fewer expensive field tests. In this work, a virtual model, i.e., digital twin, of the test site in OuluZone (situated outside of Oulu, Finland) was constructed, enabling comparative testing between real and virtual worlds. The goal was to develop a model that is both accurate, and contains enough details for physical simulation as well as sensor simulation (including GPS, cameras and lidar). On the other hand, the model needs to be as lightweight as possible to be able to provide real-time simulation for the vehicle, while also providing enough details for visual sensors. A semi-automatic way of creating a digital twin of the test environment has been made. Some steps are already automated, i.e., software processing of the data, while some need manual work (i.e. using a CAD-program to cluster or clean up data).

The modeling process includes the following steps:

1. Collecting multi-modal data from the site.
2. Processing data by filtering and classifying objects in the data.
3. Reconstructing a set of 3D objects to be used in the simulator.

In the data collection phase, A UAV drone (DJI Matrice 300 RTK) with a Zenmuse L1 lidar and a RGB camera, in separate gimbals, was used for collecting image and point cloud data. Additionally, a Leica BLK360 laser scanner was used for collecting high resolution, high accuracy, colored point cloud data to verify the photogrammetry results. Important aspects of the data collection were the time of the year (and day), as well as various weather conditions that affect the quality of the data. In direct sunlight the colors are the best, but the shadows are too sharp. By contrast, in cloudy conditions shadows disappear, but the color quality is not as good. Also, rain prevents data collection due to noise in images and lidar data.

In step 2, image segmentation, using the DeepLab V3+ network (Ariram et al., 2021), was done on 2D-images for detecting, segmenting, and classifying: grass, asphalt, buildings, people, etc., to separate layers.

In step 3, the classification information was used to create separate 3D-models from segmented clusters, providing separate 3D-meshes for grass, asphalt, etc. Depending on the type of object and the role in the simulation, a simplified version of the mesh is generated to make the simulation faster. Further, based on the classification, some additional parameters are attached to the models, for example terrain softness, friction, etc. Also, large areas like the parking lot, are divided into several smaller areas. This makes the simulation faster since the non-required areas can be omitted in the simulation. The knowledge about the object types is also used for leaving out dynamic objects like cars, people, etc. from the simulation model.

For exteroceptive sensor emulation, the used simulation software Mevea can run in co-simulation with the game engine Unity (Unity), for inclusion of ready-made sensor emulation models and higher visual fidelity. If necessary, Unity can be run on a separate computer. A view from the scanned virtual environment can be seen in Fig. 6 below.

Fig. 6. View from the scanned virtual environment

To improve the accuracy of the GPS-coordinate system in the virtual environment, a well-defined point in the environment was chosen as a reference point. The GPS-coordinates for this point were then measured with high accuracy in the physical world, using a handheld RTK GPS sensor.

2.7 GPS sensor

The physical vehicle was equipped with a dual-antenna RTK GPS sensor (Leica iCG82 with Leica iCON gps 80 Dual GNSS Standard Heading). Therefore, the digital twin needed to be equipped with an identical sensor. The used software had support for emulation of GPS sensors of different types. However, at the time of the modeling, only single-antenna GPS sensors were supported (dual-antenna GPS sensors have later become supported). Therefore, only one of the antennas was modeled, taking into account its position on the vehicle. Thereafter, a Python script was created to calculate the position of the other antenna, based on the position of the first antenna and the orientation of the rigid body that the antennas were attached to. A Python script was also created to handle the TCP/IP communication between the virtual GPS sensor and the tracking script of the vehicle. This communication code needed to be carefully written to avoid disrupting the real-time simulation, e.g. by blocking the simulation while waiting for a message.

3. Results

3.1 Articulated joint flow curve

The flow driving the articulated joint of the digital twin during a full turn (-40° to +40°), at full turning speed, is shown in Fig. 7. The turn was performed at standstill, on a completely flat ground surface.

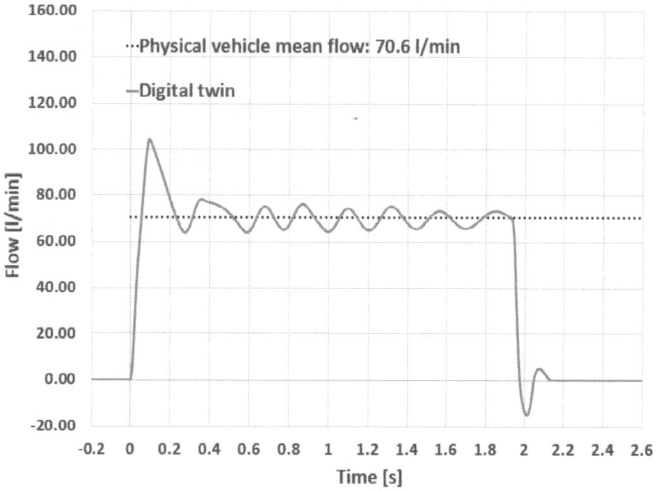

Fig. 7. Flow driving the articulated joint of the digital twin during a full turn

The dynamics of the flow seen in Fig. 7 appear realistic, and the full turn takes 1.98 s. Since the volumes of the A and B-side of the articulated joint cylinders were known, as well as the time required for the physical vehicle to do a corresponding full turn (1.98 s), the mean flow driving the articulated joint of the physical vehicle could be calculated, and was found to be 70.6 l/min (the dotted line in Fig. 7). Therefore, it seems reasonable that the simulated flow stabilizes around 70 l/min, which is close to the calculated mean flow of the physical vehicle. Future validation work is planned to include comparisons between the full flow curves of the physical vehicle and the digital twin. That validation will further increase the understanding of the level of realism of the simulated flow curve.

3.2 Hydraulic motor circuit level of detail impact on articulated joint calibration

A hybrid Vehicle-in-the-Loop (ViL) test, with the physical vehicle coupled to wheel-hub dynamometers, was planned. In the hybrid ViL test, the physical hydraulic motor circuit was to be used, while the rest of the vehicle, including the articulated joint, were to be simulated. Therefore, the calibration of the simulated articulated joint (see section 2.5) needed to be accurate enough, for this articulated joint to correctly overcome the internal resistance of the physical hydraulic motor circuit and the rolling resistance of the simulated wheels, while turning.

As shown in Fig. 4., the articulated joint hydraulics do not have a direct hydraulic dependency on the hydraulic motors, since they are connected to separate pumps. But they are still connected indirectly (mechanically) since the wheel-ground interface (see section 2.4) will cause the wheel motors to turn when the articulated joint turns the vehicle.

A simplified overview of the hydraulic motor circuit is shown in Fig. 8.

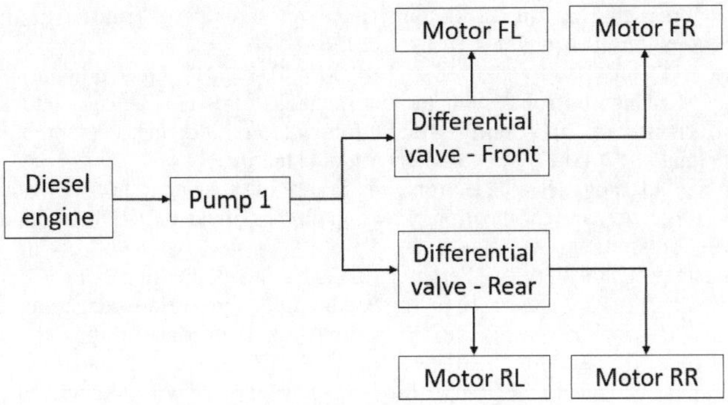

Fig. 8. Simplified overview of the hydraulic motor circuit

To determine the impact of the level of detail of the hydraulic motor circuit model, three different calibrations of the articulated joint were made. These calibrations used vehicle models with an increasing level of detail of the hydraulic motor circuit.

In the first, most simple model, no hydraulic motors were used at all, so the wheels could turn completely freely. Because the differential valves of the physical vehicle (see Fig. 8) allow oil to flow relatively freely between the wheel motors, this could still be seen as a reasonable approximation, although the resistance when turning can be expected to be lower than for the physical vehicle (see section 2.4).

The second, "simple" model, had only the hydraulic wheel motors, directly connected to the same hydraulic volume coming from the pump, without any differential valves in between, or other surrounding hydraulic components. This allowed the motors to exchange oil completely freely when the wheels were turning.

In the third model the hydraulic motor circuit was modeled in as high level of detail as possible, including representations of the differential valves, charge pump, oil cooling, etc.

Each calibration of the digital twin's articulated joint was done in two steps. Firstly, the additional disc brakes were calibrated, to account for any missing resistance in the driveline, see *2.4 Tyres and rolling resistance*. Secondly, the calibration of the L90LS valve of the digital articulated joint was made, by performing a turning test at standstill as explained in section 2.5. to determine Δp_{nom}.

For the most simple model, without any hydraulic motors, the brake calibration described in section 2.4 could not be performed. This resulted in no added disc brake force, which makes comparison to this model a bit unfair. However, this model still serves a purpose as an example of calibrating the L90LS valve, in calibration step 2, using a vehicle model with too low resistance. Calibration of the L90LS valve using the most simple model, yielded a semi-empirical flow rate coefficient C_v (see section 2.3 and Eq. 3) that was 10% lower than the value achieved by using the most detailed vehicle model.

It was noted that for the second "simple" model, the disc brake force needed to be increased by 50% in the first calibration step, compared to what was used for the detailed model (the detailed model is the one described in section 2.4). This showed that the simple model had significantly lower internal resistance in its hydraulic motor circuit. With the 50% increase in disc brake force however, the simple model resulted in a calibration of the L90LS valve that didn't differ significantly from the one achieved using the detailed model.

The results from the three calibrations are shown in Table 2.

Table 2
Summary of calibration results, detailed model is used as reference

Model	Hydr. motor circuit	Disc brake force (calibr. 1)	L90LS Flow rate coefficient C_v (calibr. 2, Eq. 3)
Detailed model	Detailed	100%	100%
Simple model	Simple	150%	101%
Most simple model	None	None	90%

This shows that the level of detail of the hydraulic motor circuit affects the resistance during turning, which in turn affects the calibration of the articulated joint hydraulics. However, it seems possible to compensate for missing internal hydraulic resistance, by substituting with added disc brake force, for this calibration purpose.

Future physical and ViL tests will show whether the detailed hydraulic motor circuit accurately captured the internal resistance in the physical hydraulic motor circuit, or not. They will also show how accurate the digital calibration of the L90LS valve was.

3.3 Test run

The digital twin of the vehicle on a test run in the scanned virtual environment can be seen in Fig. 10 below.

Fig. 10. The digital twin of the vehicle on a test run in the scanned virtual environment

The test run in Fig. 10 consisted of using the emulated GPS-sensor and the same tracking script as the vehicle, to follow a GPS-path that the physical vehicle had previously been following, while logging various vehicle and sensor data. The simulation was possible to run in real-time using a time step of 1 ms for the main simulation and 33 µs for the hydraulics. This was done on a laptop with an Intel i7-10750H processor (6 cores, 12 threads), a Nvidia RTX 2070 MaxQ graphics card, and 16 GB of RAM. The most critical hardware attribute for running this simulation, was the single-core CPU performance. Changes to the simulation, for example using a more complex environment, visual sensor emulation, larger timesteps, etc., could make other components, such as the graphics card, the new bottleneck.

4. Conclusions and future work

A HiL-capable multi-physics digital twin of a forestry vehicle was created. The digital twin appeared to have a high level of realism and was possible to run in real-time in a scanned virtual environment, following a real GPS-track, on a laptop.

Future work includes validation against the physical vehicle to determine and improve on the accuracy of the model. Fully physical, as well as hybrid ViL tests, in the scanned OuluZone environment, are planned to aid in this validation. The hybrid ViL tests are planned to be performed using wheel-hub dynamometers in the NUVE-Lab in Oulu, Finland.

5. Acknowledgements

This research was conducted within the project Nordic Platform for Development of Autonomous Utility Vehicles (NUVE) which was funded by the EU, through Interreg Nord, as well as by multiple additional stakeholders.

6. Declaration of competing interest

The authors declare that they have no known competing financial interests or personal relationships that could have appeared to influence the work reported in this paper.

7. References

Ariram, S., Pennanen, T., Tikanmaki, A., Roning, J., 2021. Garbot - Semantic Segmentation for Material Recycling and 3D Reconstruction Utilizing Robotics, in: 2021 IEEE International Conference on Mechatronics and Automation, ICMA 2021. Presented at the 2021 IEEE International Conference on Mechatronics and Automation, ICMA 2021, pp. 1255–1260.

Baharudin, M.E., Rouvinen, A., Korkealaakso, P., Mikkola, A., 2014. Real-time multibody application for tree harvester truck simulator. Proceedings of the Institution of Mechanical Engineers, Part K: Journal of Multi-body Dynamics 228, 182–198.

Choi, H.S., Crump, C., Duriez, C., Elmquist, A., Hager, G., Han, D., Hearl, F., Hodgins, J., Jain, A., Leve, F., Li, C., Meier, F., Negrut, D., Righetti, L., Rodriguez, A., Tan, J., Trinkle, J., 2021. On the use of simulation in robotics: Opportunities, challenges, and suggestions for moving forward. Proceedings of the National Academy of Sciences of the United States of America 118.

Dabbiru, L., Goodin, C., Scherrer, N., Carruth, D., 2020. LiDAR Data Segmentation in Off-Road Environment Using Convolutional Neural Networks (CNN). SAE Technical Papers.

Dallas, J., Cole, M.P., Jayakumar, P., Ersal, T., 2021. Terrain adaptive trajectory planning and tracking on deformable terrains. IEEE Transactions on Vehicular Technology 70, 11255–11268.

Engelmann, D., Unger, M.R., Geimer, M., 2017. Testing and benchmarking a powertrain with independent wheel control for heavy machinery, VDI Berichte.

Fathy, H.K., Filipi, Z.S., Hagena, J., Stein, J.L., 2006. Review of hardware-in-the-loop simulation and its prospects in the automotive area, in: Proceedings of SPIE - The International Society for Optical Engineering. Presented at the Proceedings of SPIE - The International Society for Optical Engineering.

Hružík, L., Bureček, A., Vašina, M., 2017. Mathematical simulation and measurement of expansion of hydraulic hose with oil. Tehnicki Vjesnik 24, 1905–1914.

Kelber, C.R., Reis, B.R.R., Figueiredo, R.M., 2016. Improving functional safety in autonomous guided agricultural self propelled machines using hardware-in-The-loop (HIL) systems for software validation, in: IEEE Conference on Intelligent Transportation Systems, Proceedings, ITSC. Presented at the IEEE Conference on Intelligent Transportation Systems, Proceedings, ITSC, pp. 1438–1444.

Lee, C.J., Jeon, C.W., Han, X.Z., Kim, J.-H., Kim, H.-J., 2017. Electro-hydraulic steering control system for automated steering of an agricultural tractor, in: 2017 ASABE Annual International Meeting. Presented at the 2017 ASABE Annual International Meeting.

Li, S., Lideskog, H., 2021. Implementation of a system for real-time detection and localization of terrain objects on harvested forest land. Forests 12.

Lideskog, H., Karlberg, M., Bergsten, U., 2015. Development of a Research Vehicle Platform to Improve Productivity and Value-extraction in Forestry, in: Procedia CIRP. Presented at the Procedia CIRP, pp. 68–73.

Lindroos, O., La Hera, P., Häggström, C., 2017. Drivers of advances in mechanized timber harvesting - a selective review of technological innovation. Croatian Journal of Forest Engineering 38, 243–258.

Mevea. Available at: https://mevea.com (accessed 20 June 2023).

Peiret, A., Karpman, E., Kovács, L.L., Kövecses, J., Holz, D., Teichmann, M., 2021. Modelling of off-road wheeled vehicles for real-time dynamic simulation. Journal of Terramechanics 97, 45–58.

Rahikainen, J., Mikkola, A., Sopanen, J., Gerstmayr, J., 2018. Combined semi-recursive formulation and lumped fluid method for monolithic simulation of multibody and hydraulic dynamics. Multibody System Dynamics 44, 293–311.

Raikwar, S., Jijyabhau Wani, L., Arun Kumar, S., Sreenivasulu Rao, M., 2019. Hardware-in-the-Loop test automation of embedded systems for agricultural tractors. Measurement: Journal of the International Measurement Confederation 133, 271–280.

Tsolakis, N., Bechtsis, D., Bochtis, D., 2019. Agros: A robot operating system based emulation tool for agricultural robotics. Agronomy 9.

Unity. Available at: https://unity.com (accessed 20 June 2023).

Watton, J., 1989. Fluid power systems: Modeling, simulation, analog, and microcomputer control, first ed., Prentice Hall International, Hemel Hempstead.

Wolf, P., Groll, T., Hemer, S., Berns, K., 2020. Evolution of robotic simulators: Using UE 4 to enable real-world quality testing of complex autonomous robots in unstructured environments, in: SIMULTECH 2020 - Proceedings of the 10th International Conference on Simulation and Modeling Methodologies, Technologies and Applications. pp. 271–278.

Zheng, Y., Cheng, B., Huang, Q., Liu, J., 2018. Research on Virtual Driving System of a Forestry Logging Harvester. Wireless Personal Communications 102, 667–682.

4456 | Modeling soil-tool interaction of a cultivator sweep using DEM

Peeyush Soni [a,*], Sumit Parmar [b], Prakhar Patidar [a]

[a] *Department of Agricultural and Food Engineering, Indian Institute of Technology Kharagpur, Kharagpur, India*

[b] *Former M. Tech Student, Department of Agricultural and Food Engineering, Indian Institute of Technology Kharagpur, Kharagpur, India*

* *Corresponding author: soni@iitkgp.ac.in*

ABSTRACT

The discrete element method (DEM) is a powerful tool for predicting soil disturbance and cutting forces in soil cutting tools, aiding in design optimization. In this study, we calibrated DEM input parameters to model a sandy loam soil using the hysteretic spring contact model, linear cohesion model, and a nominal particle radius of 5 mm. The DEM simulations were validated using experimental lab results, considering the effects of speed and depth. Some soil properties were measured in the laboratory, while others were obtained from literature and used as input parameters in Altair EDEM.

We constructed a 3-D model of the cultivator tyne, and by combining the implement design with soil parameters and implement parameters (depth of operation, various speeds), we performed simulations in Altair EDEM software. Prior to the soil bin test, essential sensors such as load cells and torque transducers were calibrated. Overall, the DEM results closely aligned with the experimental findings, demonstrating similar trends. Comparing the measured and predicted tillage forces under various conditions, we achieved correlation coefficients ranging from 0.8 to 0.85. This demonstrates a highly accurate prediction of tillage forces.

Keywords
DEM
Soil-tool interaction
Cultivator tyne
Soil bin

1. Introduction

Tool geometry is an essential factor for a tillage tool to be able to perform according to its desired criteria. The criteria used for the design of the tillage tool typically include draught to operate the tool, the volume of soil loosened, and total energy requirements. Due to the tool activity, the state of the tilled soil depends on the mechanical actions of the soil, its initial condition, and the function of the tool operating parameter. A significant parameter in deciding tool size and shape is the soil pressure on tillage tools and their distribution with respect to tool wear.

Tool interaction with agricultural soil deals with soil cutting in order to achieve suitable crop conditioning. The quality of tillage output is calculated in terms of draft or energy input. Tillage tool design optimization requires minimization of the input energy. Tillage is estimated to account for approximately half of the energy used in crop production. The draft criterion for causing soil failure to the desired tilth depends on the pattern of soil failure, which is a combined effect of the parameters of the soil, tool, and system parameters. The power requirements of a tillage tool depend on several factors related to soil type and conditions, the design of the tool, and operating parameters.

Many studies on soil–tool interaction have been carried out in the past to improve the design of engaging soil tools. Soil–tool interactions are usually characterized by two phenomena: forces arising at the soil–tool interface (draught and vertical forces) and displacement of soil particles (soil disturbance).

Numerical methods, including the finite element method and discrete element method, are used to simulate the interaction between soil and tillage apparatuses. These methods are described to be not only capable of calculating tool forces but to simulating soil loosening.

The discrete element model (DEM) used to model the behavior of granular materials is suitable for modeling soils and interactions between soil and rigid or flexible bodies. The non-linearity of the soil behavior and soil–implement interaction can be easily depicted using this model. Consequently, this method seems to be a promising way to model granular soil–implement interaction without the limitation of the methods mentioned above, and may serve as a tool for optimization of the design process.

The overall objective of this research was to model the sweep-soil interaction using Altair® EDEM. The specific objectives included a) creation of a 3D model of the cultivator sweep, b) determination of soil properties towards model simulation, c) simulation of the performance of the sweep using Altair EDEM, and d) experimental validation of the simulated results by laboratory tests.

2. Material and methods

2.1 *Determination of DEM input parameter for soil-sweep interaction*

In this study, two categories were used for the DEM parameters: material properties and interaction properties. A combination of measurements and data from existing literature, as given in Table 1, was used to obtain the material properties. The interaction properties, including the coefficient of restitution for soil-soil, coefficient of restitution for soil-steel, coefficient of friction for soil-steel, and coefficient of rolling friction for soil-steel, were determined using a similar combination of measurements and literature data, as given in Table 1. The interaction properties of coefficient of friction for soil-soil and

https://doi.org/10.56884/BLLT5880
ISBN 978-1-942112-55-6 (pdf)

coefficient of rolling friction for soil-soil, along with the integration time step, were calibrated to account for the particle size used in the simulations, following the methodology described by (Ucgul et al. 2014), to ensure accurate simulation results. The parameters were adjusted to match the simulation results with the measured outcomes of angle of repose as shown in Fig. 1 and angle of inclination test as shown in Fig. 2 and bulk density conducted on sandy clay loam soil. It is important to note that the validation of actual tillage results was not considered during parameter adjustment and was separate from the calibration process.

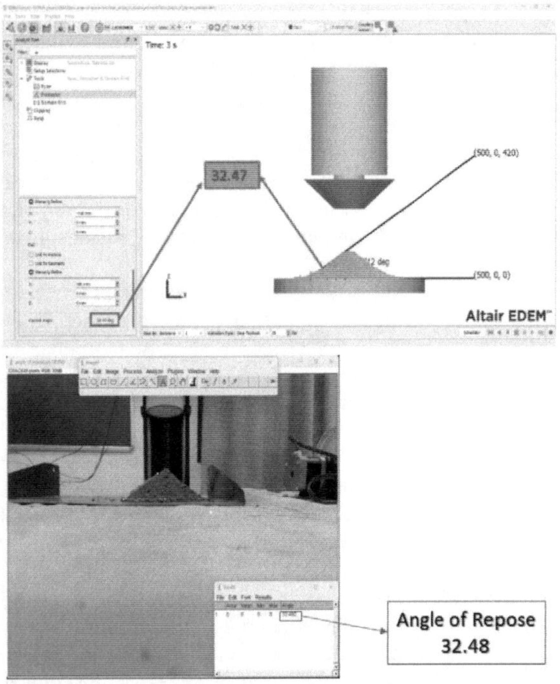

Fig. 1. Angle of repose test in laboratory setup and DEM

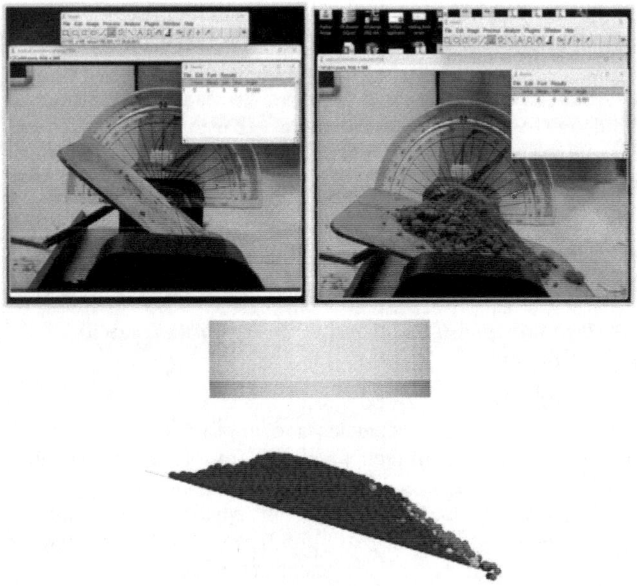

Fig. 2. Angle of inclination test in laboratory setup and DEM

Table 1
DEM parameters used in simulations

Property	Values	Source
Poisson's ratio of soil (v)	0.25	Ucgul et al. (2014a)
Particle radius (mm)	5	Selected based on available computation time
Particle size distribution	0.95– 1.05	Ucgul et al. (2014a)
Bulk density of soil (ρ), kg/m^3	1427±9	Measured
Shear Modulus (G), Pascals	3×10^7	calibrated
Co. of restitution of soil	0.6	Barr et al. (2018)
Co. of Static Friction of soil-soil	0.8	calibrated
Co. of Rolling friction of soil-soil	0.4	calibrated
Poisson's ratio of high carbon steel (v)	0.3	Horabik and Molenda (2016)
Solid density of high carbon steel (ρ), kg/m^3	7865	Budynas and Nisbett (2012)
Shear Modulus of steel (G), Pascals	7.9×10^{10}	Budynas and Nisbett (2012)
Co. of restitution of soil-tool	0.6	Barr et al. (2018)
Co. of Static Friction of soil-tool	0.7	Calibrated
Damping factor	.05	EDEM (2022)
Stiffness factor	0.95	EDEM (2022)
Yield strength of soil (MPa)	2.8	calibrated
Yield strength of steel (MPa)	380	Budynas and Nisbett (2012)
Co. Rolling friction of soil-tool	0.2	calibrated

2.2 *Laboratory Experiment*

Experiments were conducted in a soil bin setup (Fig. 3) comprising a soil processing trolley, linear motion transmission system, control panel, and data acquisition (DAQ) system. For evaluate the performance of the developed test rig under controlled conditions, a rectangular soil bin with dimensions of 15.0 m in length, 1.8 m in width, and 0.6 m in depth was utilized. The sandy clay loam soil inside the bin had a solid density of 2.20 g/cm^3.

To facilitate the movement of the soil processing trolley and test rig along the track, two side rails were installed on top of each side of the bin wall. The soil processing trolley consisted of a rotary tiller, levelling blade, and roller, which were responsible for tilling, levelling, and compacting the soil, respectively. The trolley was employed to prepare the test bed for different soil compaction levels. A gear reduction unit with eight-speed options enabled the back-and-forth motion of the soil processing trolley on the soil bin track at the desired forward speed by selecting the appropriate gears. The gear reduction unit was connected to the input shaft of a rotating drum, which, in turn, was attached to the soil processing trolley using stainless steel rope and a threaded wire rope tensioner. The control panel, located outside the bin, operated the soil processing trolley, test rig, and hydraulic system. The test rig (Fig. 3) comprised a vertical frame on which a sweep The operating depth of the sweep could be adjusted and controlled using a rack and pinion mechanism. A guiding frame connected the main frame to the bin's side rails through rollers, allowing for movement of the test rig within the soil bin. Four rollers on the guiding frame ensured that the test rig followed the intended path, while six rollers in the C-section of the side rails balanced the upward load and maintained the desired operating depth during motion. The main frame also featured two cylindrical side rollers pressed against the rails to counteract side thrust, secure the main frame, and prevent it from slipping out of the rail. The forward speed was measured using a proximity switch attached to the developed test rig. A 2000 kg-capacity S-type load cell was horizontally mounted between the soil processing trolley and the test rig to measure the draft force. A Spider-8 DAQ system with eight channels was utilized to record the output from various sensors in the soil bin.

Fig. 3. Experimental setup in soil bin

2.3 *Virtual soil bin set up and 3D models of the sweep*

For this study, a virtual soil bin was established with dimensions of 2000 mm length, 700 mm width, and 300 mm depth (Fig. 5a). The width of 600 mm was chosen to account for lateral soil movement observed in the virtual soil bin. The depth of 90 mm was set to accommodate the operation of a sweep up to a maximum depth of 7 cm. The length of 1500 mm was selected to ensure adequate data collection from the sweep and to capture soil movement as it interacts in the virtual soil bed. The virtual soil bin was filled with particles and packed to attain the desired cone index and bulk density. A CAD model and soil-sweep interaction of the sweep was shown in shown in Fig. 4a It was then imported into DEM software in igs format. The interaction between soil and the sweep in the virtual soil bin can be observed in Fig. 4b.

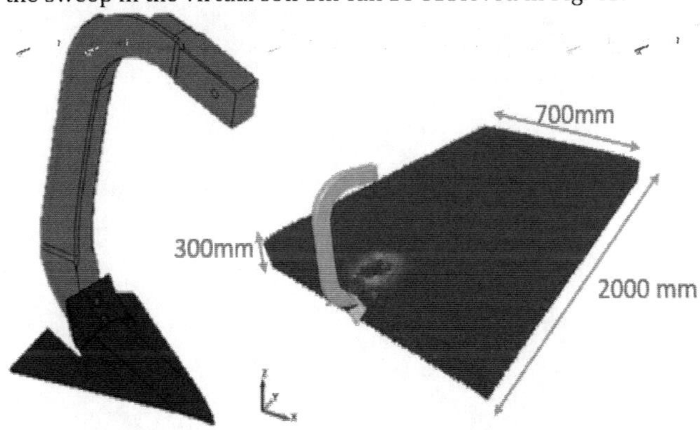

(a) CAD model of sweep (b) DEM Soil–sweep interaction

Fig. 4. CAD Model and soil-sweep interaction in virtual soil bin

3. Results and Discussion

3.1 *Effect of forward velocity on draft of sweep at constant depth of operation*

The effect of forward velocity on the vertical force requirement at a depth of 10 cm is shown in Fig. 5. The results indicated that the vertical force was not significantly affected by changes in forward velocity. The mean vertical forces were found to be 47, 50, and 59 at 3, 4, and 5 kmph, respectively. The increase in vertical force with depth was not linear and was virtually independent of speed. Ucgul et al. (2015) also observed similar trends regarding the effect of forward speed on vertical forces.

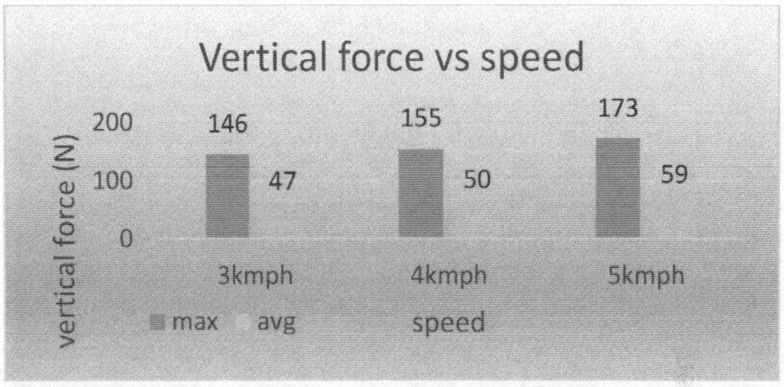

Fig. 5. Effect of forward velocity on the vertical force requirement at a depth of 10 cm

3.2 *Effect of depth on draft requirement of sweep at constant speed*

The mean draft of the sweep tyne for an average operating depth of 10, 12.5, and 15 cm was found to be 286, 388, and 521 N, respectively, for a constant forward speed of 4 kmph as shown in Fig. 6. The draft force increased linearly with depth due to an increase in the volume of soil being handled.

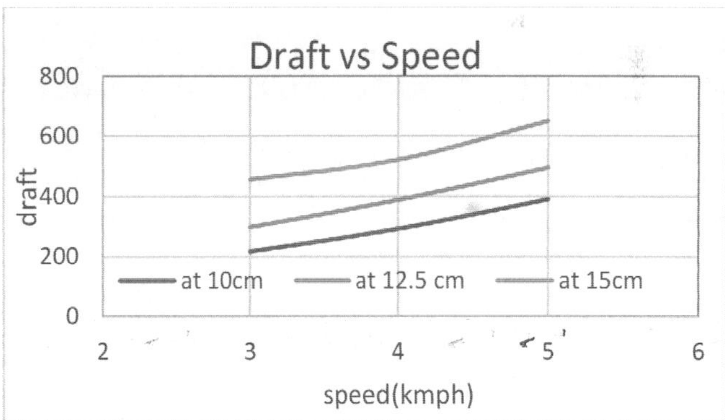

Fig. 6. CAD Model and soil-sweep interaction in virtual soil bin

3.3 *Virtual soil bin set up and 3D models of the sweep*

The vertical force measured in the soil bin of the sweep tyne at depths of 10, 12.5, and 15 cm is shown. The mean vertical force of the sweep tyne for an average operating depth of 10, 12.5, and 15 cm was found to be 40.23, 73.69, and 116.32, respectively, for a constant forward speed of 4 kmph. As shown in the Fig, 7, vertical forces increase with increase in depth operation due more soil reaction force on sweep at higher depth of operation.

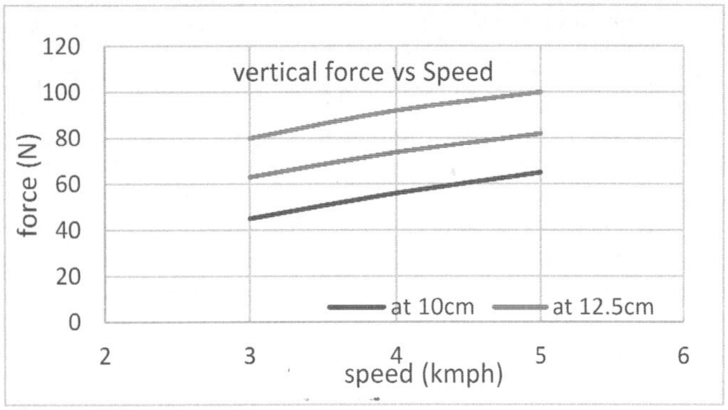

Fig. 7. Vertical load virtual soil-bin results

4. Model validation

The Comparison of the simulated and measured draft for varying depth (10 cm, 12.5 cm) have been shown earlier. The simulated result for the draft (R^2 = 0.85 at 10 cm depth and R^2 = 0.84 for 12.5 cm depth) were closely correlated to experiment results.. These good relations give confidence that the use of the hysteric spring contact model with consideration of soil cohesion can model actual sweep interaction for different operating conditions in a sandy clay loam soil.

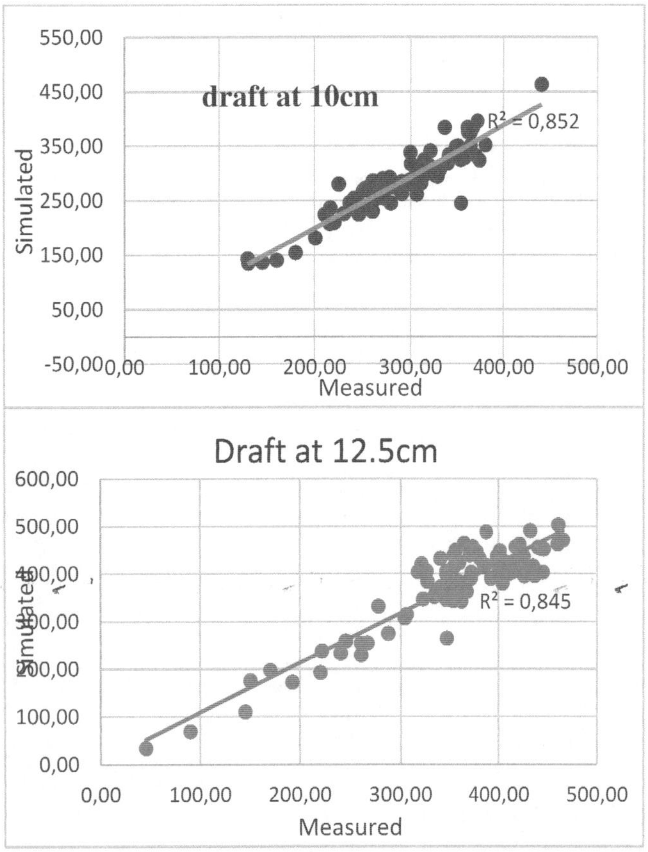

Fig. 8. Correlation between the measured and predicted draft (a) depth = 10 cm and (b) 12.5 cm

5. Conclusions

The influence of the range of operating conditions on the sweep in sandy clay loam soil was simulated using a discrete element method (DEM) with a hysteretic spring contact model. The DEM input parameters were obtained in this study by comparing simulation results to actual tests of the angle of repose and angle of inclination for a sandy clay loam soil. By comparing simulated and measured tillage forces for various combinations of depth and speed for the sweep, correlation values between 0.84 and 0.85 were obtained. The good correlation provides a reason to encourage further research into the application of the hysteretic spring contact model for a broader variety of soil conditions and tillage tool types.

This study found that an increase in speed and depth of operation results in an increase in the draft requirement for the sweep due to the larger volume of soil being handled at greater depths. Overall, the study demonstrates the potential of using 3D DEM to simulate various agricultural equipment operating under different conditions, thereby avoiding the need for repetitive tests in soil bins and saving time and resources required for testing developed tools. These findings have significant implications for the design and optimization of agricultural machinery, particularly for agricultural operations in sandy clay loam soil.

6. References

Barr, J.B., Ucgul, M., Desbiolles, J.M.A., Fielke, J.M., 2018. Simulating the effect of rake angle on narrow opener performance with the discrete element method. Biosystems Engineering. 171, 1-15.

Budynas, R.G., Nisbett, K.J., 2012. Shigley's Mechanical Engineering Design, 9th ed. McGraw-Hill Education, New York.

Horabik, J., Molenda, M., 2016. Parameters and contact models for DEM simulations of agricultural granular materials. Biosystems Engineering. 147, 206–229.

Ucgul, M., Fielke, J.M., Saunders, C., 2014. 3D DEM tillage simulation: validation of a hysteretic spring (plastic) contact model for a sweep tool operating in a cohesionless soil. Soil and Tillage Research. 144, 220–227.

Ucgul, M., Fielke, J. M., & Saunders, C. 2015. Defining the effect of sweep tillage tool cutting edge geometry on tillage forces using 3D discrete element modelling. Information Processing in Agriculture. 2(2), 130-141.

4595 | Evaluation and comparison of driving performance of lunar exploration rover wheel in different soils

Keisuke Takehana [a,*], Shino Kizaki [a], Kentaro Uno [a], Tomomi Tanaka [a], Gentaro Suda [a], Kazuya Yoshida [a]

[a] *Department of Aerospace Engineering, Tohoku University, Sendai, Miyagi, Japan*

* *Corresponding author: keisuke.takehana.s1@dc.tohoku.ac.jp*

ABSTRACT

Wheeled mobile robots, rovers, are highly effective in lunar exploration as they can move the Moon's surface for detailed exploration. However, the lunar regolith, soft soil with fine powdery sand, can cause wheel slippage, resulting in an inability to travel for the rover.

To analyze the driving performance of a rover wheel, a single-wheel testbed is usually used. In our experiment, force sensors mounted on the wheel measure the reaction force, which allows for computing the wheel's traction performance evaluation criteria.

This system can control the rotation and translation of the wheels separately, realizing experiments in any slippage condition. This method can measure accurate traction data under more precise slip ratios than the conventional single-wheel tests, where a traction load is applied along with wheel rotation. Moreover, this testbed can conduct experiments using regolith simulant with a cohesive property that has similar mechanical properties to those of lunar regolith, in addition to Toyoura sand, which is non-cohesive sand collected from the earth.

This paper presents the results of a driving test on two types of loose soil: Toyoura sand and regolith simulant (FJS-1). The wheel used in the experiment is the preliminary version of the actual flight model of a 10 kg class lunar exploration microrover. The results reveal that the traction performance on both sands improves as the slip ratio increases. The performance did not depend on velocity but on vertical load. It should be noted that the cohesive simulant shows a higher difference in traction performance than Toyoura sand. These findings, measured in detail from the low-slip to the high-slip range, contribute to the actual driving operation of the rover missions.

Keywords
Terramechanics
Lunar exploration rover
Single-wheel test
Grouser wheel
Driving performance
Lunar regolith simulant

1. Introduction

In recent years, the importance of lunar exploration has grown significantly, particularly in the context of discovering water ice resources and establishing manned bases. Lunar exploration can greatly benefit from the utilization of rovers, which are wheeled mobile robots capable of conducting detailed explorations by traversing the lunar surface. However, the lunar surface consists of soft soil composed of fine powdery sand deposits known as regolith. When a rover moves on such soft soil, the wheels tend to sink and slip, and excessive slippage can hinder the rover's forward movement, ultimately impeding its ability to travel. To ensure the safe driving of rovers on soft soil, it is crucial to comprehend the mechanical interaction between the wheels and the soil and to regulate the driving process accordingly.

The study of wheel-soil interaction, commonly referred to as Terramechanics, has seen the development of a fundamental wheel mechanics model by Bekker and an improved traction force model by Wong (Bekker, 1956, 1960, 1969; Wong, 2008; Wong and Reece, 1967). Based on the Terramechanics theory, the distribution of vertical and shear stresses on the wheel contact surface was experimentally measured (Hettiaratchi and Reece, 1975), and the effect of low gravity environment on driving performance was evaluated (Niksirat et al., 2019). Such conventional analysis of rover wheel performance typically utilizes a single-wheel testbed.

Our research group has developed a single-wheel testbed capable of measuring the reaction force exerted on the wheels during driving, using force sensors attached to the wheels (Higa, 2017). This testbed facilitates the calculation of wheel traction performance. Notably, (Sutoh et al., 2013) and (Nagaoka et al., 2020) have investigated the influence of the grouser shape on driving performance, while (Higa et al., 2015, 2019) have measured the stress distribution on the grouser part and wheel surface of a grouser wheel, and quantitatively assessed the reaction force at the grouser part. Additionally, (Nakamura et al., 2020) have employed the image particle measurement method to study the sand flow phenomenon beneath the wheel, thereby elucidating the relationship between sand flow and traction force.

Conventional research has typically involved experiments with free-slip systems, applying a traction load concurrent with wheel rotation. While this approach enables the measurement of nearly constant traction force, inaccuracies arise due to sliding resistance in the travel direction, and the slip ratio depends on the applied traction load. However, our system allows for independent control of wheel rotation and translation, enabling experiments under various slip conditions. Consequently, more precise traction data can be obtained, surpassing the capabilities of conventional free-slip experiments where the traction load is applied during wheel rotation. Furthermore, aside from Toyoura sand, non-cohesive sand collected from Earth, the testbed can also be utilized for experiments with lunar-regolith simulant, cohesive sand possessing mechanical properties similar to those of lunar regolith. Although differences in driving performance due to varying mechanical properties between soils, particularly differences in cohesion, are anticipated, few studies have quantitatively evaluated these discrepancies.

This paper presents the findings of driving tests conducted on two types of loose soil: Toyoura sand and regolith simulant (FJS-1). The wheel employed in the experiments corresponds to the actual flight model of a 10 kg class lunar exploration micro-rover. To comprehensively evaluate the wheel's traction performance, the traction force, and sinkage at each slip ratio ranging from 0 % to 50 % were measured. Through these driving experiments, the

https://doi.org/10.56884/JSFS4189
ISBN 978-1-942112-55-6)pdf)

disparities in driving characteristics arising from distinct soil properties have been examined.

2. Single-wheel testbed

This section presents the details of the single-wheel testbed employed in the experiments, as illustrated in Fig. 1. This has overall dimensions of 2.5 m in length, 0.3 m in width, and 1.5 m in height, with one wheel moving in a fixed direction on the sand. The testbed consists of the following components: (1) wheel drive unit, (2) passive vertical moving unit, (3) translational drive unit, (4) wheel, and (5) sandbox. Each component is described in further detail below.

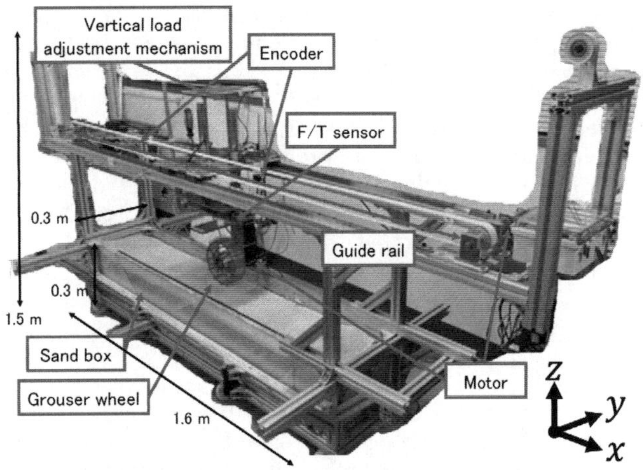

Fig. 1. Single-wheel testbed apparatus

2.1 Wheel Drive unit

This component controls the wheel's rotation and measures the reaction force exerted on the wheel by the soil. It incorporates a Maxon DC motor, a planetary gear head with a reduction ratio of 16:1, and an encoder as the wheel drive motor. The motor's output shaft is connected to a harmonic drive with a reduction ratio of 50:1, resulting in a final reduction ratio of 800:1. The motor is linked to an STM32 microcontroller via a Hibot motor driver, enabling wheel speed control by the microcontroller. The wheel's speed command value can be specified from the operation workstation, allowing control of the wheel's rotational motion at a maximum angular speed of 1 rad/s. The wheel drive is powered by the motor's rated voltage of 18 V, and the current value is set to 3.59 V, which corresponds to the motor's maximum continuous current and is supplied to the microcontroller. An upper section of the wheel drive unit is equipped with a 6-axis force sensor manufactured by Leptrino, enabling the measurement of traction force and vertical force acting on the axle.

2.2 Passive vertical moving unit

This unit is designed to passively float in response to the reaction force from the sand surface. Positioned between the wheel drive unit and the translational drive unit, this mechanism consists of a linear shaft, a linear encoder, and a vertical load adjustment unit. The connection of each unit via a linear shaft allows the wheel to freely slide in the vertical direction, simulating the floating and sinking of a wheel during operation. The linear encoder measures the amount of wheel sinkage after the start of driving, with a maximum allowable sinkage of 70 mm for a wheel diameter of 200 mm. The vertical load can be adjusted as needed by placing weights on the vertical load adjustment section, facilitating experiments with a maximum vertical load of 70 N.

2.3 Translational drive unit

This unit serves as a mechanism to forcibly transport the wheel unit in the direction of wheel travel. It comprises a motor for wheel transportation, a linear guide, a timing belt, and a linear encoder. Fixed to the vertical mechanism through a linear shaft, this unit is equipped with a Maxon DC motor, a planetary gear head with a reduction ratio of 231:1, and an encoder for wheel transportation. Similar to the wheel drive unit, the motor is connected to an STM32 microcontroller, allowing for motor speed control by the microcontroller. The vertical drive and this mechanism are interconnected to the transport motor via timing belts, timing pulleys, and shaft couplings. By regulating the speed of this motor, the guide rail moves forward, enabling the wheels to translate at the desired speed, up to a maximum of 0.1 m/s. The amount of translational movement is measured using a horizontal linear encoder.

2.4 Wheel

The wheel employed is a grouser wheel of the same model as the wheels utilized in the development of the Rashid Rover by MBRSC (Almaeeni et al., 2021; Emirates Lunar Mission 2022, 2023). Figure 2 illustrates the wheel used, which features a diameter of 200 mm, 14 grousers, a grouser height of 20 mm, and a weight of 1.18 kg.

Fig. 2. A test wheel prepared based on MBRSC rover design. (Image credit: MBRSC)

2.5 Sandbox

A sandbox serves as the designated test field. It has dimensions of 1.6 m in length, 0.3 m in width, and 0.3 m in depth, and allows for a maximum wheel travel distance of 1 m. For the experiments, separate sandboxes are prepared, one filled with Toyoura standard sand and the other with regolith simulant (FJS-1). Table 1 provides information on the mechanical properties and Fig. 3 displays an enlarged view of each soil (Higa et al., 2019; Ryu et al., 2018; Toyoura keiseki kogyo co., ltd, 2023).

Table 1
Sand mechanical properties

Type	Toyoura Standard Sand	Regorith Simulant (FJS-1)
Particle size (μm)	106-300	7-4000
Specific gravity (g/cm^3)	2.63	2.94
Angle of Repose (°)	36.0	41.0
Friction Angle (°)	38.0	39.4
Cohesion (kPa)	0.0	8.0

Fig. 3. Soils: Toyoura standard sand (left) and regolith simulant (right)

Toyoura standard sand is obtained by sieving natural silica sand collected from Earth. Due to its natural occurrence, the sand grains are rounded, facilitating excellent flowability. These inherent characteristics contribute to reduced errors in experiments and measurements, ensuring stable and reliable results. Regolith simulant (FJS-1) replicates the shape, chemical composition, and mechanical properties of lunar regolith. Compared to Toyoura sand, the regolith simulant tends to exhibit variations in soil hardness based on sand conditions and depth, primarily due to its sharp particle shape and cohesive nature.

3. Experimental methods and procedures

This section describes in detail the methods and procedures of the experiment using the single-wheel testbed, and the experimental conditions.

3.1 Vane shear test

To ensure consistent driving conditions and account for the soil condition before driving, a preliminary assessment of soil hardness was conducted. The shear force of the soil was determined through a shear test utilizing a hand vane equipped with blades at the end of a rod, as depicted in Fig. 4. By inserting the rod with the vane blade into the soil and rotating it, the shear moment was measured. Tests were performed to ensure that the shear moment remained within the range of 0.015 Nm to 0.020 Nm. If the shear moment deviated from the specified range, the soil was tilled again, followed by subsequent shear tests. The average shear moment values obtained from the vane shear tests are 0.015 Nm for the Toyoura standard sand and 0.017 Nm for the regolith simulant.

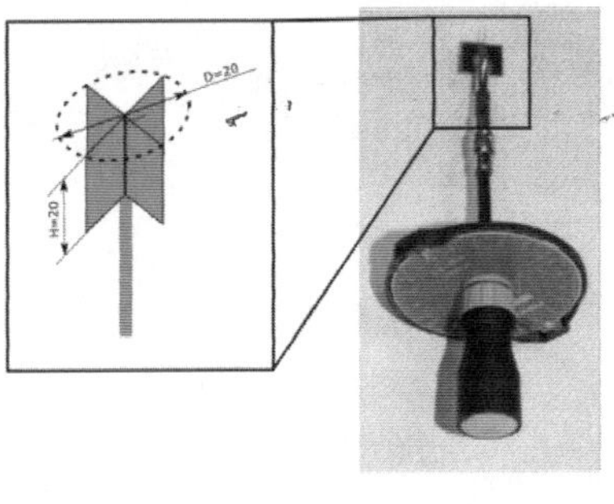

Fig. 4. Hand vane

3.2 Procedure

The experimental procedure for the driving test is outlined as follows:

1. Level the sand surface using a flat plate to achieve a constant, horizontal height.
2. Assess the soil hardness using a hand vane.
3. Configure the wheel's rotational and translational velocities based on the desired slip ratio, and initiate driving with controlled constant velocity motion.
4. Capture force data acting on the wheel through force sensors and encoders during the driving process.
5. Upon completion of the measurement, return to step 1 and repeat the experiment three times under each condition to obtain data.

3.3 Physical quantities to be measured

Given that the system allows independent adjustment of rotation velocity and translational velocity, driving experiments with varying slip ratios can be conducted. The slip ratio: s is expressed using the wheel radius: r, angular velocity: ω, and translational velocity: v_x, which are measured by linear encoders installed on the wheel drive unit and the translational unit, respectively.

$$s = \frac{r\omega - v_x}{r\omega} \tag{1}$$

The traction force: F_x and vertical load: F_z acting on the wheel are measured by force sensors mounted on the top of the wheel. The traction coefficient: μ is derived from these two forces, serving as the primary physical quantity indicating the wheel's driving performance.

$$\mu = \frac{F_x}{F_z} \tag{2}$$

At the beginning of the driving process, the wheel already experiences a certain amount of sinking from the sand surface due to its own weight. This sinking distance is referred to as static sinkage. When driving commences from this state, the wheel's rotation scrapes away the sand, causing the wheel to sink even deeper. This sinking attributed to driving is known as dynamic sinkage. In this experiment, the dynamic sinkage: z_d is measured using a linear encoder mounted on the passive vertical mover.

The traction coefficient and dynamic sinkage for each slip ratio are the measured physical quantities in this experiment.

3.4 Experimental conditions

In this experiment, two parameters, namely the slip ratio and the vertical load applied to the wheels, were varied by adjusting the mass of the weight on the passive vertical moving unit. Table 2 presents the conditions of the driving experiment, where the wheel rotation velocity was set to 0.02 m/s and 0.04 m/s, and the slip ratio ranged from 0 % to 50 % in increments of 10 %. This velocity range is chosen because the planned speed of the subject Rashid rover was a few 0.01 m/s, which is slowly driving. In addition, the vertical load was set to 25 N as a standard, 40 N, and 65 N. Note that this criterion is due to the fact that the Rashid rover is a 4-wheel rover weighing 10 kg, so the load on one wheel is assumed to be 25 N.

Table 2
Test conditions

rω [m/s]	F_z [N]	s [%]					
0.02	25	0	10	20	30	40	50
	40	0	10	20	30	40	50
	65	0	10	20	30	40	50
0.04	25	0	10	20	30	40	50
	40	0	10	20	30	40	50
	65	0	10	20	30	40	50

4. Data capture and analysis

This section elucidates the analysis methodology applied to the time-series raw data acquired from the force sensors and encoders during the driving experiment. Time-series data represent measured values recorded throughout a single driving experiment. As an illustrative example, Fig. 5 portrays the time-series data captured under the following conditions: vertical load of 25 N, rotation velocity of 0.02 m/s, and translational velocity of 0.014 m/s. Over time, after an initial transient phase, the sinkage stabilizes at a constant value, indicating the attainment of steady wheel driving. Concurrently, the average reaction force acting on the wheel also reaches a steady state.

Hence, it is established that a steady state is achieved following a transient period from the initiation of driving, and only this steady-state interval is extracted for subsequent analysis. The steady-state section used for analysis satisfies the following three conditions:

1. Sinkage variation is below 1.5 mm/s.
2. Slip ratio falls within ±5 % of the specified experimental parameter value.
3. Vertical load remains within ±10 % of the specified experimental parameter value.

The oscillation period of the sinkage data is consistent with the period of the number of grousers, indicating that the oscillation is due to the grouser. To mitigate the oscillatory nature of the values, a moving average of the time-series raw data is employed during the analysis to smooth the data. The steady-state interval's raw data obtained from this analysis is segmented into approximately 0.5 s intervals for the 0.02 m/s driving condition and 0.25 s intervals for the 0.04 m/s condition. Moving averages are computed to derive multiple traction coefficients and sinkage data for each interval. Since three driving experiments were conducted for each condition, the number of data samples tripled. In the subsequent graphs, these data sets are averaged and organized based on the slip ratio. The error bars in the graphs represent standard deviations, while the dashed lines approximate polynomial curves.

5. Test results

This section presents the outcomes of the driving experiments conducted to elucidate the influence of soil properties with distinct characteristics on the driving behavior of the wheels, specifically evaluating traction performance and sinkage in relation to the slip ratio. Firstly, the results of experiments performed on two different soil types with varying properties are summarized individually for each soil. Subsequently, the driving performance of the different soils under identical conditions is compared, and the disparities in driving performance are examined.

5.1 Toyoura standard sand

Figure 6 (top) illustrates the traction coefficient results for each slip ratio obtained from the driving experiment conducted on Toyoura standard sand. The traction coefficient exhibits a consistent increase with an escalating slip ratio, up to the maximum 50 % slip ratio achieved in this experiment. This behavior can be attributed to the maintenance of traction facilitated by the grouser wheel. Conversely, beyond the 50 % slip ratio, it is anticipated that the traction coefficient will reach its peak and plateau.

In Fig. 6 (top left), a graph assessing the velocity dependency under the same vertical load condition is presented. Notably, the two approximate curves align closely, indicating the absence of velocity dependency in the measured range within the low-speed region of this experiment.

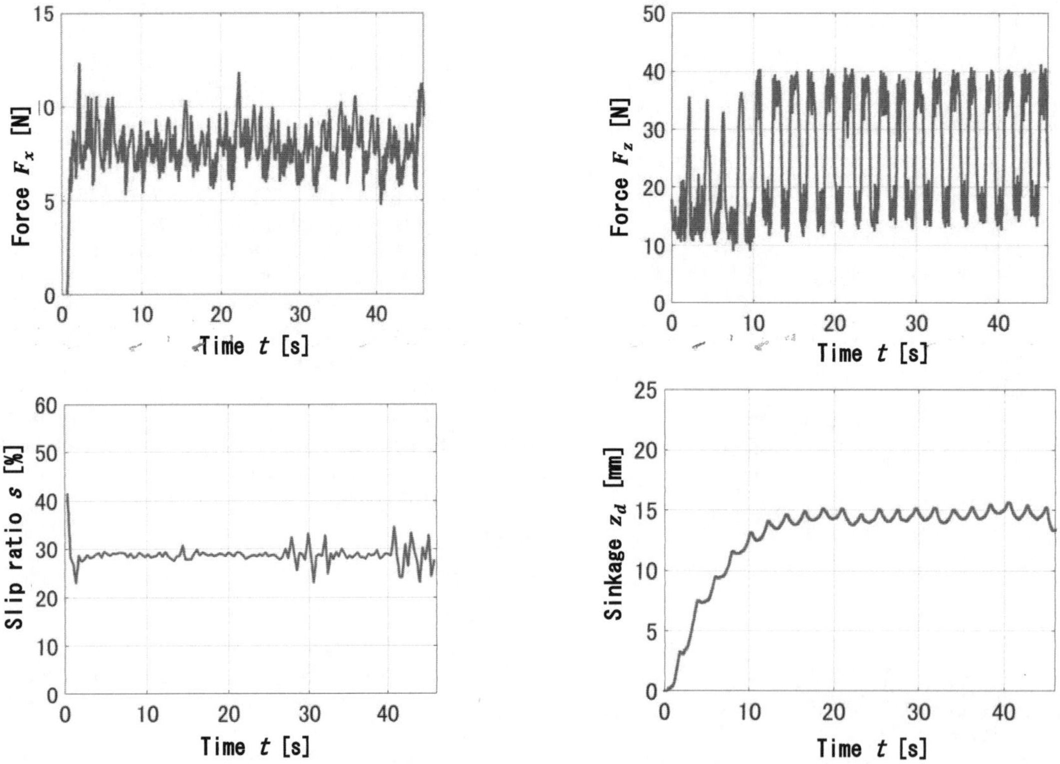

Fig. 5. Representative time-series data obtained in our experiment using single wheel testbed (Toyoura standard sand, Vertical load: 25 N, wheel driving velocity: 0.02 m/s and translational velocity: 0.014 m/s, that realize slip ratio: 30 %)

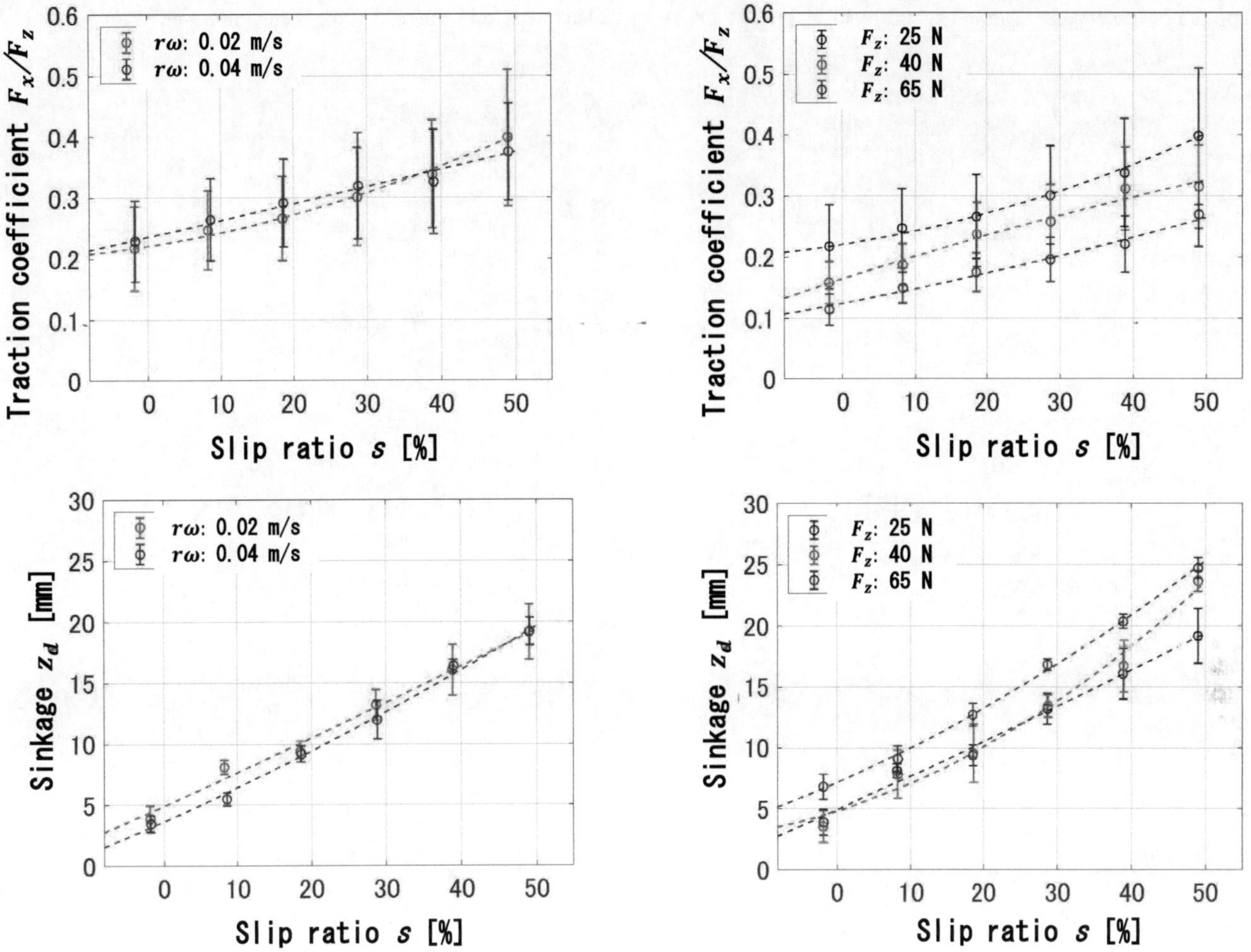

Fig. 6. Toyoura standard sand results: Traction coefficient and sinkage with different driving speeds when vertical load is 25 N (left), and Traction coefficient and sinkage with different vertical loads when driving speed is 0.02 m/s (right)

Furthermore, Fig. 6 (top right) portrays a graph evaluating the load dependency at a fixed driving velocity. The three approximate curves do not coincide, signifying a decrease in the traction coefficient as the load increases, thus demonstrating the presence of load dependence.

Subsequently, Fig. 6 (bottom) depicts the sinkage at various slip ratios. It is evident that as the wheel slip increases, the wheel sinks deeper. Figure 6 (bottom left) indicates that there is no variation in sinkage with increasing velocity, suggesting the absence of velocity dependency within the measured range. Conversely, Fig. 6 (bottom right) displays an upward shift in the sinkage curve as the vertical load increases, underscoring the presence of load dependence. Notably, the trend of this dependence aligns with that of the traction coefficient.

5.2 Regolith simulant (FJS-1)

Figure 7 (top) presents the traction coefficient results for each slip ratio obtained from the driving experiment conducted on regolith simulant. Similar to Toyoura standard sand, the traction coefficient displays an increasing trend with an escalating slip ratio. Figure 7 (top left) depicts the results for different driving velocity conditions at each load, revealing no discernible change in the traction coefficient with increasing velocity. Conversely, Fig. 7 (right) illustrates the results for different loading conditions at each driving velocity, manifesting a clear load dependence, with the traction coefficient decreasing as the load increases.

Next, Fig. 7 (bottom) presents the sinkage results at various slip ratios. This trend also mirrors the observations with Toyoura standard sand, as there is no significant difference in sinkage with increasing velocity, while the sinkage exhibits an upward trend with increasing load.

5.3 Comparison in each soil

As described in Sections 5.1 and 5.2, both types of soils exhibit an increase in traction coefficient and sinkage with an escalating slip ratio. Furthermore, they share the common characteristic of being independent of velocity but dependent on vertical load. Firstly, let's discuss this qualitative phenomenon. The consistent rise in traction coefficient with increasing slip ratio can be attributed to the augmentation of soil reaction force caused by the wheel's scraping action, resulting in increased sinkage. Additionally, as the vertical load increases, the vertical stress exerted by the wheel on the soil intensifies, leading to a larger area of soil disturbance and facilitating sinkage. On the other hand, when the velocity is increased, there is no variation in the vertical stress applied by the wheel to the soil, resulting in a constant area of soil disturbance and no change in sinkage. In fact, (Ono et al., 2021) conducted a Particle Image Velocimetry (PIV) analysis of soil deformation, reporting that the sand flow area in Toyoura standard sand also increases with an increase in vertical load, while the influence of wheel rotation angular velocity on sand flow area is nearly negligible. Moreover, the increase in vertical load and sinkage is believed to enlarge the soil contact area, subsequently elevating the driving resistance and reducing the traction force, ultimately leading to a decrease in traction coefficient, which significantly contributes to the load dependence of driving performance.

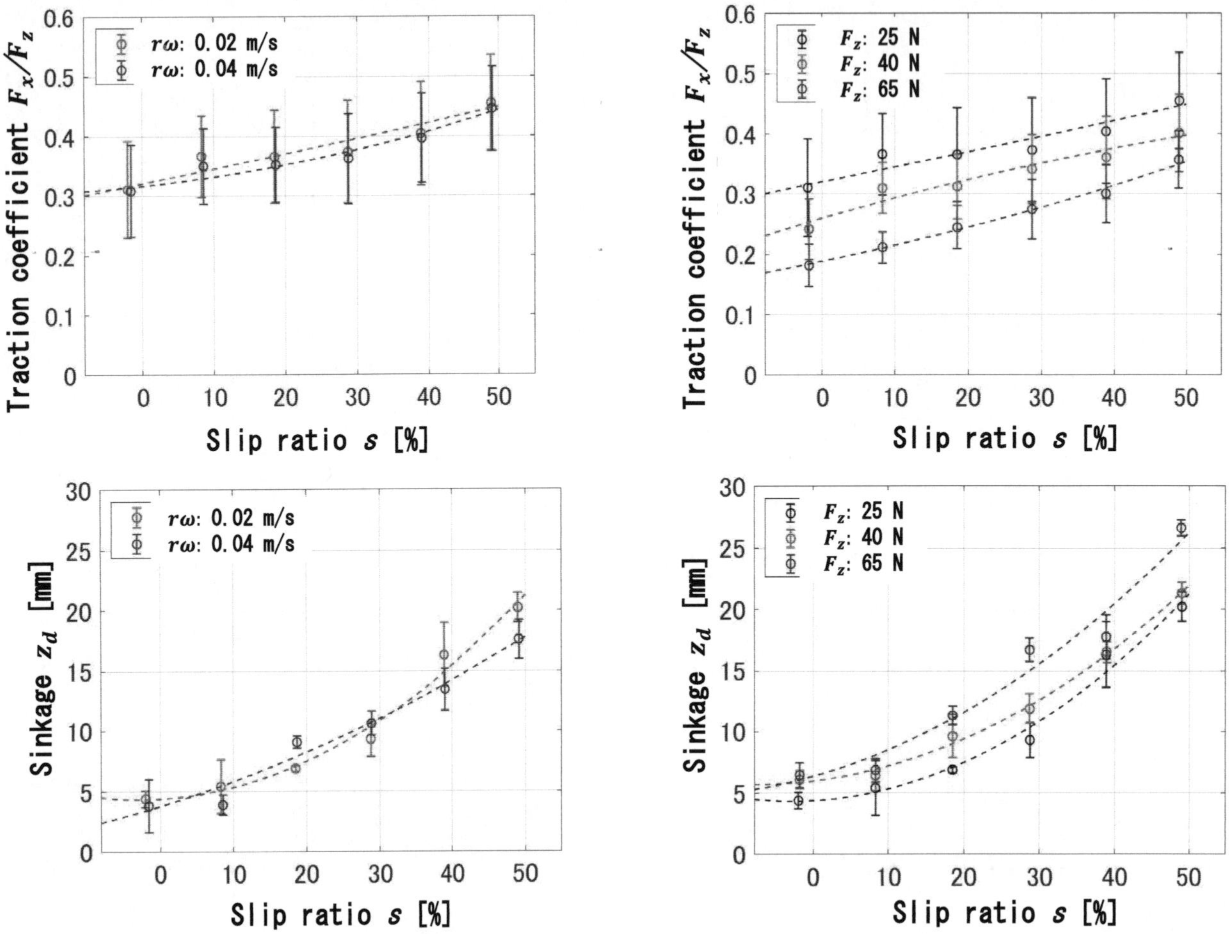

Fig. 7. Regolith simulant results: Traction coefficient and sinkage with different driving speeds when vertical load is 25 N (left), and Traction coefficient and sinkage with different vertical loads when driving speed is 0.02 m/s (right)

Next, Fig. 8 presents a comparison of the traction coefficient and sinkage between Toyoura standard sand and regolith simulant when driven on the same soil under identical conditions. The results demonstrate that the traction coefficient of the regolith simulant surpasses that of Toyoura standard sand in all conditions. Although there is no significant difference in sinkage among the different soils, the sinkage of Toyoura standard sand is consistently 2 mm to 3 mm greater than that of the regolith simulant. Ono et al. (2021) reported that Toyoura standard sand exhibits a larger sand flow area under the same conditions. This larger flow area causes the wheel to scrape out more sand, resulting in increased sinkage. Furthermore, this effect can be attributed to the cohesion of the regolith simulant. It is plausible that the regolith simulant generates a greater reaction force when the same amount of sand is scraped out. Additionally, due to its denser nature and higher specific gravity, the regolith simulant's weight amplifies the reaction force generated by the same amount of sand, thereby enhancing its traction performance.

Lastly, Fig. 9 displays the possible slope angles calculated from the traction coefficient within the range of results obtained in this experiment. When considering the motion of an object on a frictional slope, the maximum slope angle: θ can be determined by an equation based on the equilibrium of forces.

$$\theta = \tan^{-1}(\mu) = \tan^{-1}\left(\frac{F_x}{F_z}\right) \quad (3)$$

The results indicate that for both Toyoura standard sand and regolith simulant, as the vertical load increases, the slope angle decreases. Furthermore, the regolith simulant exhibits higher angles for all vertical loads, enabling it to ascend steeper slopes.

6. Conclusion and future work

This study aimed to assess the driving performance of the grouser wheel through driving experiments conducted on a single-wheel testbed, covering slip ratios ranging from 0 to 50 %. The traction coefficient and sinkage were measured for each slip ratio, and two soil types, namely Toyoura standard sand and regolith simulant, were evaluated in terms of their traction performance. The findings revealed a shared load dependence in the traction coefficient of both Toyoura standard sand and regolith simulant, irrespective of velocity. This dependency can be attributed to soil failure resulting from an increase in soil contact pressure. Notably, the regolith simulant exhibited a higher traction coefficient than Toyoura standard sand across all conditions, primarily due to the cohesive properties of the regolith simulant, which yield a greater soil reaction force.

Furthermore, the experiments conducted at low speeds of 0.02 to 0.04 m/s revealed that both the traction coefficient and sinkage were not influenced by velocity. For future investigations, it is recommended to conduct similar experiments at higher speeds to gain a more detailed understanding of the driving performance of grouser wheels.

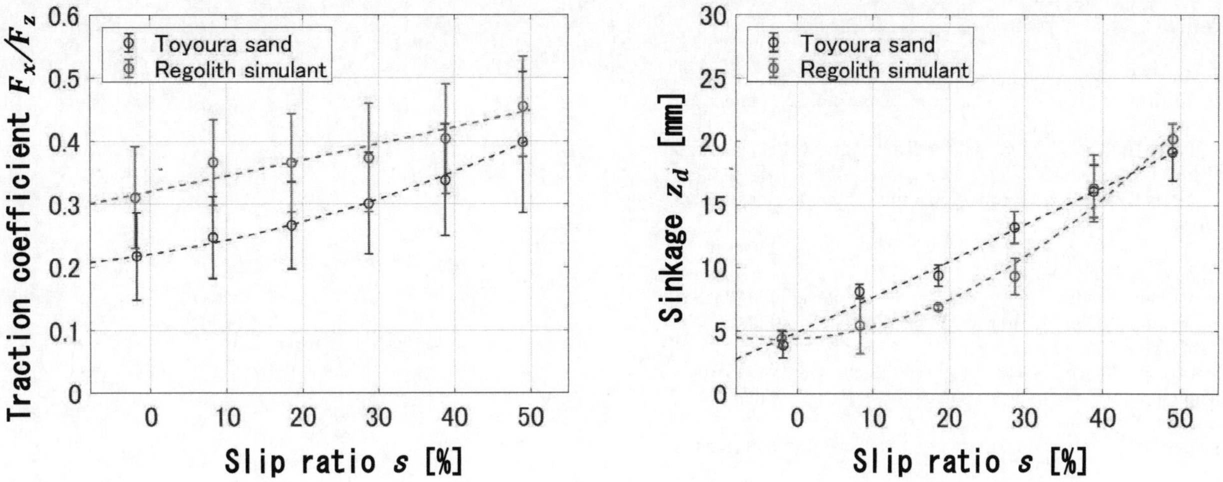

Fig. 8. Comparison of traction coefficient (left) and sinkage (right) of Toyoura standard sand and regolith simulant (Vertical load: 25 N, wheel driving velocity: 0.02 m/s)

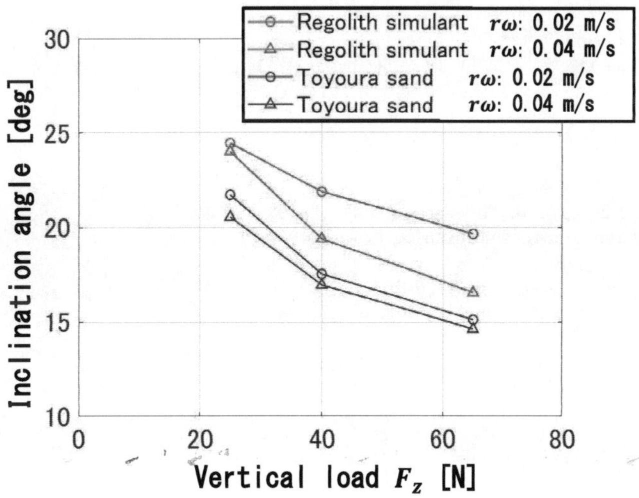

Fig. 9. Comparison of slope angle of Toyoura standard sand and regolith simulant

7. Acknowledgements

We are very grateful that this project was helped by Dr. Sebastian Els, Mohammed Bin Rashid Space Centre (MBRSC) to supply important information about the wheel parameter of the Rashid rover.

8. Declaration of competing interest

The authors have no known competing financial interests or personal relationships that could have appeared to influence the work reported in this paper.

9. References

Almaeeni. S., Els, S.G., Almarzooqi, H., 2021. To a dusty Moon: Rashid's mission to observe lunar surface processes close-up. 52nd Lunar and Planetary Science Conference 2021.

Bekker, M.G., 1956. Theory of land locomotion. University of Michigan Press, Ann Arbor.

Bekker, M.G., 1960. Off-the-road locomotion. University of Michigan Press, Ann Arbor.

Bekker, M.G., 1969. Introduction to terrain-vehicle systems. University of Michigan Press, Ann Arbor.

Emirates Lunar Mission 2022 / Rover "Rashid" https://forum.nasaspaceflight.com/index.php?topic=52020.0 (accessed on 31/5/2023).

Hettiaratchi, D.R.P., Reece, A.R., 1975. Boundary Wedges in two-dimensional passive soil failure. Geotechnique. 25, 197–220.

Higa, S., Nagaoka, K., Nagatani, K., Yoshida, K., 2015. Measurement and modeling for two-dimensional normal stress distribution of wheel on loose soil. J. Terramech. 62, 63–73.

Higa, S., Nagaoka, K., Yoshida, K., 2017. Reaction force/torque sensing wheel system for in-situ monitoring on loose soil. 19th International and 14th European-African Regional Conference of the International Society for Terrain-Vehicle Systems (ISTVS).

Higa, S., Nagaoka, K., Yoshida, K., 2019. Stress distributions of a grouser wheel on loose soil. J. Terramech. 85, 15–26.

Nagaoka, K., Sawada, K., Yoshida, K., 2020. Shape effects of wheel grousers on traction performance on sandy terrain. J. Terramech. 90, 23–30.

Nakamura, H., Nagaoka, K., Yoshida, K., 2020. Soil flow analysis for grouser wheels based on a particle image velocimetry method. J. Terramech. 91, 233–241.

Niksirat, P., Skonieczny, K., Nassiraei, A.F., 2019. Characterizing the effects of reduced gravity on rover wheel-soil interactions using computer vision techniques. 2019 International Conference on Robotics and Automation (ICRA).

Ono, S., Namikawa, S., Yoshida, K., 2021. Analysis of soil flow and traction mechanics for lunar rovers over different types of soils using particle image velocimetry. J. Terramech. 95, 89–100.

Ryu, B.H., Wang, C.C., Chang, I., 2018. Development and geotechnical engineering properties of kls-1 lunar simulant. J. Aerospace Eng. 31 (1).

Sutoh, M., Nagaoka, K., Nagatani, K., Yoshida, K., 2013. Design of wheels with grousers for planetary rovers traveling over loose soil. J. Terramech. 50 (5–6), 345–353.

Toyoura keiseki kogyo co., ltd, about toyoura sand. http://www.toyourakeiseki.com/english/about.html (accessed on 31/5/2023).

Wong, J.Y., 2008. Theory of ground vehicles. John Wiley & Sons Inc, Hoboken, New Jersey.

Wong, J.Y., Reece, A., 1967. Prediction of rigid wheel performance based on the analysis of soil-wheel stresses part i. performance of driven rigid wheels. J. Terramech. 4, 81–98.

4681 | Simplified models of terrain-vehicle interaction for real-time applications

Dávid Körmöczi [a], Péter Kiss [b,*]

[a] *Doctoral School of Mechanical Engineering, Hungarian University of Agriculture and Life Sciences, Gödöllő, Hungary*

[b] *Department of Vehicle Technology, Hungarian University of Agriculture and Life Sciences, Gödöllő, Hungary*

* *Corresponding author: kiss.peter@uni-mate.hu*

ABSTRACT

Accurate modeling or simulation of the vehicle-terrain interaction is critical for effective off-road vehicle navigation. While several high-accuracy methods exist (for example FEM simulation) they typically require computational capacity that exceeds what can be installed in a vehicle. Therefore, they are not applicable for real-time off-road vehicle navigation purposes, where computer hardware capacity is limited by the need for onboard installation. To address this challenge, simplified and less detailed models must be developed for real-time applications. This paper compares three different two-dimensional static terrain-vehicle models, considering accuracy and computational capacity requirements. Results of the comparison provide insights into the suitability of each model for real-time navigation of off-road vehicles.

Keywords
Off-road vehicle navigation
Vehicle-terrain interaction
FEM simulation

1. Introduction

A fundamental part of navigation of vehicles in an off-road environment is to determine whether the vehicle is able to traverse a given part of the terrain. In the absence of a human driver, an autonomous vehicle has to evaluate the surrounding area, and plan the optimal route for the vehicle. For planning of the optimal route, several factors can be taken into account (such as travel time and fuel consumption), however - considering the complexity of such navigation tasks – it is often deemed sufficient to find the shortest route along which the vehicle can move, while avoiding all non-traversable obstacles of the terrain (Liu et al., 2017; Hong et al., 2021). These parts of the operational area are usually categorized as "go" or "no go", referring to whether the vehicle can move through a given part of the area or not (Hua et al., 2022).

The terrain effects the dynamics of the vehicle in several ways. An obstacle can be non-traversable if the vehicle contacts the obstacle through any part other than the wheels (namely, nose-in failure and hang up failure, see Fig. 1) (Laib, 2002; Lukács and Vég, 2022), or when the gravity force acting on the center of mass falls outside the ground contact area (loss of stability, see Fig. 2) (Liu and Paul, 1999). Both of these factors depend on the geometrical position of the vehicle, which can be described using the coordinates of the wheel centers. For this reason, modeling the deformation of the tire and the terrain (soil) is very important. Apart from the stability and obstacle negotiation capability of the vehicle, the tire deformation effects the rolling resistance (Karaftath, 1988), and in agricultural applications the pressure applied to the ground through the contacting area of the tire can also have an unfavorable effect on the soil from the perspective of seed growth (Tardif-Paradis et al., 2015).

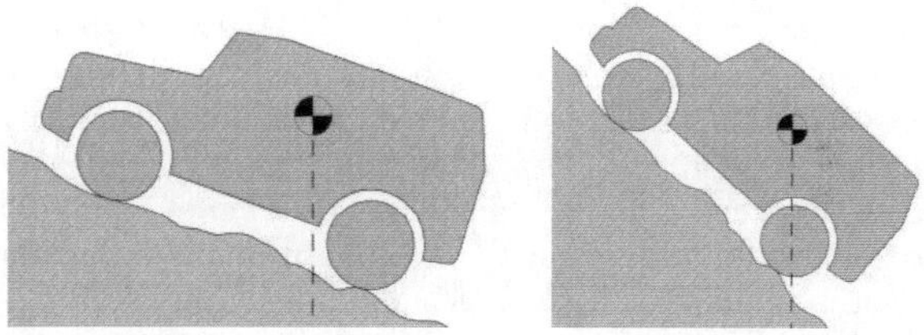

Fig. 1. Longitudinal stability of the vehicle: (a) stable position; (b) unstable position

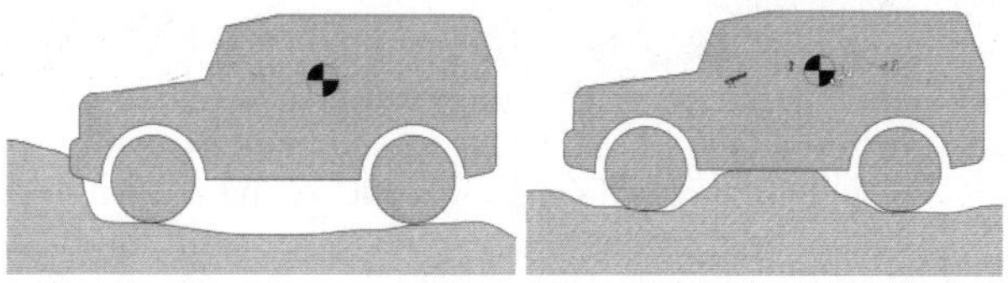

Fig. 2. Typical types of untraversable obstacles: (a) nose-in failure; (b) hang up failure

Many already available and widely used methods are applicable to examine the interaction of the vehicle and the terrain, from the perspective of the tire-soil deformation. On one side, there are several models to characterize the deformation of the soil and the tire, and thus determine the total displacement as a function of the wheel or axle load. The most widely used models are the Bernstein model (Bernstein, 1913):

$$p = k \cdot z^n \tag{1}$$

https://doi.org/10.56884/BPGE9748
ISBN 978-1-942112-55-6 (pdf)

the Saakyan model (Saakyan, 1959):

$$p = k \cdot \left(\frac{z}{D}\right)^n \tag{2}$$

and the Bekker model (Bekker, 1969):

$$p = \left(\frac{k_c}{b} + k_\phi\right) \cdot z^n \tag{3}$$

where:

p - ground pressure [kPa],
z - sinkage [mm],
n - sinakge coefficient [-],
D - diameter of the contact area [mm],
b - the smaller characteristic dimension of the contact area [mm],
k_c - cohesion coefficient [-],
k_ϕ - internal friction coefficient [-].

Most of these models are, at least partially, based on empirical data [He at al., 2019]. On the other hand, universal modeling methods such as finite element method (FEM) are applicable for this purpose (Xia, 2011), or a combination of finite and discrete element method (DEM) can be used to model the characteristics of the rubber tire and the soil, respectively (Kotrocz, 2019).

However, the applicability of these models is limited. The load-displacement equations are only applicable in case of a flat undeformed ground surface, and not generally (Ding et al., 2014). Considering that most of these models were created from an agricultural perspective, where the vehicles move on a surface with little to no macroscopic obstacles, general applicability was not an important aspect. Computer simulations, such as FEM and DEM would be applicable for this purpose, however they are highly limited by the available computational capacity. While this may not be an issue during scientific research, in real time applications the available timeframe makes such simulation methods inapplicable, or at the very least, requires a remarkably less detailed model. It has also to be taken into account, that the available on-board hardware capacity - considering the operational conditions of an off-road vehicle, such as temperature and vibrations - is way less than of a comparable desktop computer or server. Data cloud based remote computing theoretically could be a solution for this, but in many application fields, such as military or emergency services, the navigation cannot rely on the availability of a high-speed network.

For these reasons, it is necessary to select an optimal modeling method (or create one if none of the available models meets all criteria). An optimal model:

- is applicable in a general case;
- need computational capacity that do not exceed the performance of a readily available on-board hardware;
- is based only on theoretical methods and not empirical data, to be easily adapted for different conditions.

Determining the accuracy of a model is not only necessary to select the most suitable modeling method. With a known accuracy, the error can be easily compensated by applying a safety factor, thus lower accuracy models can also be used reliably - even though the introduction of a safety factor partially reduces the theoretically traversable areas.

In this paper, 3 different modeling methods (including sub-types, 7 models in total) are compared to a FEM reference model, and both the accuracy and the computational time of the models is compared. The reference FEM model (which is considered to be orders of magnitude more accurate than the other models) is created based on data available in the literature, and the less computationally intensive models are compared to this reference model.

2. Materials and Methods

2.1 Parameters of the vehicle and terrain

To compare the different models, a reference terrain profile was created. The terrain profile is based on a photogrammetric 3D scan of the agricultural test field (Fig. 3) at the Hungarian University of Agriculture and Life Sciences (Szent István Campus). Considering that the longitudinal obstacle negotiation capabilities of off-road vehicles are far more important compared to lateral, a 2-dimensional terrain section was used. For the 3-dimensional FEM models, the same profile was used as a constant cross section 3D solid. The position of the vehicle has been determined at the same 50 points of the terrain by each of the following modeling methods.

Fig. 3. Point cloud 3D model of the test area

To describe the mechanical properties of the soil, the stress-deformation characteristics shown on Fig. 4 have been used, where the mechanical properties of the rubber and the soil (composed from the typical values for a soft clay soil) can be seen, respectively (Das, 2011; López et al., 2012). A soft soil was selected for the reason to examine the results at a relatively higher deformation, and thus larger absolute error. As it can be seen on the figure, the rubber tire has a linear stress-deformation relationship, as it is expected that the deformation does not reach the point where a significant non-linear characteristic must be taken into account.

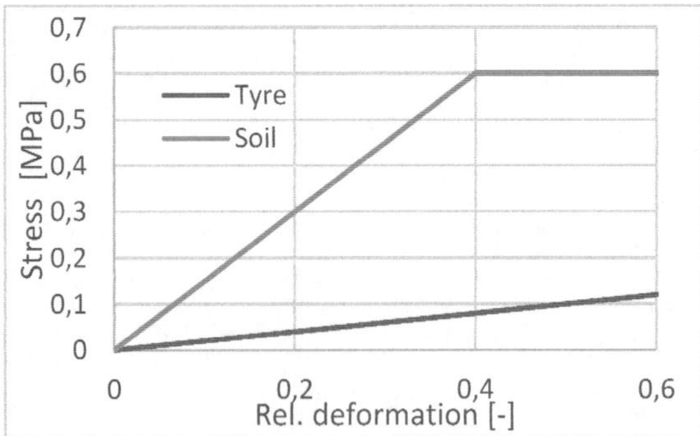

Fig. 4. Stress-displacement characteristic of the simulated soil and tire

Even without further examination of the effect of the elastic parameters on the model accuracy, it is trivial that a more rigid material pairing would mean smaller absolute error in the results. The hypothetical vehicle used in the simulations was a 4 wheeled vehicle with a 1.2 meter wheelbase, and 600 mm x 200 mm x 250 mm (external diameter, width and rim diameter) solid rubber tire with a linear characteristic (with a Youngs modulus of 0,2 MP) [Schaefer, 2010], and a total mass of 500 kg, comparable to a typical small unmanned ground vehicle used in many application fields. The tread pattern of the tire is neglected, as the research of Zeng et al. (2022) has shown that the thread profile only causes a constant offset in the displacement. It has to be noted that while these parameters are partially hypothetical, the aim of the research is not to model the interaction of a specific vehicle and terrain, but to compare modeling methods.

2.2 Reference model

As a reference, a finite element model was used. The model takes into account the deformation of the terrain and the rubber tire, other parts of the vehicle are modeled as rigid bodies, as their only purpose in this model is to distribute the load to the tires. The model was created in Ansys Workbench R1 (2023). For better visualization, only a single wheel and rim is shown on Fig. 5, a full model consists of a pair of these. The model created was a static model, meaning that the effect of rolling of the tire is not taken into account, instead the deformation of the tire and the terrain is modeled as a static state in each position, with only a vertical displacement of the wheel center allowed by the constrains. This is a necessary simplification, because if a rolling tire model have been used, the comparison to the other, simplified models (which do not take the effect of rolling into account) would not be applicable.

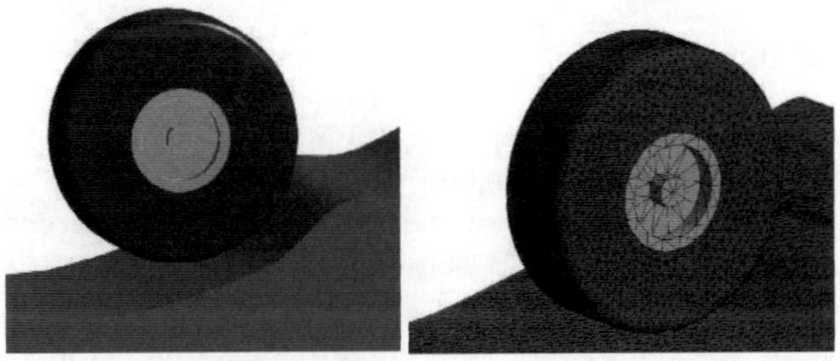

Fig. 5. Reference model: (a) geometrical model of the wheel on a terrain section; (b) tet-rahedral meshing of the model

The reference model was created with a 20 mm meshing element size (triagonal) for the rubber tire, and an adaptive element size triagonal meshing for the terrain. The width and height of the modeled terrain section is chosen so that the effect of the load does not appear at the boundaries of the terrain section, thus equivalent to an infinite half space. The wheel load was determined ad each point based on the load shifting between the axles due to the pitch of the vehicle, with a 1250 N default load. The connection of the rigid rim and flexible tire was modeled as "Bonded", the interaction of the tire and terrain as "No separation". The meshing of the FEM model and the results of a simulation can be seen in Fig 5. (b). and 6, respectively.

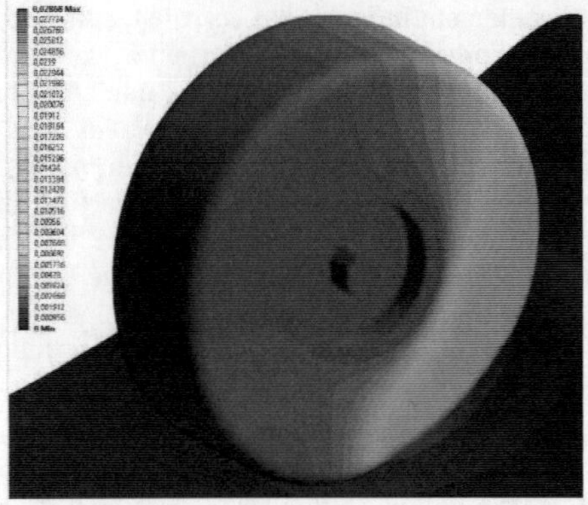

Fig. 6. Results of the reference finite element model, showing the displacement under a 1250 N wheel load

2.3 Finite element models

Apart from the one used as reference, two other, less detailed finite element models have been examined and compared to the reference model. The FEM 1 and FEM 2 models have a decreased meshing resolution of 5 and 15 cm, respectively. From the other perspectives, these models are identical to the reference model, so they will not be discussed in detail.

2.4 Linear rheological model

The simplest model which can be used assumes that the displacement of the axle is the same, at any part of the terrain, as it would be on a perfectly flat, horizontal terrain section, under the same load. This means that, compared to the position of the axle aster the deformation can be derived from the theoretical non deformed position, and a load dependent constant, which constant can be derived from a look up table created by using the reference model. It was found that -within the load interval examined here – the axle displacement is linear to the vertical force, with a theoretical stiffness of 22.6 N/mm.

2.5 Parallel element model

The rheological behavior of the soil and the rubber tire can be described by many various models. On Fig. 7 different soil rheological models can be seen. These models are different in many perspectives, however, considering that in this study only the static, steady state deformation is examined, the velocity dependent dashpot elements of the models can be neglected. With this modification, all mentioned soil models could be simplified to a single elastic element, or to a serially connected elastic and frictional element. The latter will be used in this study to describe the behavior of the soil. For similar reasons, the behavior of the rubber tire (which is usually characterized by the parallel connection of an elastic and a speed dependent dashpot element) will be characterized by simple linearly elastic elements.

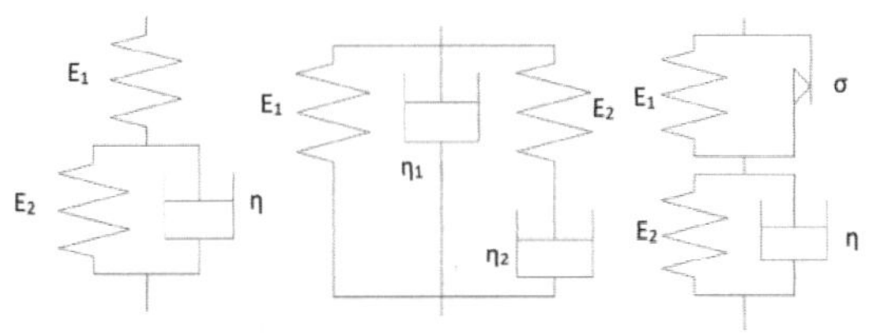

Fig. 7. Different rheological models of soil [Józefiak and Zbiciak, 2017; Komamura, 1974; Shao-wu et al., 2005]

The parallel element models combine the features of the finite element method and the conventional rheological models. Both flexible bodies are divided into a finite number of separate elements, however - contrary to finite element models - the elements of a body have no effect on each other. As Fig. 9 shows, the model can be considered a set of parallel, vertical rheological elements with the same characteristics. The rheological parameters of the tire and the soil elements have been identified using the reference FEM model in two steps, by using a theoretically non-flexible tire to identify the soil parameters, and a non-flexible terrain to identify the tire parameters, as it can be seen in Fig. 8. With using a horizontal resolution of 1 cm, are the following: ktire=25.53 N/cm, ksoil=168.9 N/mm, Fsoil=92.3 N. it also has to be taken into account, that the critical force of the soil models' frictional part is only exceeded in a very small area of the terrain, which makes the parameter identification of the frictional elements critical force inaccurate, but it also means that this parameter has only minor effects on the results.

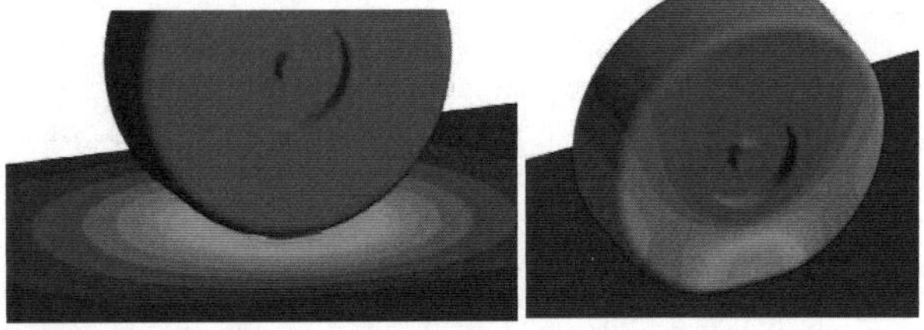

Fig. 8. Parameter identification using the reference model: (a) identification of the soil parameters using a rigid tire mode; (b) identification of the tire parameters on a rigid, perpendicular surface.

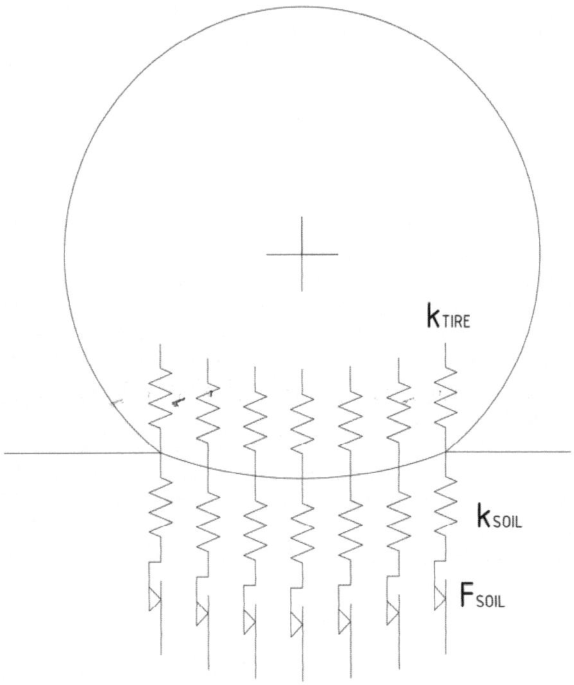

Fig. 9. Parallel element rheological model

The model was applied using 3 different values of longitudinal elements size, 10 mm; 25 mm and 50 mm. These models are referred to as parallel element 1.1; 1.2 and 1.3, respectively.

2.6 Parallel element model 2

The second type of the parallel element model used is similar to the first model in many ways. The main difference is that in the second model the rheological elements are not vertical, but perpendicular to the (undeformed) surface of the terrain and the tire, as it can be seen in Fig. 10. The software implementation of both parallel element models was done in Python 3.

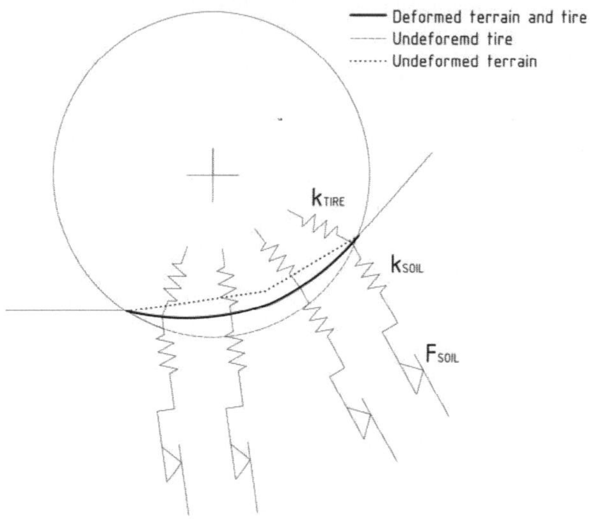

Fig. 10. Parallel element rheological model

From all other perspectives, the model is identical to the parallel element 1.1 model, and the same rheological parameters have been used.

3. Results

On the selected terrain profile, all 5 of the earlier discussed have been used at the selected points of the terrain profile. it was assumed, that the finite elements model used as reference is perfectly accurate in a sense that, while the error of the model is not zero, it is negligible compared to the errors of other, less detailed models. Since the exact computational time required for each simulation method would be hard to interpret, a percentage value (compared to the least capacity-intensive linear model) has been used. The computing time only refers to the runtime of the simulation itself and does not include the importation of the initial data, identification of the model parameters, and the exportation and visualization of the results.

Table 1
Results of the model comparison

Model	Avg. accuracy [mm]	Rel. computational time [-]
Reference	-	5199
FEM 1	3.352	2172
FEM 2	6.133	641
Linear	17.802	1
Parallel element 1.1	10.764	12.38
Parallel element 1.2	11.550	7.12
Parallel element 1.3	11.684	4.23
Parallel element 2	9.628	20.86

The results show that the less computational capacity a modeling method needs, the less accurate it is. It is important to note that this statement is not true in general, just for the models compared in this study. Modeling methods that require higher computational capacity and still are less accurate can also exist.

By examining the results, it can be seen that there is no "best" modeling method, as the more accurate models require consistently more computational capacity. However, there are models that are inferior from the perspective of their cost-performance ratio. For example, it can be seen that if a parallel element 1 type model is used, it is not necessary to use a higher resolution than 50 mm, as it would not significantly improve the accuracy.

As mentioned earlier, the deformation of the tire and the soil affects the mobility and stability of the vehicle in many ways. The accuracy of the examined models can be interpreted differently, depending on the purpose of the modeling. The error of even the less accurat model is small comparable to the external dimensions of the vehicle, and for example only has a minor effect on the change in the load distribution between the axles due to the pitch of the vehicle. However, from the perspective of the obstacle negotiation capabilities of the vehicle, even these inaccuracies are significant. Considering that a typical small UGV, comparable to the one used in the simulations, usually has a ground clearance of no more than around 150-250 mm, even a 20-30 mm safety clearance would have a significant effect on the obstacle negotiation capabilities. Based on the results of this study, it would be possible to select an optimal modeling method for a specific application, making a compromise between the accuracy and the computational capacity. Apart from selecting an optimal method, the results can be useful to determine what safety margins should be used for the mobility models. The safety margin should be large enough to compensate all of the possible inaccuracies of the mobility models, but also as small as possible not to reduce the theoretically traversable parts of the operational area more than what is absolutely necessary.

4. Conclusion

In this study, a total of 8 different terrain-tire interaction models (3 model types with different rasolutions) have been compared, both from a perspective of accuracy and computational capacity. based on the results, it can be said that all of the examined models may have a use in some circumstances. It is not possible to select a "perfect" model, as a compromise has to be made between the accuracy of the model and the available computational capacity. The results of this study can be used to select the optimal model for a specific application. Also, it is planned to continue the research and create an iterative method combining some of these models, adaptively using the more detailed (and thus, computational capacity intensive) models only for the critical parts of the terrain, while less detailed models can be used on the other parts. In this study, it was not thoroughly examined whether a mathematical connection between the accuracy of the models and the terrain profile can be found. However, it was experienced that the error of the models is larger in points where the terrain profile has a high microscopic roughness and/or has a high slope. Further research of this phenomena could provide important results, which could be essential to create the earlier mentioned combined models, as it could provide a method to determine which models should be used at ani given part of the terrain.

5. Acknowledgements

Project no. 2022-2.1.1-NL-2022-00012 has been implemented with the support provided by the Ministry of Culture and Innovation of Hungary from the National Research, Development and Innovation Fund, financed under the 2022-2.1.1-NL funding scheme.

The drone-based 3D mapping of the test area was carried out with the contribution of the Department of Geodesy, Budapest University of Technology and Economics.

6. Declaration of competing interest

The authors declare that they have no known competing financial interests or personal relationships that could have appeared to influence the work reported in this paper.

7. References

Bekker, M.G., 1969. Introduction to terrain-vehicle systems. part I: The terrain. part ii: The vehicle. Michigan Univ Ann Arbor.

Bernstein, R., 1913. Probleme zur experimentellen Motorpflugmechanik. Der Motor-wagen.

Das, B.M., 2011. Geotechnical engineering handbook. J. Ross publishing.

Ding, L., et al., 2014. New perspective on characterizing pressure–sinkage relationship of terrains for estimating interaction mechanics. Journal of Terramechanics. 52, 57-76.

He, R., et al., 2019. Review of terramechanics models and their applicability to real-time applications. Journal of Terramechanics. 81, 3-22.

Hong, Z., et al., 2021. Improved A-Star algorithm for long-distance off-road path planning using terrain data map. ISPRS International Journal of Geo-Information. 10.11, 785.

Hua, C., et al., 2022. A Global Path Planning Method for Unmanned Ground Vehicles in Off-Road Environments Based on Mobility Prediction. Machines. 10.5, 375.

Józefiak, K., Zbiciak, A., 2017. Secondary consolidation modelling by using rheological schemes. MATEC Web of conferences. Vol. 117. EDP Sciences.

Karaftath, L.L., 1988. Rolling resistance of off-road vehicles. Journal of construction engineering and man-agement. 114.3, 458-471.

Komamura, F., Rodney J.H., 1974. New rheological model for soil behavior. Journal of the Geotech-nical Engineering Division. 100.7, 807-824.

Kotrocz, K., 2019. Talaj diszkrét elemes modellezése talaj-kerék kölcsönhatásának elemzéséhez. Dissertation, Budapest University of Technology and Economics (Hungary).

Laib, L., 2002. Terepen mozgó járművek. Szaktudás Kiadó Ház, Budapest.

Liu, J., and Paul D.A., 1999. Off-road vehicle rollover and field testing of stability index. Journal of agri-cultural safety and health. 5.1, 59.

Liu, Q., et al., 2017. Global path planning for autonomous vehicles in off-road environment via an A-star al-gorithm. International Journal of Vehicle Autonomous Systems. 13.4, 330-339.

López Bravo, E., et al., 2012. Determinación de las Propiedades Mecánicas en un Suelo Arcilloso como Función de la Densidad y el Contenido de Humedad. Revista Ciencias Técnicas Agropecuarias. 21.3, 05-11.

Lukács, T.J., Vég, R.L., 2022. Az autonóm terepjáró eszközök. Műszaki Katonai Közlöny. 32.2, 107-116.

Saakyan, S.S., 1959. Vzaimodeistrie vedomogo kolesa i pochvi.

Schaefer, R.J., 2010. Mechanical properties of rubber. Harris' Shock and Vibration Handbook, Sixth edition, A. Piersol, T. Paez (Eds), McGraw-Hill Companies Inc. 33-1.

Shao-wu, G., et al., 2005. Solution of a rigid disk on saturated soil considering consolidation and rheology. Journal of Zhejiang University-Science A. 6, 222-228.

Tardif-Paradis, C., et al., 2015. Effect of planter and tractor wheels on row and inter-row weed populations. Crop Protection. 71, 66-71.

Xia, K., 2011. Finite element modeling of tire/terrain interaction: Application to predicting soil compaction and tire mobility. Journal of Terramechanics. 48.2, 113-123.

Zeng, H., et at., 2022. Experimental investigation of tire traction performance on granular terrain. Journal of Terramechanics. 104, 49-58.

4816 | Effects of humidity on the emissions of the diesel engines

Daniel Szőllősi [a,b,*], Peter Kiss [c]

[a] Doctoral School of Mechanical Engineering, MATE, Gödöllő, Hungary

[b] IBIDEN Hungary Technical Center, Exhaust system evaluation, IBIDEN Hungary Ltd., Dunavarsány, Hungary

[c] Department of Vehicle Technology, Institute of Technology, MATE, Gödöllő, Hungary

* Corresponding author: szollosi.daniel1@phd.uni-mate.hu

ABSTRACT

Above any common perspective, for heavy duty and off-road conditions at least, the usage of the internal combustion engines will continue for decades because of the advantages on this area, like fuel storing, weight and range. During the operation of the internal combustion engines, the main contributors for the burning process is the fuel and air. The fuel component possibilities and effects widely investigated, but almost no attention for the other contributor, the air. It is considered as a fix parameter but at least one component, the humidity can widely change depending on environmental circumstances and can have great effect on the engine operation. It can be considered as the oldest and simplest emission treatment technic what we using – even unnoticedly – since the usage of the internal combustion engine. It is essential to have a proper overview on this area, it should be summarized and clearly understand the effects of air humidity. This knowledge is a baseline for lot of further investigation in this area. The water effect is significantly important for the diesel engine, where excess air going into the combustion chamber. This paper gives us insight into the different levels of humidity induced performance and emission differences. It reveals the connection between the particle number and gas emission in relation to the water introduced to the combustion chamber. To answer today's most urging questions, in this survey, data also provided connected to the ultrafine particles (sub 23nm), which emerging with the new emission regulations.

The paper provides guidelines and possibilities to show the potential of artificially increased water content in the combustion air and the potential of water-using techniques.

Keywords
Emission reduction
Humidity
Water fumigation
Ultrafine (sub. 23 nm) particles
Nitrogen oxides (NO_x)

1. Introduction

1.1 Emissions of diesel engines

Many toxic and less harmful compounds are emitted as a result of diesel engine combustion. Unburned fuel is the source of hydrocarbon (HC) emissions. It forms at the lower-temperature portions of the combustion chamber, where their breakdown products and oxygen-containing derivatives may contain partially or fully non-oxidized hydrocarbons. The reaction for complete combustion is depicted in Eq. 1:

$$C_xH_y + (x + y/4)\,O_2 \rightarrow x\,CO_2 + (y/2)\,H_2O \tag{1}$$

Diesel oil does not completely burn when there is insufficient air or oxygen, resulting in incomplete combustion. Hydrogen and carbon oxidized during this process, and as a result, produces carbon monoxide, carbon, and water. Because diesel engines have a substantial air surplus, CO (carbon monoxide) production is limited. At high combustion temperatures of around 2,000 °C, combustion mostly produces CO, not CO_2. The balance has shifted in favor of CO creation in such cases Eq. 2. It only changes direction during the expansion phase and when it cools down after passing through the exhaust (Kalló, 2000).

$$CO + \tfrac{1}{2}\,O_2 \leftrightarrow CO_2 \tag{2}$$

Particulate matter (PM) in diesel engines resulted from improper combustion. They are formed by the agglomeration of various particles along the exhaust system. The particles are extremely complicated since they contain a wide range of components like carbon, partially burned gasoline, partially burned lubricating oil, ash from fuel and lubricating oil, water, and metal particles (Agarwal, 2007). These components are frequently adsorbed onto a central carbon core to create agglomerates ranging in size from 15 to 40 nm, with 90% of them being less than 1 μm (Burtscher, 2005). The first step in the creation of particulates is nucleation, and the nucleation particles are subsequently agglomerated. The nucleation occurs in the engine cylinder (carbon, ash) and the exhaust system (hydrocarbons, sulfuric acid, water) via homogeneous and heterogeneous nucleation mechanisms. Depending on its aerodynamic diameter, ambient particulate matter divided into following categories which is calculated by measuring the diameter of a sphere with a density of 1 g/cm3 and an air velocity equal to that.

- Coarse particles or PM_{10}—particulates of an aerodynamic diameter ≤ 10 μm,
- Fine particles or $PM_{2.5}$—particles with a diameter ≤ 2.5 μm,
- Ultrafine particles of diameters below 0.1 μm or 100 nm,
- Nanoparticles, characterized by diameters of less than 50 nm (Dieselnet, 2002).

Airborne nitrogen does not interact with oxygen in a typical combustion. However, due to the inhomogeneous combustion characteristic of diesel engines, nitrogen already mixes with oxygen at a starting temperature of 1600 °C to form NO_x emissions. The two basic components of NO_x emissions are NO and NO_2. There are three primary methods for NO to be created during diesel combustion. Before the thermal mechanism can occur, NO is swiftly created in the flame zone of the laminar premixed flames in the quick (or Fenimore) process (Fenimore, 1972). In case of the nitrous oxide mechanism, oxygen radicals (O) and volatile nitrogen species like hydrogen cyanide or nitrous oxide oxidized during combustion nitrogen interacts to create intermediaries like N_2O and is oxidized into NO. Low temperature and lean fuel (0.8)

https://doi.org/10.56884/THMG6024
ISBN 978-1-942112-55-6 (pdf)

are the conditions that this mechanism operates in (Turns, 2000). The third and most important mechanism is the thermal (or Zeldovich) mechanism. Take happen at temperatures substantially higher than 1300 °C. It begins with the two reactions described by Zeldovich, but later adds a third phase (Rente, 2004):

$$N + O_2 \leftrightarrow NO + O \quad (3)$$

$$O + N_2 \leftrightarrow NO + N \quad (4)$$

$$N + OH \leftrightarrow NO + H \quad (5)$$

Under normal air conditions, the NO that predominates in NO_x inside the combustion chamber slowly oxidizes to NO_2. Measurements performed and data evaluated with the procedure detailed in the experimental chapter. Results can confirm the effects of the humidity varies in one year.

1.2 Emission regulations

In response to health concerns, new and more stringent emission limits for new diesel engines have been introduced worldwide. These force continuous improvements and a system approach, where advanced engine control must be used simultaneously and combined with various sophisticated exhaust after-treatment technologies.

In this paper light duty engine tests performed for investigating humidity effects. The light duty regulations covered by the new EURO 7 criteria, and will came to force from 2025 along with the passenger car limits. As shown on Fig. 1 for tailpipe CO_2 fleet average should be decreased to Zero till 2035.

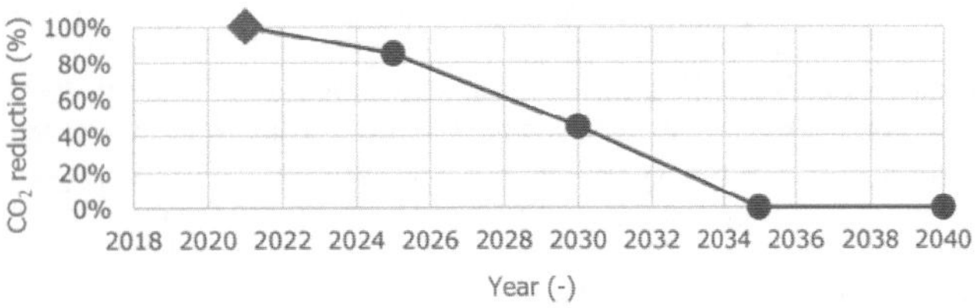

Fig. 1. EU targets for Light-Duty vehicles CO_2, tailpipe emissions, fleet average reference to 2021 (AVL Techday Emission Conference, 2023)

For EURO 7 further requirements will be applied to the engine emissions, like 10 nm emission, new component (total hydrocarbon - THC, Non-methane hydrocarbons - NMHC, ammonium - NH_3) measurements, along with testing conditions changing, broadening and became more severe (European Commission, 2022).

For related non-road machinery regulations, in the US, the Tier 4 standards is in force, while in Europe the latest regulation in force is the Stage V. They use wide variety of different categories for regulating the non-road vehicles from several hundred cubic centimeter to marine vessel engines, or depending on engine power. The regulated components are PM (particle mass) and engine gaseous emissions. In the latest EU Stage V, PN (particle number) regulation limits appeared also for non-road vehicles. As a result, application of DPF (Diesel particulate filter) became necessary, while in US the regulation not necessitate the usage of DPF. As this also shows, above Tier 3/4 and the corresponding Stage III/IV, the harmonization between the different party regulation (EU and USA) has been largely lost. Engine and equipment makers have put regulatory authorities in the EU, the United States, and Japan under pressure to standardize global emission regulations (Dieselnet, 2023, 2021; European Commission, 2023).

1.3 Water/humidity effect

Numerous scientific papers confirmed that water added into the combustion chamber has a considerable effect on its emission. These impacts are described further below.

The addition of water increases the mass of the air-fuel mixture and the associated air volume for the same delivered energy and unit time, resulting in improved atomization and mixing (Sheng et al., 1995). Ignition delay is consistently observed in all diesel engines (Sawa and Kajitani, 1992; Miyamoto et al. 1995; Janousek, 2010). Improved combustion rejection, higher peak pressure (and heat release maximum) and higher pressure rise rate (Tsukahara et al., 1989). From the point of view of NO_x control, increased heat release rate and peak pressure are counterproductive effects. However, the addition of water also results in reduced peak combustion temperatures, which is the primary effect responsible for NO_x reduction in all methods of adding water (Andrews et al., 1988). Combustion temperatures are reduced due to the heat sink effect of water, both because of the increase in gas mass and specific heat (Ishida and Chen, 1994) (an effect that closely resembles NO_x reduction mechanisms via exhaust gas recirculation - EGR). The improved atomization and mixing results in faster combustion speed, which leads to a reduction in the overall combustion duration. This helps reduce NO_x formation from the diffusion flame. Chemical reactions can also occur, such as accelerated water gas reaction due to increased partial water pressure (Ishida and Chen, 1994; Allen, 2001). Around maximum power, the fuel enrichment strategy can be replaced by water injection, resulting in a reduction of PM emissions and fuel consumption (Rohit et al., 2017; Fan et al., 2020).

Water can cause PM emission decreasing, through better oxidation of carbon in the premixed flame due to improved mixing and a higher than stoichiometric (equivalence ratio) oxygen content in the flame, and inhibition of soot precursors (Langer et al., 2001). However, the mechanisms of PM reduction by water are generally less understood than those of NO_x reduction.

2. Experimental

2.1 Measurement system, layout

A medium duty engine (EURO V/EEV compatible inline 6 cylinder) was chosen for humidity data investigation. It has a wide range of on-road and non-road applications, including harvester, piste machine, and public transportation vehicles. The chosen engine model does not use EGR, which allows us to avoid the overlapping effect of EGR and humidity (Ishida and Chen, 1994; Wei et al., 2022), to view simply the water generated differences.

The tests performed on a Horiba HT460 test cell equipped with fluid and air conditioners. Horiba MEXA 7500 is utilized for nitrogen oxides (NO/NOx), carbon oxides (CO/CO_2), oxygen, and hydrocarbon (THC - total hydrocarbon) gas components data collection. Particle number measurements were carried out with AVL APC489 for 10 nm measurements and CPC488 for 23 nm and above measurements. We measured raw exhaust gas during the testing and utilized dilution factor 20000 to avoid particle counter clogging.

The stationary cycle was chosen for the experiments because it provided better comparability. The test protocol includes a precondition (WHSC – world harmonized stationery cycle - mode 9 - 125 sec), a five-minute soak, and the WHSC cycle. The humidity, fuel ratio and gas components are averaged from each cycle and the particle number is summed for the corresponding cycle.

2.2 Humidity calculations

Relative humidity, temperature and pressure was logged in the engine intake and from that absolute humidity was calculated. To calculate absolute humidity, first saturation vapor pressure needs to be calculated as can be seen in Eq. 3 (Wagner and Pruss, 2002).

$$ln(P_{ws} / P_c) = T_C / T \cdot (a_1 \cdot \vartheta + a_2 \cdot \vartheta^{1.5} + a_3 \cdot \vartheta^3 + a_4 \cdot \vartheta^{3.5} + a_5 \cdot \vartheta^4 + a_6 \cdot \vartheta^{7.5}) \quad (3)$$

P_{ws}: saturated water vapor pressure (Pa)
P_c: critical pressure equal to 22.1 MPa
T_c: critical temperature equal to 647.1 K
T: ambient temperature (K)
ϑ: 1- (T/T_c)
$a_1, a_2, ..., a_6$: empirical constants

From the saturation vapor pressure, water vapor pressure can be calculated Eq. 4 (Lawrence, 2005).

$$P_w = RH \cdot P_{ws} / 100 \quad (4)$$

RH: relative humidity (%)
P_w: water vapor pressure (Pa)

Ideal gas law can be adapted to water vapor, and with that absolute humidity can be calculated Eq. 5 (Wagner and Pruss, 2002).

$$AH = (P_w / (R_w \cdot T)) * 1000 \quad (5)$$

AH: absolute humidity (g/m^3)
Rw: specific gas constant for water vapor 461.5 (-)

The majority of water injection publications give the injected water volumes in terms of water fuel ratio (Hadidi et al., 2022; Pamminger et al., 2020; Wei et al., 2022). Additional calculations are required to compare the results with these. To accomplish this, we should execute the following calculations:

$$P_d = P - P_w \quad (6)$$

P_d: pressure of dry air (Pa)
P: ambient pressure (Pa)

After calculating the pressure of dry air, as a next step the air density needs to be calculated. We need to simplify the ideal gas equation and rearrange like below:

$$\rho = P_d / (R_d \cdot T) + P_w / (R_w \cdot T) \quad (7)$$

ρ: air density (kg/m^3)
Rd: specific gas constant for dry air 287.1 (-)

With these values, water fuel ratio can be calculated:

$$WFR = \left(\frac{MAF \cdot (AH / \rho)}{m_f}\right) \cdot 100 \quad (8)$$

WFR: water fuel ratio (%)
MAF: intake air mass flow (g/s)
m_f: fuel flow (g/s)

3. Results and discussion

Measurements performed during the year of 2022 and 2023. This allows to track the humidity conditions that can occur during the year in the temperate zone, in a mix of continental, oceanic and Mediterranean climates typical of Hungary.

Emissions measured in reflection to the characteristic emission of the diesel engine like the particulates and nitrogen oxides. All the results show clear tendencies:

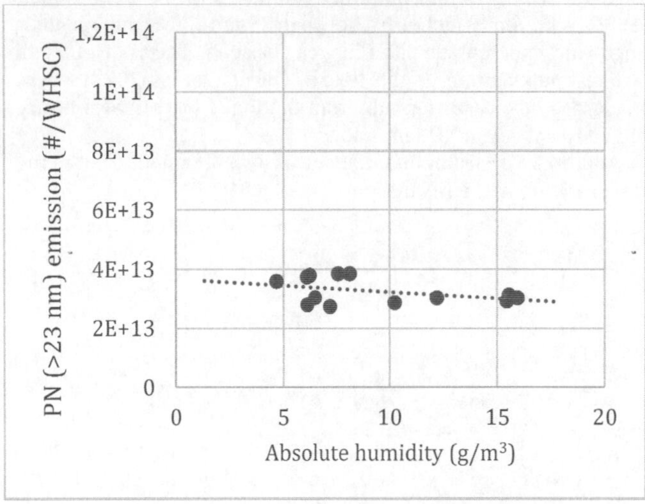

Fig. 2. 23 nm and above particle number results (with CPC488)

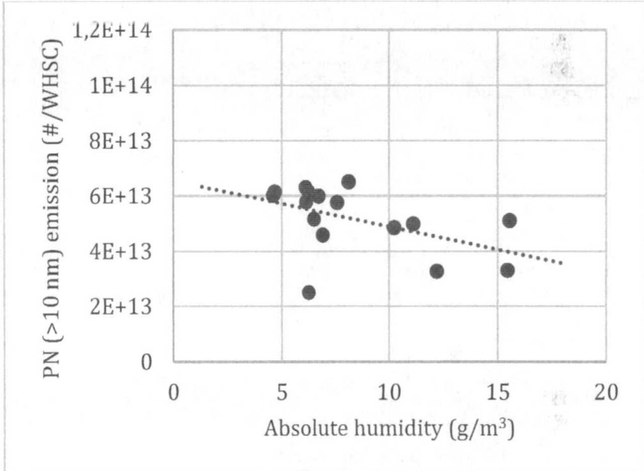

Fig. 3. 10 nm and above particle number results (with APC489)

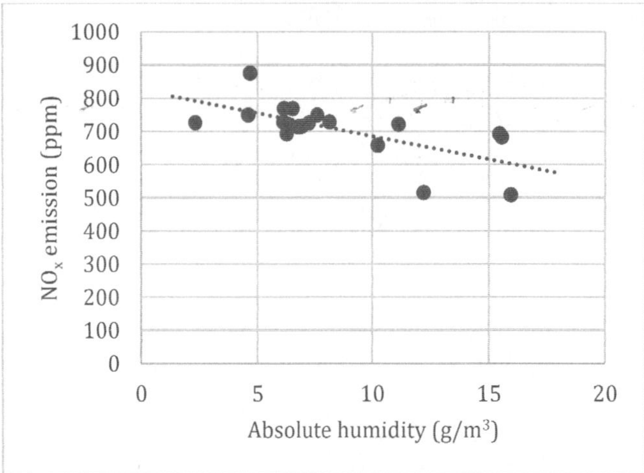

Fig. 4. NO_x gas emission results (with MEXA 7500)

As can be seen on Fig. 2 and Fig. 3, emissions are decreasing with increasing humidity. Comparing the two figures, the emission of 10 nm particles decreases at a faster rate (~30% along the scale of ~5-15 g/m³ absolute humidity compared to the 23nm result ~15%) as humidity increases. Between 23 nm and 10 nm results a roughly fifty percentage difference can be seen. This is in line with the latest publications on this field by (Bíró and Kiss, 2023) where an average of 45% observed. From the Fig. 4 we can also notice steep decreasing on NO_x emissions.

Additional gas component data was also obtained to aid in the interpretation of the results.

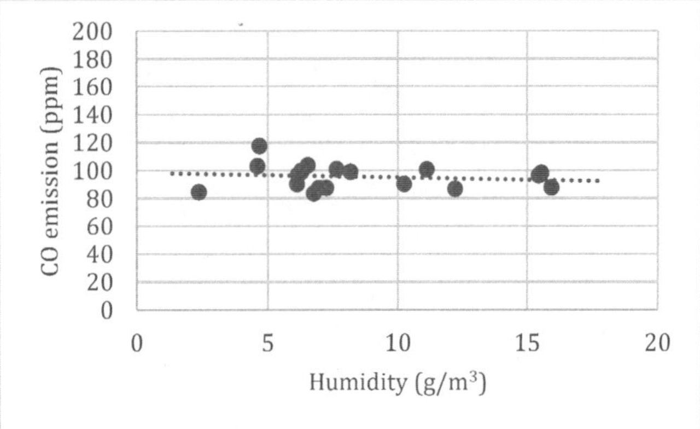

Fig. 5. CO gas emission results (with MEXA 7500)

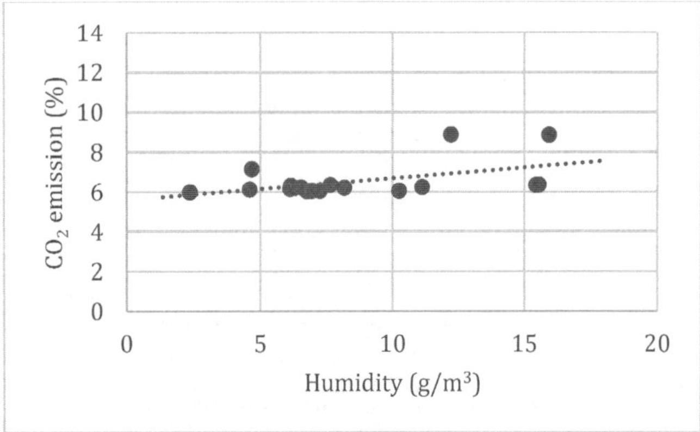

Fig. 6. CO_2 gas emission results (with MEXA 7500)

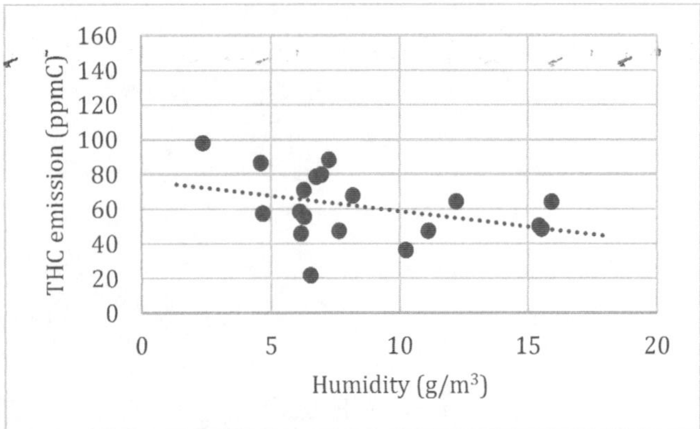

Fig. 7. Total hydrocarbon gas emission (with MEXA 7500)

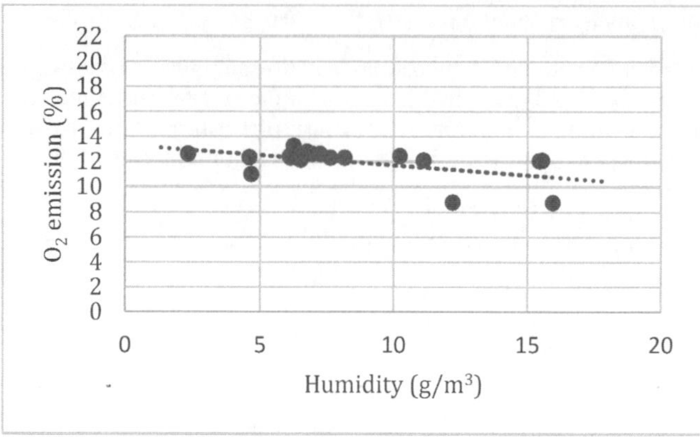

Fig. 8. O_2 gas emission results (with MEXA 7500)

The most noteworthy result from the carbon dioxide emission results shown in Fig. 6 is a noticeable increase in CO_2 emissions as humidity increases. In terms of CO emission, less then ~5% dropping in emission can be seen in Fig. 5.

The standard deviation is larger for hydrocarbons (see Fig. 7), but the downward trend is also visible.

Figure 8 show ~15% decreasing of the oxygen in the engine out emissions.

Conclusions and future work

In overall, from the results of the exhaust gas components we can see the effects of the air as one of the main contributor on the burning process.

From emission reduction point of view, these results are the following:

- The Nitrogen oxide results shows decreasing tendency as can be expected from literature data.
- Previous research has found that the mechanism for reducing PM emissions is not well understood. The particle number emission shows a declining trend in this investigation. When the emission of 10 nm and 23 nm particles is compared, the emission of the 10 nm particles decreasing at a higher rate as humidity increases. This indicates that water has a greater effect on smaller particles. There is a possibility water can catalyze the agglomeration process in the nanoparticle size range.
- The carbon monoxide level did not vary significantly, which is appropriate given the diesel engine air surplus and its overall low CO emission.

  CO_2 results however shows clear increasing. Considering the Eq. 1, and that there is no changing in fuel injection, we can conclude that the oxidation and burning is more complete.
- With bigger scattering, but the THC emissions also decreasing. This result is consistent with and supports the idea of more complete oxidation, which we may assumed during the CO_2 data evaluation also.
- Oxygen level slowly decreasing in connection to the CO_2 increased emission.

Calculating these results to the water to fuel ratio (Eq. 6-8) and then add the scale used by previous studies (Hadidi et al., 2022; Pamminger et al., 2020; Wei et al., 2022) gives the result shown in Fig. 9.

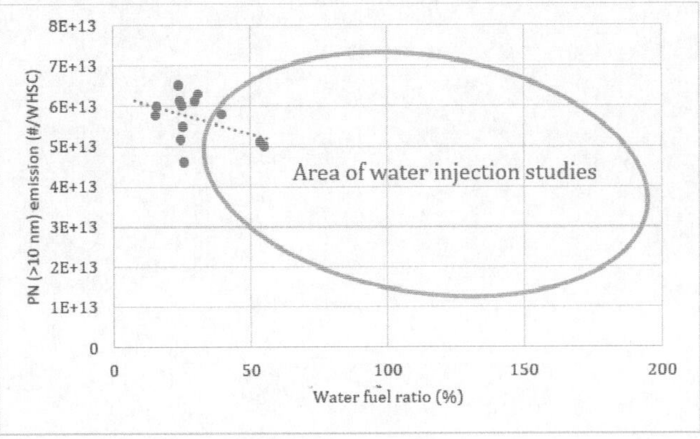

Fig. 9. Particle number results as a function of WFR ratio, also showing characteristic range of water injection studies

Even in the rather dry climate of Hungary we can see a water fuel ratio of 10-50% which, compared to the 25-150% WFR range commonly studied in publications, can cause quite a significant distortion in the results.

Related to the new EURO 7 legislation good results can be seen how water effect on the 10 nm size range emission. Cold starting tests must be performed in order to continue the investigations in accordance with the regulations. Also, in order to expand the general knowledge, it may be worth doing the tests with a gasoline engine as well.

4. Acknowledgements

Project no. C1767774 has been implemented with the support provided by the Ministry of Culture and Innovation of Hungary from the National Research, Development and Innovation Fund, financed under the KDP-2021 funding scheme.

5. Declaration of competing interest

The authors declare that they have no known competing financial interests or personal relationships that could have appeared to influence the work reported in this paper.

6. References

Agarwal A.K., 2007. Biofuels (alcohols and biodiesel) applications as fuels for internal combustion engines. Progress in Energy and Combustion Science vol. 33, 233–271.

Andrews G.E., Bartle K.D., Pang S.W., Nurein A.M., Williams P.T., 1988. The Reduction in Diesel Particulate Emissions Using Emulsified Fuels. SAE Technical Paper 880348, doi:10.4271/880348

Bíró N., Kiss P., 2023. Internal combustion engine's sub and over 23 nm solid particle measurement. MDPI. Sensors 2023. 23(2), 590; https://doi.org/10.3390/s23020590

Burtscher H., 2005. Physical characterization of particulate emissions from diesel engines: a review. Journal of Aerosol Science vol. 36, 896–932.

Dieselnet 2023. United States: Nonroad Diesel Engines, Revision 2023.03, https://dieselnet.com/standards/us/nonroad.php, date of download 2023.05.30

Dieselnet 2023. EU: Nonroad Engines, Revision 2021.07, https://dieselnet.com/standards/eu/nonroad.php, date of download 2023.05.30

European Commission 2022. Emissions from non-road mobile machinery, https://single-market-economy.ec.europa.eu/sectors/automotive-industry/environmental-protection/non-road-mobile-machinery_en, date of download 2023.05.29

European Commission 2023. Commission paves way for mobile machinery to move easier on public roads across the EU, https://ec.europa.eu/commission/presscorner/detail/en/ip_23_2046, date of download 2023.05.29

Fan Y., Wu T., Xiao D., Xu H., Li X., Xu M., 2020. Effect of port water injection on the characteristics of combustion and emissions in a spark ignition direct injection engine, Elsevier, ISSN: 0016-2361, e-ISSN: 1873-7153, DOI: https://doi.org/10.1016/j.fuel.2020.119271

Fenimore C.P., 1971. Formation of nitric oxide in premixed hydrocarbon, Symposium (International) on Combustion Volume 13, Issue 1, Pages 373-380

Ishida M., Chen Z., 1994. An Analysis of the Added Water Effect on NO Formation in D.I. Diesel Engines, SAE Technical Paper 941691, doi:10.4271/941691

Janousek Grant S., 2010. Evaluation of Ethanol and Water Introduction via Fumigation on Efficiency and Emissions of a Compression Ignition Engine Using an Atomization Technique, Biological Systems Engineering--Dissertations, Theses, and Student Research. 9., University of Nebraska, pp. 124, https://digitalcommons.unl.edu/biosysengdiss/9, search engine: Google. keywords: Water fumigation, diesel, date of download: 2021.05.22.

Jaraha A., 2001. Using Flame Quenching to Reduce NOx Emissions and High Temperature Corrosion, 9th Annual North American Waste-to-Energy Conference May 5–9, 2001 Miami, Florida, pp. 97-102, ISBN: 0-7918-3525-1, https://doi.org/10.1115/NAWTEC9-108

Kalló D., 2000. Katalitikus eljárások a környezetvédelemben, University of Veszprém

Kurt Engeljehringer 2023. The way to Euro 7: Legislation Introduction, AVL Techday Emissions, presentation, pp 29, 9.-10.5.2023

Langer, D.A., Petek N.K., Schiferl E.A. 2001. Maximizing the Effectiveness of Water Blended Fuel in reducing Emissions by Varying Injection Timing or Using After-Treatment Device, SAE Technical Paper 2001-01-0513, doi:10.4271/2001-01-0513

Lawrence M.G., 2005. The Relationship between Relative Humidity and the Dewpoint Temperature in Moist Air: A Simple Conversion and Applications, Bulletin of the American Meteorological Society Volume 82 Issue 2, p 225-234, https://doi.org/10.1175/BAMS-86-2-225

Miyamoto N., Ogawa H., Wang J., Ohashi H., 1995. Significant NOx Reductions with Direct Water Injection into the Sub-Chamber of and IDI Diesel Engines, SAE Technical Paper 950609, doi:10.4271/950609

Pamminger M., Wang B., Hall C.M., Vojtech R., Wallner T., 2020. The impact of water injection and exhaust gas recirculation on combustion and emissions in a heavyduty compression ignition engine operated on diesel and gasoline, International Journal of Engine Research Vol. 21, 1555-1573, DOI: 10.1177/1468087418815290

Rente T., 2004. Injection strategies for Heavy Duty DI Diesel Engines. Doctoral thesis, Chalmers University of Technology

Rohit A., Satpathy S., Choi J., Hoard J., Surnilla G., Hakeem M., 2017. Literature survey of water injection benefits on boosted spark ignited engines (No. 2017-01-0658). SAE Technical Paper; 2017.

Sawa N., Kajitani S., 1992. Physical Properties of Emulsion Fuel (Water/Oil-Type) and Its Effect on Engine Performance under Transient Operation, SAE Technical Paper 920198, doi:10.4271/920198

Sheng H., Chen L., Wu C., 1995. The Droplet Group Micro-Explosions in W/O Diesel Fuel Emulsion Sprays, SAE Technical Paper 950855, doi:10.4271/950855

Tsukahara M., Yoshimoto Y., Murayama T., 1989. W/O Emulsion Realizes Low Smoke and Efficient Operation of DI Engines without High Pressure Injection, SAE Technical Paper 890449, doi:10.4271/890449

Turns S.R., 2000. An Introduction to Combustion, McGraw-Hill (Second edition), 168-171

Wagner W., Pruss A., 2002. The IAPWS Formulation 1995 for the Thermodynamic Properties of Ordinary Water Substance for General and Scientific Use, Journal of Physical and Chemical Reference Data 31, 387–535, https://doi.org/10.1063/1.1461829

Wei X., Qian Y., Meng S., Hua Y., Qiu L., Sun Y., Wan J., Zhang X., 2022. Effects of coupling port water injection and EGR on the spray water evolution, combustion and emission of a premixed stoichiometric natural gas engine: A numerical study, The Science and Technology of Fuel and Energy volume 324, part A, Elsevier, Article 124861

5552 | New approach to controlled vehicle drawbar pull testing

Orian Welling [a,*], Michael Stearns [a], Zach Tomberg [a]

[a] US Army ERDC-Cold Regions Research and Engineering Laboratory (CRREL), New Hampshire, United States

* Corresponding author: orian.z.welling@erdc.dren.mil

ABSTRACT

In this work, we show a new vehicle drawbar test method that allows simple regulation of a constant vehicle speed through the test while requiring relatively little training or practice from the test vehicle driver. The improved method relies on a capstan winch mounted on a fixed stable point with a controlled release of cable over a 180*m* distance. The winch and cable are capable of sustaining loads up to 88*kN* at a constant cable speed. The load on the cable is measured with a vehicle mounted load cell. In the paper / presentation, we show a comparison of test data from drawbar tests utilizing a separate holdback vehicle and holdback driver with tests using the new methodology and show that measurement stability is improved and that the target vehicle speed is held much closer to the target point, eliminating error or need to correct for vehicle acceleration.

Keywords
Mobility Modeling
Off-road
Drawbar
Vehicle Performance
Vehicle Testing

1. Introduction

Determining the mobility of vehicles across offroad surfaces is a critical capability for Army. Over the years, a substantial amount of research has been conducted by ERDC and others to measure vehicle performance parameters on different surfaces, and for different vehicle configurations, in order to support efforts for accurate vehicle mobility modeling. These data support models in use by the Army including the NATO Reference Mobility Model (NRMM), GeoWATCH, and other stand-alone implementations. At the same time, our armed forces, as well as those of our adversaries, are continually developing new vehicles and vehicle technologies to improve mobility, and conflicts continue to shift to new parts of the world where having a strong understanding of vehicle performance over the local terrain is critical. As a result, it is anticipated that there will be an ongoing need to measure vehicle performance parameters of many vehicles, and across many terrain types, for years to come.

Due to the focus on evaluating different terrain types, most military testing of vehicle performance parameters has evolved around portable approaches where equipment can easily be brought and tested in new locations (e.g. to multiple locations across multiple military training grounds in the US). Traditionally, drawbar testing, which aims to determine the tractive force a vehicle can produce on a certain terrain, is conducted with two vehicles - a test vehicle and a holdback vehicle. While this approach is highly portable, it relies on well trained test drivers, and in challenging terrain it can be difficult to conduct well controlled tests even with seasoned drivers. In this paper, we describe the importance of the drawbar testing approach, the different tests and basic calculations associated with the tests and introduce a new testing approach that reduces the need for well-trained drivers or for identifying well paired holdback vehicles for each test vehicle.

2. Background

When estimating the mobility of a vehicle or developing a computational mobility model for a conceptual vehicle, it is important to consider which key parameters can be easily scaled relative to known / empirically measured vehicles. At the highest level, the two most important vehicle performance parameters are the vehicle load ($\boldsymbol{F_{Load}}$, the force exerted on the ground normal to the ground plane) and the drawbar force ($\boldsymbol{F_{DB}}$, the force exerted on the ground parallel to the ground plane, and in line with the direction of travel). These two forces are linked in many ways. For example, a higher load can lead to deeper sinkage into soft soil or snow through an increased ground pressure (the load divided by the tire or track contact area), which can in turn lead to higher resistive force in the direction of vehicle travel as the vehicle pushes through the terrain. Conversely, a higher load also results in a higher maximum friction force between the tire or track and the ground, potentially increasing the maximum achievable tractive force before slipping occurs. When considering turning, or side slope capabilities, the lateral force ($\boldsymbol{F_{Lat}}$, the force exerted on the ground plane, parallel to the ground and perpendicular to direction of travel) is also important.

https://doi.org/10.56884/TISO6782
ISBN 978-1-942112-55-6

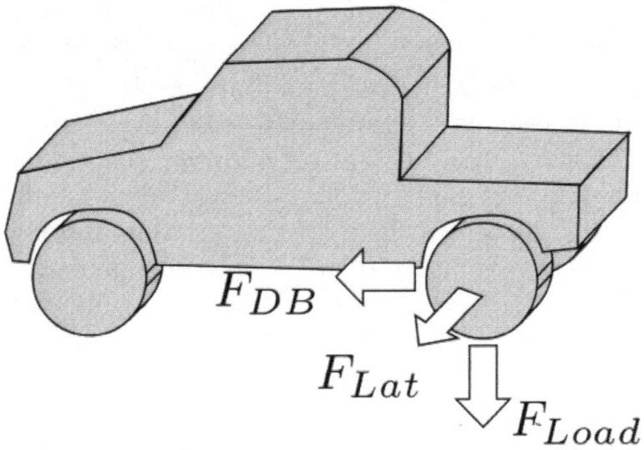

Fig. 1. Basic vehicle performance parameters include Drawbar Force (***F_{DB}***), Load Force (***F_{Load}***), and Lateral Force (***F_{Lat}***)

A benefit of the performance parameters discussed here as a basis for describing and modeling vehicle mobility is that they are both relatively simple descriptors of what is in reality a highly complex vehicle-terrain interaction, and yet they describe the actual vehicle behavior well in most circumstances. Many more complex models are available to describe the interaction forces between the vehicle and the ground, from full discrete element models evaluating forces between a tire or track and soil or snow particles, to analytical approaches such as use of Mohr-Coulomb stress analysis of soil under a wheel. In each case, it is possible to simplify the more complex models back to the performance parameters simply by summing the forces across the surface of the tire in the direction normal to the ground plane (load force), the direction parallel to the vehicle travel along the ground plane (drawbar force) and the direction perpendicular to the vehicle travel (lateral force). The forces are summed specifically at the interacting surface between the tire and the ground since it is this interface specifically where the vehicles tractive force is generated (Nowatzki and Karafiath, 1974).

While these three parameters can describe well the state of a particular vehicle, a few manipulations are often helpful to enable scaling to other vehicles. First, it is often helpful to split the "net" drawbar forces described above into a tractive force component (the locomotive component of force exerted on the ground by the powered running gear - the tractive force is often referred to as "gross drawbar force" ***F^*_{DB}***) and a resistive component (forces from compression of the soil, pushing over of vegetation, gravitational forces climbing a hill, or wind resistance - these forces are often referred to in aggregate as the motion resistance force ***F_{MR}***, or in parts as the e.g. the terrain motion resistance or vegetation motion resistance). The relationship between these forces is:

$$F^*_{DB} = F_{DB} + F_{MR} \tag{1}$$

The gross drawbar pull ***F^*_{DB}*** is typically higher than the net drawbar pull (though a vehicle driving down a smoothly paved hill for example could have a higher net drawbar). Conceptually it is useful because it removes the variability of things like slope, sinkage, and vegetation to push over that aren't closely tied to the vehicle, and instead more closely captures the interaction between the vehicle running gear and the ground itself. For a given vehicle and a given surface (e.g. grass, gravel, snow), the maximum ***F^*_{DB}*** will be fairly constant regardless of hills or varying firmness of the ground, while the maximum ***F_{DB}***, which is quite literally the maximum pulling capacity of the vehicle, will be lower when traveling uphill, or sinking deeper into the soil, or when the vehicle is pushing over shrubs. For a given vehicle and surface, an improvement in maximum ***F^*_{DB}*** is nearly always tied to some technological improvement (better tires, reduced tire pressure, etc.). Further, it can be helpful to normalize the maximum ***F^*_{DB}*** by the weight of a vehicle - this normalized value is known as the drawbar coefficient ($C_{DB} = max(F^*_{DB})/Weight_{veh.}$), and is often similar between different vehicles of the same general vehicle class and weight (e.g. ***C_{DB}*** would be expected to be similar between different models of 4x4 light pickup on a given terrain). Using a known ***C_{DB}*** for a similar vehicle and scaling by vehicle weight is often a reasonable first approximation to estimate the performance of a new vehicle of interest (at least to the degree that the performance is limited by tire slip rather than engine power).

2.1 *Measurement of Drawbar Pull*

Since it is most closely related to vehicle traction, the gross drawbar force ***F^*_{DB}***, or the associated drawbar coefficient, is often the target measurement for drawbar testing. However, since it is typically not possible to directly measure the interaction forces at the ground (not practical except small scales where entire test surfaces can be build on a load measuring table), the ***F^*_{DB}*** is instead calculated from measurements on three standard tests. These standard tests are:

- Net Drawbar Force test (***F_{DB}***): Measures pulling force of the vehicle on the surface of interest.
- Total Motion Resistance test (***F_{MR}***): Measures the force required to tow the vehicle across the surface of interest.
- Internal Motion Resistance test (***$F_{MR-Internal}$***): The force required to tow the vehicle across an ideal surface (often referred to as "hard surface rolling resistance", and typically conducted on a surface such as asphalt which is close to ideal in that deformation or roughness of the surface contribute negligible resistive force on the vehicle relative to the tractive forces of interest).

In a **Net Drawbar Force** test, a tow cable a tow cable and load cell can be connected to the back of the vehicle and connected to some holdback load (e.g. a sledge, a trailer, or another vehicle) - the value measured on the load cell is the net drawbar force. Tests are usually conducted on flat ground to avoid the need to back out gravitational forces. Also to avoid errors from acceleration forces or correction thereof, tests are typically conducted at a constant speed (at least to the degree the test drivers can maintain a constant speed during the test). In order to measure the maximum FDB, it is common to use a second "holdback" vehicle of similar or slightly greater weight to the test vehicle. A second driver applies carefully brakes in the holdback vehicle, keeping the vehicle traveling at the target speed for the test, while the test vehicle driver ramps up application of the throttle until the test vehicle drive wheels start to slip. For most surfaces, the maximum FDB is achieved right at the point the drive wheels begin slipping.

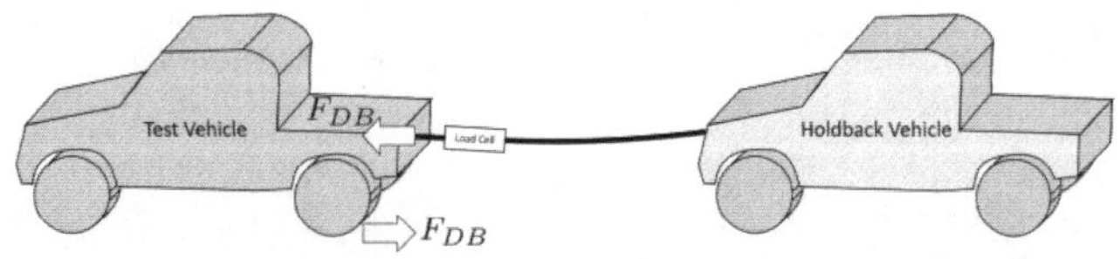

Fig. 2. Basic drawbar test configuration - at constant velocity the net drawbar force exerted by the vehicle on the ground is equal to and balanced by the measured tow force

From Eq. 1, we now have the net drawbar force, and can calculate the gross drawbar force we would like to know if we know the motion resistance on the target terrain. The **Total Motion Resistance** test helps us get this value, it is conducted by towing the test vehicle, with a load cell connected to the tow cable to measure the force required to tow the vehicle. The test mimics the drawbar as closely as possible so the measured $F_{MR\text{-}Total}$ value is as close as possible to the same resistive force acting on the vehicle during the net drawbar force test (which we couldn't measure directly in that test) – this includes conducting the test at the same vehicle speed, on the same slope (if a no-slope test location could not be found), with the same vegetation loading, wind loading, and so on. Also, the running gear (tires or tracks) should be configured to be as close to "free running" as possible - if major components of the driveline are left engaged (vehicle transmission, locking differentials, or even the track assembly or very soft balloon tires), the internal friction of those systems will result in an internal resistive force FMR–Internal which will be balanced by a resistive tractive force at the vehicle-terrain interface which is not pertinent to the maximum gross drawbar pull value of interest *[Wong]* - in the maximum F^*_{DB} case, we are looking for the maximum tractive force the tire or track can exert on the terrain before "slipping out", and are not concerned with the internal workings of the vehicle beyond their delivering sufficient power to the running gear to create slip (else we would note the maximum F^*_{DB} is power limited - a case where we cannot effectively measure the traction limits of the vehicle).

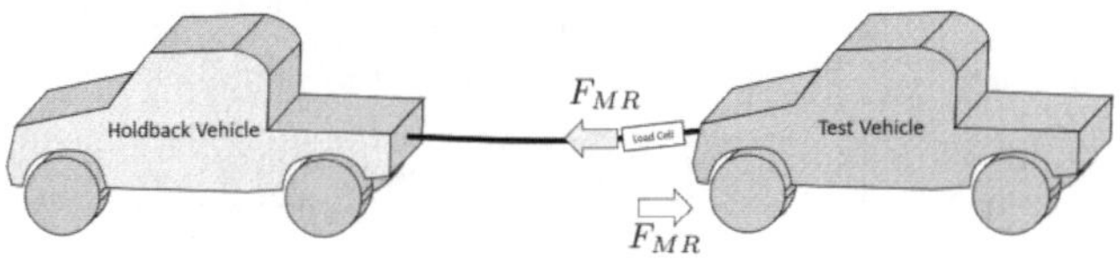

Fig. 3. Basic total motion resistance test - at constant velocity the measured towing force is equal to and balanced by the motion resistance force acting on the vehicle terrain interface, which includes the external terrain resistance as well as a tractive resistance as the tire resists turning due to internal resistances in the vehicle drivetrain; the test setup for the internal motion resistance test is the same, but a very low resistance surface such as asphalt or concrete is used for the test

For vehicles that can be configured into a highly "free running" state for the total motion resistance test , it may be reasonable to assume that the internal motion resistance is negligible and to calculate F^*_{DB} directly from Eq. 1 using $F_{MR\text{-}Total}$ (e.g. a vehicle where all hubs can be fully unlocked, and with standard or high pressure tires, so only the minimal hub bearing friction and minimal tire deformation resist motion). However, many vehicles, including nearly all tracked vehicles, have a sufficiently large $F_{MR\text{-}Total}$, even in their most "free running" configuration, that conducting one more test to be able to remove it from the calculation is merited. The Internal Motion Resistance test is conducted very similarly to the total motion resistance test, but instead of being conducted on the test surface, it is conducted on as ideal (e.g. non-resistive to motion) surface as possible, such as blacktop or concrete (as a result, this test is also often referred to as a hard surface rolling resistance test). The test should be conducted with the same "free running" vehicle configuration (e.g. neutral with differentials unlocked) as the motion resistance test, and at the same speed and same slope (ideally flat) as the motion resistance test. While some small external motion resistance forces act on the vehicle even in this case (e.g. wind resistance, some terrain resistance based on roughness or stickiness of the blacktop), they can generally be made sufficiently small relative to the larger drawbar forces that they can be assumed negligible, and $F_{MR\text{-}Internal}$ can be assumed to be the same as the measured towing force for the test.

Based on these three tests, Eq. 1 can be re-written as:

$$F^*_{DB} = F_{DB} + F_{MR} = F_{DB} + (F_{MR-Total} - F_{MR-Intrenal}) \quad (2)$$

Where each of the forces is input as the positive scalar measurement taken from the load cell during the respective tests (not vectors where signs are flipped based DB and MR forces acting in opposite directions - this convention is used here since load cell force measurements in both tests are always positive due the similar towing configurations).

2.2 Existing Methods

Today, the standard net drawbar force test is usually conducted with use of holdback vehicle that is slightly heavier than the test vehicle, and a holdback driver applies brakes during the test to keep the two linked vehicles traveling at the target speed. Many factors go into the ability of the holdback driver to maintain a speed close to the target point, including driver skill, availability of adequate speed readouts (may need readouts with accuracy beyond what a normal speedometer would provide, such as 10ths of miles per hour), terrain variability, relative traction of the holdback vehicle vs. the test vehicle, and touchiness of the brakes. Pairing and configuring holdback vehicles, and training drivers, is often a time-consuming component of testing new vehicles, and even in the best of pairings, it is not uncommon to throw out several test runs due to variable or out-of-target speeds.

Alternative methods include outfitting test vehicles with loadcells on each wheel. This can provide high fidelity data including details of the lateral forces on the wheels that are not measured with a traditional drawbar approach, but it is limiting due to the cost of instrumenting each vehicle with custom loadcells on each axle. For broad testing across multiple vehicle types (often a desire for developing mobility models), it is impractical to use integrated loadcells. Another alternative method which has been discussed in the field, but not widely used or published, is use of controlled winches or other devices to enable constant speed testing. This paper describes a new approach based on this concept for drawbar testing of vehicles up to 9,000kg.

3. New Measurement Setup

Fundamentally, the role of a holdback vehicle and driver is to provide a variable holdback force such that the speed of the test vehicle remains constant regardless of the force with which the test vehicle is pulling. Such a goal can be easily met with simple control algorithms, provided the physical equipment is capable of sustaining the required force. A medium tactical vehicle may have a F_{DB} in the range of 50kN - for a drawbar test conducted at 8km/h this equates to approximately 110kW of power that must be provided or dissipated by the test system. This is a large amount of power to provide to a mobile test setup for testing of vehicles at different terrains (though not impossible - large, trailered diesel generators are available up to about 200kW), however it is a relatively modest amount of energy to dissipate, for short periods, with common hydraulic brake systems.

For this approach, we designed and built a capstan with a structural dynamic load capacity of 88kN, affixed a brake rotor to the capstan driveshaft with an electric over hydraulic actuator powering the brake calipers, and programmed a PID control system based on the angular velocity of the capstan shaft to control the capstan brake to the achieve the target setpoint speed regardless of load applied. We used a 180*m*, 38*mm* diameter dynema cord rated to 400*kN* with the capstan, such that the test vehicle could drive to a length of up to 180*m* for drawbar testing. Figure 4 shows the design of the capstan and braking system.

Fig. 4. Electric over hydraulic braking system inside the capstan is controlled with PID system to maintain a constant rope speed

Initial testing with the capstan drawbar test system with an MRZR Alpha test vehicle confirmed the system could smoothly control the speed of the vehicle through the course of a test, first limiting the vehicle at the target speed after acceleration, then allowing spin up past loss of traction and up to a slipping wheel speed approximately 2x the vehicle speed. Figure 5 shows the test setup in action during initial testing.

Fig. 5. Capstan secured to trailer or fixed platform for testing. Rope from capstan is secured to load cell at rear of test vehicle

4. Conclusions and Next Steps

The initial testing clearly showed the potential value of the equipment to achieve more controlled testing over a range of vehicles while also reducing the need to find well paired holdback vehicles for each new vehicle tested. Additionally, the system can be easily set to a target speed, or speed profile, and then operated with minimal training, reducing the need to train drivers or hire trained drivers for testing.

As a next step, we plan to conduct parallel tests on several snow surfaces with the capstan testing system, and a traditional holdback driver approach. Testing staff will be interviewed to determine the relative ease of each approach, tests will be timed to determine efficiency of both methods, and the data will be reviewed to determine if the data collected with the automated capstan system is truly more controlled for speed (or other issues) through the course of the test.

5. References

Nowatzki, E., Karafiath, L., 1974. General yield conditions in a plasticity analysis of soil-wheel interaction. Journal of Terramechanics. 11(1), 29–44.

Wong, J., 2009. Terramechanics and Off-Road Vehicle Engineering, Elsevier.

6027 | Digital precision planning tool for autonomous forest regeneration of mixed tree species

Erik Arvidsson [a,*], Magnus Karlberg [a], Karin Hjelm [b], Håkan Lideskog [a]

[a] *Department of Engineering Sciences and Mathematics, LTU, Laboratorievägen 14, Luleå, SE-971 87, Sweden*

[b] *Southern Swedish Forest Research Centre, SLU, P.O. Box 190, Lomma, SE-234 22, Sweden*

* *Corresponding author: erik.arvidsson@ltu.se*

ABSTRACT

Forest management planning often neglects within-site variation, focusing on decisions for entire forest stands, which impedes effective forest regeneration. Incorporating mixed species stands can mitigate risks associated with monocultures and offer numerous benefits. However, implementing mixed stands present challenges such as higher costs and complex management. This research aims to develop a method for precise and efficient forest regeneration planning in Swedish silviculture. The study developed digital tools to be used in forest regeneration planning and makes it suitable for deployment on autonomous vehicles. Traditional regeneration methods involving labour-intensive processes and soil disturbance have limitations, while autonomous vehicles show promise in improving energy efficiency and reducing soil damage. The aim is to increase site productivity by determining the spatial distribution of varied species in mixed stands using density constraints set by the user.

The Digital Precision Planning Tool (DiPPT) developed in this study utilises existing site index maps to enhance biomass growth and select the most suitable species for each location. The tool considers species-specific site productivity and density constraints to create the planting map. Two approaches were developed: the Max function, which selects the species with the highest site productivity for each location, and an iterative process that optimises species-specific factors to meet user-defined density constraints.

The DiPPT was validated using test sites provided by the forest company Sveaskog. Results show that the tool's approach outperforms manual planning in terms of site productivity. The average site productivity using DiPPT is 10% higher compared to manual planning. The tool demonstrates its potential for optimising species placement and promoting species diversification in forests.

Keywords
Site index
Site preparation
Site productivity
Precision forestry

1. Introduction

Forest management planning typically do not consider within-site variation and instead focuses on decisions for the entire forest stand, which hinders effective and precise forest regeneration. The process of mapping and identifying microsite variations within a stand is time-consuming, making it challenging to implement different management options on the same site. However, understanding and considering within-site variation enables more precise and efficient forest regeneration strategies (Nordin, 2023).

In 2021, 452 million forest plants were delivered from Swedish nurseries where 96% of the plants were Scots pine and Norway spruce (Fürst, 2022). Although these species grow well, relying mainly on two species introduce risks, for e.g., pests and pathogens. Replacing monoculture pine or spruce stands by mixing in birch has numerous benefits, including increased biodiversity, financial security, and improved water quality (Felton et al., 2016). Additionally, mixed stands are less susceptible to pests and pathogen damage. However, this approach also comes with drawbacks such as higher logging costs, more complex management, and higher risk for browsing damage (Felton et al., 2016).

By establishing mixed species stands, the spread of the fungi Heterobasidion spp., which causes root rot, is reduced. The fungi are generally spread in monocultures of spruce through root-to-root contact (Huuskonen et al., 2021). Mixed stands may also be more resistant to global warming than monocultures (Fahlvik et al., 2011).

Traditional methods of regeneration are labour intensive and disturb much soil. Today, forest regeneration is typically conducted by mechanised site preparation followed by manual planting of seedlings.Instead, the use of autonomous vehicles can improve the energy efficiency of forest regeneration, by minimising the area of prepared soil. By using smaller and less heavy forest vehicles, ground damage can be reduced. Additionally, it simplifies the inclusion of multiple species due to one assumption made by this study, which is that the autonomous forest machine can carry several different species of seedlings, which enables autonomous planting of mixed forests in one operation.

There are different soil preparation techniques used today in the Swedish forestry sector, of which the most common are mounding and disc trenching, which disturbs 37% respectively 53% of the soil surface area (Krekula et al., 2018; Sikström et al., 2020). Approximately 50% of the attempts to create mounds on clearcuts are unsuccessful due to obstacles, and of those failed attempts, more than 30% are specifically attributed to encountering stumps (Lideskog and Karlberg, 2014). If there were better methods that could pinpoint good spots to scarify, and which species has the highest probability to grow in that spot, fewer faulty attempts would be needed and therefore would reduce the disturbed soil area.

Saksa et al. (2021) has researched the possibility to use harvester data to predict local growth for each cell in a raster map over the site. With that data, they calculate delineations of micro-stands and then predict whether Scots pine or Norway spruce should grow in those areas. The article concluded that this method could be used as a decision support tool during the regeneration phase of forest management. Overall, precision forestry has the potential to increase forest productivity and sustainability.

Similarly, the Forestry Research Institute of Sweden (Skogforsk) conducted a work study on precision forestry by analysing harvester data together with map data to make smarter decisions in forest regeneration. For example, planning mixed

https://doi.org/10.56884/SDFR7512
ISBN 978-1-942112-55-6 (pdf)

forests in areas infected by the fungus Heterobasidion could reduce the spread of the fungi. The study showed satisfactory results, with a digital model corresponding to a manual model manufactured on-site. The researchers believe that the model can be used as a decision-making tool during on-site planning of the regeneration map (Friberg et al., 2019).

When collecting site index properties from harvester data, it is necessary to cut down the forest to obtain the data. However, by using a non-destructive method of collecting the site index properties, as presented in this paper, it becomes possible to plan the regeneration before the harvest takes place. This approach enables the modification of the harvesting plan, allowing for specific trees to be preserved to produce seeds for a natural regeneration process.

There is currently a research gap in how to optimise the placement of certain species for mixed stands, based on density constraints set by the user, which is the focus of this study. To address this gap, the objective is to develop a tool that can increase the site productivity based of the placement of species in mixed stands using density constraints set by the user. By addressing this research gap, the tool developed in this study can help spreading risks and promote diversification in forests.

1.1 *Author contributions*

E.A. was the main contributor by collecting and processing the data, creating the necessary tools, and conducting the analysis of the results. E.A. also took responsibility for the writing of the paper. The work was supervised by M.K., H.L., and K.H., who provided guidance and support throughout the process. They reviewed the work and approved the final version of the paper.

2. Materials & methods

The overall idea was to make use of existing data, contributing to prediction of enhanced biomass growth, based on choosing the best species for a certain location. The development process involved a literature review and benchmarking of available tools and to determine which data was available to enable better decision-making in forest regeneration. Based on existing data, a prototype tool was developed and verified, and the performance and stability of subsystems were tested. Validation of the tool was then conducted with test sites provided by Sveaskog, which is a forest company owned by the Swedish state. The tool developed during this study was named DiPPT after Digital Precision Planning Tool.

2.1 *DiPPT*

A common parameter in silviculture is site index, which is an estimate of a certain stand height at a certain age (Skovsgaard and Vanclay, 2008). For instance, if a pine stand has a site index of twenty using the H_{100} system, it indicates that the height of the dominant pines is expected to reach twenty meters after 100 years. However, certain species like silver birch use the H_{50} system, which measures the dominant height after 50 years instead. To establish a justifiable comparison, all site indexes in this study were transformed into the H_{100} system according to (Bogghed, 2010). Another parameter is site productivity **SP**, which is how much biomass a site produces, measured in cubic meter per hectare and year $m^3 fo/(ha \cdot yr)$.

Currently, the Swedish University of Agricultural Sciences (SLU) has created digital site index raster maps for Scots pine and Norway spruce only. The map is based on aerial laser scanning, latitude, altitude, and soil moisture. The resolution of the raster cells in the map are 12.5m x 12.5m. Each species has their own unique map, and by comparing data from the different maps in a certain coordinate, an analysis can be made of what the most suitable species is for that coordinate.

The study involved the use of site index raster maps **SI** to generate a site productivity map by converting the maps from tabular data (Bogghed, 2010). The total number of species studied was denoted by *n*. Due to the limited supply of high-resolution site index maps, the ability to support custom site index maps was made. This was done by entering a static site index that covered every cell of the raster map with the same value. By comparing the site productivity of each species *i* for every raster cell indexed by *x* and *y* in the field, a probability approximation matrix **P** was developed for each species

$$P(SP_i, x, y) = \frac{SP_i(x,y)}{\sum_i^n SP_i(x,y)}. \quad (1)$$

Two different approaches were developed during this study. One approach which is called the Max function, maximising site productivity by selecting the most suitable species for each location based on the available data. The probability approximations from Eq. 1 were then used to create a planting map. For each raster cell, the species with the highest site productivity was selected. By applying a Gaussian filter to the site index data, local variability was reduced, decreasing the distortions of the site productivity map. In cases where multiple species had the same site productivity, they were alternated when adding them to the planting map.

In the other approach, the plant density for multiple species $\boldsymbol{\rho_i}$ across the entire stand are prescribed by the user. The function tunes species-specific factors $\boldsymbol{k_i}$ through an iterative process until the error $\boldsymbol{\Delta_i}$ is lower than the threshold for all species. The error is calculated as

$$\Delta_i = \frac{\sum_{x=1}^{x_{max}} \sum_{y=1}^{y_{max}} (k_i \cdot P(SP_i,x,y) > k_{\neq i} \cdot P(SP_{\neq i},x,y))}{x_{max} \cdot y_{max}} - \frac{\rho_i}{\sum_{i=1}^n \rho_i}. \quad (2)$$

The species-specific factors are multiplied by the probability approximation for each respective species. If a particular species has the highest probability, a value of one was added to its corresponding Boolean matrix, while the Boolean matrices of the other species received a value of zero. The Boolean matrix values were then summed for each species, and the resulting sum was divided by the total number of elements in the matrix.

The species-specific factors were then updated iteratively with the error multiplied with a learning rate α to avoid overshooting. When dealing with multiple species, the species with the largest error had its species-specific factor updated first. Thereafter, the factors of the other species were adjusted in the opposite direction. This approach ensured that the boundary condition was met, where the sum of all species-specific factors equalled one.

$$\begin{cases} k_{i_{new}} = k_{i_{old}} - \alpha \cdot \Delta_i \\ \sum_{i=1}^n k_i = 1. \end{cases} \quad (3)$$

To evaluate the tool, Sveaskog sites were randomly selected from a large inventory of different sites distributed over Sweden. The location of the testing sites is shown in Fig. 1, while Table 1 provides the corresponding coordinates in SWEREF 99 TM projection, which is the spatial reference system utilised throughout this study.

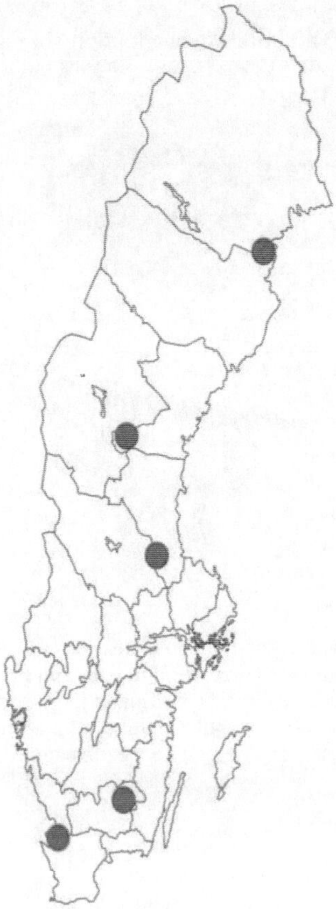

Fig. 1. A map over Sweden with the location of the test sites marked with red coloured dots

Table 1
The name of the different test sites and their corresponding coordinates in SWEREF 99 TM projection.

Test site name	Coordinate	Area [ha]
Piteå	N 7249370, E 784131	12.8
Snöberg	N 6934648, E 517575	3.7
Ovansjö	N 6730560, E 574877	2.3
Linnebjörke	N 6312031, E 508494	5.0
Oxhult	N 6248755, E 378739	11.2

The mean site productivity of the species Scots pine and Norway spruce on the test sites can be seen in Table 2.

Table 2
Mean site productivity of pine and spruce monocultures of the test sites.

Test site	SP Spruce [m^3 *fo/(ha yr)*]	SP Pine [m^3 *fo/(ha yr)*]	Diff.
Piteå	3.67	3.85	4.9%
Snöberg	5.64	4.32	-23.4%
Ovansjö	7.09	4.91	-30.7%
Linnebjörke	10.24	5.26	-48.6%
Oxhult	12.58	6.39	-49.2%

3. Results

This chapter presents the results obtained from the analysis and evaluation of the test sites using DiPPT. The initial case involved different test sites using pine and spruce. The second case, custom density constraints were implemented, requiring an equal density of pine, spruce, and birch.

3.1 Case 1: Comparisons between manual and digital planning

The test sites were evaluated using the DiPPT with the two approaches. During the first approach, the Max function was applied, which displays the species with the highest site productivity in each location. With the second approach, Sveaskog's planting strategy was applied as a density constraint. This allowed for a comparison of site productivity between Sveaskog's planting strategy and the Max function. Due to data availability, this analysis focused specifically on Norway spruce and Scots pine.

The results, from both Sveaskog's manual planning and the Max function are presented in Table 3, which provides the different test sites with their corresponding results. Column no. 2 in Table 3 presents at which density Sveaskog has planned their regeneration (seedlings per hectare), which was later sent to their planting contractors. The third column displays the occurrence of the species that the tool recommends having an increase of mean site productivity, using the Max function. Note that the Max function outputs the occurrence of the species, and not a density.

Table 3
Comparisons between manual planning of Sveaskog and planning with DiPPT Max function.

Test site	Sveaskog planning	DiPPT Max
Piteå	2500 seedlings/ha pine	92.7% pine, 7.3% spruce
Snöberg	1800 seedlings/ha pine	100% spruce
Ovansjö	2500 seedlings/ha spruce	100% spruce
Linnebjörke	2500 seedlings/ha pine	100% spruce
Oxhult	2500 seedlings/ha spruce	100% spruce

To further analyse the results from Table 3, the mean site productivity of each test site was extracted from DiPPT, using both approaches. The mean site productivity of each site is displayed in Table 4, of both Sveaskog's planned regeneration and using the Max function. The diff column shows the percentage difference of using the DiPPT instead of manual planning. Note that the site productivity of Table 4 corresponds to the species occurrence data in Table 3.

Table 4
Comparisons of site productivity between manual planning of Sveaskog and planning with DiPPT Max function.

Test site	SP Sveaskog [m^3 *fo/(ha yr)*]	SP Max [m^3 *fo/(ha yr)*]	Diff.
Piteå	3.85	3.86	0.4%
Snöberg	4.32	5.64	30.5%
Ovansjö	7.09	7.09	0.0%
Linnebjörke	5.25	10.24	94.9%
Oxhult	12.58	12.58	0.0%

Figure 2 provides an example of a raster map, illustrating the site productivity of the Snöberg test site. The two maps showcase the different functions, the upper utilising the Sveaskog density constraint and the lower utilising the Max function.

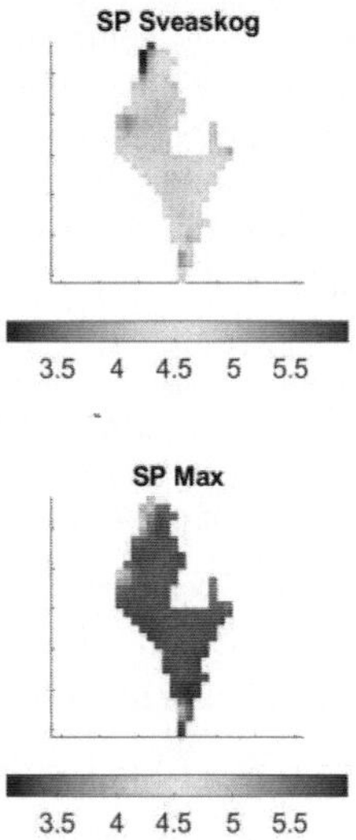

Fig. 2. The site productivity of the Snöberg test site is represented in two plots. The upper plot displays the site productivity raster based on Sveaskog's planned regeneration, consisting entirely of pine. The lower plot illustrates the site productivity raster generated using the Max function, consisting entirely of spruce. The raster cell size is 12.5 x 12.5 meters

By taking the area of each site into account, the average site productivity of all test sites was 7.07 *m³fo/(ha·yr)* on Sveaskog's planning and DiPPT were 7.76 *m³fo/(ha·yr)*, which results in a predicted increased biomass of 9.76 %.

3.2 Case 2: Equal seedling density of pine, spruce, and birch

In this case, the different test sites were set to have a density constraint of eight hundred seedlings per hectare for each species, spruce, pine, and birch, totalling a seedling density of 2400 seedlings per hectare on the site. Since digital site index maps were not available for birch, an assumed static site index according to Table 5 was assigned to all raster cells in the birch site index map. The values of the birch site index were set to a value slightly lower than the mean site index of the dominant species, ensuring that the value does not govern the site.

Table 5 shows the result of mean site productivity of the test sites using the Max function referred as SP Max, as well as density constraints referred as SP case. The table also shows the difference between the results.

Table 5
Comparisons of site productivity between Max function and the case with density constraints.

Test site	Birch SI $[H_{50}]$	SP Max $[m^3 fo/(ha\,yr)]$	SP Case $[m^3 fo/(ha\,yr)]$	Diff.
Piteå	16	3.93	3.89	-1.1%
Snöberg	16	5.65	4.99	-13.3%
Ovansjö	22	7.11	6.58	-8.1%
Linnebjörke	26	10.24	8.18	-25.2%
Oxhult	26	12.59	9.32	-35.1%

The results from test site Piteå is presented as the species placement in Fig. 3 and the site productivity in Fig. 4.

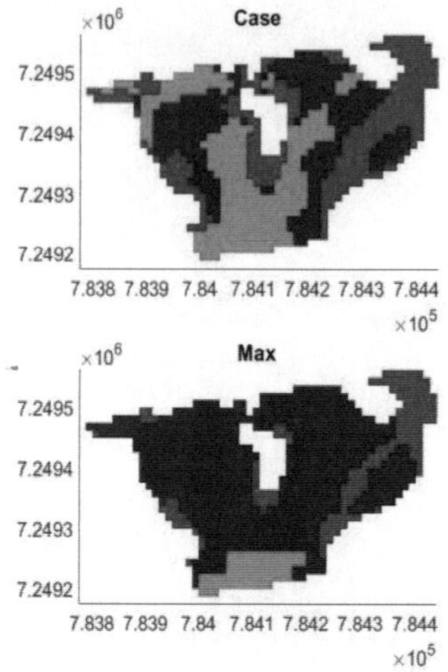

Fig. 3. The species placement of the Piteå test site, red cells indicate the selection of birch, blue cells correspond to pine, and green cells correspond to spruce. The upper plot displays the case with density constraints and the lower plot shows the species placement utilising the Max function. The raster cell size is 12.5 x 12.5 meters

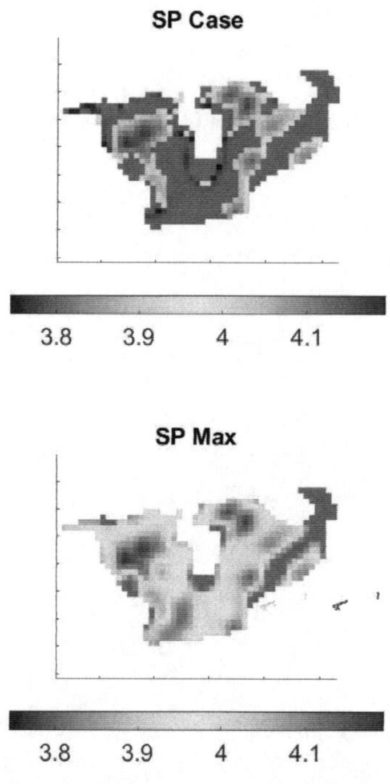

Fig. 4. Site productivity of test site Piteå. The upper plot displays the case with density constraints and the lower plot shows the species placement utilising the Max function. The raster cell size is 12.5 x 12.5 meters. The site productivity raster is corresponding to the species pattern in Fig. 3

A pattern from Fig. 3 and 4 can be seen that the highest site productivity was in areas covered by pine and the lowest were covered with spruce and birch. Additionally, the plots revealed a pattern where areas of the lower plot with the lowest site productivity were transformed into spruce or birch at the upper plot. While areas with high site productivity values for pine were designated for pine reservation. This approach ensures a high mean site productivity while fulfilling the planting density requirements regulated by the boundary conditions.

4. Discussion

The reliability of the site index maps used in this study has not been evaluated yet, therefore the results from this study can deviate from reality. Although if the results should deviate, the DiPPT tool provides a methodology on how site index data can be used to enhance the placement for varied species. The tool can also input more site index maps than those used in this study, e.g., site index maps created from harvester data.

The manual regeneration plans from the test sites can also be unreliable. This is since regeneration planning processes usually are conducted by experience and the easiest choice for the site, and maybe not the correct choice for optimal growth. This can lead to a sub-optimal decision being made out in the field. Therefore, growth experiments using this tool is to recommend.

One notable advantage of the tool's constraint function is its ability to mitigate the impact of biased data. For instance, if the site index data exhibits a bias towards spruce, but the forest owner's prior experiences indicate that birch thrives best in that area, the constraint functions capability allows for overriding the bias.

Another key factor is the forest owner's desire to introduce a species that deviates from the species with optimal growth, possibly driven by personal interest. In such cases, the tool assists the owner in identifying the most suitable areas of the site for planting the specific seedling, in terms of biomass optimisation. Just as the data presented in Table 5, the forest owner can examine how the average site productivity may change depending on the chosen plant density. Afterward, the forest owner can assess whether this decision is advantageous or not. For example, on the Piteå test site, the average site productivity only decreased by -1.09%. This suggests that opting for an equally mixed tree forest could be a favourable option.

The difference between pine and spruce in Table 2 becomes more pronounced as you go down the table. One plausible reason for this is the increasing gap in how well pine and spruce grow in various locations. In general, pine is more common in the northern parts of Sweden, while spruce is more common in the southern parts. It is plausible that the similar trend of the diff column observed in Table 5 is a result of the same underlying factor since the densities of the species were set as a constraint.

5. Conclusions and future work

In this study, a methodology for increasing within-site variation, while keeping a high average site productivity, was developed by using digital site index maps. Implementing precision forestry techniques, e.g. through smaller autonomous vehicles, can enhance energy efficiency and reduce soil disturbance. Applying the methodology presented in this paper will provide a species placement plan before entering the site. This plan can then be used not only to select a route minimising the wheel to ground damage, but also to do precise scarification operations minimising the ground damage due to the site preparation equipment. Further research is needed to optimise the local placement of species in mixed stands, and the development of tools like DiPPT shows promise in enhancing decision-making and site productivity in forest regeneration.

The DiPPT methodology can be extended beyond Sweden and applied to other regions in Fennoscandia, provided that digital site index data for the relevant species is available for those locations. In this study, the convenience of utilising the Swedish national projected coordinate system, SWEREF 99 TM, was acknowledged. However, if a three-dimensional coordinate system is preferred, adjustments to the methodology would be necessary.

To make a better usage of DiPPT, a local planner for an autonomous vehicle needs to be developed, that utilises the results generated. The local planner needs to conduct precise decision-making during planting and be able to manage different obstacles in the terrain.

The result from this study was only compared to test sites and not evaluated with growth experiments, therefore the tool needs to be validated, on its performance in predicting plant growth and productivity. This will provide valuable insights into the accuracy and reliability of the tool's predictions, further validating its effectiveness.

There is also potential to extend its optimisation capabilities beyond biomass. The exploration of optimising for additional factors, such as recreation, lumber values, and biodiversity, could be considered. Embracing this broader approach to optimisation will foster more comprehensive and sustainable forest management practices, effectively addressing diverse stakeholder interests and ecological considerations.

Limiting the analysis to only the H_{100} system, as performed in this study, presents certain drawbacks. By converting a H_{50} species to the H_{100} system, an assumption regarding the average growth over the period from year 0 to 100 is made. However, it should be noted that the species in the H_{50} system typically reaches its harvestable age within 50 years. With the conversion, we assume an additional 50 years of growth at a stalled pace, which contributes minimally to the overall biomass growth. To have a better transparency to the forest owner, it is recommended to develop a more comprehensive life cycle analysis model for each stand, which incorporates the consideration of multiple stand ages and site index systems.

6. Acknowledgements

This research was conducted within Trees For Me, a centre of excellence supported by the Swedish Energy Agency and almost fifty stakeholders, including Sveaskog.

7. Declaration of competing interest

The authors declare that they have no known competing financial interests or personal relationships that could have appeared to influence the work reported in this paper.

8. References

Bogghed, A., 2010. Produktionsförmåga vid olika ståndortsindex och konvertering av ståndortsindex mellan trädslag. Swedish, Accessed 2023-05-09.

Fahlvik, N., Agestam, E., Ekö, P.M., Lindén, M., 2011. Development of single-storied mixtures of Norway spruce and birch in southern Sweden. Scandinavian Journal of Forest Research 26, 36–45. doi:10.1080/02827581.2011.564388.

Felton, A., Nilsson, U., Sonesson, J., Felton, A.M., Roberge, J.M., Ranius, T., Ahlström, M., Bergh, J., Björkman, C., Boberg, J., Drössler, L., Fahlvik, N., Gong, P., Holmström, E., Keskitalo, E.C.H., Klapwijk, M.J., Laudon, H., Lundmark, T., Niklasson, M., Nordin, A., Pettersson, M., Stenlid, J., Sténs, A., Wallertz, K., 2016. Replacing monocultures with mixed-species stands: Ecosystem service implications of two production forest alternatives in sweden. Ambio 45, 124 - 139. doi:10.1007/s13280-015-0749-2.

Friberg, G., Jacobson, S., Möller, J.J., Bhuiyan, N., Willén, E., 2019. Regeneration planning using harvester information and geodata. Skogforsk ARBETSRAPPORT 1002-2019. Swedish, [Link].

Fürst, M., 2022. Statistik från Skogsstyrelsen - levererade skogsplantor 2021. Swedish, [Link].

Huuskonen, S., Domisch, T., Finér, L., Hantula, J., Hynynen, J., Matala, J., Miina, J., Neuvonen, S., Nevalainen, S., Niemistö, A., Piri, T., Siitonen, J., Smolander, A., Tonteri, T., Uotila, K., Viiri, H., 2021. What is the potential for replacing monocultures with mixed-species stands to enhance ecosystem services in boreal forests in Fennoscandia? Forest Ecology and Management 479, 118558. doi:10.1016/j.foreco.2020.118558.

Krekula, B., Bergqvist, J., Fries, C., Gällerspång, J., Reisek, J., Ringagård, J., Sollander, E., Svensson, L., Wågström, K., 2018. Föreskrifter för anläggning av skog - Regeringsuppdrag. Technical Report RAPPORT 2018/13. Skogsstyrelsen. Swedish, [Link].

Lideskog, H., Karlberg, M., 2014. Automatic clearcut obstacle identification using a time-of-flight camera, in: A Methodology for Automation of Mechanized Forest Regeneration, pp. 97–100.

Nordin, P., 2023. Regeneration measures in time and space: Site preparation, planting and digital tools. Ph.D. thesis. Linnaeus University, Department of Forestry and Wood Technology & Skogforsk, Sweden. doi:10.15626/LUD.489.2023.

Saksa, T., Uusitalo, J., Lindeman, H., Häyrynen, E., Kulju, S., Huuskonen, S., 2021. Decision support tool for tree species selection in forest regeneration based on harvester data. Forests 12. doi:10.3390/f12101329.

Sikström, U., Hjelm, K., Hanssen, K.H., Saksa, T., Wallertz, K., 2020. Influence of mechanical site preparation on regeneration success of planted conifers in clearcuts in Fennoscandia – a review. Silva Fennica 54. doi:10.14214/sf.10172.

Skovsgaard, J., Vanclay, J., 2008. Forest site productivity: A review of the evolution of dendrometric concepts for even-aged stands. Forestry 81, 13 - 31. doi:10.1093/forestry/cpm041.

6630 | Real-time measurement of tire sinkage using stereo cameras

Hannah White [a,*], Corina Sandu [a,*], Jyotirmoy Mukherjee [a,b], Andrea L'Afflitto [b], David Gorsich [c], Michael Cole [c]

[a] Terramechanics, Multibody and Vehicle Systems Laboratory (TMVS) Lab, Virginia Tech, Blacksburg, VA, USA

[a] Advanced Control Systems Lab, Virginia Tech, Blacksburg, VA, USA

[c] The U.S. Army DEVCOM Ground Vehicle Systems Center, Warren, MI, USA

* Corresponding authors: hnwhite@vt.edu, csandu@vt.edu

ABSTRACT

There is a need for autonomous vehicles in off-road situations, such as in military and farming, but most established autonomous vehicle models are created for on-road conditions only and thus, do not account for the terrain properties. The terrain parameters and terrain-vehicle interaction parameters are essential for autonomy in off-road conditions. These conditions must be measured or estimated in real-time to be included in autonomy models. One of the most important aspects in estimating the performance of a vehicle on deformable terrain is tire sinkage. This parameter affects the slip-sinkage relationship, which, in turn, has a significant influence on the tractive performance of the vehicle. For real-time applications, it is reasonable to assume that the rut left behind has the same value as the sinkage. This paper presents part of a larger study focused on the mobility of autonomous vehicles in off-road conditions, and it details one of the methods employed for real-time sensing of vehicles and vehicle-terrain parameters. Specifically, this paper shows how image processing can be employed to assess the sinkage of a tire on a deformable surface. The approach presented here requires collecting the depth data of a tire rut in real-time using a set of two stereo cameras. The first camera measures the profile of the undisturbed terrain in front of the tire. The second camera measures the depth profile of the tire rut. These profiles are compared to deduce the final rut depth. The measurements from the two cameras are also used to compute the tire's entry and exit angles; these parameters can further be employed in the computation of the shear stresses along the contact patch of the tire.

Keywords
Terrain Sensing
Real-Time
Tire Sinkage
Terramechanics
Stereo Cameras

1. Introduction

As the use of autonomy is growing, most of the work focuses on on-road terrain. These results, however, are not easy to adapt to off-road environments, where the terrain varies and its parameters are unknown. There are many applications that can benefit from the use of off-road autonomy, such as the military and agriculture. Thus, there is a need for classical terramechanics tools to be adapted for autonomous navigation.

Tire sinkage is one of the most important parameters for vehicle-terrain modeling. It is used in the fundamental Bekker-Wong pressure sinkage relationship, as seen in Eq. 1 (Wong, 2022).

$$p = \left(\frac{k_c}{b} + k_\phi\right) z^n. \qquad (1)$$

The sinkage can also be used to find the entry and exit angles of the tire in the terrain. Chan and Sandu (2014) define the *vertical soil position* at any point along the tire's surface as

$$z = R(\cos\theta - \cos\theta_e), \qquad (2)$$

and the sinkage as

$$z_0 = R(1 - \cos\theta_e). \qquad (3)$$

Equation (3) can then be rearranged to solve for the entry angle

$$\theta_e = \cos^{-1}\left(1 - \frac{z}{R}\right). \qquad (4)$$

Chan and Sandu (2014) also define the *unloaded terrain depth behind the tire* as

$$u = R(1 - \cos\theta_r). \qquad (5)$$

In real-time applications, to save computational time, it can be assumed that since the soil is relatively inelastic, the rut depth is equal to the sinkage. With this assumption, the exit angle will equal 0°.

A challenge when adding autonomy to off-road vehicles is adding sensors that are easy to access and do not interfere with the integrity of the tires. This paper proposes a method of measuring tire sinkage using a set of two stereo cameras that are mounted on the body of the vehicle, in this case, a Terramechanics Testing Rig.

1.1 Literature Review

There is not an abundance of research on tire rut depth or sinkage measurement in real-time. Botha et al. (2019) did a study on the feasibility of measuring 3D rut profiles using cameras in real-time using digital image correlation. Digital image correlation uses the disparity between objects in two images to produce the distance.

Another method of measuring rut depth is photogrammetry, which is the process of using overlapping images to create 3D profiles. Though photogrammetry produces very accurate profiles, it is not well suited for real-time applications due to the computation time required. Kenarsari et al. (2017) used digital photogrammetry to obtain the 3D profiles of both a static tire print and a continuous rut. Pierzchała (2016) measured the rut of the tracks left at a logging site by tracked vehicles.

Lidar is another method that has been employed for measuring tire rut depth. Salminvaara et al. (2018) mounted lidar sensors on the backs of trucks. The data was post-processed to get the profile of both ruts left by the truck.

https://doi.org/10.56884/TOVX2833 — DISTRIBUTION STATEMENT A. Approved for public release; distribution is unlimited. OPSEC #7702
ISBN 978-1-942112-55-6 (pdf)

Naranjo et al. (2014) used optical sensors mounted on the rim inside the tire to measure the tire deflection, a string potentiometer to measure the vertical wheel displacement and an ultrasonic sensor to measure the undisturbed soil in front of the tire. These three measurements were used to find the tire sinkage. This method is difficult to implement due to the sensors being mounted inside the tire.

Many experiments measure the rut depth by using a ruler. He et al. (2019) used a ruler to measure the depth of a static tire test at several points in the rut. While this obviously cannot be used in real-time, it has the advantage of being very inexpensive, compared to other methods.

This paper focuses uses a real-time method that is able to measure the rut depth in real-time, which has rarely been done before. This paper also takes measuring the rut depth a step farther to be able to calculate the entry angle of the tire.

2. Experiment

2.1 Theory

In this experiment, the rut depth is found by using two stereo cameras to measure the depth profile in front of and behind the tire, as it can be seen in Fig. 1. This experiment relies on the assumption that the rut left behind by the tire is equal to the sinkage.

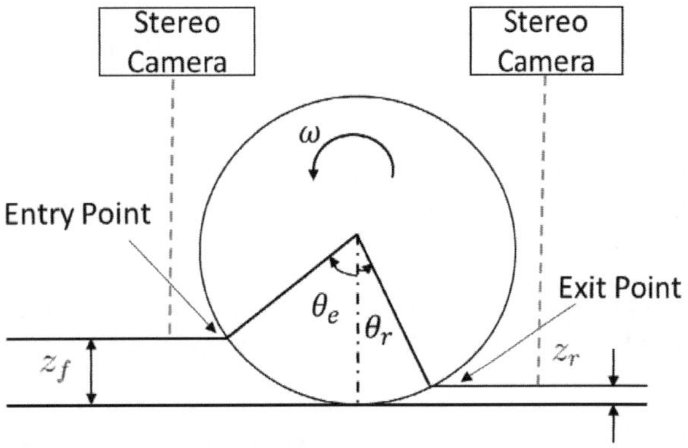

Fig 1. Diagram of camera setup

The frames provided by the cameras are trimmed to approximately the middle eight of the rut, to be able to obtain the area of maximum sinkage. Each depth frame is then averaged and compared to get the final depth.

2.2 Terramechanics Test Rig

The experiment was performed on the Terramechanics Testing Rig at the Terramechanics, Multibody, and Vehicle Systems Laboratory (TMVS) at Virginia Tech. The rig was filled with fine dry sand. The soil was uniformly prepared before the tests were performed. This preparation process began by loosening the sand with a shovel to eliminate the compacted sand. The Terramechanics Test Rig has a leveling blade on its frame which is then used to level the soil.

The rig can vary camber, toe, slip, and normal load (Sandu, et al., 2008), (Sandu, 2015). For every test, the toe was set to 0° and the camber setting was held constant. The tests were run at 0%, 5%, 30%, 40%, 50%, 60%, -5%, -15%, -25%, -40%, and -60% slip. The tests were run twice at each slip value. The normal force applied was 2000 Newtons. The 3D profiles of the ruts were captured with an Einstar 3D scanner. These scans were used to find the maximum sinkage at each slip. This was done by forming a plane using the undeformed portion of the scan and measuring the distance to the surface geometry of the scan (Sonalkar et al., 2023). This was used to validate the data collected by the cameras. The tire used was an AT 171 manufactured by Balkrishna Industries.

The effective rolling radius (R_e) of the tire was found experimentally before testing. This process entailed measuring the force in the longitudinal direction for various angular velocities while keeping the linear velocity constant. The angular velocity with the lowest longitudinal force was assumed to be the zero slip. The *effective rolling radius* was then computed as

$$R_e = \frac{V}{\omega_0}. \tag{5}$$

The effective rolling radius was 0.3556 meters The angular velocity for other values of slip was then calculated using the basic equation for slip

$$S = \frac{R_e\omega - V}{V} * 100. \tag{6}$$

2.3 Cameras

Two Intel® RealSense™ D405 stereo cameras (Intel® RealSense™ Depth Camera D405, 2023) were used to measure the rut depth in this experiment. The cameras were attached to the frame of the Terramechanics Testing Rig, one in front of the tire and one behind it. The cameras attached to the rig are shown in Fig. 2. The cameras were leveled and mounted at the same distance from a reference height and mounted in the same lateral position.

Fig. 2. Terramechanics Testing Rig with the stereo cameras mounted

The code to collect the data was programmed in C++. The cameras collected images at 30 frames per second. The frames provided a depth value at each pixel. Fig. 3 shows an image of the colored depth values and the corresponding RGB picture of the rut.

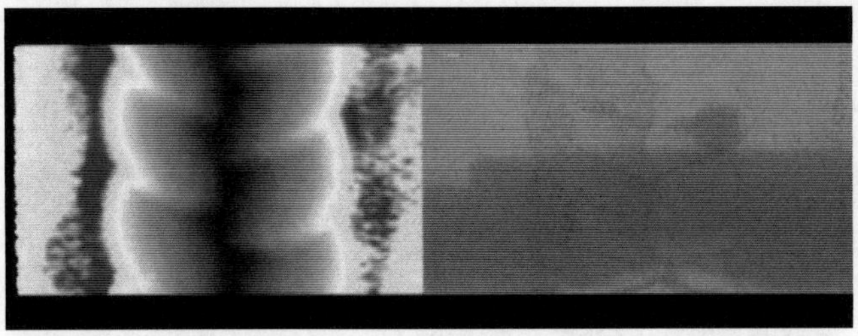

Fig. 3. Colored depth values and the RGB image of the corresponding rut.

3. Results

The results of the experiment were imported into MATLAB for post-processing and plotting purposes.

3.1 *Sinkage*

The graphs for the experimental runs with positive sinkage values are shown in Fig. 4 and negative sinkage values are shown in Fig. 5. These values were found using the theory discussed in the previous section. The sinkage values from -40% to 40% slip can be observed to have a large peak before settling. The sinkage values at 50%, 60%, -50%, and 60% have a sharp rise at the beginning and then peak at their steady-state values. The values of both runs tend to follow very similar paths, with the exception being a 40% slip. The maximum value provided by the scanner can be observed to be slightly higher than the steady-state value. Since the rut depth has a large variation, with the center being much deeper due to the nature of the sand, the average value from the cameras is smaller than the true maximum value. Overall, the results converged to a true sinkage value. It can also be observed that most of the values of slip ratio have the same wavy pattern in the depth values. This wavy pattern ("ripples") has been observed before by He et al. (2019), and by Irani et al. (2010).

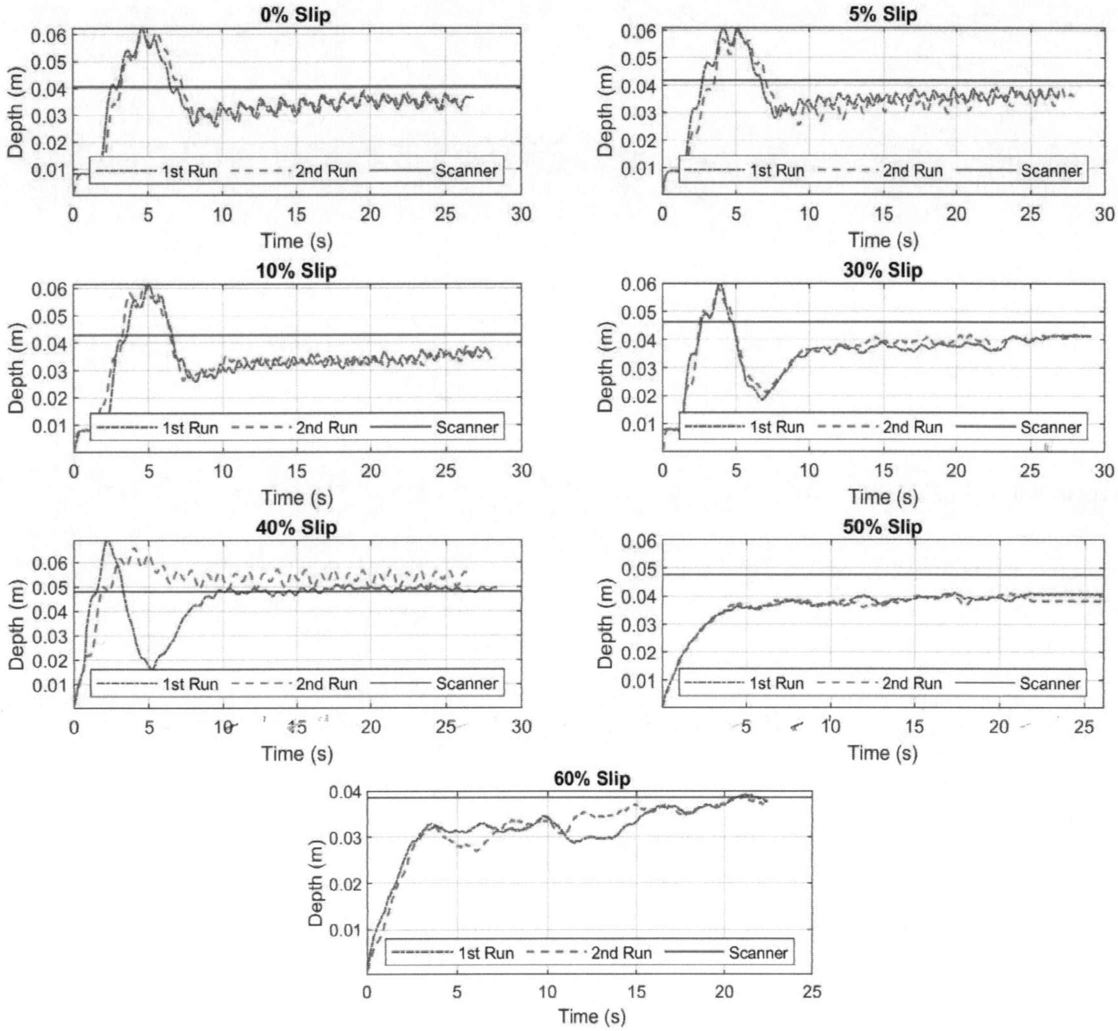

Fig 4. Tire sinkage for runs with positive slip values

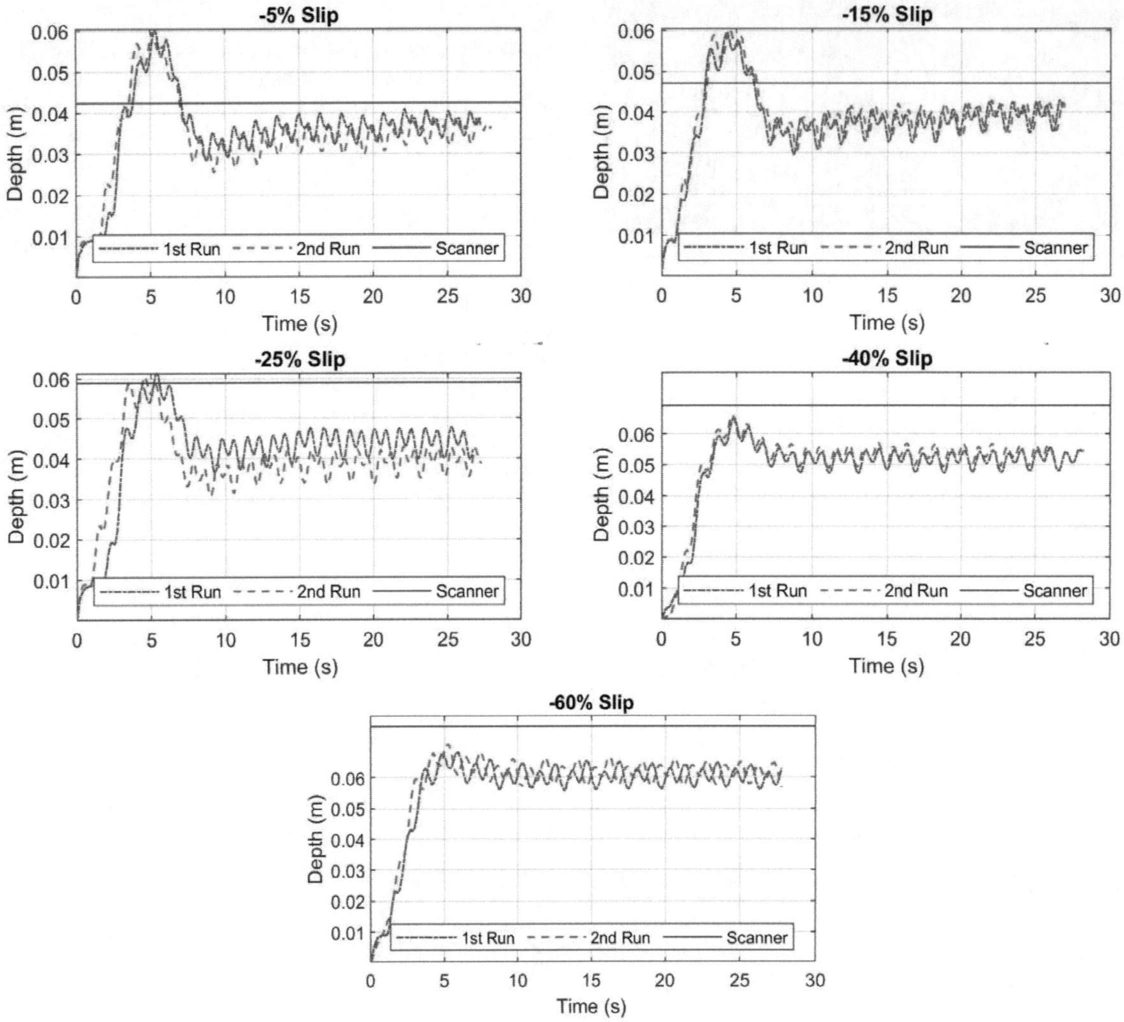

Fig 5. Tire sinkage for runs with positive slip values

3.2 Entry Angle

The entry angles are shown in Fig. 6 and Fig. 7. These angles are found using Eq. 3 mentioned in the Introduction. The angles from the positive slip values can be seen to stay very close together, between 20° and 30°. The only exception is at 40% slip, where the entry angle is higher. The entry angles for the runs at the negative slip seem to be more evenly distributed according to the slip ratio.

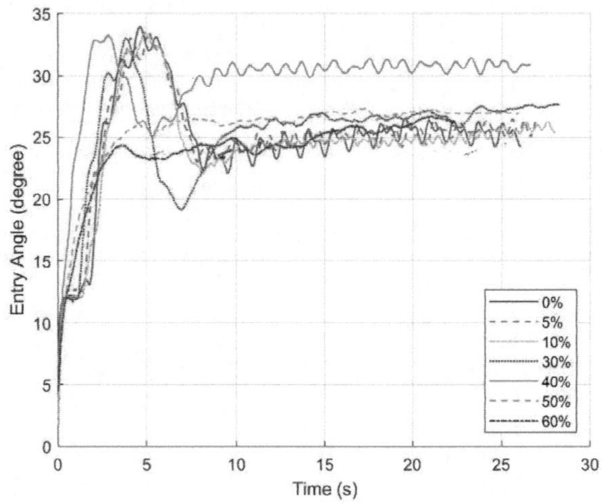

Fig 6. Entry angles of positive slip

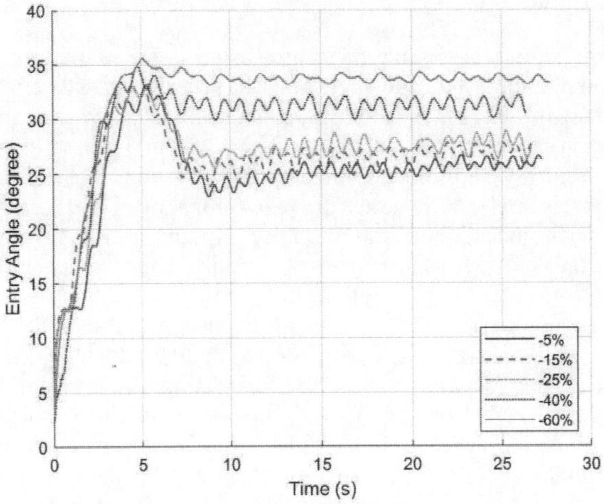

Fig 7. Entry angles of negative slip

4. Conclusions and future work

The research presented in this paper used two Intel RealSense D405 cameras to measure the rut depth created by an agricultural tire in real-time on a Terramechanics Testing Rig. The depth measurements were then used to calculate the approximate entry angle of the tire in the sand. These measurements have been proven to be accurate and repeatable.

The next step in this research project will be to conduct further testing using different types of terrain. Another sensor will be employed to measure the tire deflection so its effects can be considered for both a real-time estimation of the effective rolling radius and the entry angle. After further testing in the lab, additional studies will be conducted by implementing this methodology on a full vehicle for field testing. Finally, this will be integrated with a full autonomy stack.

5. Nomenclature

b	Tire width	[m]
k_c	Pressure-sinkage parameter	[kN/m^{n+1}]
k_φ	Pressure-sinkage parameter	[kN/m^{n+2}]
n	Sinkage exponent	[]
p	Pressure	[kPa]
R	Tire radius	[m]
R_e	Effective rolling radius	[in]
S	Tire slip ratio	[%]
u	Unloaded terrain height	[u]
V	Linear velocity	[m/s]
z	Vertical Tire Position	[m]
Z_0	Tire sinkage	
θ	Angle of soil particle	[°]
θ_e	Tire entry angle	[°]
θ_r	Tire exit angle	[°]
ω	Angular tire velocity	[deg/s]
ω_0	Angular tire velocity at zero slip	

6. Acknowledgements

This study has been supported by the Automotive Research Center (ARC) under the Cooperative Agreement W56HZV-19-2-0001 U.S. Army DEVCOM Ground Vehicle Systems Center, as well as by the TMVS Lab at Virginia Tech.

7. Declaration of competing interest

The authors declare that they have no known competing financial interests or personal relationships that could have appeared to influence the work reported in this paper.

8. References

Botha, T., Johnson, D., Els, S., Shoop, S., 2019. Real-time rut profile measurement in varying terrain types using digital image correlation. J. Terramech. 82, 53-61.

Chan, B. J., Sandu, C., 2014. Development of a quasi-steady state tyre model for on-road and off-road vehicle dynamics simulations: Part II – off-road rigid wheel model. Int. J. Vehicle Systems Modelling and Testing 9, 107-136.

He, R., Sandu, C., Osorio, J., 2019. Systematic tests for study of tire tractive performance on soft soil: Part I – Experimental data collection. J. Terramech. 85, 59-76.

Irani, R.A., Bauer, R.J., Warkentin, A., 2010. Modelling a single-wheel testbed for planetary rover applications. In: ASME 2010 Dynamic Systems and Control Conference. Cambridge, Massachusetts, USA, pp. 181–188

Intel® RealSense™ Depth Camera D405. https://www.intelrealsense.com/depth-camera-d405/ (accessed 29 May 2023)

Kenarsari, A. E., Viton, S. J., Beard, J. E., 2017. Creating 3D models of tractor tire footprints using close-range digital photogrammetry. J. Terramech. 74, 1-11.

Naranjo, S. D., Sandu, C., Taheri, S., Taheri, S., 2014. Experimental testing of an off-road instrumented tire on soft soil. J. Terramech. 56, 119-137.

Pierzchała, M., Talbot, B., Astrup, R., 2016. Measuring wheel ruts with close-range photogrammetry. J. For. Res. 89, 383-391.

Salmivaara, A., Miettinen, M., Finér, L., Launiainen., S., Korpunen, H., Tuominen, S., Heikkonen, J., Nevalainen, P., Sirén, M., Ala-Ilomäki, J., Uusitalo, J., 2018. Wheel rut measurements by forest machine mounted LiDAR sensors – accuracy and potential for operational applications?. Int. J. For Eng. 29, 41-52.

Sandu, C., Taylor, B., Biggans, J., Ahmadian, M., 2008. Building Infrastructure for Indoor Terramechanics Studies: The Development of a Terramechanics Rig at Virginia Tech, in: Proceedings of the 16th ISTVS International Conference, Turin, Italy. Turin, Italy.

Sandu, C., 2015. Chapter 2. Severe Environmental Conditions and Situations. 2.1. Experimental and Modeling Terramechanics Studies, in: Vantsevitch, V. V., Blundell, M. V., (Eds.), Advanced Autonomous Vehicle Design for Severe Environments, IOS Press BV, NATO ASI, Vol. 44 of NATO Science for Peace and Security Series — D: Information and Communication Security (29), pp. 40-68.

Sonalkar, C., Ravichandran, N., Vilsan, A., Sandu, C., Lin, T., 2023. In, Proceedings of the 16th European-African Regional Conference of the ISTVS, Lublin, Poland.

Wong., J. Y., 2022. Theory of Ground Vehicles, fifth ed., John Wiley & Sons, New York.

6690 | Benchmarking of methods to evaluate penetration resistance for Clegg hammer and cone penetrometer

Mohit N Shenvi [a,*], Corina Sandu [a], Costin Untaroiu [b]

[a] *Terramechanics, Multibody and Vehicle Systems (TMVS) Lab, Virginia Tech, Blacksburg, Virginia, United States*

[b] *Center for Injury Biomechanics (CIB), Virginia Tech, United States*

* *Corresponding author: mshenvi@vt.edu*

ABSTRACT

The computation of sinkage and of penetration resistance in deformable terrains are important aspects for the evaluation of the terrain properties from a terramechanics perspective. The Clegg Impact Hammer and the Cone Penetrometer are two widely used devices for evaluating the penetration resistance of soils. Variants of the cone penetrometer technique that have been used for snow evaluation are the Rammsonde and the Russian Snow Penetrometer, and very few studies exist for using the Clegg Impact Hammer technique for snow. For terrain such as highly compacted snow, like the one used in ASTM F1805 traction testing of tires, the process of measuring the terrain resistance becomes very complex.

This work describes the working methodologies and underlying assumptions of the Clegg Impact Hammer and an in-house developed device inspired by the Russian Snow Penetrometer. Furthermore, these devices are employed to measure the properties of compacted snow in a compressive regime. This work focuses on the relative evaluation of the sinkage and penetration resistance parameters based on the classical methods widely used in literature and two new proposed hypotheses for the evaluation of the outputs of these two devices respectively. The results show that the consideration of the surface area of the cone for evaluation of resistive pressure applied by the terrain amounts in a huge change to the applied pressure as well as the Bekker sinkage modulus values. In comparison, the newly developed formulations for the evaluation of Clegg hammer modulus and sinkage values from the Clegg Impact Hammer do not show a major change. Possible reasons for this behavior in both cases is discussed.

Future work in this field would be an evaluation of sinkage and penetration resistance of different terrain types to validate the proposed hypothesis and/or provide correction factors to the hypothesis based on the field testing results for these types of terrain.

Keywords
Compacted snow
Clegg Impact Hammer
Cone penetrometer
Penetration resistance
Sinkage

1. Introduction

Many researchers have been interested in snow mechanics; however, compressed snow mechanics has received less attention (Shenvi et al., 2022b). Theoretically, the density of snow can range from fresh snow (70–150 kg/m³) to ice (917 kg/m³) (Russell-Head and Budd, 1989). For almost 50 years, organizations like the Cold Regions Research and Engineering Laboratory (CRREL) of the U.S. Army have focused on the mobility of automotive tires on compacted snow and snow-covered roads, that have densities ranging from 370 kg/m³ to 600 kg/m³ (Shoop et al., 2010).

In North America, winter tire testing is performed according to the ASTM F1805-20 standard (ASTM, 2020), which specifies methodologies for straight-line traction testing of passenger car and light truck tires on snowy or icy surfaces. The type of surface used in such tests prescribed by the standard is in the compacted snow domain. Prior to tire testing, a CTI penetrometer (recommended by the standard) is used to characterize the compacted snow, but using the corresponding data in snow modeling is challenging. Additionally, the majority of the currently available devices do not perform well in the compacted snow domain (Shoop et al., 2019).

An in-house device was developed and used simultaneously along with the Clegg Impact Hammer and the CTI penetrometer to study the compressive properties of compacted snow (Shenvi et al., 2022a). To evaluate the Bekker parameters for the Clegg hammer and the in-house device, classical formulations found in the literature were utilized. However, these formulations have their inherent assumptions, which may or may not be applicable in compacted snow. Thus, an attempt is made to revisit the development of the formulations from a physics-based perspective and a comparison of the differences in the major outputs of the two devices is discussed.

Section 2 of this work provides an introduction to the two devices and prior attempts discussed in the literature for their use in snow and/or compacted snow. Section 0 provides explanations of the formulations found in the literature for the device outputs and our proposed changes for these formulations, as well as the reasons for these changes. Section 4 discusses the results from the two methods for both devices followed by interpretation and effects on the analysis from a terramechanics perspective. Section 5 summarizes the conclusions while highlighting the possible avenues for future development and validation of the proposed methodologies.

2. Description of testing devices

In-situ testing of snow for characterization purposes is primarily of compressive or shear nature. In compressive testing, vertical penetration of the snow using a plate-type or cone-type surface is performed; however, the majority of the snow testing devices and their standards by organizations (e.g., ASTM) have their roots in soil testing methodologies (Shenvi et al., 2022b). On the other hand, the available shear testing devices have their own drawbacks for use in compacted snow field testing (Shoop et al., 2019). The development of a device for fast field-testing methodology that will provide snow characterization data in compressive and shear domains is investigated in this study. The work presented in this paper focuses on the characterization of compressive behavior of snow.

https://doi.org/10.56884/ZDXV8929
ISBN 978-1-942112-55-6 (pdf)

2.1 Clegg Impact Hammer

The Clegg Impact Hammer, invented by Dr. Baden Clegg for evaluating the pavement base has a flat-faced solid shaft with an attached accelerometer to measure the deceleration experienced by the solid shaft on contact with the terrain (Clegg, 1976). The device is available commercially in 5 different sizes ranging from 0.5 kg up to 20 kg. The device is widely used for evaluation of pavement bases, turfs, sports grounds, etc. To the best knowledge of the authors, the device is manufactures in Australia, the UK, and the US; however, based on personal correspondence with one of the manufacturers, the differences in the electronic parts and signal processing algorithms is unknown. A schematic of the Clegg Impact Hammer is presented in Fig. 1.

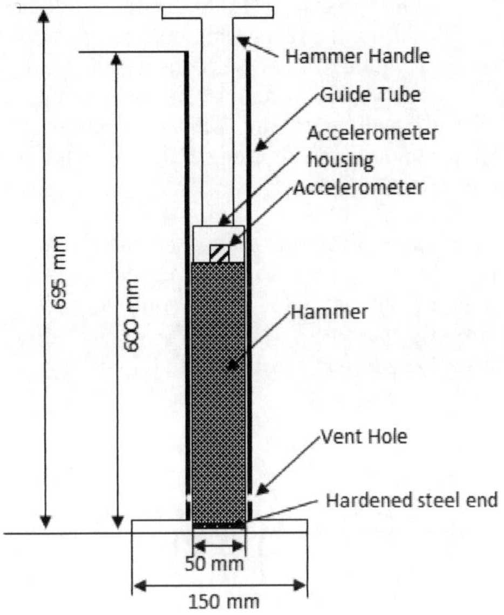

Fig. 1. Schematic of Clegg Hammer (Shenvi et al., 2022b)

The standard governing the use of the Clegg hammer is ASTM D5874 (ASTM, 2016). The 4.5 kg variant is considered the "standard" variant according to ASTM and was used in this work. The operating procedure of the device consists of dropping the hammer consecutively 5 times at a given location and using the Clegg Impact Value (CIV) or the maximum deceleration faced by the hammer in the 4th drop (S D Instrumentation Ltd., 2022). The remote attached to the Clegg hammer displays the highest deceleration faced by the hammer in terms of CIV; however, the exact filtering algorithm is not found in literature and depends on the manufacturer of the unit.

The Clegg hammer has been widely used for soil applications but its use in snow applications has been primarily performed by CRREL. CRREL has used the device for evaluation of strength of snow roads from as early as 1989 (Lee et al., 1989) with latter studies showing a comparative analysis of the applicability of different size Clegg hammers on compacted snow in Antarctica (Shoop et al., 2010; Shoop et al., 2012). The results of these studies postulated that the 4.5 kg hammer might be too heavy for compacted snow evaluation purposes and the 2.25 kg variant might be better suited for this scenario. However, the 2.25 kg variant is comparatively less used and thus is difficult to procure for testing purposes. In this work, the focus is on the assumptions and derivations of formulations that provide the sinkage, the Clegg Hammer Modulus (dynamic Young's modulus) and the penetration resistance force.

2.2 In-House Developed Device

The Rammsonde Penetrometer was initially developed as a dynamic cone penetrometer for use in snow (Abele, 1963). Over time, its applicability and utility in compacted snow was questioned, as the device has its inherent drawbacks (Niedringhaus, 1965; Shoop et al., 2019). This led to the development of the Russian Snow Penetrometer (RSP) for compacted snow evaluation that difffer in a few features from the Rammsonde, especially the cone properties (Shenvi et al., 2022b).

The working procedure of both devices consists of dropping a fixed weight from a fixed height on a guide tube causing the cone and the device to penetrate the snow. A record of the number of drops required to attain a certain penetration/sinkage is kept, which allows one to do a vertical profiling of the snow column. An in-house device was developed in our laboratory, having as inspiration the Russian Snow Penetrometer. The changes in the design were based on learnings from literature about different devices and their pros and cons (Shoop et al., 2020; Shenvi et al., 2022b). The in-house device features a 1.5 kg drop weight, base diameter of 20 mm and the maximum drop height of 400 mm. The device has gradations to allow a variable drop height study to evaluate effectiveness of change in drop height and is shown in **Errore. L'origine riferimento non è stata trovata.**.

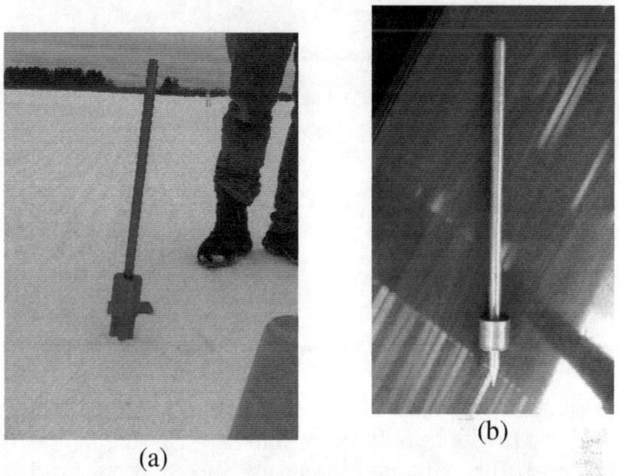

(a) (b)

Fig. 2. In-house developed device at Virginia Tech (a) when in use in snow, (b) Picture of entire device with gradations on guide tube (Shenvi et al., 2022a)

3. Description of Formulations and Data Collection Methodologies

The Clegg Impact Hammer has been employed in past studies described in the literature to evaluate the sinkage, the Clegg Hammer Modulus and the penetration resistance force, which can be utilized to find the pressure-sinkage relationship. The in-house developed device works on the principle of the cone penetrometer and is based on the working principle of the Rammsonde penetrometer. The main assumption of the formulations found in literature are described and discussed in this section.

3.1 In-house developed device

The in-house device is based on the working principle of the Rammsonde and RSP and has the following assumptions:

- The friction between the guide tube and the hammer is negligible.
- There is no energy loss on impact.

- The instrument uses a single type of cone, thereby considering the size effect to be negligible, i.e., independent evaluation of k_c and k_ϕ is not possible.

The testing procedure consists of dropping the hammer five times while measuring the height from the top of the device to a metal plate placed adjacent to it on the snow. This helps to evaluate sinkage due to each drop which was used with Eq. 1 to evaluate the penetration resistance force offered by the snow. This penetration resistance force has classically been assumed to be resisting the base area of the cone to evaluate the pressure due to a certain sinkage, as shown in Eq. 2 (Wong and Irwin, 1992). However, practically speaking, the snow is resisting the penetration due to the entire conical column, as shown in Fig. 3, and, thus, the resistive force is also acting on the surface area of the cone, not just its base area as depicted in Eq. 3. This would cause the forces to give a normal resistance effect perpendicular to the slant height of the cone and a frictional shear resistance effect along the slant height of the cone when the force directions are resolved.

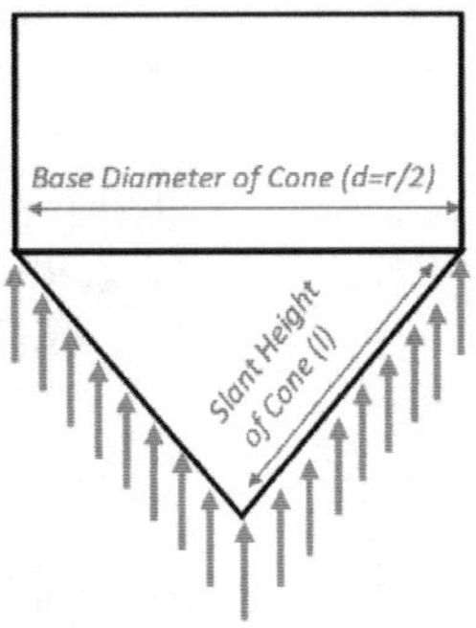

Fig. 3. Resistive forces act on the conical column instead of the base

In this work, the relative difference in the pressure evaluations is considered to further evaluate the sinkage exponent (according to Eq. 4) and the sinkage modulus (using Eq. 5). In the terramechanics literature we frequently see the assumption of the pressure being derived dependent on the base area of the cone; however, we postulate this assumption to be true in cases where the ratio of the surface area to the base area of the cone is not very high. In the case of the in-house device this ratio is about 3.864 which is much higher than for other commercially available devices, primarily due to its smaller cone angle.

$$R = \frac{W*h*n_{blows}}{z} + W + Q \quad (1)$$

$$p = \frac{R}{\pi*r^2} \quad (2)$$

$$p = \frac{R}{\pi r^2+\pi rl} \quad (3)$$

$$n = \frac{\Sigma p^2 \Sigma p^2 ln(p) ln(z) - \Sigma p^2 ln(p) \Sigma p^2 ln(z)}{\Sigma p^2 \Sigma p^2 (ln(z))^2 - (\Sigma p^2 ln(z))^2} \quad (4)$$

$$k = \frac{\Sigma p z^n}{\Sigma z^{2n}} \quad (5)$$

$$p = kz^n \quad (6)$$

where h is the height of drop in mm, z is sinkage in mm, R, W, and Q represent the resistance of snow to penetration, the weight of drop hammer, and the weight of penetrometer in Newtons respectively, n_{blows} is number of blows required to achieve sinkage z, r is radius of base of the cone in mm, l is the slant height of cone in mm and p is pressure in MPa.

A total of 6 readings (the initial one and one after each drop) were taken as part of the measurement methodology used, which involved measuring the height of a fixed point on the device in relation to the terrain. This allowed the evaluation of 5 sinkage values for the hammer drops at each location in each dataset. The number of blows in the evaluation must be adjusted appropriately in order to apply Eq. 1 since the second and subsequent sinkages must be treated as cumulative sinkages.

3.2 *Clegg Impact Hammer*

The Clegg Impact Hammer calculates the deceleration caused by the terrain when a hammer of certain weight is dropped from a predetermined height. Among the features of the Clegg hammer important from a terrain characterization perspective are the ease to convert to California Bearing Ratio (He et al., 2020) and the possibility of calculating the Clegg Hammer Modulus (Shoop et al., 2019). The Clegg Hammer Modulus is considered analogous to the Young's modulus however it should be considered more of a dynamic Young's modulus in uniaxial compression. The formulations pertaining to the Clegg Hammer are derived with some assumptions (Clegg, 1994) as:

- The clearances between the guide tube and the hammer have no influence, as the hammer drop is vertical.
- The hammer's impact with the terrain doesn't cause any energy loss.
- The retardation curve is constant with time (Fig. 4).

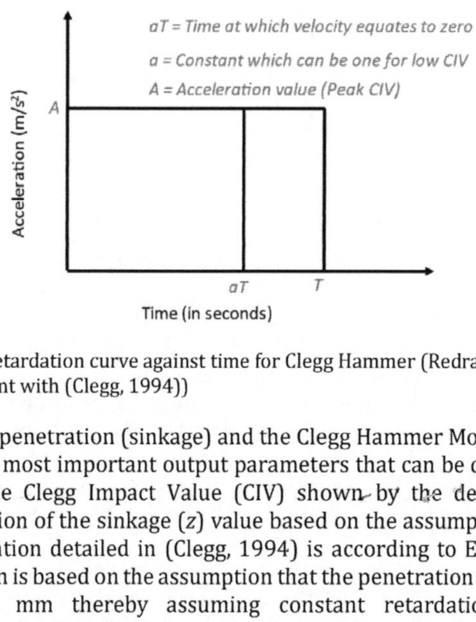

Fig. 4. Retardation curve against time for Clegg Hammer (Redrawn in agreement with (Clegg, 1994))

The penetration (sinkage) and the Clegg Hammer Modulus are the two most important output parameters that can be calculated from the Clegg Impact Value (CIV) shown by the device. The calculation of the sinkage (z) value based on the assumptions and a derivation detailed in (Clegg, 1994) is according to Eq. 7. This equation is based on the assumption that the penetration is greater than 1 mm thereby assuming constant retardation curve (Dentamaro, 1984). However, one thing to note is that the numerator in Eq. 7 will change with every drop as the working mechanism of the device only allows a measured drop height (h) for the first drop. The sinkage evaluated during a drop should be added to the drop height to evaluate sinkage of the next drop. This updated Eq. 8) is not found in literature and is of scientific interest in this work as this causes the numerator of the sinkage evaluation formulation to be variable instead of a constant value which is the real condition in field.

$$z = \frac{h}{10*CIV} \quad (7)$$

$$z_i = \frac{h+\Sigma z_{i-1}}{10*CIV} \quad (8)$$

Based on the elastic theory for a rigid bearing plate and with a Poisson's ratio assumed to be 0.5 on impact, the penetration can also be calculated by Eq. 9, where p is peak contact stress (in MPa), r is radius of the hammer (in mm) and E is the Elastic Modulus (MPa). The peak contact stress can also be calculated using Eq. 10, where m is the mass of the hammer and g is the gravitational acceleration. Equating Eqs. 7 and 9 and using Eq. 10 we get Eq. 11 (Clegg, 1994). Upon inputting the values of mass, gravitational acceleration, and radius of the hammer (Clegg, 1995), Eq. 12 is obtained. Dr. Clegg further substitutes the value of drop height in Eq. 12 and converts to sinusoidal waveform assumption (Millar, 1977) to obtain the classical formulation to evaluate the Clegg Hammer Modulus in MPa (Eq. 14). However, as discussed earlier, the value of the drop height changes for every drop except for the first one (Eq. 8) thus the assumption of a constant drop height to calculate the elastic modulus may not be completely accurate. The assumption of a square wave is more appropriate in low strength levels, however for compacted snow, it makes more sense to assume sinusoidal waveform (Dentamaro, 1984). Thus, we derive a revised equation for calculating elastic modulus in MPa (Eq. 13).

$$z = \frac{1.18*(p)*(r)}{E} \tag{9}$$

$$p = \frac{m*CIV*10*g}{\pi r^2} \tag{10}$$

$$E = \frac{1.18*m*(CIV)^2*100*g}{\pi r h} \tag{11}$$

$$E = \frac{66358.089*(CIV)^2}{h} \tag{12}$$

$$E_i = \frac{0.03984*(CIV)^2}{h+\sum z_{i-1}} \tag{13}$$

$$E = 0.088*(CIV)^2 \tag{14}$$

$$R = \sqrt[0.48]{\frac{(0.24*CIV+1)^2}{1.44}} \tag{15}$$

Further, the calculation of the ram resistance force in kg_f is performed according to Eq. (15) that is based on the conversion between the CIV and the California Bearing Ratio. As the correlation might be based on empirical data, this work continues to use the same formulation for the ram resistance force. However, the evaluation of the correlations for compacted snow evaluation would be of scientific interest.

4. Results

The field measurements done as a part of this work were conducted at the Smithers Winter Test Center in March 2022 and the data was analyzed statistically based on classical formulations (Shenvi et al., 2022a). The results of this section focus on the data collected during day 3 of the testing regime which had a total of 5 datasets collected at an interval of 2 hours with the first dataset being approximately 2 hours since the grooming process to create an ASTM F1805 compatible test track. Six locations were arbitrarily chosen in the area of the track that did not obstruct the regular operations of winter tire testing.

4.1 *In-house Developed Device*

The in-house device was used for height measurements using a laser distance sensor from a fixed position, producing five sinkage datapoints for each test, roughly every two hours. The analysis procedure described in section 3.1 was utilized to evaluate the variation that the pressure faced by the device as depicted in Table 1 and the change in the sinkage exponent and sinkage modulus values as shown in Table 2.

Table 1
Evaluation of pressure for in-house device using classical and proposed methods based on data of day 3

Location	Mean of pressure by classical method (MPa) (Eq. 2)	Mean of pressure by proposed method (MPa) (Eq. 3)	Percentage Difference
A	2.33	0.48	-79.44
B	2.49	0.51	-79.44
C	2.44	0.50	-79.44
D	1.93	0.40	-79.44
E	1.92	0.40	-79.44
F	2.09	0.43	-79.44

Table 2
Evaluation of sinkage modulus and sinkage exponent by classical and proposed methods based on data of day 3

Location	Mean by classical method		Mean by proposed method		Percent Difference	
	k	n	k	n	k	n
A	0.182	1.030	0.038	1.030	-79.44	0
B	0.208	1.043	0.043	1.043	-79.44	0
C	0.192	0.846	0.040	0.846	-79.44	0
D	0.133	1.289	0.027	1.289	-79.44	0
E	0.132	1.475	0.027	1.475	-79.44	0
F	0.154	1.191	0.032	1.191	-79.44	0

Based on the observations from these results, it can be seen that the change in the area being considered for the resistive force application has a huge impact on the pressure being calculated from the process. Furthermore, this solely seems to affect the sinkage modulus evaluation by the same amount in terms of percentage difference but does not affect the sinkage exponent at all. One of the possible reasons for this could be the neglect of the size effect due to usage of a single cone size.

4.2 *Clegg Impact Hammer*

The Clegg Impact Hammer output values were used to evaluate the Clegg Hammer Modulus and the sinkage using the classical methods found in the literature, as well as the proposed methods in section 3.2. The data presented in this section (Table 3-Table 5) is based on data of day 3 during the testing regime for the 6 locations.

Table 3
Mean of CHM for Clegg Hammer Device using classical and proposed methods based on data of day 3

Location	Mean of CHM by classical method (MPa) (Eq. 14)	Mean of CHM by proposed method (MPa) (Eq. 13)	Percentage Difference
A	22.26	22.40	0.60
B	22.05	22.04	-0.02
C	19.52	19.39	-0.66
D	18.09	17.85	-1.33
E	20.93	20.49	-2.06
F	18.51	18.00	-2.73

Table 4
Mean of sinkage of 3-5 drops for Clegg Hammer Device using classical and proposed methods based on data of day 3

Location	Mean of sinkage from 3-5 drops by classical method (MPa) (Eq. 14)	Mean of sinkage from 3-5 drops by proposed method (MPa) (Eq. 13)	Percentage Difference
A	2.85	2.85	0
B	2.89	2.91	0.64
C	3.08	3.12	1.28
D	3.27	3.33	1.98
E	2.93	3.01	2.71
F	3.28	3.39	3.36

Table 5
Mean and Standard Deviation of resistance force (R) evaluated using Clegg data of day 3

Location	Mean of R (N)	Std. Dev. of R (N)	Coefficient of Variation (COV)
A	3250	880	0.271
B	3270	1490	0.456
C	2660	1050	0.395
D	2400	1100	0.458
E	2920	520	0.178
F	2600	1880	0.723

The comparison of the results using the classical methods with results found in literature was published previously (Shenvi et al., 2022a). Based on the minor difference between the classical and the proposed methods, as shown in Table 3 and Table 4, it can be deduced that the outputs are similar. This is due in part to the contribution of a highly compacted snow layer, whereas the classical formulations were developed based on assumptions of a softer layer that yields more than 1 mm of sinkage, which may or may not be the case in compacted snow. Since the formulation calculating the resistance force was not changed in this work, the results in Table 5 are reproduced from (Shenvi et al., 2022a).

5. Conclusions and Future work

In this work, an attempt is made to propose new formulations to be used for the determination of useful outputs from a Clegg Impact Hammer and an in-house device that is similar to a dynamic cone penetrometer.

As part of this attempt, new formulations were derived to evaluate the pressure and the Bekker parameters for the in-house device. It was found that this revised formulation led to an 80% decrease in the pressure calculated and the sinkage modulus values at all locations. Mathematically, this makes sense, as basically the denominator of the pressure formulation changes and that has a direct effect on the sinkage modulus. There was no effect seen on the sinkage exponent value. For the Clegg Impact Hammer, the effect of the sinkage in a previous drop on the drop height of a subsequent drop was used to revise the formulation of the sinkage evaluation and of the Clegg Hammer modulus evaluation. The revised formulations do not have a major impact on the sinkage and CHM values.

From a future work perspective, the developed formulations for the in-house device pressure calculations should be utilized in a similar manner to see the effect of different cone sizes on different terrains. This will help to check the extent of the effect on different terrains and also to find the ratio of surface area to base area of cone beyond which the surface area of cone must be utilized for pressure evaluations. Using different cone sizes would also help in evaluating the cohesion and internal friction modulus of Bekker's pressure-sinkage equation, thus helping to find the extent to which each of them is affected due to a larger cone size. For the Clegg Impact Hammer formulations, it would be beneficial to re-evaluate the resistance force formula, as the change in the drop height for every drop will have some effect on that formula as well.

6. Nomenclature

r	Radius of cone/Clegg hammer	[mm]
l	Slant Height of cone	[mm]
E	Young's Modulus/Clegg Hammer Modulus	[MPa]
h	Drop Height	[mm]
nblows	Number of blows required to achieve sinkage z	[]
p	Pressure	[MPa]
Q	Weight of Penetrometer	[N]
R	Penetration resistance of the snow	[N]
W	Weight of drop hammer	[N]
z	Sinkage/Penetration	[mm]

7. Acknowledgements

This study was supported by the Center for Tire Research (CenTiRe), an NSF I/UCRC (Industry University Cooperative Research Center) at Virginia Tech. The authors wish to thank the project mentors and the members of the Industrial Advisory Board of CenTiRe for their support and guidance and Smithers Winter Test Center for providing access to the test site. The authors would also extend gratitude to Mr. Jim Crandell of Dr. Baden Clegg Pty Ltd., Australia for sharing classical literature pertaining to the Clegg Hammer.

8. Declaration of competing interest

The authors declare that they have no known competing financial interests or personal relationships that could have appeared to influence the work reported in this paper.

9. References

Abele, G, 1963. A correlation of unconfined compressive strength and ram hardness of processed snow. Hanover, New Hampshire, U.S. Army Cold Regions Research and Engineering Laboratory. 1-23.

ASTM, 2016. D5874-16: Standard Test Method for Determination of the Impact Value (IV) of a Soil. West Conshohocken, PA, ASTM International: 1-9.

ASTM, 2020. F1805-20: Standard Test Method for Single Wheel Driving Traction in a Straight Line on Snow. West Conshohocken, PA, ASTM International: 1-13.

Clegg, B., 1976. An impact testing device for in situ base course evaluation. Australian Road Research Board Conference Proc.

Clegg, B., 1994. Technical Note 1: Calculation of Penetration and Elastic Modulus from CIV. Jolimont, Western Australia 6014, Australia, Dr Baden Clegg Pty Ltd.

Clegg, B., 1995. Technical Note 2: Hammer Modulus and Application of 20 kg Hammer. Jolimont, Western Australia 6014, Australia, Dr Baden Clegg Pty Ltd.

Dentamaro, A.L., 1984. Correlation of the Clegg Impact Tester to the California Bearing Ratio for CBR 20% or Less. Nedlands, Western Australia, Australia, Department of Civil Engineering, University of Western Australia.

He, R., Sandu, C., Mousavi, H., Shenvi, M.N., Braun, K., Kruger, R., Els, P.S., 2020. "Updated standards of the international society for terrain-vehicle systems." Journal of Terramechanics. 91, 185-231.

Lee, S.M., Haas, W.M., Brown, R.L., Wuori, A.F., 1989. Improving snow roads and airstrips in Antarctica. Hanover, New Hampshire, U. S. Army Corps of Engineers, Cold Regions Research and Engineering Laboratory: 1-24.

Millar, L.R., 1977. Application of the Soil Impact Tester to Road Pavement Construction. Department of Civil Engineering. Nedlands, Western Australia, Australia, University of Western Australia. Honours.

Niedringhaus, L., 1965. Study of the Rammsonde for Use in Hard Snow. Hanover, New Hampshire, U.S. Army Cold Regions Research and Engineering Laboratory: 23-23.

Russell-Head, D.S. and Budd W.F., 1989. Compacted-Snow Runways: Guidelines for their Design and Construction in Antarctica. Hanover, New Hampshire, U.S. Army Cold Regions Research and Engineering Laboratory.

S D Instrumentation Ltd., 2022. CIST/882: 4.5 Kg Clegg Impact Soil Tester (IV Display): Operating Manual. Top Farm, Tellisford, Bath, England: 1-30.

Shenvi, M.N., Sandu, C., Untaroiu, C., 2022a. Benchmarking of Compression Testing Devices in Snow. 11th Asia-Pacific Regional Conference of the ISTVS. Harbin, China, ISTVS: 1-7.

Shenvi, M. N., C. Sandu and C. Untaroiu (2022b). "Review of compressed snow mechanics: Testing methods." Journal of Terramechanics 100: 25-37.

Shoop, S., Knuth, M., Crandell, J., 2012. Using a Clegg Impact Hammer to measure snow strength. Cold Regions Engineering 2012: Sustainable Infrastructure Development in a Changing Cold Environment Quebec City, Canada, American Society of Civil Engineers.

Shoop, S., Phetteplace, G., Wieder, W.L., 2010. Snow Roads at McMurdo Station , Antarctica. Hanover, New Hampshire, U.S. Army Cold Regions Research and Engineering Laboratory: 1-112.

Shoop, S.A., Wieder, W.L., Elder, B., 2020. Mechanical Property Measurements on Various Snow Surfaces. Hanover, New Hampshire, U.S. Army Cold Regions Research and Engineering Laboratory.

Shoop, S.A., Wieder, W.L., Elder, B.C., Beal, S.A., Deeb, E.J., 2019. Assessment of Field Methods for Measuring Mechanical Properties of Snow. Hanover, New Hampshire, U.S. Army Cold Regions Research and Engineering Laboratory.

Wong, J.Y., Irwin, G.J., 1992. Measurement and Characterization of the Pressure-Sinkage data for Snow obtained using a Rammsonde. Journal of Terramechanics. 29(2), 265-280.

6736 | Ride comfort comparison between 4-poster and straight-line driving simulations

Cor-Jacques Kat [a,*], P. Schalk Els [a]

[a] *Vehicle Dynamics Group, Department of Mechanical and Aeronautical Engineering, University of Pretoria, Pretoria, South Africa*

* *Corresponding author: cor-jacques.kat@up.ac.za*

ABSTRACT

Difference thresholds of whole-body vibration is important to determine perceptibility of changes in a vehicle's dynamics. Difference thresholds can be used to guide ride comfort improvements. Difference thresholds have been estimated for vertical and multi-axial seat vibration in laboratory settings. It is postulated that estimated difference thresholds in a laboratory setting should be applicable to real-world driving conditions given that the stimuli are similar. This paper considers the stimulus associated with the vehicle's dynamic response i.e. vibration. The vibration stimulus characteristic of a vehicle on a 4-poster test rig and straight-line driving is investigated through simulation. A non-linear SUV vehicle model is simulated on a 4-poster test rig and driven in a straight-line over three ISO road profiles that range from on- to off-road conditions i.e. class B, C and D. The resulting translational and rotational vibration for all three axes at the driver position make up the vibration stimulus. Differences in the vibration characteristics exists between the 4-poster test rig and straight-line driving simulations. Further investigations are needed to determine whether differences are only present in simulations or also practically and the influence of the difference in vibration stimuli to difference thresholds.

Keywords
Ride comfort
Vehicle vibration
Difference thresholds

1. Introduction

Difference thresholds of whole-body vibration is important to determine perceptibility of changes in a vehicle's dynamics. It can be used to guide ride comfort improvements. Difference thresholds can be defined as the minimum change in the magnitude of whole-body vibration required for the seated person to perceive the change in magnitude (Mansfield and Griffin, 2000). Studies have estimated difference thresholds in lab environments to random vibration stimuli e.g. vertical stimulus on a seat (Mansfield and Griffin, 2000) and multi-axial stimulus in a vehicle on a 4-poster test rig (Gräbe et al, 2020). It is however not yet clear how well these laboratories estimated difference thresholds translate to real-world driving. It is postulated that estimated difference thresholds in a laboratory setting should be applicable to real-world driving conditions given that the stimuli (vibration, noise, visuals, cognitive load, etc.) are similar. This paper considers the stimulus associated with the vehicle's dynamic response, specifically vibration.

Utilizing 4-poster test rigs for difference estimation has advantages over performing these experiments on a proving ground. An obvious advantage is with respect to the relative ease with which the magnitude of vibration and duration between signals in a trial can be defined. Other advantages include: cost, repeatability, no need for full driving prototype, and it is not limited by terrain profiles available on a test track. However, the 4-poster might pose frequency band limitation with respect to excitation due to the actuators and valves. Therefore, the replication of the vibration stimulus characteristic of a vehicle driving in a straight-line on the 4-poster test rig, should be considered.

The aim of the paper is to compare the ride comfort, quantified by the weighted vibration magnitude of the seat vibration, between 4-poster test rig and straight-line driving simulations.

2. Simulation model

The MSC.ADAMS/View non-linear vehicle model is of a Land Rover Defender 110. The model is based on a previously validated version of the model [Els et al, 2007]. The vehicles' standard spring and damper is replaced by a hydro-pneumatic system. The hydro-pneumatic suspension system allows for the selection of various suspension settings including a handling and ride mode. The ride mode is used in this study. The Michelin LTX 235/85 R16 tires are modeled using FTire (Bosch et al, 2016).

The vehicle model will be simulated in two environments: 1) on a 4-poster test rig and 2) driving in a straight line. The 4-poster will use a 2-dimensional flat road on each corner of the test rig and the contact with the four tires. Each corner of the 4-poster test rig is actuated and its vertical displacement controlled by the desired road profile that corresponds to a constant speed. For the straight-line driving environment, the vehicle will drive at a constant speed and in a straight-line over a 3-dimensional road.

The following test conditions are considered. Three ISO 8608 (1995) road profiles are considered i.e. class B, C, D, and three vehicle speeds i.e. 60, 80 and 100 km/h. The 100 km/h on the class B road is relatively high and would not typically be part of normal driving.

The simulations are set up such that for the 4-poster test rig there is 1.5 seconds at the start of the simulation with no actuator inputs to the tires. This allows any start-up transients to dissipate before the 4-poster excites the vehicle. Similarly, for straight-line driving a flat road section is added to allow the vehicle to speed up and

https://doi.org/10.56884/NQUR9670
ISBN 978-1-942112-55-6 (pdf)

reach the desired speed, which is then kept constant whilst driving over the desired road profile.

The 3-dimensional road profile is generated from the power spectral density of the specific ISO 8608 (1995) road class. The road is generated such that the longitudinal and lateral direction of the road corresponds to the desired ISO road class. The 3-dimensional grid is then used to generate the Curved Regular Grid (CRG) road type that is used to define the road in MSC.ADAMS. The vertical displacement inputs used to drive the four corners of the 4-poster test rig is obtained from the 3-dimensional road by extracting a track (a single line in the longitudinal direction) from the grid for the left and right-hand track of the vehicle. The time delay between the front and rear actuators of the 4-poster test rig is determined from the vehicle speed and wheelbase.

3. Results

The translational and rotational vibration at the driver position is weighted according to BS 6841 (1987). BS 6841 (1987) used instead of the newer ISO 2631 (1997) to benefit from other published data on difference thresholds (Mansfield and Griffin (2000), Gräbe et al, 2020). It should be noted that the weighted vibration has been frequency weighted and multiplied by the applicable multiplication factor.

The weighted vibration magnitude of the six vibration components and the difference in the magnitude between the two environments are presented in Table 1.

It is observed that the magnitude of vibration in the vertical direction for straight-line driving is larger than on the 4-poster test rig. This is also observed for yaw, with the only exception being for 60 km/h on the class B road. Vibration in the other four directions (Fore-aft, Lateral, Roll and Pitch) tends to be higher on the 4-poster test rig than for straight-line driving. The absolute value of the difference in vibration magnitude between the 4-poster test rig and straight-line driving ranges from 1.7% to 61.5%. The larger differences are associated with the yaw direction which ranges from 16.7% to 61.5%. Excluding yaw, the absolute value of the differences range from 1.7% to 29.2%.

The weighted vibration magnitude is also considered for the following three frequency bands: Primary ride 0.5 - 5 Hz, Secondary ride 5 – 30 Hz and Harshness 30 – 80 Hz. These are based on the frequency bands as defined by SAE Standard J670 (2008). Note that the standard defines harshness from 30 to 100 Hz. In this paper, harshness is only up to 80 Hz as content above this is discarded based on ride comfort frequency weightings in BS6841.

The weighted vibration magnitude for the three frequency bands of interest and the difference in the magnitude between the two environments are presented in Table 2, Table 3 and Table 4 for the three roads, respectively. It is observed that the vibration magnitude in secondary ride tends to be higher for straight-line driving than the 4-poster test rig for all directions (the only exceptions are yaw for 80km/h over class B and D, and pitch for 80km/h over class C and D). No other consistent trends are observed. The absolute value of the difference in vibration magnitude between the 4-poster test rig and straight-line driving ranges from 0.5% to 95.8%. There does not seem to be a specific direction or frequency band having larger differences than others.

Table 1
Vibration magnitude (weighted) at driver seat

		Vibration magnitude [m. s.$^{-2}$, r. m. s.]						Difference [%]		
		4-Poster			Straight-line driving			$\frac{4Poster - Straight-line\ driving}{Straight-line\ driving} \times 100$		
		60 km/h	80 km/h	100 km/h	60 km/h	80 km/h	100 km/h	60 km/h	80 km/h	100 km/h
ISO B	Fore-aft	0.09	0.09	0.09	0.09	0.09	0.10	0	0	-10.0
	Lateral	0.02	0.02	0.02	0.02	0.02	0.02	0	0	0
	Vertical	0.42	0.44	0.41	0.48	0.51	0.45	-12.5	-13.7	-8.9
	Roll	0.65	0.61	0.67	0.60	0.58	0.64	+8.3	+5.2	+4.7
	Pitch	2.27	2.36	2.31	2.23	2.24	2.23	+1.8	+5.4	+3.6
	Yaw	0.05	0.05	0.05	0.04	0.06	0.13	+25	-16.7	-61.5
ISO C	Fore-aft	0.18	0.20	0.21	0.17	0.20	0.22	+5.9	0	-4.6
	Lateral	0.03	0.03	0.03	0.03	0.03	0.03	0	0	0
	Vertical	0.88	0.90	0.86	0.97	1.03	0.95	-9.3	-12.6	-9.5
	Roll	1.13	1.25	1.27	0.91	1.06	1.13	+24.2	+17.9	+12.4
	Pitch	4.80	5.16	5.34	4.47	4.87	5.23	+7.38	+6.0	+2.1
	Yaw	0.06	0.06	0.05	0.08	0.12	0.11	-25.0	-50.0	-54.6
ISO D	Fore-aft	0.42	0.46	0.46	0.37	0.44	0.46	+13.5	+4.6	0
	Lateral	0.07	0.08	0.09	0.06	0.07	0.08	+16.7	+14.3	+12.5
	Vertical	1.85	1.78	1.67	1.97	2.01	1.89	-6.1	-11.4	-11.6
	Roll	2.48	3.07	3.63	1.92	2.39	3.28	+29.2	+28.5	+10.7
	Pitch	11.04	11.91	11.91	9.83	11.04	11.71	+12.3	+7.9	+1.7
	Yaw	0.11	0.12	0.12	0.20	0.24	0.17	-45.0	-50.0	-29.4

Table 2
Vibration magnitude (weighted) at driver seat over ISO class B road

		Vibration magnitude $[\mathrm{m.\ s.^{-2},\ r.\ m.\ s.}]$						Difference [%]		
		4-Poster			Straight-line driving			$\frac{\text{4Poster} - \text{Straight} - \text{line driving}}{\text{Straight} - \text{line driving}} \times 100$		
		60 km/h	80 km/h	100 km/h	60 km/h	80 km/h	100 km/h	60 km/h	80 km/h	100 km/h
Fore-aft	P	0.083	0.087	0.084	0.076	0.083	0.091	+9.2	+4.8	-7.7
	S	0.023	0.032	0.039	0.037	0.036	0.042	-37.8	-11.1	-7.1
	H	0.002	0.002	0.002	0.006	0.005	0.004	-66.7	-60.0	-50.0
Lateral	P	0.015	0.014	0.013	0.011	0.010	0.013	+36.4	+40.0	0
	S	0.009	0.010	0.012	0.012	0.013	0.013	-25.0	-23.1	-7.7
	H	0.001	0.001	0.001	0.001	0.001	0.001	0	0	0
Vertical	P	0.228	0.118	0.122	0.145	0.104	0.131	+57.2	+13.5	-6.9
	S	0.353	0.421	0.388	0.455	0.498	0.428	-22.4	-15.5	-9.4
	H	0.033	0.036	0.031	0.028	0.033	0.037	+17.9	+9.1	-16.2
Roll	P	0.532	0.505	0.520	0.414	0.375	0.473	+28.5	+34.7	+9.9
	S	0.335	0.340	0.413	0.420	0.434	0.415	-20.2	-21.7	-0.5
	H	0.046	0.046	0.047	0.052	0.053	0.056	-11.5	-13.2	-16.1
Pitch	P	2.202	2.246	2.135	2.108	2.114	2.038	+4.5	+6.2	+4.8
	S	0.531	0.713	0.878	0.703	0.742	0.916	-24.5	-3.9	-4.2
	H	0.034	0.034	0.035	0.031	0.034	0.039	+9.7	0	-10.3
Yaw	P	0.050	0.040	0.039	0.036	0.053	0.130	+38.9	-24.5	4
	S	0.018	0.021	0.026	0.024	0.020	0.025	-25.0	+5.0	0
	H	0.002	0.002	0.003	0.002	0.002	0.003	0	0	0

P – Primary ride, S – Secondary ride, H – Harshness

Table 3
Vibration magnitude (weighted) at driver seat over ISO class C road

		Vibration magnitude $[\mathrm{m.\ s.^{-2},\ r.\ m.\ s.}]$						Difference [%]		
		4-Poster			Straight-line driving			$\frac{\text{4Poster} - \text{Straight} - \text{line driving}}{\text{Straight} - \text{line driving}} \times 100$		
		60 km/h	80 km/h	100 km/h	60 km/h	80 km/h	100 km/h	60 km/h	80 km/h	100 km/h
Fore-aft	P	0.176	0.188	0.193	0.154	0.181	0.201	+14.3	+3.9	-4.0
	S	0.047	0.068	0.082	0.073	0.074	0.094	-35.6	-8.1	-12.8
	H	0.003	0.003	0.003	0.011	0.009	0.008	-72.7	-66.7	-62.5
Lateral	P	0.026	0.028	0.026	0.015	0.017	0.016	+73.3	+64.7	+62.5
	S	0.018	0.020	0.022	0.022	0.025	0.027	-18.2	-20.0	-18.5
	H	0.002	0.002	0.002	0.002	0.002	0.002	0	0	0
Vertical	P	0.489	0.257	0.288	0.315	0.216	0.278	+55.2	+18.9	+3.6
	S	0.724	0.861	0.811	0.919	1.001	0.901	-21.2	-14.0	-10.0
	H	0.052	0.059	0.063	0.039	0.051	0.071	33.3	+15.7	-11.3
Roll	P	0.910	1.032	0.946	0.474	0.623	0.647	+92.0	+65.7	+46.2
	S	0.635	0.690	0.806	0.782	0.842	0.913	-18.8	-18.1	-11.7
	H	0.058	0.059	0.063	0.065	0.071	0.078	-10.8	-16.9	-19.2
Pitch	P	4.668	4.909	4.999	4.240	4.627	4.763	+10.1	+6.1	+5.0
	S	1.097	1.539	1.872	1.416	1.512	2.170	-22.5	+1.8	-13.7
	H	0.061	0.058	0.065	0.053	0.057	0.079	15.1	+1.75	-17.7
Yaw	P	0.057	0.048	0.035	0.075	0.109	0.104	-24.0	-56.0	-66.4
	S	0.029	0.031	0.037	0.041	0.037	0.046	-29.3	-16.2	-19.6
	H	0.004	0.005	0.005	0.003	0.004	0.006	33.3	-25.0	-16.7

P – Primary ride, S – Secondary ride, H - Harshness

Table 4
Vibration magnitude (weighted) at driver seat over ISO class D road

		Vibration magnitude [m. s.$^{-2}$, r. m. s.]						Difference [%]		
		4-Poster			Straight-line driving			$\frac{\text{4Poster} - \text{Straight} - \text{line driving}}{\text{Straight} - \text{line driving}} \times 100$		
		60 km/h	80 km/h	100 km/h	60 km/h	80 km/h	100 km/h	60 km/h	80 km/h	100 km/h
Fore-aft	P	0.404	0.434	0.431	0.346	0.407	0.416	+16.8	+6.6	+3.6
	S	0.095	0.142	0.161	0.137	0.155	0.192	-30.7	-8.4	-16.2
	H	0.006	0.006	0.007	0.018	0.016	0.016	-66.7	-62.5	-56.3
Lateral	P	0.054	0.069	0.080	0.034	0.041	0.060	+58.8	+68.3	+33.3
	S	0.036	0.040	0.040	0.044	0.053	0.056	-18.2	-24.5	-28.6
	H	0.003	0.003	0.004	0.003	0.003	0.005	0	0	-20.0
Vertical	P	1.081	0.570	0.684	0.729	0.499	0.670	+48.3	+14.2	+2.09
	S	1.493	1.680	1.514	1.830	1.939	1.760	-18.4	-13.4	-14.0
	H	0.107	0.135	0.138	0.083	0.133	0.149	+28.9	+1.5	-7.4
Roll	P	2.050	2.603	3.178	1.047	1.511	2.491	+95.8	+72.3	+27.6
	S	1.383	1.561	1.720	1.597	1.842	2.074	-13.4	-15.3	-17.1
	H	0.101	0.110	0.123	0.112	0.137	0.154	-9.82	-19.7	-20.1
Pitch	P	10.788	11.417	11.269	9.445	10.580	10.826	+14.2	+7.9	+4.1
	S	2.291	3.315	3.821	2.700	3.139	4.463	-15.2	+5.6	-14.4
	H	0.133	0.133	0.168	0.106	0.123	0.161	+25.5	+8.1	+4.4
Yaw	P	0.094	0.097	0.088	0.187	0.229	0.153	-49.7	-57.6	-42.5
	S	0.058	0.071	0.080	0.064	0.064	0.080	-9.4	+10.9	0
	H	0.009	0.012	0.014	0.007	0.010	0.014	+28.6	+20.0	0

P – Primary ride, S – Secondary ride, H - Harshness

4. Discussion

From the results, differences in vibration magnitude are observed in the weighted vibration magnitude of each of the six seat directions between the 4-poster test rig and straight-line driving. There are also differences observed when considering the frequency bands associated with primary ride, secondary ride and harshness.

Whether these differences are significant in influencing the applicability of difference thresholds estimated on the 4-poster test rig to real-world straight-line driving needs further investigation. It is postulated that if the stimulus characteristic is similar between environments, then there should be a high probability that the difference thresholds be applicable across environments.

There could be several explanations for the observed differences between the environments. It is well known that differences exist in tire dynamics between non-rolling and rolling tires. The differences in interaction between the tire on the 2-dimensional flat road and the 3-dimensional road may result in significant discrepancies in tire model behavior. These effects might be mitigated by considering alternatives in determining the vertical motion of the four actuators of the 4-poster test rig. In this paper the vertical displacement of the actuators was defined by extracting a track from the 3-dimensional road profile. This is similar to what would be done practically, where a single line measurement of the vertical height of the road is used. A possible alternative would be to perform a signal reconstruction process to determine the vertical displacement of each of the four actuators of the 4-poster rig that would minimize the error between the vehicle vibrations of interest. Alternatively, the accelerations of the unsprung mass can be measured and used to drive actuators that are directly connected to the wheel hub, by-passing the tire-road contact. A possible disadvantage of the alternatives is that measurements in the straight-line environment would be needed before the 4-poster simulations, or experiments, can be performed.

5. Conclusions and future work

Differences in the vibration characteristics exists between the 4-poster test rig and straight-line driving simulations. Further investigations are needed to determine whether differences are only present in simulations or also practically. Full validation of the vehicle model, tire model and environments (i.e. 4-poster test rig and straight-line driving) needs to be performed. Investigations are also required into the influence of the difference in vibration stimuli to difference thresholds. This would give guidance on how close vibration stimuli between environments need to be in order for difference thresholds to be applicable across them.

We have developed a suspension system that can change stiffness and damping characteristics of each corner allowing for the manipulation of the vehicle's dynamic response, enabling this research to be performed easily on both a 4-poster test rig and real-world driving.

6. Declaration of competing interest

The authors declare that they have no known competing financial interests or personal relationships that could have appeared to influence the work reported in this paper.

7. References

Bosch, H.B., Hamersma, H.A. and Els, P.S., 2016. Parameterization and validation of an all-terrain SUV FTire model, Journal of Terramechanics, Vol 67, pp.11–23.

British Standards Institution, BS 6841 1987. British Standard Guide to measurement and evaluation of human exposure to whole-body mechanical vibration and repeated shock, London.

Gräbe, R.P., Kat C.-J., van Staden, P.J. and Els, P.S., 2020. Difference thresholds for a vehicle on a 4-poster test rig, Applied Ergonomics, vol. 87, 103115, doi: 10.1016/j.apergo.2020.103115.

International Organization for Standarization, ISO 2631-1 1997. Mechanical vibration and shock - evaluation of human exposure to whole-body vibration, Geneva.

International Organization for Standarization, ISO 8608 1995. Mechanical vibration – Road surface profiles – Reporting of measured data, Geneva

Mansfield, N.J., Griffin, M.J. 2000. Difference thresholds for automobile seat vibration, Applied Ergonomics, vol. 31, no. 3, pp. 255-261.

SAE International 2008. Vehicle Dynamics Terminology. SAE Standard J670, Rev. 2008-01-24.

Els, P.S., Theron, N.J., Uys, P.E. and Thoresson, M.J., 2007. The Ride Comfort vs. Handling Compromise for Off-Road Vehicles, Journal of Terramechanics, Vol. 44, pp. 303–317.

6861 | Researches on airplane performance on a grass airfield

Jarosław Pytka [a,*], Paweł Tomiło [b],
Anna Zalewska-Tytłak [a], Mateusz Klepka [a],
Sławomir Tarkowski [a]

[a] *Faculty of Mechanical Engineering, Lublin University of Technology, Lublin, Poland*
[b] *Faculty of Management, Lublin University of Technology, Lublin, Poland*
* *Corresponding author: j.pytka@pollub.pl*

ABSTRACT

The paper presents experimental researches on airfield performance of an aircraft operating from a grass runway. For this research, a methodology of field measurements was developed and measurements of take-off and landing distance were carried out on a grass airfield, taking into account important factors, such as soil cone index, soil moisture, vegetation density and atmospheric factors. The method of measuring the distance of take-off and landing ground roll based on neural network models turned out to be very convenient to use, the advantage is cheap hardware and ease of use.

On the basis of the results of *CI* and *MC* measurements on two test airfields, it was concluded that the ground performance of aircraft obtained at grass airfields with different soil substrate may differ significantly. Therefore, it is necessary to distinguish grass runways in terms of soil type. Flight test results showed that the impact of green vegetation on the runway is more significant than the impact of soil strength, expressed by means of soil *CI*.

The overarching goal is to implement the measurement method and obtained results into the GARFIELD system – an online information system on grass airfield conditions.

Keywords
Grass airfield
Wheel-grass interaction
Takeoff and landing distance
Cone index
Soil moisture
Green vegetation

1. Introduction

Grassy airfields were the first aerodromes for airplanes. Over the years, grassy airfields have been gradually replaced by paved runways, but some are still available for small, private airplanes and gliders, for light transport aircraft in less populated regions, and for use as emergency airstrips. General and business aviation use of airfields with grass runways is becoming more important (Berster et al. 2011, Reynolds-Feighan and Maclay, 2016) and there are even certain types of business or fighter jets that were designed to operate from grass. One drawback of the grass surface, however, is that this kind of runway is very sensitive to weather conditions, which influences trafficability of landing gear wheels and can lead to problems during takeoff and landing.

Literature analysis of the topic this indicates that the problem of flying from grass airfields is not fully explored. Magazines for pilots publish articles of a popular, not scientific nature, containing practical tips on how to determine the take-off distance (Krawcewicz, 2011) or instructional tips for pilots using dirt or grass airfields (Mauch and Kutschke, 2016, CAA of New Zealand, 2011), but they are not content of scientific and technical importance. In particular, there is no universal model of interactions between the tire and the surface of a grass runway. Partly, the issue was the subject of research for the safety of driving a car in areas where the paved road is replaced only by tracks with sod surface (Cenek et al, 2005). Another source of methods and knowledge in the field subject are aircraft design handbooks, in which simplistic models can be found, being a modification of the models used for paved runway, in which values of the rolling resistance coefficient are introduced as for soil. Predicting aircraft airfield performance based on the above models may turn out to be unreliable, especially in the case of extreme runway conditions (e.g. tall grass, high soil moisture).

The GARFIELD system uses a database of general aviation airplanes with technical data required to calculate airfield performance as well as a soil database covering the European Union territory. Finally, the system is linked with an online weather service to have past, current, and future meteorological conditions to simulate the mechanical strength of grass runway on a given airfield that, in turn, is needed to predict wheel performance and calculate takeoff or landing distances. A general purpose of the system is to incorporate grass airfields into a regular air transportation system by determining their suitability for flight operations and by disseminating that information in text or coded format to airmen (Pytka et al, 2019a, Pytka et al 2019b).

Therefore, the authors undertook to develop a model of airplane performance on a grass runway and the purpose of this particular paper is the development of a methodology and experimental measurements of aircraft performance on a grass runway.

The contribution in this paper is as follows:

- elaboration and selection of field measuring methods, including measuring devices;
- performing a flight test campaign, in which airfield performance of an airplane was measured at various conditions;
- analysis of the results, with particular emphasis on the impact of potentially significant factors.

https://doi.org/10.56884/EVDP1196
ISBN 978-1-942112-55-6 (pdf)

2. Methodology

The basic airfield performance of the aircraft is the length of the take-off distance (from starting point to the lift-off) and the landing distance (from touchdown to stop). When flying from a grass runway, these performances depend on the tire-turf interaction strongly. From literature, it's known that this interaction is affected by several soil/turf parameters, such as: soil moisture, vegetation and soil strength (Shoop et al, 2015). Of course, other factors are also of importance: wind, density altitude, piloting technique, but they will not be addressed here as "non-terramechanical".

In the light of the above, methods for measuring the following quantities characterizing the airfield performance of the aircraft have been developed:

- ground roll of take-off and landing;
- soil cone index;
- soil moisture;
- green vegetation.

An important element of the research program was the choice of the site and time of the measurements. The idea was to obtain results under specific soil conditions (moisture content, strength), which directly resulted from weather factors.

2.1 Take-off and landing ground roll distance

The essence of the measurement method is the identification of the moment of lift-off (during take-off) and touchdown (during landing) on the basis of continuous measurement of the aircraft's vertical acceleration. It is obvious that at the moment of take-off, when the aircraft loses contact with the runway, the amplitude of the vertical acceleration changes (decreases). On the other hand, at the moment of touchdown, there is a short-term acceleration peak, as a consequence of landing gear hitting the ground. Identification of these points in conjunction with the method of reckoning allowed for determining the distance of the take-off and landing of the aircraft.

A significant improvement was the use of artificial neural networks, which enabled the automatic recognition of lift-off and touchdown points. An important assumption was the ability to measure the performance of the aircraft without the participation of third parties (ground observers), using an on-board device in the aircraft. A measuring device IMUMETER was developed, which consists of the following components:

- single - board computer;
- triaxial accelerometer
- triaxial magnetometer
- GPS sensor;
- wireless communication interface;
- power supply.

Figure 1 shows the schematic of the IMUMETER device. The sensors included in the device had the following parameters:

- GPS RTK sensor: 25 Hz navigation rate with declared horizontal position accuracy of 2.5 m and 0.01 m (with and without RTK respectively);
- Accelerometers: configurable measuring range ± 2 / ± 4 / ± 8 / ± 16 g, refresh rate up to 952 Hz;
- Magnetometers: configurable measuring range of ± 4 / ± 8 / ± 12 / ± 16 Gauss, refresh rate up to 80 Hz;

A handheld device, depicted in Fig. 1, was with a test engineer onboard and was used to monitor and control of measurements as well as for viewing the results. A RTK base station was installed in a building, approx. 5000m apart from the test airfield. The base station consisted of the following devices:

- antenna which is capable of receiving the L1/L2 bands for GPS, GLONASS, Galileo, and BeiDou constellations;
- GNSS reciever (GPS, GLONASS,Galileon, BeiDou);
- server (PC).

Communication between the IMUMETER device installed on the test plane and the base station was made via the Internet using the NTRIP server.

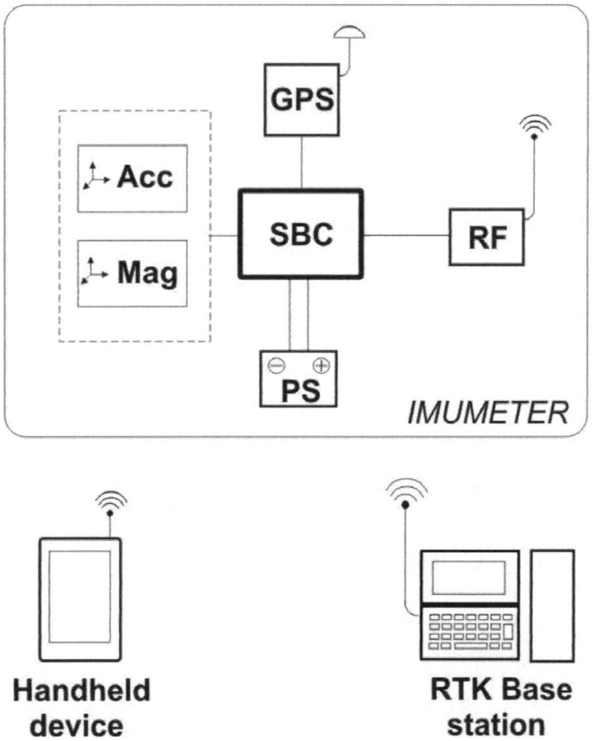

Fig. 1. Schematic of the IMUMETER device for takeoff and landing distance measurements. SBC = single board computer, RF = radio frequency device, PS = power supply, Acc=accelerometer, Mag=magnetometer

More details of the IMUMETER device can be found in the cited works (Tomiło, 2023, Pytka, 2023). The photograph in Fig. 2 shows the measuring device, which is the hardware platform of the method, installed in the test aircraft.

The operation of the IMUMETER device consists in the fact that during the flight the sensors continuously measure selected quantities, which are analyzed and classified in an artificial neural network model. The device starts counting the distance when data specific to the take-off roll appear. The counting of the distance continues until the moment of lift-off, which is recognized based on measured data. Similarly, when landing, the appearance of a set of touchdown data triggers the command to start counting the distance by the GPS sensor until the aircraft comes to a stop.

An important component of the IMUMETER measuring device is the artificial neural network model (NNM) mentioned above. It was developed on the basis of multiple flight tests, taking into account a variety of factors that affected the accuracy of the classification of measurement data sets. In detail, the neural network model is described in reference (Pytka et al, 2022).

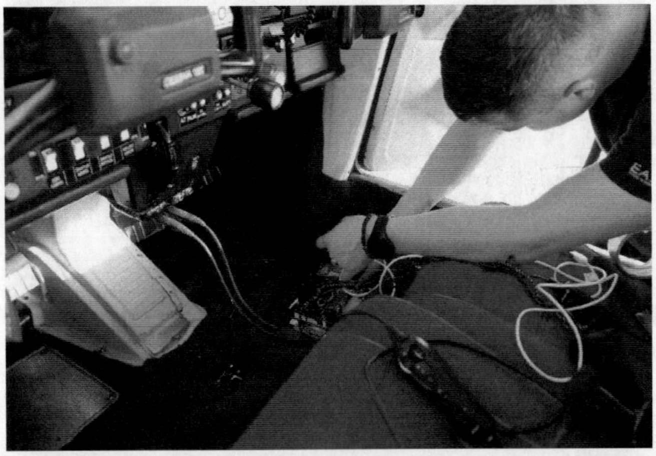

Fig. 2. Installation of the IMUMETER device in a test airplane on a floor panel of the passenger seat (upper photo) and the device ready for measurement (lower photo)

2.2 Soil cone index

To measure the soil cone index, a commercially available, typical measuring device was used - a cone penetrometer with an electronic measurement and recording system. The measurement was carried out in accordance with the user's manual of the device, typically 10 - 12 repetitions within the part of the runway, which was actually in use. Since the runway was covered with green vegetation (grass), the measurement point was prepared by removing the green parts of the grass from where the measurement cone was placed. Vertical distribution of the soil *CI* was measured down to a depth of approx. 25 cm.

2.3 Soil moisture

The measurement of soil moisture was carried out using a hand-held device operating on the principle of time-domain reflectometry (TDR). The method uses the phenomenon of changing the permittivity of the soil depending on soil moisture and allows for a quick and non-invasive measurement in situ (Wilczek et al. 2016, Pytka et al, 2019)

The measuring device used has a two-electrode probe, which should be placed in the soil. The measurement result is available almost immediately. Typically, 10-12 repetitions were performed on the part of the runway in use, similar to the CI measurement.

2.4 Green vegetation

When developing a method for measuring green vegetation (grass), various solutions were considered. One of them was a method based on opto-electronic measurement of the height of grass stems. The measuring device, described in detail in Pytka et al, 2019 consists of an emitter and a light detector, spaced approx. 0.5m apart, positioned so that there are grass stems in between. Both the emitter and the detector have pairs of optoelectronic elements arranged vertically along the scale every 10mm. The presence of grass stalks at a given height from the soil surface breaks the optoelectronic circuit so that the height of the grass can be determined. However, the method does not allow to determine the density of grass growth, and this parameter is important in terms of tire-turf mechanics. Therefore, a different method was finally used, described by Shoop et al 2015, which consists in measuring the mass of green (above-ground) parts of grass taken from a unit area. Simply cut the green parts of grass from a 1,0x1,0m square, then weighed. The index obtained in this way was called unit mass vegetation and marked m_G.

2.5 Test airfields with grass runway

The research was carried out at three grass airfields, located in south-east Poland, Lubelszczyzna region. They are: Radawiec airfield located close to Lublin city, Świdnik airfield, adjacent to the LUZ Lublin Airport located in Świdnik and Depułtycze Królewskie airfield, near the city of Chełm. Radawiec airfield is based on a loess- sandy soil. Vegetation is a mixture of various grasses, fescue, ryegrass, panacea, and burrow together with low meadow vegetation. The runway of the Radawiec airport is 1200m long, heading 290 and is rolled in order to level out unevenness, that arises as a result of inhabitation by small rodents. The grass is mowed 3 times a year. On the runway, the grass is partially degenerated, wheel trakcs along the runway are significant. There is a separate runway for model airplanes, on which the grass is approx. 3 cm high, and the unevenness is much smaller than on the main runway. The soil substrate of the Świdnik airfield is very similar to that in Radawiec. Vegetation in Świdnik is a fairly dense mixture of grasses and meadow plants. The grass on the runway is rather rarely growing, partially degenerate.

Depułtycze Królewskie airport is located approx. 70km apart from Lublin and was built on marly soil. This airfield has two runways, namely a bitumen and a grass one. The grass runway, overgrown with grass of a homogeneous composition, with the predominant species of field grass, resistant to creasing, is fertilized and renovated once a year.

2.6 Test program and procedures

Measurements were carried out during the summer (June - September 2022). The test airfield was located on a sandy loam soil. It was assumed that the measurements should be carried out in conditions ensuring the variability of the tested parameter (e.g. soil cone index) and the stability of other parameters (green vegetation and soil moisture). This turned out to be problematic, which is why the collected research material is quite modest, despite about 300 test flights over 30 flying hours. For each case, a minimum of 5 repetitions (test flights) have been performed.

A typical test flight consisted of a four-turn pattern. The pilot tried to use the same piloting technique, especially during take-off and landing. The flights took place mainly in the morning to eliminate the effect of the wind.

Various aircraft were used in the tests, mainly to check how the design of the airframe, landing gear, etc. affects the functioning of the NNM model. However, results obtained with one type of aircraft have been selected for this study. It was the PZL 110 Koliber 150, a four-seat low-wing aircraft with nose wheel landing gear. The aircraft is allowed to operate from grass airfields and has a landing gear with tires measuring 350 x 120 mm (diameter x width). Figure 3 shows the Koliber aircraft during a test flight.

Fig. 3. The PZL 110 Koliber 150 aircraft during a test flight, just prior to touchdown

3. Results

The results of this research include grass runway characterization by means of cone index and soil moisture, effect of grass upon tractive parameters of tire - runway system and effect of grass runway on airfield performance of the test airplane.

3.1 Grass runway characterization

Characterization of the grass airfield surface was carried out on the basis of the soil *CI* index and soil moisture. Both parameters were measured as part of full-season studies, i.e. throughout one year, with a 1-week interval. Figure 4 presents the values of the soil cone index at 10cm depth, with averaged results for individual months of the year for clarity.

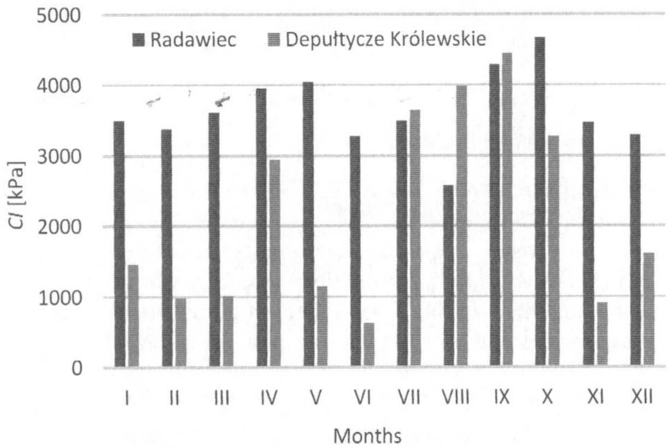

Fig. 4. Values of soil cone index CI at a depth of 10 cm for grass runway on airports in Radawiec and Depułtycze Królewskie

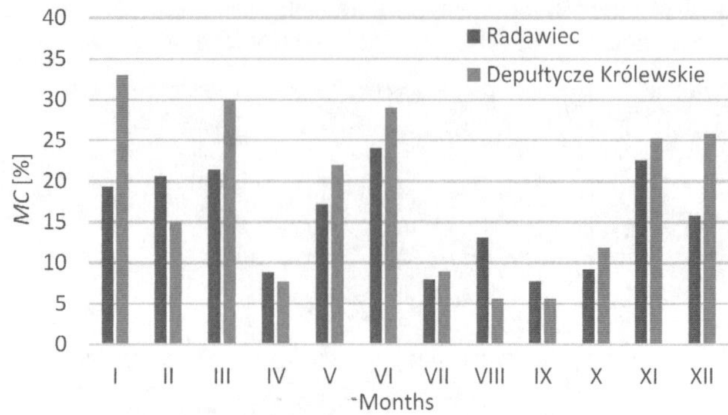

Fig. 5. Values of soil moisture content MC for grass runway on airports in Radawiec and Depułtycze Królewskie

It is noticeable that the *CI* values for the runway in Radawiec are higher for most months of the year than in Depułtycze Królewskie. Especially in the winter and late autumn months, they are even more than three times higher. This is due to the high content of chalk (marl) in the soil of the airport in Depułtycze Królewskie, and thus, significant sensitivity to humidity.

Figure 5 contains corresponding soil moisture content readings for the two airfields.

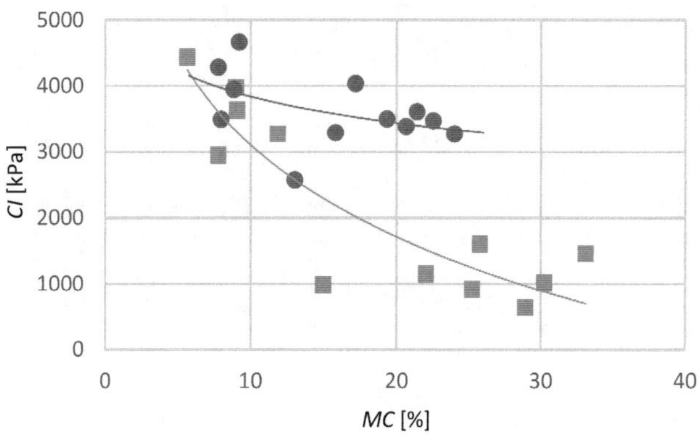

Fig. 6. Relationship between the soil cone index and soil moisture content for both airports: Radawiec - blue line, Depułtyczne Królewskie - orange line

Figure 6 shows the correlation of the soil cone index *CI* with soil moisture for both airports studied. There is quite a clear difference: the *CI* values for the airport in Depułtycze Królewskie are almost 3 times lower, while the humidity range for this airport is larger and covers values between approx. 6% and almost 35%. Soil moisture at the Radawiec airport ranges from 8 to 24%.

3.2 Effect of grass upon tractive parameters

One of the main traction parameters is the rolling resistance coefficient, k_{RR}. The value of k_{RR} determines i. a. on the dynamics of acceleration and braking (rolling down) of the aircraft. Figure 7 shows the values of the k_{RR} coefficient for the wheel of the test aircraft, obtained on several variants of the surface of the Radawiec airfield. In this case, the green vegetation was not quantified, and only the individual variants of the grass surface were presented descriptively. The values are in the range of 0.025 - 0.1, so the difference between the largest and smallest value does not exceed one order of magnitude.

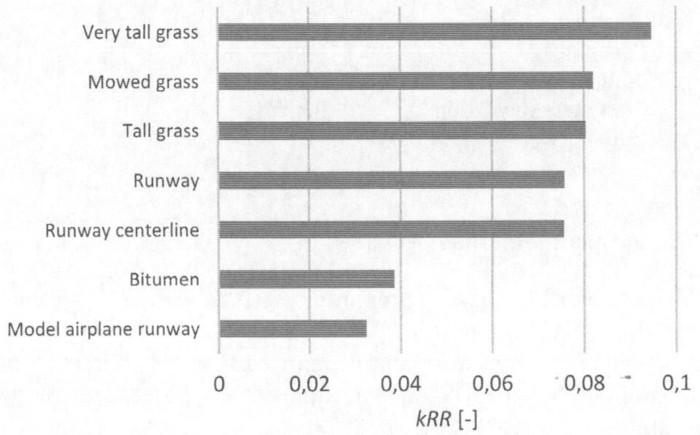

Fig. 7. Values of the rolling resistance coefficient k_{RR} of the test aircraft wheel for various variants of the grass surface of the Radawiec airfield

The dependence of the rolling resistance coefficient of the test aircraft wheel on the vegetation is shown graphically in Fig. 8. The trend line describing the characteristics is of the logarithmic type, which indicates that the increase in the value of the coefficient for the growing unit mass of vegetation is of the regressive type. Results for Radawiec airport.

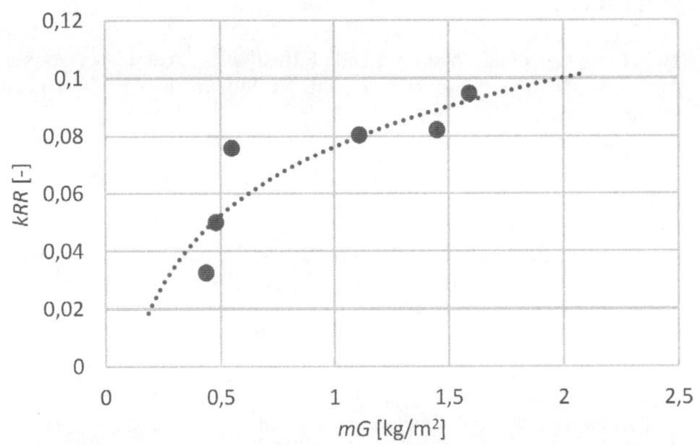

Fig. 8. Dependence of the rolling resistance coefficient k_{RR} of the test aircraft wheel on the unit vegetation mass m_G for the Radawiec airport

On the basis of the above presented results, it is concluded that the ground performance of aircraft obtained at these airfields may differ significantly. In particular, in high humidity conditions, due to lower *CI* values, the aircraft's take-off performance at the airport in Depułtycze Królewskie will be worse than at Radawiec airport. This applies to the autumn-winter period, when meteorological conditions cause high soil moisture. Landing performance, i.e. the length of the landing roll, will be improved, but the reduced *CI* values imply the risk of the wheels sinking into the soft ground of the grass runway, which may result in a nose-down, especially in the case of an aircraft with a tail wheel. The general conclusion is that different grass airfields cannot be treated in the same way in terms of aircraft ground performance, as is the case in handbooks, as well as in numerous guides, manuals or technical articles (Gudmundsson, 2014). There, the grass runway is most often treated as one separate type of surface. It is therefore necessary to distinguish grass runways in terms of soil type.

3.3 *Effect of grass runway on airfield performance*

The most important result from the practical point of view is the obtained dependence of the take-off and landing ground roll distances on the soil *CI* index and the unit mass vegetation m_G. Figures 9 and 10 show the relationships obtained from the results of flight tests carried out at the airfield in Radawiec. The *CI* values on the horizontal axis of Figure 9 are on a logarithmic scale. A strong dependence of the length of the roll-out of the aircraft on *CI* can be noticed, with a significant increase in the distance to stop above *CI*=3000kPa. Then the mechanical strength of the runway surface is high, the deformation is small, which results in low rolling resistance and a long landing ground roll. An important note here: the maximum *CI* value of 10000 kPa was assumed for the bitumen runway as a reference point. The opposite situation occurs during take-off, because low rolling resistance translates into shorter take-off ground roll, but the nature of this dependence is significantly different, the decrease in length is much less intense than the increase in landing distance. It should be noted that the presented results were obtained in measurements carried out on a runway covered with short grass, the value of the m_G index was about 0.5 kg/m2.

Figure 10 presents the relationships of take-off and landing ground roll distances with the unit mass vegetation. As in the case of the soil *CI* index, also in the case of m_G, two different relationships are visible, namely the increase in m_G results in the extension of the take-off ground roll distance and the shortening of the landing ground roll. The intensity of changes in the distances is similar in both cases. This is logically understandable: the more green vegetation on the runway surface, the longer the take-off ground roll and the shorter landing rollout due to the increased rolling resistance. The greatest changes (influence of m_G on aircraft performance) are noticeable for the range of low values of unit mass vegetation, more or less up to $m_G = 1$ kg/m^2.

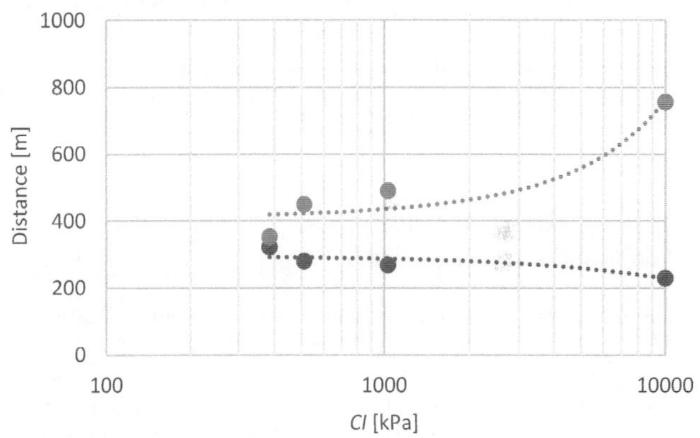

Fig. 9. Impact of *CI* on the ground roll distances of take-off and landing. Orange line - landing (length of the landing roll), blue line - take-off (length of the take-off roll)

At this point, it is worth analyzing the range of measured distance values, taking into account the range of changes of the *CI*. It should be emphasized that the flight tests took place in the summer-autumn season, when the range of *CI* values was quite narrow (385m - 1030kPa). Well, the length of the take-off ground roll of the aircraft varied from 269m to 323 m, which is an increase of 20%. The length of the landing ground roll decreased from 491m to 353 m, which is a change of 28%. These results are close to the recommendations given in pilot's manuals or flight manuals regarding the difference in performance between a paved and flexible runway.

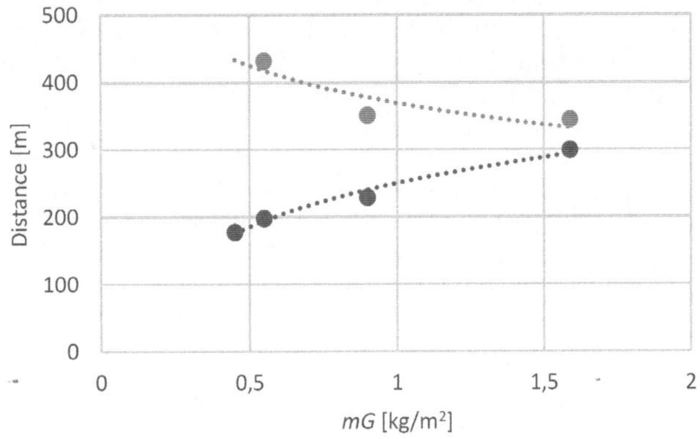

Fig. 10. Influence of m_G on the length of take-off and landing. Orange line - take-off, blue line – landing

As for the influence of vegetation, the range of take-off ground roll extends from 177m for the lowest value of m_G to 299m for the maximum value of unit mass vegetation. The percentage increase in length was therefore 69%, which is a very big change, much bigger than typical practical recommendations. This is very important from the point of view of safety, because it is during the take-off that most accidents related to short runway or, strictly speaking, to the extension of the take-off distance due to vegetation on the runway occur (Garrison, 2013, Landsberg, 2016, Landsberg, 2017). Analogous values for the landing ground roll distance: 432m for the lowest m_G and 344m for the highest, the relative change was 18%.

It should be expected that the range of changes in the length of both take-off and landing ground roll of the aircraft in the months outside the summer-autumn season may be greater.

Finally, it can be concluded that the impact of green vegetation on the runway is more significant than the impact of soil strength, expressed by means of soil *CI*.

4. Conclusions and future work

The paper describes experimental studies of aircraft ground performance at an airport with a grass runway. Research methods and devices were presented. The cone penetrometer and TDR methods allowed characterization of the runway surface and may be a tool for calibrating the GARFIELD information system. The method of measuring the distance of take-off and landing ground roll turned out to be very convenient to use, the advantage is cheap hardware and ease of use.

On the basis of the results of *CI* and *MC* measurements on two test airfields, it is concluded that the ground performance of aircraft obtained at grass airfields with different soil substrate may differ significantly. Therefore, it is necessary to distinguish grass runways in terms of soil type.

Flight test results showed that the impact of green vegetation on the runway is more significant than the impact of soil strength, expressed by means of soil *CI*.

Future work would include the development of a tire - grass interaction model, which will take into account both unit mass vegetation as a global parameter of grass runway but also selected detailed parameters of the surface covered with green vegetation, which may affect the performance of the tire as well as the entire airplane.

5. Nomenclature

CI	Cone Index	[kPa]
k_{RR}	Rolling resistance coefficient	[-]
m_G	Unit mass vegetation	[kg/m^2]
MC	Soil moisture content	[%]

6. Acknowledgements

This work was prepared within the project PM/SP/0065/2021/1 entitled "Method for airplane takeoff and landing distance measurement", financed by the Ministry of Education and Science (Poland) as a part of the Polish Metrology Programme.

7. Declaration of competing interest

The authors declare that they have no known competing financial interests or personal relationships that could have appeared to influence the work reported in this paper.

8. References

Berster, P., Gelhausen, M.C., Wilken, D., 2011. Business aviation in Germany: An empirical and model-based analysis. Journal of Air Transport Management 17, 354 – 359.

Cenek, P.D., Jamieson, N.J., McLarin, M.W., 2005. Frictional characteristics of roadside grass types. Proc. International Surface Friction Conference, 2005, Christchurch, New Zealand

Garrison, P., 2013. Sunday Drive. Flying Magazine, 8, 40 – 42.

Gudmundsson, S., 2014. General Aviation Aircraft Design. Applied Methods and Procedures. Butterworth-Heinemann/Elsevier, 1st edition, Amstedram, The Netherlands

Krawcewicz, K., 2011. Czy samolot wyjdzie? (Will an airplane takeoff?). PLAR Przegląd Lotniczy Aviation Revue, 8/2011 (204), pp. 40-43.

Landsberg, B., 2016. Soft field, soft thinking. Who is responsible for a takeoff gone wrong? AOPA Pilot, 9 (2016), pp. 20

Landsberg, B., 2017. Margins to live by. Beautiful places, challenging runways. AOPA Pilot, 8 , pp. 80 - 84

Mauch, H and Kutschke, E., 2016. Tiefe Boden – hohe Kunst? Fliegermagazin, 1, 64-67.

Pytka J., Budzyński P., Józwik J., Łyszczyk T., Gnapowski E., Laskowski J., 2019. GARFIELD information system – old problems and new perspectives. In Proceedings of the 6th International Workshop on Metrology for Aerospace (MetroAeroSpace), Torino, Italy.

Pytka J., Budzyński P., Józwik J., Łyszczyk T., Gnapowski E., Laskowski J., 2019. GRASSTAM – an idea of a notice on grassy runway condition. In Proceedings of the 6th International Workshop on Metrology for Aerospace (MetroAeroSpace), Torino, Italy.

Pytka, J, Budzyński, P., Kamiński, M., Łyszczyk, T., Józwik, J., 2019. Application of the TDR Moisture Sensor for Terramechanical Research. Sensors, Vol. 19, No. 9, p. 2116.

Pytka J., Budzyński P., Tomiło P., Michałowska J., Błążejczak D., Gnapowski E., Pytka J.D., Gierczak K., 2022. Measurement of aircraft ground roll distance during takeoff and landing on a grass runway, Measurement 195, 111130.

Pytka J., 2023. Airplane Performance on Grass Airfields. CRC Press, Taylor&Francis Group, Boca Raton, USA.

Reynolds-Feighan A., Maclay P., 2016. Accessibility and attractiveness of European airports: A simple small community perspective, Journal of Air Transport Management 12, 313–323.

Shoop, S.A., Coutermarsh B., Cary T., Howard H., 2015. Quantifying vegetation biomass impacts on vehicle mobility. Journal of Terramechanics, 61 (2015) 63-76.

Tomiło P., 2023. Metoda pomiaru osiągów naziemnych samolotu z wykorzystaniem algorytmów atencji oraz ekscytacji w modelu q-kształtnej sztucznej sieci neuronowej (A method of measuring aircraft ground performance using attention and excitement algorithms in a q-shaped artificial neural network model). Ph D Thesis, Faculty of Mechanical Engineering, Lublin University of Technology, Lublin, Poland.

Wilczek A, Szypłowska A, Kafarski M, Skierucha W. 216. A Time-Domain Reflectometry Method with Variable Needle Pulse Width for Measuring the Dielectric Properties of Materials. Sensors (Basel). 16(2):191. doi: 10.3390/s16020191.

Civil Aviation Authority (CAA) of New Zealand, 2011. Takeoff And Landing, Good Aviation Practices Series, , Wellington.

7198 | Analysis of aircraft dynamics while overcoming terrain obstacles

Jerzy Józwik [a,*], Jarosław Pytka [b], Andrzej Łukaszewicz [c], Sylwester Korga[d], Kamil Żyła[d]

[a] Department of Production Engineering, Faculty of Mechanical Engineering, Lublin University of Technology, Lublin, Poland, j.jozwik@pollub.pl

[b] Department of Motor Vehicles, Faculty of Mechanical Engineering, Lublin University of Technology, Lublin, Poland, j.pytka@pollub.pl

[c] Institute of Mechanical Engineering, Faculty of Mechanical Engineering, Bialystok University of Technology, Bialystok, Poland, a.lukaszewicz@pb.edu.pl

[d] Department of Computer Science, Faculty of Electrical Engineering and Computer Science, Lublin University of Technology, Lublin, Poland, s.korga@pollub.pl, k.zyla@pollub.pl

* Corresponding author: j.jozwik@pollub.pl

ABSTRACT

Flying from grass airfields is associated with the risk of surface roughness and their impact on safety and comfort during take-off and landing. This paper presents the measurement method and experimental results of a study investigating the effect of terrain obstacle geometry on the aircraft landing gear dynamics. Three different models of terrain obstacle geometry were tested: rectangular, triangular and semi-circular. Experiments were conducted using the GOM PONTOS measuring system, which made it possible to perform a dynamic analysis of the landing gear of a Cessna 152 airplane passing over the obstacles. The study investigated dynamic parameters of three elements of the aircraft: landing gear wheel and leg as well as the fuselage, for which accelerations and velocities were measured when the aircraft was passing over the obstacles. The paper includes details of the measurement method as well as the results and their analysis.

Keywords
Grass airfield
Obstacle geometry
Tire measurement model
Tire deflection
Dynamic test
Airplane

1. Introduction

In the provisions of the Aviation Law, objects are considered aviation obstacles occurring in the strip of the runway, regardless of their height or simply those that pose a potential threat to the movement of aircraft. In case of airfields with a grass runway, one speaks of terrain obstacles (off-road unevenness), tall grasses, bushes, molehills, stones, sticks and others.

The risk associated with the presence of terrain obstacles on the surface of the runway results from the possibility of generating, forces acting transversely to the direction of the aircraft's motion during take-off, roll-out or taxiing. The result is vibrations of the aircraft structure as well as the veer off, which can lead to significant damage or even a catastrophe. Significant horizontal forces acting on the wheels of the landing gear may cause the landing gear leg to break or bend, which in extreme cases leads to a collision of the fuselage with the ground and to the plane's fire.

An example of the effect of unevenness is shown in the photo, Fig. 1. A small 2-seater plane was damaged as a result of hitting a terrain obstacle - unevenness of the runway surface. The front landing gear leg was damaged.

Fig. 1. A small airplane with damaged nose landing gear as a result of passing over unevenness on a grass runway

But it's not just small planes flying from grass airfields that are at risk from uneven natural surfaces. An example related to the risk of driving over bumps, especially the grassy lane, is the consequence of a runway excursion. On November 3rd, 2011, a Singapore Airlines Boeing B777-300 veered right off runway 08R while landing at Munich Airport. The aircraft completed its landing run on the grass surface. Thanks to the very well-maintained grass surface adjacent to the runway at Munich Airport, there were no damages or collisions. Figure 2 shows an aerial view of the final position of the B777-300 airplane. Figure 3 shows an aerial view of a Northrop B-2 bomber after a runway excursion, ending its landing run on grass surface adjacent to the bitumen runway.

https://doi.org/10.56884/AUDE9438
ISBN 978-1-942112-55-6 (pdf)

Fig. 2. Boeing B777-300 after an unintentional veer off maneuver on Munich Airport (courtesy Mr. Simon Hradecky, The Aviation Herald, http://avherald.com/h?article=445873f3)

Fig. 3. Northrop B2 Spirit bomber on the grass surface after a runway excursion

The event at the Maban Airstrip, South Sudan, on August 28th, 2017, took a much different course. A Coco Travel Antonov AN-26 on behalf of UNHCR, performing a freight flight from Juba to Maban Airstrip, with 3 crew, landed on Maban Airstrip in rain. The aircraft skidded and overran the runway, went across and came to a stop past the runway end.

Fig. 4. The wreckage of the Antonov An 26 after an accident of runway overrun on Maban Airstrip, Maban, South Sudan (courtesy Mr. Simon Hradecky, The Aviation Herald, http://avherald.com/h?article=4ad9d3bd)

The aircraft sustained substantial damage, fuel spilled and a fire started. The crew managed to escape with minor injuries but the aircraft and cargo burned down (Fig. 3). In this case, the aircraft's landing gear was damaged as a result of driving over an unpaved natural surface (grass/soil) and hitting the unevenness.

From the terramechanics point of view, various surface irregularities are distinguished. These are micro and macro irregularities, measuring millimeters and centimeters as well as bumps about the size of a car wheel and moguls, the dimensions of which are in the order of meters (Pytka, 2023). Following a brief survey of methods for a model description of inequalities.

In the case of repeated unevenness of the traction surface, one speaks of the terrain profile. One statistic traditionally used method has been the root mean square (RMS) of the detrended terrain-profile which is best suitable for terrains of stationary and Gaussian character. This method describes the terrain profile with only one number. Another method to characterize terrain profiles is the power spectrum density (PSD) method, which generalizes the frequency-domain shape to a sloped line which approximates the PSD. Terrain profile is obtained in the form of a frequency-domain plot with labeled slopes and breakpoints. The PSD method assumes that the terrain profile is statistically stationary and Gaussian (Brudnak et al, 2007).

The impact of unevenness on the interaction of the running element with the surface can be determined by analytical methods, among them the Multi-Body Simulation (MBS) method seems to be the most versatile and is widely used in automotive industry. Experimental methods include measurements of dynamic vehicle response, usually performed with accelerometers, inclinometers and gyroscopes installed in a test vehicle (Els et al, 2007) as well as wheel force measurements (Pytka et al, 2019a, Pytka et al 2019b).

The impact of unevenness on the driving dynamics of an off-road vehicle has been quite well described in the subject literature. On the other hand, the lack of methods and results was noticed in relation to an airplane flying from a grass airfield, where unevenness may arise as a result of the life activities of natural habitants (small rodents, wild boars) and vegetation (e.g. grass tufts). In light of the above, the main objective of this study was to develop and apply a method for analyzing the dynamic response of the aircraft landing gear to excitations arising when overcoming an obstacle in the form of surface unevenness. In particular, the focus was on the unevenness characteristic for grass airfield surfaces.

The contribution of this study was the compilation of measurement and research equipment, the preparation of model surface irregularities, the measurement of parameters describing the dynamic response of the landing gear and the preliminary analysis of these results.

2. Methods

Three different models of terrain obstacle geometry were tested: rectangular, triangular and semi-circular, each model having the same dimensions of 300x150x50mm. Experiments were conducted using the GOM PONTOS measuring system, which made it possible to perform a dynamic analysis of the suspension system of a Cessna 152 during passing the above model obstacles. All experiments were conducted for three takeoff runs of 1200, 1800 and 2700 mm for each terrain obstacle model, respectively.

For the test runs defined in this way, the approach speeds were determined, which amounted to 1,1 m/s; 1,5m/s and 1,7m/s. These speeds were determined by the GOM PONTOS coordinate measuring method (CMM) system based on the measurement of the speed of movement of the point stuck in the wheel axis, while moving in the direction parallel to the surface on which the aircraft was moving. In order to determine the estimates in the mean form, the measurements were repeated 5 times for each variant of the test.

The study investigated the following parameters: velocity v_i, acceleration a_i, where i means the direction of measurement - axes X, Y; where X – means the axis parallel to the motion plane, Y – means the axis perpendicular to the direction of motion. The analysis was performed for three crucial structural elements of the landing gear: strut, fuselage and wheel center, for which accelerations and velocities were measured when the aircraft was passing over the obstacles.

Fig. 5. Experimental setup, which includes the Cessna 152 test airplane (to the left) and GOM PONTOS measuring system (to the right)

Due to the nature of the movement along the X axis, dynamic parameters corresponding to the XY plane in the direction of the X and Y axes were analyzed. The Z axis was omitted due to small changes in this direction and the theoretical lack of movement in this axis. All measurement results were analyzed in relation to the coordinate system oriented as in Fig. 5.

2.1 Models of terrain obstacle geometry

During the tests, specially designed and fabricated regular geometric models with fixed dimensions, presented in Fig. 6, were used. The geometry of the models was taken from the observation of terrain obstacles that may occur on a typical grass runway. dimension were conventionally adopted for an average terrain obstacle. The inspiration for the obstacles developed in this way were typical surface irregularities with a triangular, circular and rectangular cross-sections. Due to the need to carry out measurements in stable environmental conditions, the tests were carried out in the laboratory of the Center of Engineering Studies of The University College of Applied Sciences in Chełm, using models specially developed for this purpose.

Taking into account the requirements of the measuring system, measuring points - markers were glued on each model. Each single marker designates a point for which the dynamic parameters of the movement (velocity and acceleration) were determined. The arrangement of the measurement points was dictated by the requirements of the analysis of three characteristic objects of the aircraft structure: strut, fuselage and wheel.

2.2 Phases of overcoming terrain obstacles

Figure 7 shows three characteristic phases of overrunning terrain obstacles. These phases include: phase I - obstacle approach phase, during which the aircraft approaches the obstacle. Phase I is used to determine the movement parameters - mainly the speed at which the aircraft will overcome the terrain obstacle.

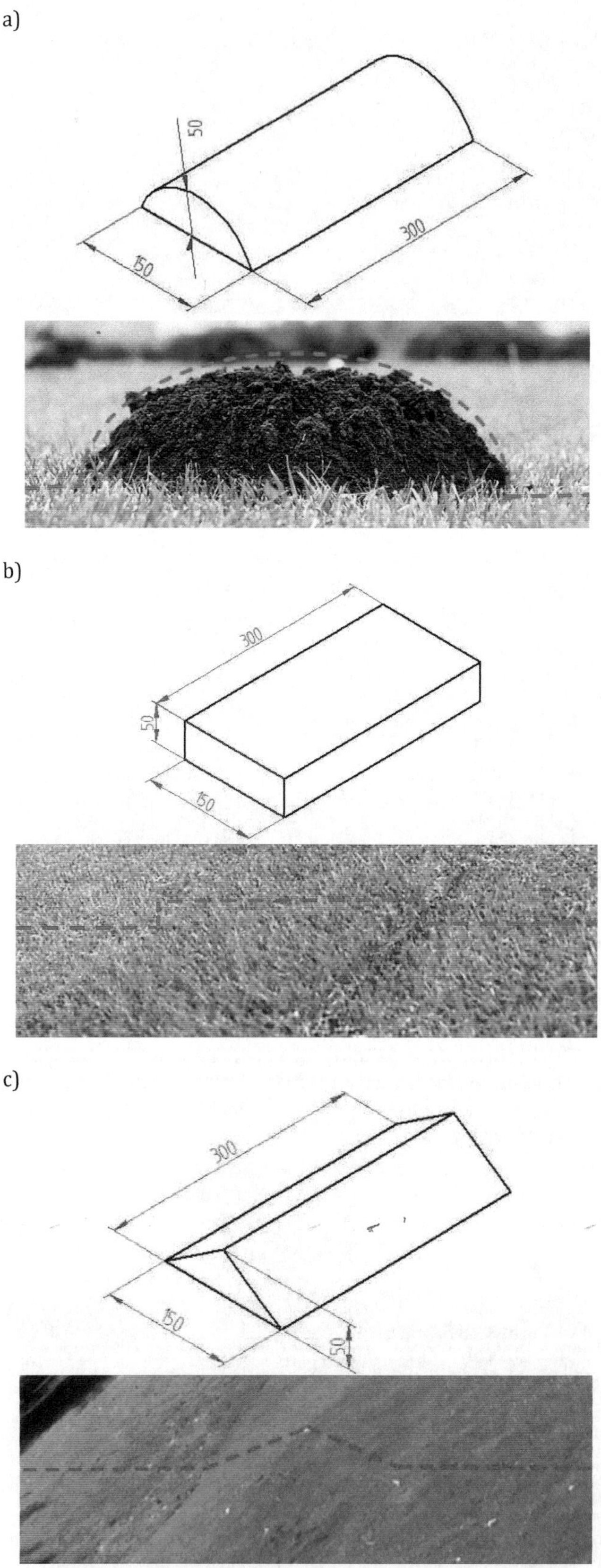

Fig. 6. Models of terrain obstacle geometry: a) circular (molehill), b) rectangular (step, threshold), c) triangular

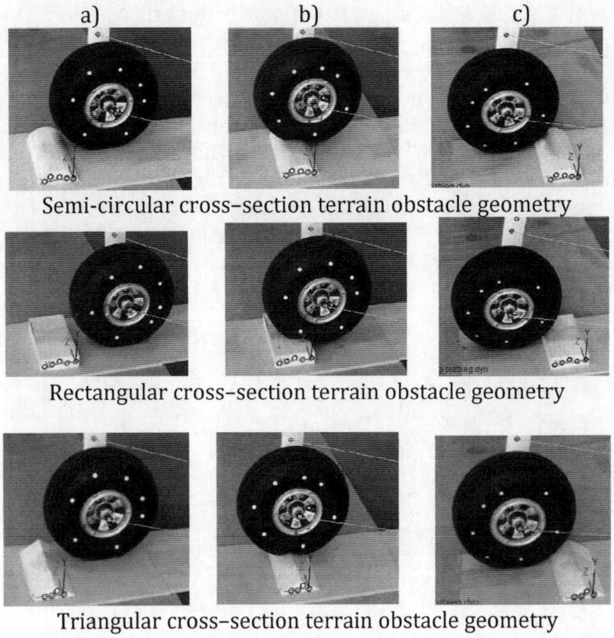

Fig. 7. Phases of overcoming terrain obstacles: a) obstacle approach phase, b) obstacle climbing phase, c) the phase of leaving the obstacle

Phase II - the obstacle climbing phase, during which the aircraft overruns and overcomes a terrain obstacle, provides valuable information related to the impact of the obstacle on the behavior of the test object and the analysis of changes in dynamic parameters analyzed in the work of the structure components (strut, fuselage and wheel). During phase II, all the dynamic characteristics of the movement change, both quantitatively and qualitatively: speed decreases, speed increases, accelerations and decelerations.

The third phase III was defined as the phase of leaving the obstacle. It allows the observation of the inertial behavior of the aircraft. All the mentioned phases of movement were analyzed for all the models of terrain obstacles adopted in the experiment (circular, rectangular and triangular cross - section terrain obstacle geometry).

The obtained results were compared with each other and presented graphically. Due to the large number of measurement points, 3 (out of 5) characteristic points related to selected elements of the aircraft structure were selected for the analysis: a strut, a fuselage and a wheel.

3. Results

3.1 Results presentation

According to the conditions and measurement parameters planned and defined in the section 2. Methods of this paper, appropriate dynamic characteristics of the system response were obtained when overcoming terrain obstacles developed for this purpose. Figure 8 presents the quantitative values of the examined dynamic parameters of motion (acceleration a_i and velocity v_i) for individual structural elements of the aircraft and selected measurement points related to them (one point on the fuselage, three characteristic points per strut and one measurement point located on the axis of the wheel in its central position (wheel center).

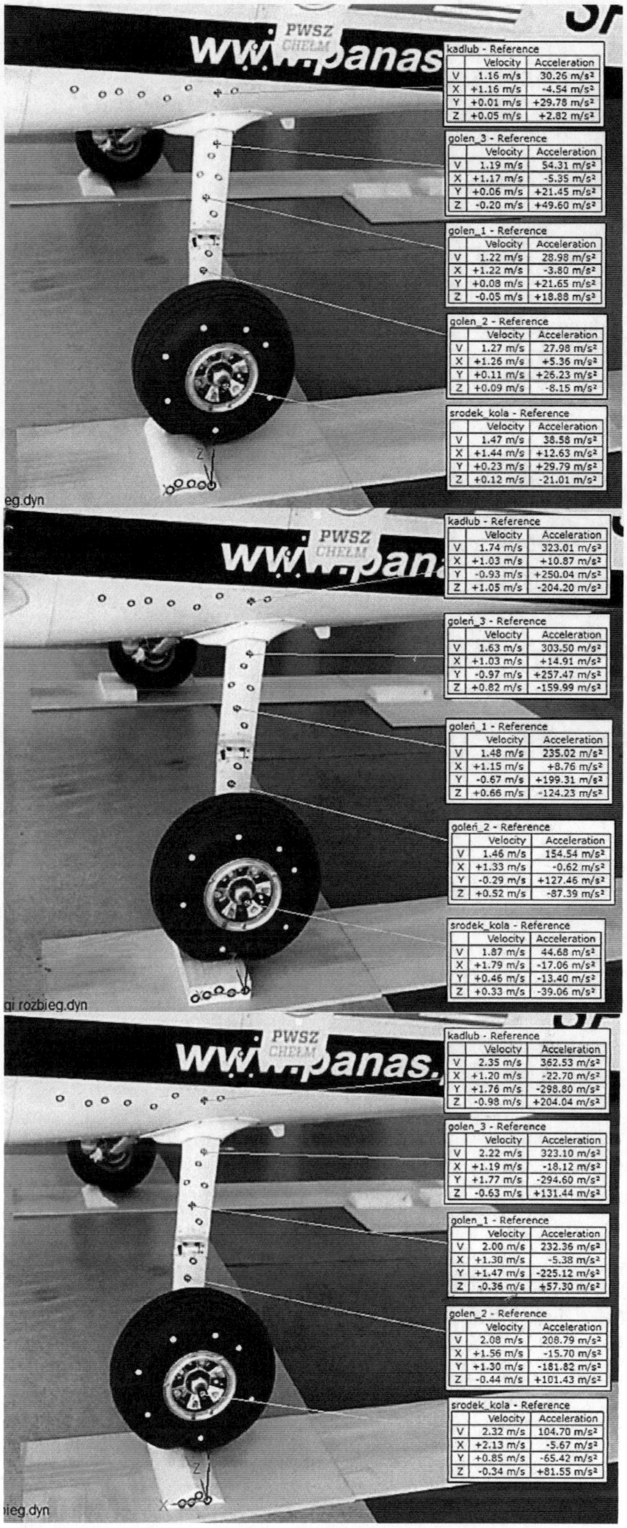

Fig. 8. Values of the velocity and acceleration of the center of the wheel and strut and fuselage, during successive phases of overcoming a terrain obstacle: a) obstacle approach phase, b) obstacle climbing phase, c) the phase of leaving the obstacle

The results presented in the figures represent exemplary, instantaneous values at a given time moment assigned to phase II of the analyzed aircraft movement, while overcoming terrain obstacles: triangular, rectangular and semi-circular. A detailed analysis of all the results of the obtained measurements indicates high values of the delay of the motion parameters when overcoming an obstacle with a rectangular (Fig. 7b) and triangular (Fig. 7c) obstacles, both in the *X* and *Y* axes, and decreases in speed, and even a change in the direction of the motion vector in *Y*-axis for struts. Changes in the direction of the *Z* axis, although they were recorded, have not been analyzed due to the lack of their significance for the research model.

In Figures 9 thru 11 the time courses of changes in the analyzed dynamic parameters (velocity and acceleration) during overcoming the terrain obstacle are shown. Figure 9 describes the character of the motion and the course of changes in acceleration a_i for the measurement point related to the center of the wheel in relation to the reference system (Fig. 5) related to the ground while overcoming a terrain obstacle.

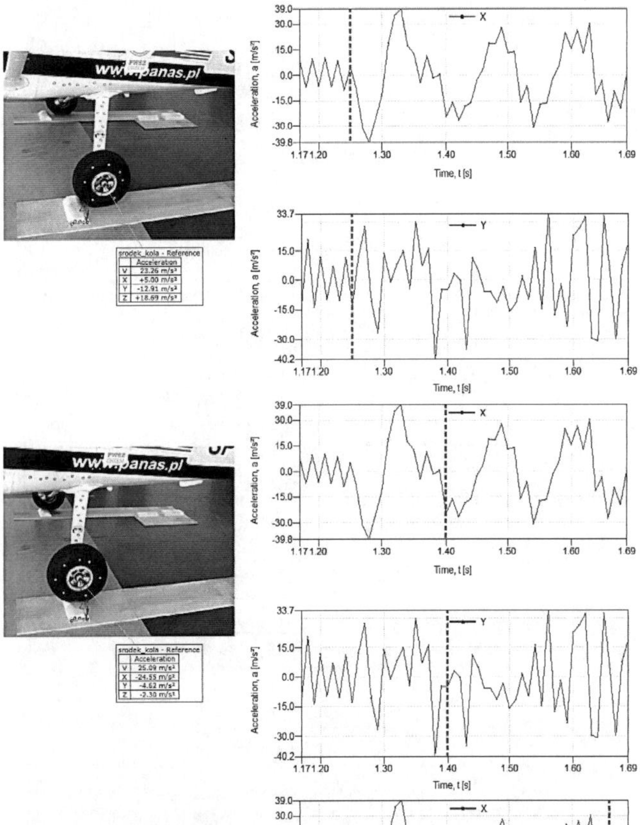

Fig. 9. Time courses of the acceleration of the center of the wheel during successive phases of overcoming a terrain obstacle with circular cross-section: a) obstacle approach phase, b) obstacle climbing phase, c) the phase of leaving the obstacle

The presented dynamic characteristics of changes in acceleration a_i concern two axes of the Cartesian coordinate system *X* and *Y*. The *X* axis was defined as the axis parallel to the velocity vector of the longitudinal (horizontal) motion on the motion surface, and the *Y* axis as the axis perpendicular to the *X* axis (vertical axis), and at the same time perpendicular to the motion surface. Figure 9 does not present changes in the value of the vi parameter, but it was included in further analyses. Figure 10 shows the characteristics of changes in the acceleration of the center of the wheel during successive phases of overcoming a terrain obstacle with rectangular cross-section (obstacle approach phase, obstacle climbing phase and the phase of leaving the obstacle).

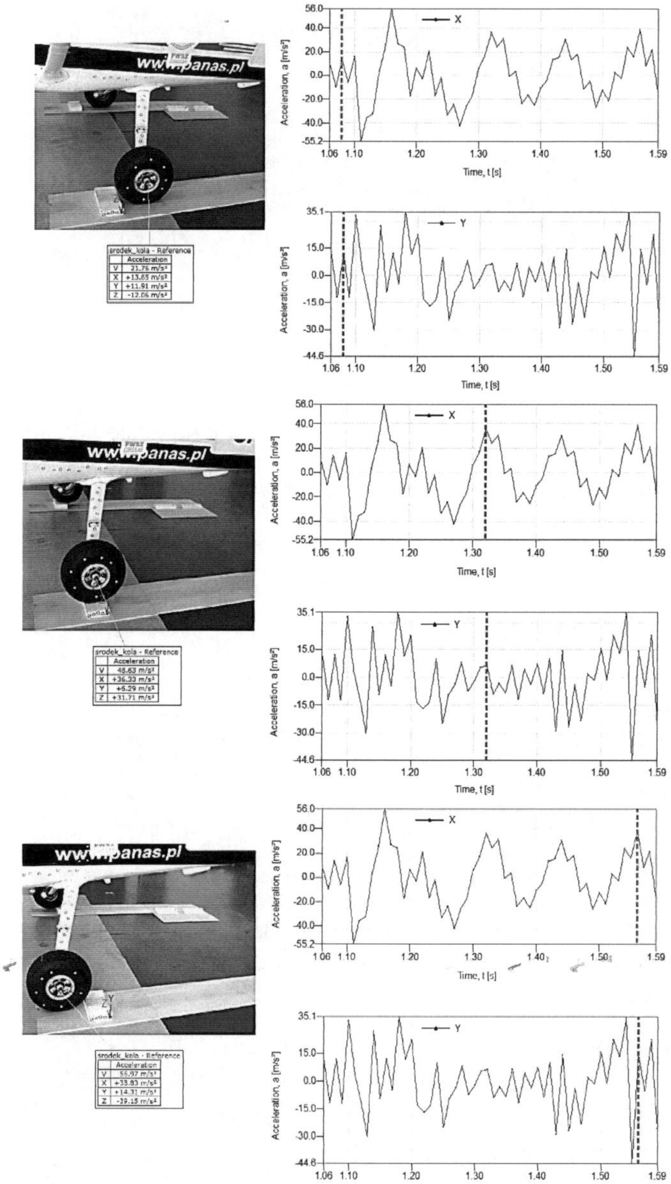

Fig. 10. Time courses of the acceleration of the center of the wheel during successive phases of overcoming a terrain obstacle with rectangular cross-section: a) obstacle approach phase, b) obstacle climbing phase, c) the phase of leaving the obstacle

The orientation of the *X* and *Y* axes in which the dynamic parameters of the movement a_i, v_i were measured, was oriented identically as in the case of terrain obstacle with circular cross-section. Similarly to the above, Figure 11 presents the characteristics of changes in the acceleration of the center of the wheel during successive phases of overcoming a terrain obstacle with triangular cross-section (obstacle approach phase, obstacle climbing phase and the phase of leaving the obstacle).

Results of a_i and v_i presented in Figures 9, 10 and 11 show a different course in the *X*-axis and the *Y*-axis. The motion parameters recorded along the *X*-axis tends to divide into phases, resulting from the rolling resistance (drag) generated by the terrain obstacles. Changes in acceleration a_i and velocity v_i obtained for the *Y*-axis are more regular and can be described by a typical cyclometric function, especially for the characteristics presented in Fig. 11.

The obtained dynamic time courses of velocity v_i and acceleration a_i made it possible to determine the values of the amplitudes of these quantities and to perform a further analysis of the cause-and-effect relationships of the impact of the type of obstacle on the behavior of individual structural elements of the aircraft.

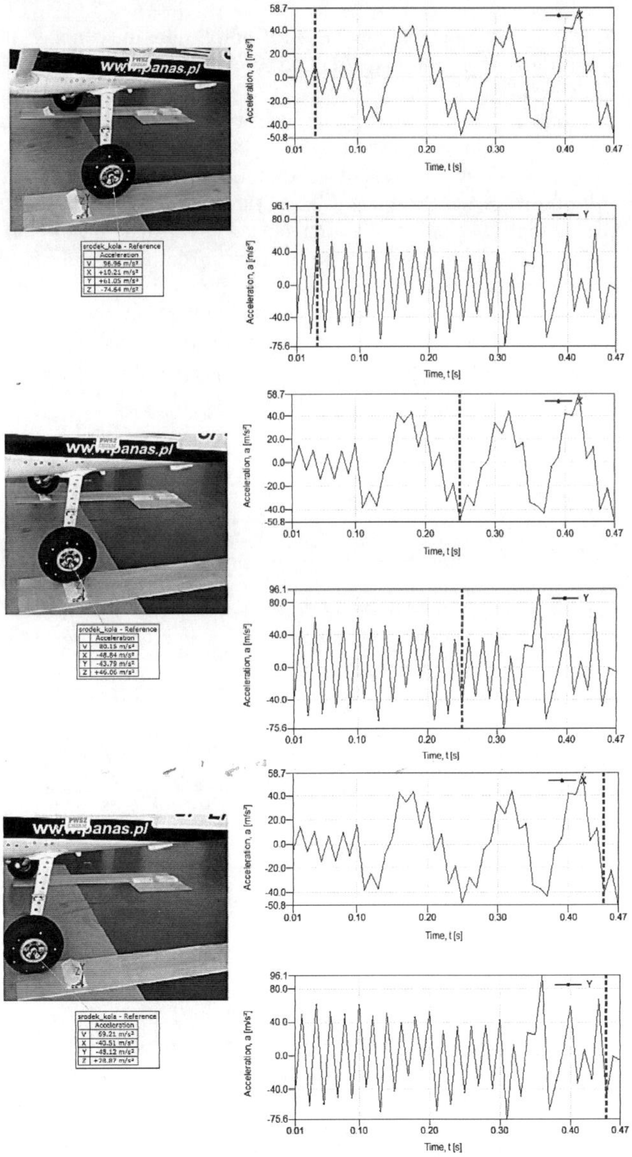

Fig. 11. Time courses of the acceleration of the center of the wheel during successive phases of overcoming a terrain obstacle with triangular cross-section: a) obstacle approach phase, b) obstacle climbing phase, c) the phase of leaving the obstacle

Tables 1 thru 3 list the values of the amplitudes of the measured values for the tested structural elements of the aircraft. Table 1 presents a comparative summary of the acceleration/deceleration values (a_i/-a_i) and the velocity vi at the wheel center measurement points in various phases (I, II, III) of overcoming a terrain obstacle. Based on the quantitative record of the tested dynamic parameters of the movement, the obtained results were visualized graphically in the graph shown in Figures 12 and 13.

Table 1
A collection of acceleration/deceleration values (a_i and -a_i) and velocity v_i of the landing gear wheel axis in various phases (I, II, III) of overcoming a terrain obstacle

Type of obstacle geometry		Wheel center			
		Acceleration amplitude [m/s²]		Velocity amplitude [m/s]	
	phase	*X*	*Y*	*X*	*Y*
Rectangular	I	+15,98 -10,16	+12,00 -12,00	+1,46 -1,36	+0.01 -0,11
	II	+55,05 -55,28	+35,13 -30,00	+1.79 -0.17	+0.76 -0.70
	III	+38.83 -44.13	+34.72 -44.69	+1.66 +0.68	+0,15 -0,30
	phase	X	Y	X	Y
Triangular	I	+14.30 -14.60	+67.05 -57.36	+1.55 +1.32	+0.30 -0.31
	II	+42.99 -48,84	+45.95 -43.79	+2.15 +0.46	+0.85 -0.57
	III	+58.79 -43.28	+58.29 -64.99	+4.24 +0.62	+0.26 -0.29
	phase	X	Y	X	Y
Semi-circular	I	+9.62 -8.93	+11.54 -12.91	+1.53 +1.43	+0.7 -0.7
	II	+39.02 -39.84	+29.79 -35.14	+1.66 +0.05	+0.76 +0.60
	III	+30.17 -27.75	+32.51 -29.94	+1.78 +0.63	+0.21 -0.40

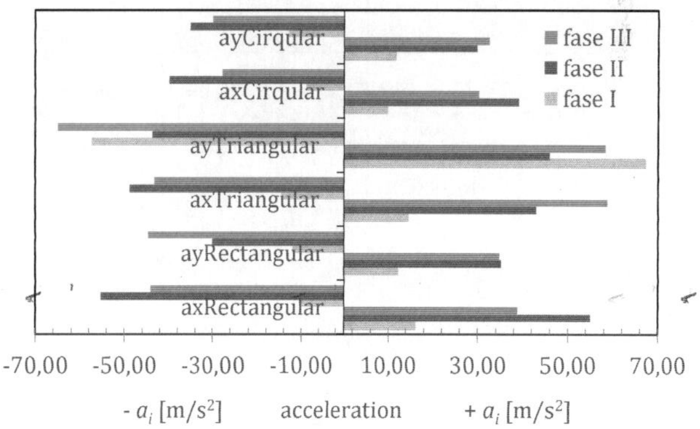

Fig. 12. Values of acceleration or deceleration of the measurement point related to the axle of the landing gear wheel during the individual phase (I, II, III) of overcoming a terrain obstacle of different geometry (rectangular, triangular and circular cross - section)

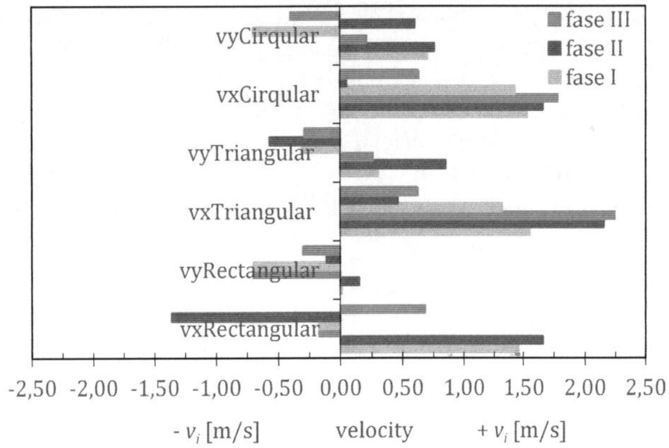

Fig. 13. Values of velocity of the measurement point related to the axle of the landing gear wheel

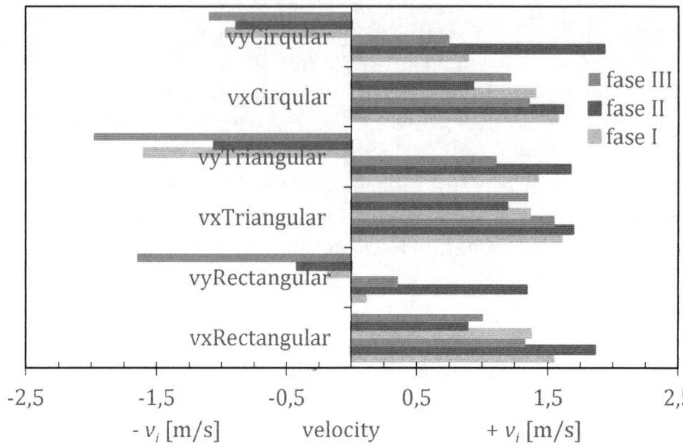

Fig. 15. Values of velocity of the measurement point related to the fuselage

Table 2
A collection of acceleration/deceleration values (a_i and $-a_i$) and velocity v_i at the aircraft fuselage measurement point in various phases (I, II, III) of overcoming a terrain obstacle

Type of obstacle geometry		Fuselage			
		Acceleration amplitude [m/s²]		Velocity amplitude [m/s]	
	fase	X	Y	X	Y
Rectangular	I	+21.83 -24.28	+277.02 -301.69	+1.54 +1.37	+0.11 -0.18
	II	+21.30 -21.69	+312.00 -298.90	+1.86 +0.89	+1.34 -0.42
	III	+23.59 -23.64	+238.79 -198.72	+1.32 +1,01	+0.35 -1.65
	fase	X	Y	X	Y
Triangular	I	+10.21 -14.24	+52.16 -50.42	+1.60 +1.36	+1.42 -1.60
	II	+42.99 -48.84	+58.79 -57.90	+1,69 +1.19	+1.67 -1.06
	III	+58.79 -50.89	+60.47 -48.12	+1.54 +1.35	+1.10 -1.98
	fase	X	Y	X	Y
Semi-circular	I	+10.25 -8.93	+11.47 -13.19	+1.57 +1.40	+0.89 -0.97
	II	+39.02 -39.84	+27.99 -27.00	+1.61 +0.93	+1.93 -0.89
	III	+32.51 -27.75	+32.51 -29.94	+1.35 +1.22	+0.74 -1.10

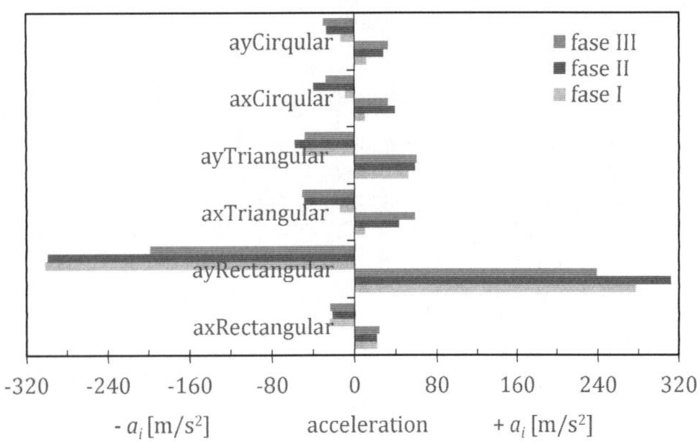

Fig. 14. Values of acceleration or deceleration of the measurement point related to the aircraft fuselage

Table 2 contains a comparison of amplitudes of changes in acceleration/deceleration (a_i/-a_i) and velocity v_i at the fuselage measurement point in various phases (I, II, III). The confirmation of the quantitative nature of changes in velocity and acceleration are shown in Figures 14 and 15.

Similarly, Table 3 presents a quantitative summary of the acceleration/deceleration (a_i/-a_i) and velocity v_i at the measurement point on the landing ger leg in various phases (I, II, III) of overcoming a given terrain obstacle. A detailed analysis of these values allows for determining the cause and effect relationships of the assessment of the impact of the terrain obstacle geometry on the dynamic parameters of the movement of individual aircraft structural elements analyzed in the article.

Table 3
A collection of acceleration/deceleration values (a_i and $-a_i$) and velocity v_i in the point on the landing gear leg

Type of obstacle geometry		Strut			
		Acceleration amplitude [m/s²]		Velocity amplitude [m/s]	
	fase	X	Y	X	Y
Rectangular	I	+18.48 -16.42	+20.79 -23.53	+18.48 -15.52	+20.79 -25.53
	II	+38.56 -39.41	+257.47 -230.03	+14.91 -12.49	+169.98 -153.66
	III	+33.78 -36.97	+194.79 -123.01	+50.18 -36.56	+194.99 -210.94
	fase	X	Y	X	Y
Triangular	I	+33.50 -18.26	+280.35 -285.56	+1.32 +1.66	+1.44 -1.37
	II	+26.46 -23.66	+249.32 -254.47	+1.63 -0,96	+1.77 -1.60
	III	+19.19 -19.54	+224.93 -248.48	+1.56 +1,35	+1.49 -1.97
	fase	X	Y	X	Y
Semi-circular	I	+14.90 -14,92	+148.79 -139.14	+1.57 +1.39	+0.76 -0.62
	II	+15.31 -12.79	+173.78 -290.44	+1.64 +0.88	+1.95 -0.95
	III	+27.36 -28.80	+173.62 -177.62	+1.30 +1.13	+0.26 -1.07

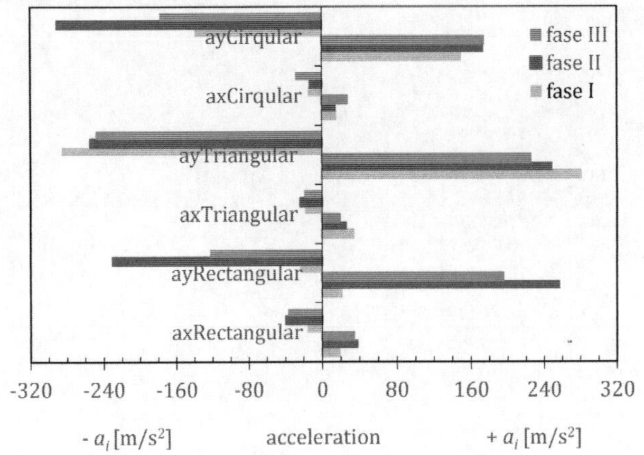

Fig. 16. Values of acceleration or deceleration of the point on the landing gear leg

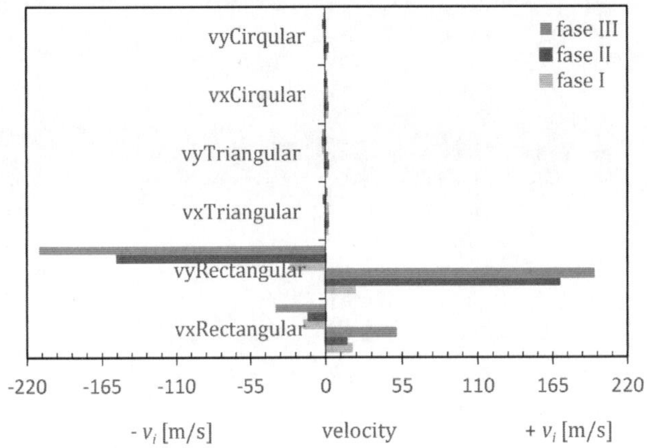

Fig. 17. Values of velocity of the point on the landing gear leg

3.2 Analysis of the results

When analyzing the above results, some trends can be observed. In the case of acceleration, the highest values of this parameter were recorded for the measurement point located on the landing gear leg (all three phases, three obstacle shapes). Similarly, great acceleration values were obtained by a point on the fuselage of the aircraft when overcoming a rectangular obstacle. Relatively small acceleration values measured for a point in the center of the wheel of the chassis result from the characteristics of the tire, which absorbs the impact energy by bending and delays the dynamic reaction.

When analyzing the velocity of the measurement points, it was found that it falls within a fairly narrow range of -2.0 to 2.0 m/s, for the measurement point located in the center of the landing gear wheel and on the aircraft fuselage. This is the case for all three phases of overcoming obstacles. However, in the case of the landing gear leg, when overcoming an obstacle with a rectangular cross-section, unexpectedly high speeds of 200 m/s were obtained. Since the result was so different from the others and it occurred in only one test variant and there is no rational explanation for this case, it was considered that the obtained result is defective (measurement disturbance, marker defect). It was noticed that the velocity of the measurement points reaches the highest values for the second phase of overcoming the obstacle. It should be taken into account that the speed of the measurement points is closely related to the speed of overcoming the obstacle, of the order of m/s.

3.3 Discussion

Based on the acceleration values of the measurement points, it is possible to determine the inertia forces acting temporarily on the aircraft element. This is one of the examples of direct practical application of measurement results. On the other hand, the analysis of the course of acceleration of individual points allows for a frequency analysis of the dynamic response of the aircraft structure to the impulse generated at the time of overcoming the obstacle.

It seems particularly interesting to determine relative velocities and accelerations for several points, e.g. a point located on the landing gear wheel in relation to the landing gear leg or the fuselage. Of course, the points on the aircraft structure elements mentioned here are only examples, it is possible to place much more markers and analyze the dynamics of other airframe elements.

On the other hand, the use of the presented method for an airplane during take-off or landing in real conditions is debatable. Probably, the measurement range as well as the limited resolution and angle of view of the GOM PONTOS camera will not allow filming the aircraft during the flight, especially since for safety reasons it will be necessary to maintain a considerable distance between the position with the measurement system and the runway. Therefore, it is necessary to strive to develop and implement measurement procedures that will enable the mapping of real conditions to the greatest possible degree of similarity.

4. Conclusions and future work

The main conclusion of this study is that the presented method is suitable for determination of dynamic response of an airplane overcoming a surface obstacle. Optical sensors of the GOM PONTOS system provided a set of data, which allowed to analyze accelerations and velocities of chosen points of aircraft's structure.

Future work would include taking into account the effects of soil and grass by performing similar tests in field conditions (on a grass airfield). The next stage of the research will be the parameterization of the test aircraft, including obtaining the values of geometric parameters, mass and inertia modules. The aim of future research will also be to develop a complete method for analyzing the impact of unevenness on the dynamics of the aircraft's motion on the ground. This will help anticipate hazardous situations during take-off, landing and taxiing at airfields with a grass runway.

5. Acknowledgements

This work was prepared within the project PM/SP/0065/2021/1 entitled "Method for airplane takeoff and landing distance measurement", financed by the Ministry of Education and Science (Poland) as a part of the Polish Metrology Programme.

6. Declaration of competing interest

The authors declare that they have no known competing financial interests or personal relationships that could have appeared to influence the work reported in this paper.

7. References

Brudnak, M.J., Gunter D.D. and Bylsma W., 2007. Modeling of Moguls on an Endurance Test Course. Proc. of the Joint North America, Asia-Pacific ISTVS Conference and Annual Meeting of Japanese Society for Terramechanics, Fairbanks, Alaska, USA.

Dubet, S., 2003. Terrain, obstacle and airport databases for electronic flight bags. Digital Avionics Systems Conference, DASC '03, Indianapolis, U.S.A., DOI: 10.1109/DASC.2003.1245956.

Els, P.S., Theron, N.J., Uys, P.E., Thoresson, M.J., 2007. The ride comfort vs. handling compromise for off-road vehicles. J. of Terramechanics, 44(4), pp.303 – 317.

Pelc-Mieczkowska, R., Tomaszewski, D., Jurgielewicz, K., 2018. Application of ALS data for GNSS terrain obstacles inventory. E3S Web of Conferences, DOI: 10.1051/e3sconf/20186300014.

Pytka, J., Józwik, J., Łyszczyk, T., Budzyński, P., Laskowski, J., Gnapowski, E., 2019. Measurement of Forces and Moments Acting on Aircraft Landing Gear Wheel. Proc. 6th International Workshop on Metrology for Aerospace (MetroAeroSpace), Torino, Italy.

Pytka J., Józwik J., Budzyński P., Łyszczyk T., Tofil A., Gnapowski E. and Laskowski J., 2019. Wheel dynamometer system for aircraft landing gear testing, Measurement (148) 2019, 106918.

Pytka, J.A., 2023. Terramechanik und Geländefahrzeuge. ATZ/MTZ Fachbuch, Springer Verlag, Wiesbaden, Germany.

Thomas, G., Vantsevich, V.V., 2010. Wheel-terrain-obstacle interaction in vehicle mobility analysis. Vehicle System Dynamics, 48: sup1, pp. 139-156, DOI: 10.1080/00423111003690496.

7477 | Evaluating the pressure performance of DPF filters using engine bench analysis

Norbert Biró [a,b,*], Péter Kiss [c]

[a] IBIDEN Hungary Kft. Technical Center, Exhaust System Evaluation, 2336 Dunavarsány, Neumann János út 1., Hungary

[b]Mechanical Engineering Doctoral School – Hungarian University of Agriculture and Life Sciences (MATE), 2100 Gödöllő, Páter Károly u. 1., Hungary

[c] Institute of Technology, Department of Vehicle Technology- Hungarian University of Agriculture and Life Sciences (MATE), 2100 Gödöllő, Páter Károly u. 1., Hungary

* Corresponding author: Norbert_Biro@ibiden.com

ABSTRACT

Passenger and commercial transportation have a substantial impact on hazardous air pollution. The progress of exhaust gas aftertreatment technology is closely aligned with global emission regulations. The upcoming Euro 7 legislation in the European Union is anticipated to be more stringent in many aspects when compared to Euro 6, although its specific requirements are yet to be determined. Consequently, new DPF technology is necessary to comply with the regulation. This study aims to compare the pressure hysteresis of the current and proposed new generation DPF substrates. The experiment will be carried out in an engine bench laboratory. For comparable results, a fixed parameter measurement layout will be introduced. The objective is to draw conclusions and determine the compliance of the upcoming Euro 7 legislation in terms of pressure hysteresis.

Keywords
Diesel Particle Filter
Euro 7
Exhaust After treatment System
Pressure Loss

1. Introduction

The mitigation of harmful emissions has emerged as a matter of tremendous significance and keen interest in both scientific and public spheres, owing to its far-reaching environmental implications. Within the Euro 27 countries, an in-depth examination of emissions patterns reveals that the transport sector is accountable for approximately 14% of the overall emissions (Eurostat, 2022). Consequently, it is imperative to diligently explore further avenues to curtail and minimize emissions. Notably, the global electricity generation in the year 2016 amounted to an astonishing 25 petawatt-hours (PWh), with fossil energy carriers contributing to a substantial two-thirds of this prodigious output. Extensive, comprehensive analyses, such as the scholarly work conducted by Gupta et al. (2017), have consistently demonstrated that internal combustion engines (ICE) powered by diesel exhibit a lower cumulative emissions profile compared to their counterparts, the battery electric vehicles (BEVs), when evaluating the complete energy life cycle, also referred to as the "well-to-wheel" analysis (Bíró et al. 2023b).

Against the backdrop of these discerning findings, it becomes evident that internal combustion engines will continue to occupy a pivotal role within the class of commercial vehicles, thereby wielding considerable influence on the economic and environmental spheres. Anticipated forthcoming regulations are poised to introduce a more rigorous regulatory framework, with a particular emphasis on particle number emissions and the frequency of regeneration. Meeting and adhering to these enhanced criteria can be accomplished through a twofold approach: augmenting the soot-holding capacity of the diesel particulate filter (DPF) or expanding the surface area available for catalytic coating. It is crucial to acknowledge that the augmentation of soot-holding capacity may engender challenges in terms of cost-effectiveness, thereby necessitating the exploration of alternative developmental pathways, such as increasing the porosity of the DPF. However, it is of paramount importance to strike a delicate balance between augmenting porosity and safeguarding other indispensable filter properties, including but not limited to pressure drop. As pressure drop exhibits a direct correlation with both engine efficiency and emission formation, it must be accorded due consideration as a pivotal aspect in all developmental endeavors.

Within the purview of this scholarly discourse, the primary objective is to undertake a meticulous and comparative evaluation of the pressure hysteresis demonstrated by a conventionally designed DPF and innovative High Porosity (HP) filter. By subjecting these filters to rigorous scrutiny and discerning their respective performances, invaluable insights can be determined, therefore propelling the advancement of cutting-edge emission control technologies for internal combustion engines.

2. Experiment

2.1 Layout

All experimental tests were conducted within the controlled environment of an engine dynamometer laboratory. The specific unit subjected to testing was a heavy-duty diesel engine with six cylinders, designed to comply with the Euro VI-d emission standards. To ensure consistency and comparability, the engine conditions of the chosen test cycles were adjusted to match the characteristics of the test engine by utilizing the emission

https://doi.org/10.56884/XSMK1643
ISBN 978-1-942112-55-6 (pdf)

automation software known as AVL PUMA Open™, as described in the study by Peng et al. (2017).

The test layout, as depicted below, provides an overview of the experimental setup:

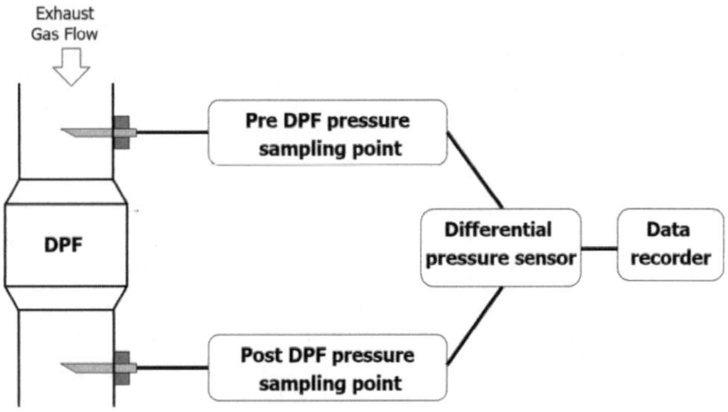

Fig. 1. Schematic sketch of the test layout (Source: N. Biró)

One notable advantage of employing an engine bench with a fully equipped conditioner is that it ensures a consistent starting condition for each test. This uniformity of the starting condition allows for better comparisons and significantly reduces uncertainty since all input variables remain the same throughout numerous repetitions. By minimizing variations in the starting conditions, the experimental setup enhances the accuracy and reliability of the results obtained.

Fig. 2. Dynamometer mounted test engine connected to test cell conditioner system

2.2 Test schedule

In order to ensure test repeatability and consistent experimental conditions, a preliminary dewaxing process is conducted on all diesel particulate filter (DPF) samples prior to each test. The dewaxing procedure involves subjecting the samples to a carefully controlled heat treatment at a temperature of 700 °C for a duration of 10 hours, as described in the study by Bíró et al. (2020).

To assess and compare the status of soot accumulation, a weight measurement of the DPF is performed after each test step that has the potential to either accumulate or burn off soot. The weight measurements of the DPF samples at different test steps are meticulously documented and recorded, as outlined in Table 1. This allows for a comprehensive evaluation of the changes in DPF weight, serving as a valuable indicator of the accumulation or removal of soot throughout the experimental procedure.

Table 1
Detailed test schedule

#	Test flow	Lead time[hr]
1	DPF preparation for test	10
2	DPF weight measurement	0.5
3	Precondition	1
4	DPF weight measurement	0.5
5	Full load curve	0.5
6	DPF weight measurement	0.5
7	Soot loading 1 g/l	2
8	DPF weight measurement	0.5
9	Full load curve	0.5
10	DPF weight measurement	0.5
11	Repeat #7 - #10, with 4 and 8 g/l	-

To ensure accurate and unaffected measurement results, it is essential to conduct the weight measurements of the DPF samples at a controlled temperature of 150 °C. This temperature selection is crucial as it minimizes the influence of ambient room air humidity on the measurements, ensuring the integrity and reliability of the results obtained. This standardized measurement condition mitigates any potential confounding factors that could arise due to variations in humidity. Prior to conducting the weight measurements, a preconditioning step is implemented, which involves the formation of a so-called "soot cake" on the DPF. This soot cake facilitates the DPF's attainment of peak filtration efficiency, enabling it to effectively capture and remove particulate matter from the exhaust gases. Bíró et al. (2023a) highlight the importance of this preconditioning step in optimizing filtration performance in the study.

During the full load curve testing, the engine is operated at its maximum load, with the engine speed rapidly ramped up from idle speed to reach a maximum of 8 revolutions per second (rpm). This rigorous testing protocol generates a vast and comprehensive dataset encompassing exhaust volume flow and pressure drop measurements of the DPF. The combination of high load and varying engine speeds provides valuable insights into the performance characteristics of the DPF under demanding operating conditions.

To simulate different levels of saturation within the DPF filter, the test schedule incorporates three distinct levels of soot loading. These levels correspond to 1 g/l for a sparsely loaded sample, 4 g/l for a moderately loaded sample, and 8 g/l for a heavily loaded sample. This deliberate variation in soot loading allows for the evaluation and comparison of the DPF's performance under different degrees of saturation, providing a comprehensive understanding of its filtration efficiency across various operating scenarios.

3. Results

In this experimental study, two distinct samples were subjected to testing: the original specification standard substrate and the newly improved high porosity substrate. Each sample underwent four Full Load Curves (FLCs), representing different operational conditions to simulate varied real-life scenarios. These conditions included no prior soot accumulation, sparsely, moderately, and heavily soot loaded conditions. By maintaining fixed conditions for each sample, the corresponding tests conducted on each sample can be effectively compared against one another, ensuring robust and meaningful analysis.

Throughout the experiments, huge amount of output data was generated, encompassing parameters such as engine speed and engine torque. These data points play a vital role in assessing and determining the comparability of the tests conducted. The evaluation process in this specific study requires the utilization of two datasets: the differential pressure over the DPF (dP_DPF) and the exhaust gas volume flow (EVF). These datasets serve as key indicators for evaluating the performance of the DPF under different conditions.

To facilitate the evaluation process and showcase the relationship between dP_DPF and EVF, a scatter point diagram was created. This diagram allows for a visual representation of the data, enabling a clear understanding of the correlation between the two parameters. Furthermore, utilizing the diagram editor feature in Microsoft Office Excel, a descriptive equation was automatically calculated to define the relationship between dP_DPF and EVF. This equation enables the extrapolation of the pressure hysteresis phenomenon observed in the specific test to the entire engine map, providing valuable insights into the overall performance of the DPF across various operating conditions.

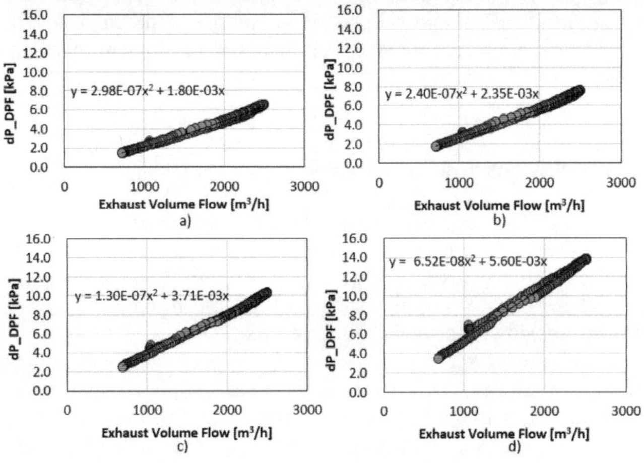

Fig. 3. Pressure and flow data of Standard substrate FLC with 0 g/l (a), 1 g/l (b), 4 g/l (c), 8g/l (d) soot amount

As depicted in Figure 3, the full load curves conducted with varying degrees of soot accumulation exhibit distinct relationships between the differential pressure and exhaust gas volume flow. This observation holds true for the Standard substrate test procedure, where four descriptive equations were derived based on the collected data. Analyzing the four diagrams, it becomes evident that the quantity of accumulated soot is directly proportional to the differential pressure experienced by the DPF.

By closely examining the relationships depicted in the diagrams, it is clear that as the soot accumulation increases, the corresponding differential pressure also increases. This direct proportionality signifies that a higher level of soot accumulation leads to a greater resistance to exhaust gas flow, resulting in an elevated differential pressure across the DPF.

These findings highlight the significant impact of soot accumulation on the DPF's performance, with the differential pressure serving as a reliable indicator of the extent of soot accumulation. The derived descriptive equations provide valuable insights into the relationship between differential pressure and soot accumulation, aiding in the evaluation and characterization of the DPF's behavior under varying operating conditions.

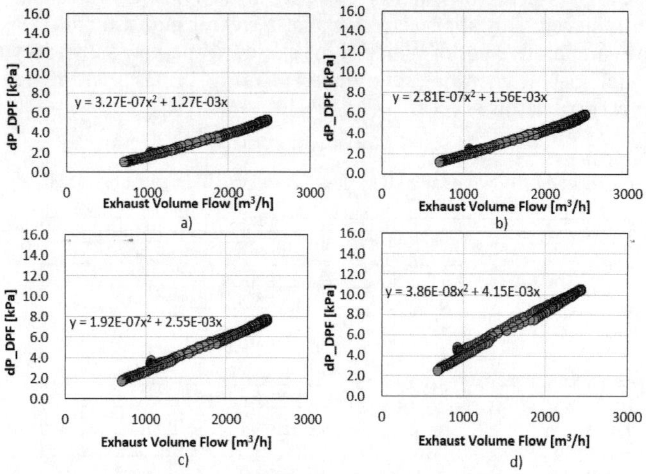

Fig. 4. Pressure and flow data of New substrate FLC with 0 g/l (a), 1 g/l (b), 4 g/l (c), 8 g/l (d) soot amount

Similar to the standard substrate, the diagrams and equations for the new high porosity substrate filter were created following the same meticulous methodology. By examining the diagrams presented in Fig. 4 for the new substrate and Fig. 3 for the standard substrate, a notable observation emerges. At equivalent levels of soot accumulation, it becomes evident that the pressure drop experienced by the new high porosity substrate is lower compared to the standard substrate.

This comparative analysis showcases a distinct advantage of the new substrate design in terms of pressure drop performance. Despite the same levels of soot accumulation, the new high porosity substrate demonstrates a reduced resistance to exhaust gas flow, resulting in a lower pressure drop across the filter. This finding underscores the superior filtration efficiency and reduced flow restriction offered by the new substrate design.

These insights provide valuable evidence supporting the effectiveness of the new substrate in mitigating pressure drop issues commonly associated with standard substrates. The lower pressure drop exhibited by the new substrate design enhances the overall performance and operational efficiency of the diesel particulate filter, making it an advantageous choice for addressing soot accumulation challenges in real-world applications.

4. Discussion

By utilizing the derived equations from Fig. 3 and Fig. 4, it becomes feasible to calculate the pressure drop at any specific exhaust volume flow point. This calculation enables a straightforward determination of the pressure drop characteristics across the diesel particulate filter (DPF) under different operating conditions. To ensure comparability, four specific exhaust volume flow points were selected: 1000, 1500, 2000, and 2500 m³/h, as indicated in Table 2. These predetermined flow rates serve as reference points for evaluating the pressure drop performance of the DPF across different scenarios. However, for the final comparison analysis, the data corresponding to the 2000 m³/h exhaust volume flow was exclusively employed. This particular flow rate was deemed representative and suitable for the comprehensive assessment and comparison of the pressure drop characteristics between the standard and new substrate designs. By focusing on a specific exhaust volume flow rate, a more precise and targeted evaluation of the pressure drop differences between the two substrates can be achieved. This approach allows for a clearer understanding of the comparative performance of the standard and new substrates under specific operating conditions, enhancing the accuracy of the final comparison.

Table 2.
Summarized test data of all FLC cycles of standard and new substrates

New substrate	dP_DPF [kPa]		Exhaust Volume Flow Rate [m³/h]			
			1000	1500	2000	2500
	Soot mass [g/l]	0.0	1.6	2.6	3.8	5.2
		0.8	1.8	3.0	4.2	5.7
		4.0	2.7	4.3	5.9	7.6
		8.0	4.2	6.3	8.5	10.6

Standard substrate	dP_DPF [kPa]		Exhaust Volume Flow Rate [m³/h]			
			1000	1500	2000	2500
	Soot mass [g/l]	0.0	2.1	3.4	4.8	6.4
		1.0	2.6	4.1	5.7	7.4
		3.9	3.8	5.9	7.9	10.1
		7.9	5.5	8.3	10.9	13.6

In Figure 5, the pressure drop values of each test conducted at an exhaust gas volume flow (EVF) of 2000 m3/h are presented alongside the corresponding soot accumulation levels. These preliminary results provide valuable insights into the relationship between soot accumulation and pressure drop for both the standard and new substrates. Upon careful examination of the data, it becomes evident that the pressure loss, or pressure drop, experienced by the new substrate is significantly lower compared to the standard substrate. This finding underscores the superior performance of the new substrate in mitigating flow restriction and reducing the resistance to exhaust gas flow. The noticeable difference in pressure drop between the two substrates supports the notion that the new substrate design effectively minimizes the negative impact of soot accumulation on the DPF. By exhibiting a lower pressure drop at equivalent levels of soot accumulation, the new substrate demonstrates enhanced filtration efficiency and improved flow characteristics. These preliminary results provide strong evidence in favor of the new substrate's ability to alleviate pressure drop issues, offering a promising solution for addressing performance limitations associated with standard substrates. Further analysis and evaluation are warranted to validate and reinforce these findings, ultimately affirming the effectiveness and benefits of the new substrate design in real-world applications.

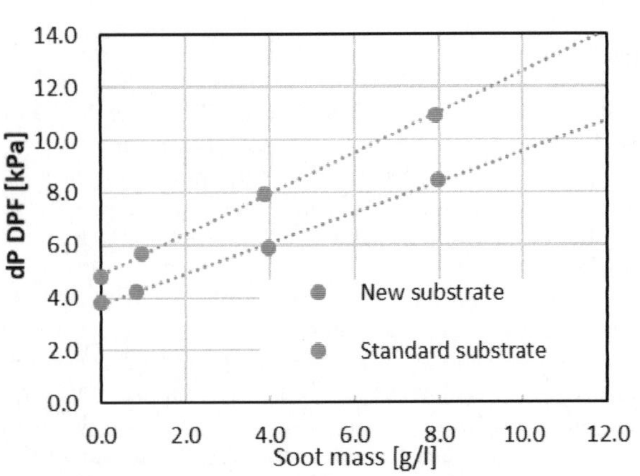

Fig. 5. Comparing new and standard substrate on 2000 m³/h exhaust volume flow

An intriguing observation arises even when simulating heavily soot-accumulated conditions at an amount of 8 g/l. The pressure drop experienced by the new substrate remains below the 10 kPa limit. This finding highlights the significant advancements of the new high porosity technology compared to the standard substrate. The ability of the new substrate to maintain a pressure drop within acceptable limits, even under challenging conditions, underscores its improved performance and efficiency. To comply with the Euro 7 legislation [European Commission, 2022] limits, the integration of a catalytic coating layer becomes necessary. This coating serves multiple purposes, including reducing the content of nitrogen dioxide (NO_2) and enabling more efficient passive regeneration. Although the technology for applying such coatings has been available since the Euro 5 legislation [EUR-Lex, 2007], the quantity of coating that can be applied is heavily influenced by the pressure drop across the DPF filter. The considerably lower pressure drop exhibited by the new substrate holds promising implications. With the new substrate's reduced pressure drop, it is hypothesized that a higher quantity of coating can be applied, potentially leading to improved passive regeneration and higher efficiency in reducing NO_2. However, further experiments are required to validate these hypotheses and ascertain the precise impact of coating quantity on regeneration efficiency and NO_2 reduction. These preliminary findings suggest that the combination of the new high porosity substrate and an increased coating quantity could lead to enhanced performance and compliance with the Euro 7 legislation requirements. Continued research and experimentation are essential to confirm and refine these hypotheses, paving the way for the development of more efficient and environmentally friendly diesel particulate filters.

5. Conclusions and future work

In conclusion, our study demonstrates the promising potential of the new high porosity substrate for developing more efficient diesel particulate filters. The observed pressure drop differences and enhanced filtration performance highlight the advantages of the new design in real-world applications.

Future research should focus on validating our findings through additional experimentation. Further investigation is needed to confirm the impact of coating quantity on regeneration efficiency and NO_2 reduction, paving the way for more optimized DPF designs. By continuing to explore the capabilities of the new substrate and conducting comprehensive assessments under various operating conditions, we can advance the development of diesel particulate filters that comply with stringent Euro 7 legislation requirements.

This research holds particular relevance to the field of terramechanics, as the pressure drop characteristics in diesel particulate filters (DPFs) play a crucial role in off-road vehicle performance and environmental impact. By exploring the superior performance of the new high porosity substrate design, we contribute to advancing terramechanics knowledge, optimizing engine efficiency, and reducing emissions, ultimately enhancing the sustainable utilization of off-road vehicles in challenging terrains.

In summary, the combination of the new high porosity substrate and improved catalytic coating integration holds significant promise in addressing performance limitations associated with standard substrates. Through continued research and experimentation, we can contribute to the advancement of cleaner and more efficient exhaust emission control technologies for a sustainable future.

6. Declaration of competing interest

The authors declare that they have no known competing financial interests or personal relationships that could have appeared to influence the work reported in this paper.

7. References

Bíró, N.; Pillinger, Gy.; Kiss, P.; Szöllösi, D.; Ohira, A. 2020. Experimental SCR system for engine dynamometer applications. Hun. Agri. Eng. 38, 56–62. http://doi.org/10.17676/HAE.2020.38.56.

Biró, N.; Kiss, P., 2023a. Emission Quantification for Sustainable Heavy-Duty Transportation. Sustainability, 15, 7483. https://doi.org/10.3390/su15097483.

Bíró, N.; Kiss, P., 2023b. Euro VI-d Compliant Diesel Engine's Sub-23 nm Particle Emission. Sensors, 23, 590. https://doi.org/10.3390/s23020590.

European Comission. Available online: https://ec.europa.eu/commission/presscorner/detail/en/ip_22_6495 (accessed on 2 April 2023).

Eurostat. Available online: https://ec.europa.eu/eurostat/statistics-explained/index.php?title=Quarterly_greenhouse_gas_emissions_in_the_EU (accessed on 21 December 2022).

EUR-Lex. Available online: https://eur-lex.europa.eu/legal-content/EN/TXT/?uri=celex%3A32007R0715 (accessed on 2 December 2022).

Gupta, M., Rafat, Y., and Alam, M., 2017. "Well to Wheel Cum Tailpipe Emission Analysis: ICE vs xEV," SAE Technical Paper, 2017-26-0107, 2-4, https://doi.org/10.4271/2017-26-0107.

Peng, Y.; Nix, A.; Li, H.; Johnson, D.; Heltzel, R. 2017. Derivation of a Representative Engine Duty Cycle from On-Road Heavy-Duty Vehicle Driving Data. J. Transp. Technol. 7, 376–389. https://doi.org/10.4236/jtts.2017.74025.

7523 | Assessing performance of light wheeled vehicles on GRC-1 using 3D scanned footprint measurements

Chaitanya Sonalkar [a,*], Nikhil Ravichandran [a], Alexandru Vilsan [a], Corina Sandu [a], Tony Lin [b]

[a] *Terramechanics Multibody and Vehicle Systems Lab, Virginia Tech, Blacksburg, Virginia, United States*

[b] *Goodyear Tire and Rubber Company, Akron, Ohio, United States*

* *Corresponding author: csonalkar@vt.edu*

ABSTRACT

In off-road conditions, the vehicle mobility is dependent on the terrain and its properties. It is necessary to characterize the soil/terrain in order to assess its traversability and to predict the performance of the tire. If not accounted for properly, it can lead to conditions where the vehicle is stuck due to excessive sinkage or is unable to satisfactorily meet its intended performance. Navigation of wheeled vehicles on fine, dry sands, such as the Lunar Soil Simulant GRC-1, has greater challenges due to very low cohesion properties. This study aims at characterizing the terrain response of such a sandy soil to different loading and slip conditions for light ground vehicles, by studying the 3D shape of the footprint/ rut left by the tire. For this, a white light- based 3D scanner will be used to collect the data. Correlations between the normal load applied, the depth and shape of the contact area, as well as the slip conditions are made to assess their impact on the drawbar pull of the tire.

Keywords
Off-road mobility
Slip-sinkage
Tire footprint
Soil compaction
3D scanning

1. Introduction

The study of off-road mobility is of utmost importance for agricultural, construction and mining, military as well as extraterrestrial applications. Mobility of vehicles in off-road environments is primarily attributed to the tire-terrain interaction and the forces arising due to it. A tire navigating an off-road terrain experiences various forces such as traction, rolling resistance, terrain resistance etc. which affects the dynamics of a vehicle. Hence study of terrain properties becomes significant for assessing vehicle performance.

The thrust developed by a wheel on a terrain is also limited due to the maximum shear strength of the soil. For any value of shear displacement, the shear stress in the failing soil layer cannot exceed the limiting value. In practical terms, if the force action on the soil due to tire threads/grousers results in shear stress values above the limiting value, the soil fails and flows underneath the tire leading to excessive sinkage and loss of traction. A tire also compacts the soil as it passes over the deformable terrain. The mechanics of soil failure leading to soil compaction beneath the surface and the soil accumulation in the front of the tire, results in compaction resistance and bulldozing resistance, respectively, experienced by the wheel.

Many theoretical, empirical, and computational approaches have been adopted in the past to determine tire and terrain response which can account for the above-mentioned phenomena for a particular loading condition. The theoretical and the empirical approaches do not hold in the entire spectrum of operating conditions due to assumptions such as homogeneity and uniformity of terrain, etc. High fidelity computational models such as the Discrete Element model (DEM) or Smoothed Particle Hydrodynamics (SPH) model show high correlation with test data but are computationally intensive and difficult to parameterize.

This paper proposes a new methodology which utilizes a white light-based 3D Scanner to scan tire footprint which is then correlated to drawbar pull measured by wheel force transducer.

2. Literature Review

For scanning and measuring the tire footprint various methods of 3-dimensional scanning were reviewed and compared by Talbot and Astrup (2021).

Photogrammetry is a method of obtaining 3D coordinates of a surface using images captured from different reference points. The accuracy of the scan is determined by the resolution and clarity of the images captured; however, the texture and the color of the specimen may also affect the accuracy.

LiDAR or Light Detection and Ranging technology measures the distance between the specimen and the light source by measuring the time required for the beam light to be reflected to the sensor. This technology is better than photogrammetry as it can be used in low lighting scenarios, but sampling rates, effective range and field-of-view can vary considerably depending on the instrument.

Ultrasound is like LiDAR, the only difference being that it uses ultrasonic waves instead of light to measure the travel times and distance, plus spatial resolution is lower due to the larger wavelength. The data measured is one dimensional, hence the sensor needs to cover the region to map it. Also, disturbances in a similar frequency range can add noise to the data.

Depth sensing cameras work on a similar principle to that of the LiDAR, but such cameras referred to in literature are used to detect motion. The sensing range of these cameras is adjustable

https://doi.org/10.56884/QBZW5768
ISBN 978-1-942112-55-6 (pdf)

making it more suitable for outside applications. 3D scanners also used depth sensing cameras to generate point cloud data of an object. Depending on the application, a depth sensing camera and a 3D scanner can be used interchangeably to obtain the best possible scanning setup.

Kenarsari et al. (2017) proposed a method of using digital photogrammetry to measure the footprint area and volume of a tractor tire. Their research included conducting experiments to measure static tire footprint as well as rut of the rolling tire. The time-lapse photography was used to take images of the rut along the tire path which were then used to generate point-cloud data and the 3D model on the rut. The 3D model was validated against a sand cone test and the error in volume estimates was around 4-7% of the results obtained from sand cone tests.

An attempt to correlate the contact volume with rolling resistance was done by (Farhadi et al., 2018). Vertical load, inflation pressure and soil moisture content were varied to collect data with different contact area and volumes. High correlation was obtained between the contact volume and rolling resistance (R=0.933).

Similarly, Ptak et al. (2022), and Ptak et al. (2023) measured the contact area using a 3D scanner and studied the variation in the contact patch with varying vertical load and inflation pressure for a radial and a bias tire in static condition.

In the above literature, the experimental tests were performed in low slip or static conditions whereas an off-road tire generally operates at higher slip ratios.

3. Experimental Equipment and Instruments

This section describes the major components required to perform the study and data acquisition.

3.1 3D Scanner

For scanning the tire footprint, a structured white light-based handheld 3D scanner (Einstar Shining 3D) was used. The 3D scanner provided depth data of tire footprint up to 0.1 mm of accuracy.

3.2 Wheel Force Transducer

A Michigan Scientific Wheel Force Transducer (LW12.8-45) was used to measure the force and moment data for the tire. This is a strain-based sensor which uses slip-ring telemetry to transmit data to the data acquisition setup. It can measure a maximum vertical load of 4082.3 kg (9000 lbs) and the traction force up to 4082.3 kg (9000 lbs).

3.3 Test Rig

The testing for this study was done on the TMVS lab Terramechanics Test Rig at Virginia Tech. This is a single wheel test rig. The major components of the rig are: (a) bed; (b) carriage assembly; (c) force hub system; and (d) two independent drivetrains.

The rig is 7.62 m (25 feet) long, 1.82m (6 feet) wide and 1.22 m (4 feet) deep. For this testing, a partition was put up to create 0.9 m (3 feet) wide and an 200mm (8 in) deep section on the upper portion of the soil bin, to achieve the required width and depth of soil.

The test rig has five degrees of freedom namely, longitudinal, and vertical translation, toe, camber, and wheel rotation. The toe angle can be varied up to ±30 degrees while the camber can vary up to 8 degrees. The test rig can apply up to 8 kN of vertical load. For this setup, the self-weight of the carriage itself was used to get the required load of 4000 N. To achieve the lower load of 2000 N, an offloading system was used, which was done by suspending gym weights using pulley-cable arrangement.

Fig. 1. Test Rig at TMVS lab, Virginia Tech

The maximum depth of soil achievable under this condition was 8 in. One of the two independent motors (carriage motor) used a belt-pulley system to give the desired longitudinal velocity to the carriage. The other (wheel motor) motor is used to provide the required wheel speed. By independently controlling these two parameters, a required constant slip ratio can be achieved.

For this setup, the self-weight of the carriage itself was used to get the required load of 4000 N. To achieve the lower load of 2000 N, an offloading system was used, which was done by suspending gym weights using pulley-cable arrangement.

3.4 Terrain details

The lunar soil simulant GRC-1 and sandy loam were two options for fine dry sands available in the laboratory. The GRC-1 soil was chosen as it has a smaller particle size and provides a very dry condition with low cohesion (1 kPa and less), which is a worst-case condition for traction.

3.5 Off-road tire and Rim

The testing for this study was done on the BKT AT 171 tire. This is an off-road tire that is designed for mud boggers. The tire has an overall diameter of 785 mm and section width of 211 mm. It has a load capacity of 450 kgs (992 lb.) at an inflation pressure of 1.4 bar (20.3 psi).

Fig. 2. Tread pattern of BKT AT 171 tire.

A modified 16×6.5-inch wheel rim was used to mount this tire on the LW12.8 system.

4. Methodology

4.1 Soil bed preparation

The method of soil bed preparation used by He et al. (2019a) and He et al. (2019b) for tests on sandy loam consisted of shoveling, tilling, leveling, and compacting. Shoveling and tilling were used to destress and declutter the soil respectively which was then levelled and compacted as required. Tilling operation was adopted as the sandy loam had a high moisture content which led to soil at lower layer to lump together. (He et al., 2019b) used a roller for the compaction process. Because of low cohesion and moisture content of GRC-1, tilling operation was not necessary. A similar process was adopted initially however it was observed that the roller bulldozed the sand along with compaction, due to which even after multiple passes, the desired compaction was not achieved. Hence it was decided to perform the experiments on uncompacted terrain as the average measured cone index gradient for the uncompacted soil (2.52 kPa/mm at 40 mm depth) seemed sufficient for the experimentation.

Another method of preparing the soil bed was described by Oravec (2009) in her dissertation. It included shoveling the soil in a specific manner and then using a tamper to compact the terrain. The tamper should be lifted to a specific height and allowed to fall on the terrain only under the influence of gravity, without any external force. This method provided consistent results but is time intensive.

For preparing the soil bed, the first step involved shoveling the soil bed thoroughly to remove residual stresses developed in the sand during the previous test run and then distribute the soil evenly on the bed.

Fig. 3. Sand bed after shoveling.

Fig. 4. Test Rig during levelling operation.

4.2 Test rig setup

The test rig is set as per the required slip ratio and load settings.

For the calculation of slip, it was necessary to estimate the value of the effective rolling radius. As defined by Grisso (2012), He et al. 2020, Kutzbach et al. (2019), Preda and Sandu (2022), Wong (2009) a pure rolling wheel has some torque applied to overcome the rolling resistance i.e. a wheel having zero drawbar pull or net tractive force. To achieve this condition some trial and error was performed by keeping the carriage velocity constant and applying varying angular velocities for the wheel. Thus, small amounts of torque were applied on wheel and the condition in which the longitudinal force measured by the wheel force transducer is negligible was considered as a zero-slip condition. From this condition, the rolling radius was calculated using the measured angular velocity of the wheel and the linear velocity of the carriage as described in Eq. 1.

$$R_r = \frac{v}{\omega} \tag{1}$$

For the given set of tests, the linear speed of the test rig was set to 0.5 km/h and was kept constant. Different slip ratios were obtained by changing the wheel rotational speed. The required rotation speed of the wheel in (rad/s) is then multiplied by a conversion factor to obtain the nominal slip ratio in ladder units (test rig control units). This nominal slip ratio was then coded in the PicPro software which controls the wheel motor.

Depending on the required wheel load, weights were added or removed from the offloading system. The normal load on the wheel center measured by the wheel force transducer was validated with normal load on wheel contact patch using a load cell.

Once the soil bed is prepared, the wheel is lowered using the winch. The data acquisition setup is then set to record, and the run is executed.

4.3 3D scanning

Once the run is executed, a section of the soil bed is chosen for 3D scanning. The scanner is connected to the laptop and calibrated.

While scanning the footprint, it was made sure that the scanner is moved at a constant velocity at a fixed height from the footprint. Once the scan was complete, a mesh with 10,000 nodes was generated from the point cloud data and exported as an .stl file (Stereolithography file format). This gave a good representation of the data and generated .stl files of reasonable size, which were then exported to CAD software for post processing.

In the CAD software, the undisturbed terrain level was identified by taking the flat portion of the soil captured in the 3D scan. A fixed area of 635 mm width and 400 mm length was considered for further analysis from the scanned model. Using necessary tools in the CAD software, certain key quantities were measured, as follows: (a) the maximum depth of rut.; (b) the volume of penetration of the tire below the undisturbed terrain; (c) the volume of soil deposited in the lateral direction, above the undisturbed soil level; and (d) The distance between consecutive tread marks in the soil. The above quantities were measured for each run and then compared.

4.4 Force Hub data processing

The data collected from the force hub was imported into MATLAB software. A Fast Fourier transform (FFT) was done considering the sampling frequency of 1000 Hz. Using the FFT output, the cutoff frequency of the Chebyshev type 2 digital filter was decided as 20 Hz. Then, the filtered data was converted from mV to N using the calibration information of the Force Hub. Offsets were also considered to account for the weight of the Force Hub system.

Due to the manual intervention required to start the rig and data recording, the two actions were not synced. For the output data obtained, the data collected can be divided into 3 sections: a) before the run; b) start of the run (transient condition). c) the steady state condition; and d) after the run ends. The section of interest for data analysis is section C, hence the rest of the data was trimmed off. This was done individually for each run and the average of this data was then calculated and considered as the output.

5. Results and Discussion

5.1 3D Scan interpretation

The scanned footprint is analyzed for the dimensions of lugs imprint and volume of the sand. All the scans shown below are for the tests performed at 2000 N normal load. As shown in Fig. 5, the longitudinal distance between two consecutive lug imprints is termed as wavelength (λ) of the footprint.

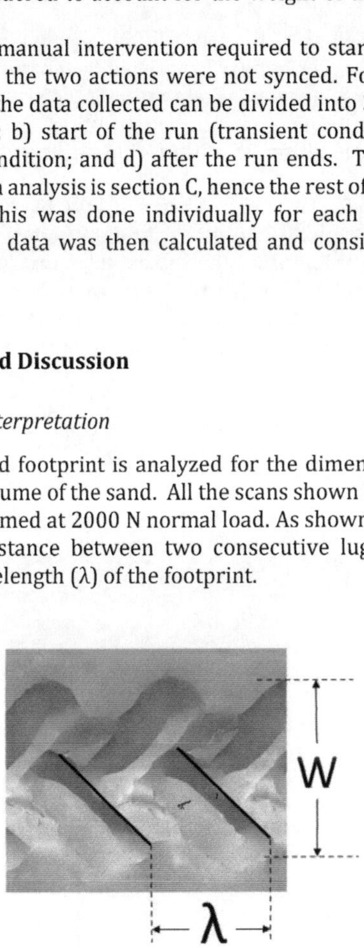

Fig. 5. Top view of scanned footprint

As seen in the front view, the rut left by the tire is 'V' shaped with two peaks at both sides. The flat area seen on the extreme right side of Fig. 6, is the undisturbed soil level. It is used as a reference for measuring the penetration volumes, the displaced volumes, and the rut depth of the soil. The volume enclosed below the reference plane and the sides of the rut in the center of the scanned footprint is termed as the penetration volume. The volume above the reference plane and the two peaks forming the shoulder of the rut as seen in Fig. 6, is termed as the displaced volume.

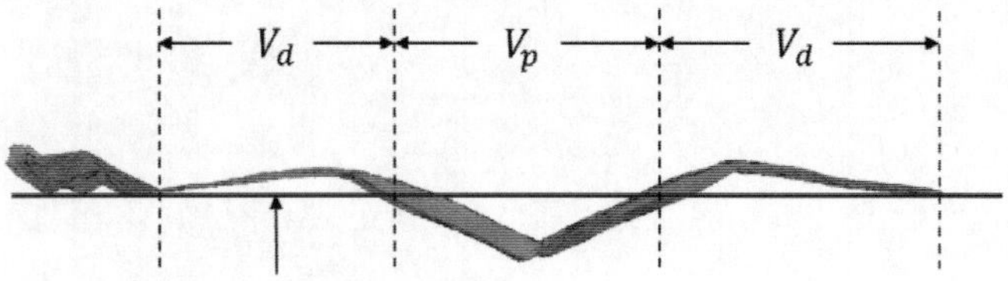

Fig. 6. Front view of scanned footprint

Fig. 7 represents the tire footprint at different values of positive slip ratios. The scanned data of each run was processed, and the images were copied next to each other in increasing order of the slip ratios. The footprint of the tire lugs, which can be prominently seen and distinguished for values of lower slip, move closer to each other as the slip increases. For slip ratios over 30%, the lug footprint blends in the wave pattern seen in the rut. The wavelength of the footprint reduces as the slip increases. Also, for negative values of slip ratios, the lug footprint elongates as the slip increases to -60%.

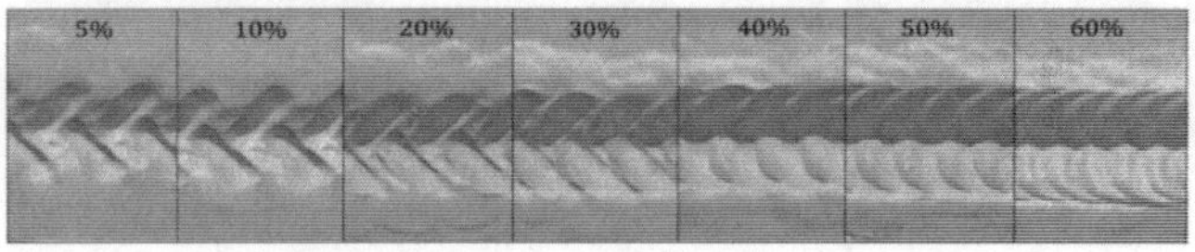

Fig. 7. Scanned tire footprint for positive slip ratios (Traction)

For lower values of slip ratios, the soil deposition along the shoulder of the footprint is considerably low. When we move further in traction or braking zones, the deposition of soil along the shoulder increases. With increasing slip (both traction and braking), more soil is excavated or scooped out and deposited along the trailing end of the tire. For positive slip ratios, it was observed that the deposition is more on one shoulder than another because of the slight camber induced due to the construction of the test rig.

For negative slip ratios, the shear failure zones (semi-circular structure at rut shoulder), can be seen at the trailing edge of the lug footprint. As the slip ratio increased on both ends, the peak along the rut shoulder became more prominent.

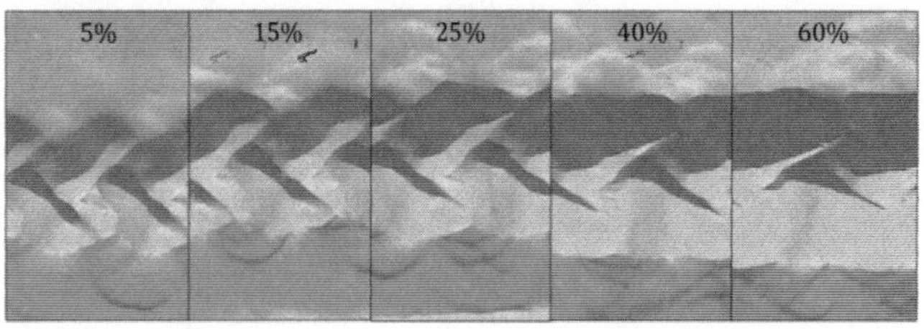

Fig. 8. Scanned tire footprint for negative slip ratios (Braking)

As evident in Fig. 6, the maximum soil sinkage occurs along the centerline of the rut profile. The rut depth or sinkage for positive slip ratios is comparatively constant while the sinkage increases for negative slips. This can be attributed to the reason that off-road tires are designed to be operated in a single direction. The increase in sinkage for negative values of slip can be observed in Fig. 9. It can be noticed that the tire generates more braking force than tractive force due to terrain resistance which in turn depends on sinkage.

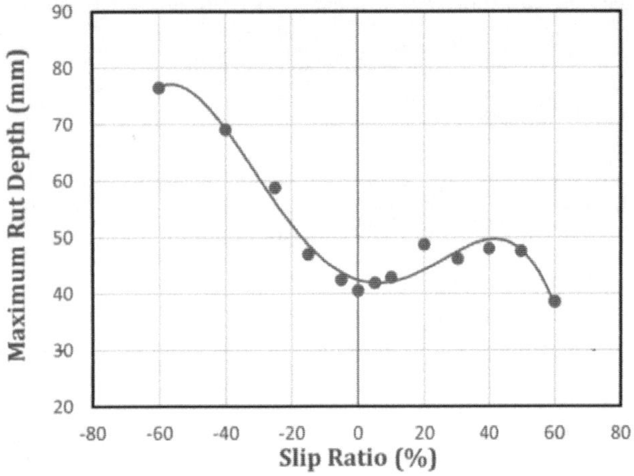

Fig. 9. Maximum Rut depth vs Slip Ratio

The displaced soil and the penetration volumes of the soil follow a similar trend as the maximum rut depth as seen in Fig. 10 and Fig. 11 as lugs of the tire are constructed obliquely to the direction of travel. During braking condition, as the soil is bulldozed by the tire lugs, the soil accumulating at the leading lug is slowly channeled and deposited along the shoulder. In case of positive slip ratio, soil is excavated by the lugs resulting in more soil being deposited at the trailing lug. Hence, more soil is deposited on the rut track for cases of positive slip ratio while in case of negative slip ratio, the soil is deposited at the shoulder of the rut track.

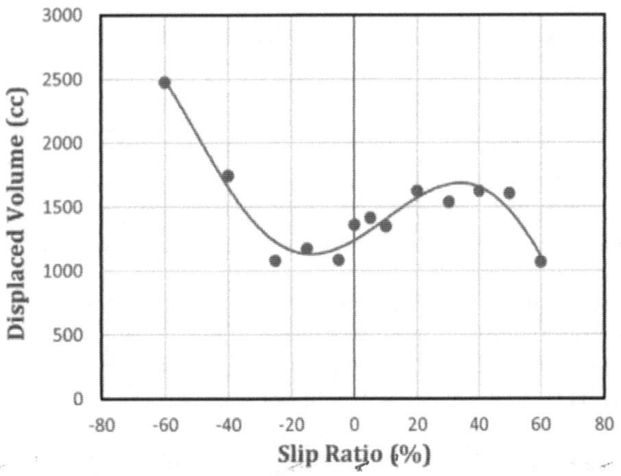

Fig. 10. Displaced volume vs Slip ratio

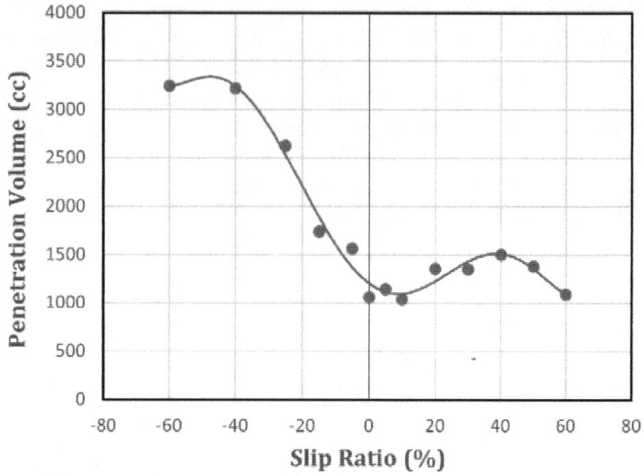

Fig. 11. Penetration volume vs Slip ratio

The wavelength of the footprint is inversely proportional to the slip ratio. As seen in fig. 12, there is a high correlation between the footprint wavelength and slip ratio. As the slip ratio increases, the lug imprints move closer to each other forming waves or ripples in the rut. This relationship will be different for each set of tires and terrain.

A similar effect was also observed by (He, 2019c) when he experimented with SRTT (Standard reference test tire) on sandy loam. At higher slip ratios, the rut shoulder protrudes above the undisturbed soil level which can be attributed to the bulldozing effect of the tire sidewall. The waves forming in the rut were observed by (He, 2019), who referred to this as the 'Ripple' effect. This primarily occurs when the tip of the local shear failure zone of the soil spreads to the surface. In this case, the shear failure zones are more evident in lower slip ratios along the rut shoulder due to taller lugs of the off-road tire used.

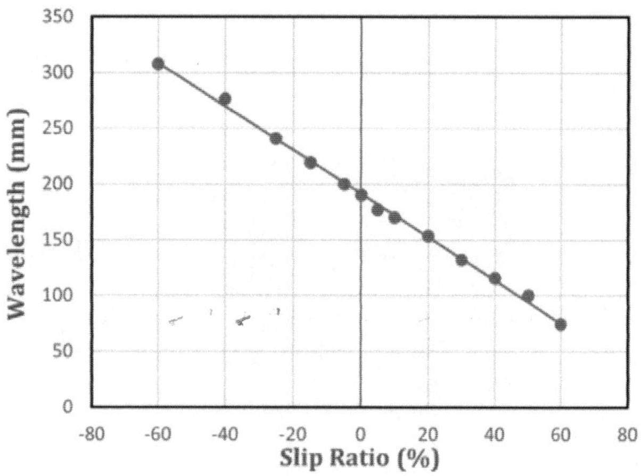

Fig. 12. Wavelength vs Slip ratio

With the above data, an evaluation of tire rut can provide us with the actual slip ratio the tire was operating at a given point which can help us evaluate the tractive force.

5.2 Drawbar pull interpretation

Initially, the tests were performed on the rated load of the tire, which was 4000 N. The following Fig. 13, shows the net traction coefficient obtained for various slip ratios. The net traction coefficient can be defined as the ratio of the drawbar pull to dynamic load.(He et al., 2020)

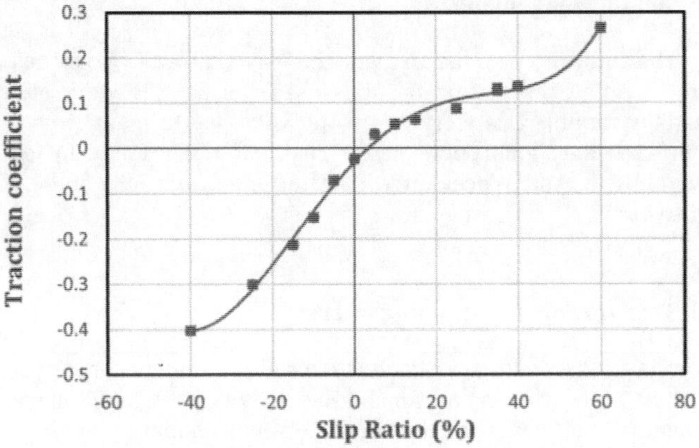

Fig. 13. Traction coefficient vs Slip (4000 N)

While a plateau was expected at the high slip ratios, a sudden increase was observed. This was most likely due to the boundary conditions effect (i.e., getting too close to the rigid bottom of the sand bed), due to high sinkage values. The general rule of thumb to be followed is that the undisturbed soil bed depth should be five times the overall sinkage and the overall width of the bin should also be five times the contact area of the tire respectively, so that the boundary effects would not be significant. This was also followed by (Oravec, 2009) for her experimentation for characterization of GRC-1. This was not the case when the tests were performed at 4000 N normal load as the sinkage obtained was more. As a result, it was decided to perform tests at a lower normal load (2000 N), to achieve lower sinkage values and eliminate the effect of boundary.

The below graph depicts the result of tests conducted at a normal load of 2000 N.

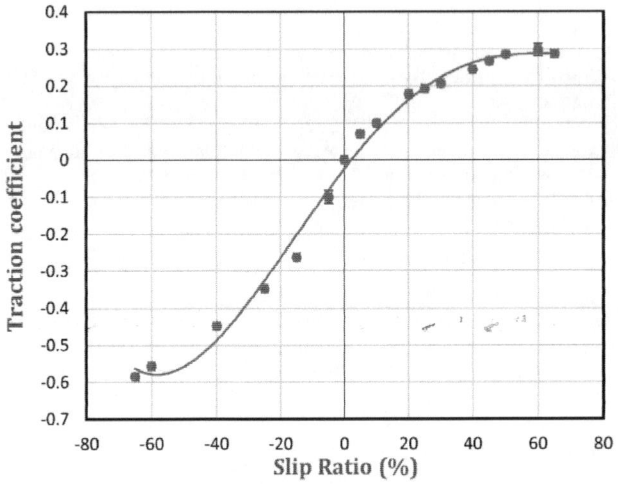

Fig. 14. Traction coefficient vs Slip ratio (2000 N)

The results of this test were more in accordance with what was expected. A plateau was observed in 60% slip. Lower sinkage values obtained for lower normal load were within the limits of rule of thumb mentioned above. This confirms the influence of boundary effect on the test results.

The higher braking force generated in the negative slip ratio zone is due to resistive force of terrain i.e., bulldozing and compaction resistance adding to the forces generated by the tire.

5.3 *Semi-empirical model*

The relationship between the net traction coefficient with the penetration volume is shown in Fig. 15.

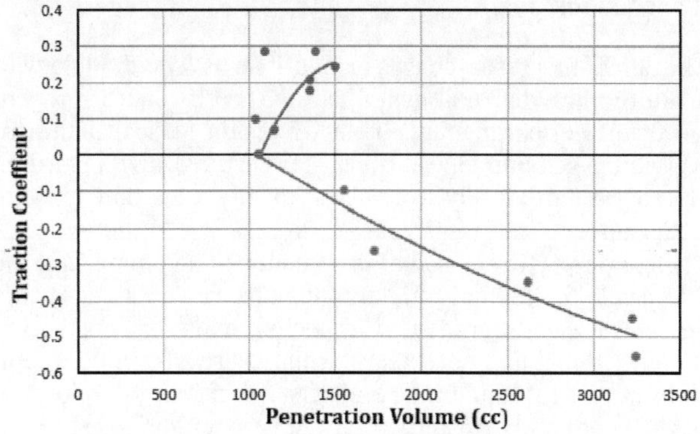

Fig. 15. Traction coefficient vs Slip ratio (2000 N)

As the graph depicts, the datapoints are concentrated about a narrow patch and a second order polynomial curve is fitted to capture all the data points. As one of the points was an outlier, it was not considered while fitting the data.

This may induce an uncertainty in predicting the traction coefficient as for a single value of penetration volume, there are two corresponding values of net traction coefficient. This is due to the reason that a single curve cannot be used to fit the data in both traction and braking conditions. Hence, although the penetration or the rut volume shows a good correlation for both traction and braking, it may not be the best quantity used to predict the traction coefficient.

A semi-empirical relationship between the traction coefficient and the penetration volume can established as follows:

$$\mu_{tr} = \begin{cases} -10^{-6} \times {V_p}^2 + 0.0034 \times V_p - 2.3983 & for\ traction \\ 3 \times 10^{-8} \times V_p - 0.0003 \times V_p + 0.3305 & for\ braking \end{cases} \quad (2)$$

The relationship between the net traction coefficient with the wavelength is shown in Fig. 16. A good correlation can be observed between the two quantities, which makes it a suitable parameter to develop semi-empirical relationships for predicting net tractive coefficient.

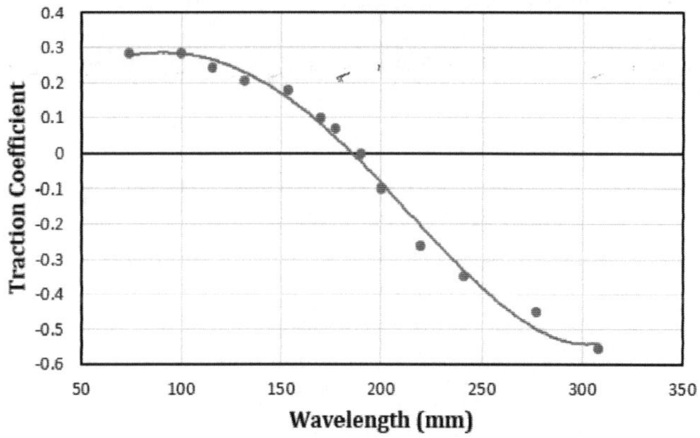

Fig. 16. Traction coefficient vs Slip ratio (2000 N)

The relationship between the net traction coefficient and the wavelength of the tire lug footprint can be given in the Eq. 3.

$$\mu_{tr} = 5 \times 10^{-10} \times \lambda^4 - 2 \times 10^{-7} \times \lambda^3 + 6 \times 10^{-6} \times \lambda^2 + 0.003\lambda + 0.0892 \quad (3)$$

6. Conclusions

The aim of this research was to identify a new methodology to correlate the drawbar pull generated by the off-road tire to tire footprint. Various parameters obtained from the 3D scan of the tire rut were analyzed and logical conclusions were drawn regarding which parameter closely correlates to the variations in the drawbar pull.

The rut depth (tire sinkage), the penetration volume, and the displaced volumes follow a similar trend. The data obtained for the above values was different than expected, however considering the unidirectional lugs of the tire, high vertical stiffness, and possible presence of boundary effects for the heavier load at higher slip ratios, the results seem realistic for the given set of experiments performed.

It was observed that the rut depth and volume of soil displaced is consistent in the traction zone, while it increases in the braking zone. This reaffirms that the tire is unidirectional and is designed to operate in a forward direction.

The wavelength of the tire footprint was observed to vary linearly with slip, increasing with negative slip and decreasing with positive slip. This can be reasoned by an increase in angular speed of the wheel, causing the lugs to interact more quickly and thus at shorter travel lengths. Due to the high correlation between wavelength and slip ratio, this was used to develop a semi-empirical model to estimate the net tractive coefficient.

7. Future work

A tire with a bi-directional thread pattern should be used for the volume study as this would provide more consistent results in traction and braking.

Further tests can be done with a greater soil depth at 4000 N to evaluate the performance of the tire under its rated load.

In the analysis of 3D scanner, there was some difficulty in determining the undisturbed soil terrain due to higher slips, as there was more deposition of soil in the lateral direction after the run. A fixed reference needs to be added to the soil bed, and that needs to be used to determine the undisturbed soil level.

Methods which can provide real time data of rut characteristics can be more useful for correlating it with the tractive forces.

8. Nomenclature

λ	Footprint Wavelength	[mm]
W	Rut Width	[mm]
V_p	Penetration volume	[cc]
V_d	Displaced Volume	[cc]
R_r	Rolling Radius	[m]
V	Longitudinal velocity of carriage	[m/s]
ω	Angular velocity of wheel	[rad/s]
μ_{tr}	Net traction coefficient	-

9. Acknowledgements

The authors would like to thank the TMVS Lab at Virginia Tech for providing the testing facilities used during this work and partially funding this work. We would also like to mention and appreciate the support of Goodyear Tire and Rubber Company for providing the soil, Wheel Force Transducer and partially funding this work.

10. Declaration of competing interest

The authors declare that they have no known competing financial interests or personal relationships that could have appeared to influence the work reported in this paper.

11. References

Farhadi, P., Golmohammadi, A., Sharifi, A., Shahgholi, G., 2018. Potential of three-dimensional footprint mold in investigating the effect of tractor tire contact volume changes on rolling resistance. J Terramech 78, 63–72. https://doi.org/10.1016/j.jterra.2018.05.003

Grisso, R.D., 2012. Traction and Tractor Performance Performance measurement of three-point mounted implement guidance systems View project.

He, R., Sandu, C., Mousavi, H., Shenvi, M.N., Braun, K., Kruger, R., Els, P.S., 2020. Updated standards of the international society for terrain-vehicle systems. J Terramech. https://doi.org/10.1016/j.jterra.2020.06.007

He, R., Sandu, C., Osorio, J.E., 2019a. Systematic tests for study of tire tractive performance on soft soil: Part II – Parameterization of terramechanics model and tire model. J Terramech 85, 77–88. https://doi.org/10.1016/j.jterra.2019.07.003

He, R., Sandu, C., Osorio, J.E., 2019b. Systematic tests for study of tire tractive performance on soft soil: Part I – Experimental data collection. J Terramech 85, 59–76. https://doi.org/10.1016/j.jterra.2019.07.004

He, R., Sandu, C., Sandu, A., Taheri, S., Kurdila, A.J., 2019c. Systematic Tire Testing and Model Parameterization for Tire Traction on Soft Soil. Blacksburg.

Kenarsari, A.E., Vitton, S.J., Beard, J.E., 2017. Creating 3D models of tractor tire footprints using close-range digital photogrammetry. J Terramech 74, 1–11. https://doi.org/10.1016/j.jterra.2017.06.001

Kutzbach, H.D., Bürger, A., Böttinger, S., 2019. Rolling radii and moment arm of the wheel load for pneumatic tyres. J Terramech 82, 13–21. https://doi.org/10.1016/j.jterra.2018.11.002

Oravec, H.A., 2009. UNDERSTANDING MECHANICAL BEHAVIOR OF LUNAR SOILS FOR THE STUDY OF VEHICLE MOBILITY.

Preda, I., Sandu Corina, 2022. CLARIFICĂRI ŞI ECHIVALENŢE: RAZELE ROŢILOR CLARIFICATIONS AND EQUIVALENCES: WHEEL RADII.

Ptak, W., Czarnecki, J., Brennensthul, M., 2022. Use of 3D scanning technique to determine tire deformation in static conditions. Journal of Agricultural Engineering 53. https://doi.org/10.4081/jae.2022.1221

Ptak, W., Czarnecki, J., Brennensthul, M., Lejman, K., Małecka, A., 2023. Evaluation of Tire Footprint in Soil Using an Innovative 3D Scanning Method. Agriculture 13, 514. https://doi.org/10.3390/agriculture13030514

Talbot, B., Astrup, R., 2021. A review of sensors, sensor-platforms and methods used in 3D modelling of soil displacement after timber harvesting. Croatian Journal of Forest Engineering. https://doi.org/10.5552/crojfe.2021.837

Wong, J.Y., 2009. Terramechanics and off-road vehicle engineering (2nd ed). Elsevier, Oxford.

7577 | Implementation of moving loads on ice in NRMM

Susan Frankenstein [a,*], Kathleen F. Jones [a]

[a] *ERDC CRREL, Hanover, New Hampshire, United States*
* *Corresponding author: susan.frankenstein@erdc.dren.mil*

ABSTRACT

Fording capabilities are common in cross country mobility models. They are limited in capability however due to their dependence on timely water level inputs and are best used for planning purposes based on monthly averages. Less common still are winter crossing considerations. To accurately predict the maximum vehicle speed on an ice cover the ice thickness, ice properties, and water depth are required. To solve for the speed an iterative process is needed which is prohibitive in mobility models. In order to implement in NRMM (NATO Reference Mobility Model) critical speed curves were generated for ice thicknesses ranging from 0.01 – 5 m and 3000 water depths spanning 0.1 – 300 m. Using these curves, two equations, one for the steady state speed and a second for the critical speed as functions of water depth and ice thickness were developed. If the water depth is known, the final speed is the minimum of the two otherwise it is set to 6.7 m/s. In NRMM, the minimum ice thickness needed to support a single non-moving vehicle is calculated. Adjustments to the minimum ice thickness can be made for weaker ice. If the ice thickness is greater than the minimum thickness, then the speed and minimum vehicle spacing are calculated.

Keywords
Fording
Winter Mobility
Off-Road

1. Introduction

The bearing capacity of ice is not a new field of study. Using frozen water bodies as winter mobility corridors has occurred for generations. One of the first investigations was conducted by Hertz (1884). Kerr (1975) presents a critical literature survey on the bearing capacity of static loads on floating ice plates. The majority of the cited studies assume that the ice and underlying water behave elastically. Solutions to these formulations to determine the minimum ice thickness given a load are either iterative or empirical for both static and moving loads and not suitable for inclusion into mobility models such as NRMM (NATO Reference Mobility Model).

In 1971 the Army Materiel Command (AMC) funded development of the AMC '71 mobility model. This model combined the efforts of WES (U.S. Army Waterways Experiment Station), TACOM (Tank Automotive COMmand) and CRREL (U.S. Army Cold Regions Research and Engineering Laboratory) (Fatherree 2006). AMC '71 was mainly used to assess off-road performance design considerations for new vehicle development. The AMC '71 model was soon followed by AMC '74. Major revisions included vehicle performance on slippery soils, muskeg and shallow snow; improved driver behavior; travel on paved roads; upgrades to the computer code (Jurkat et al. 1975b). AMC '74 was intended only as an interim until what is considered the second-generation Army Mobility Model (AMM-75) could be released (Jurkat et al. 1975a, 1975b). NATO adopted AMM-75 in 1977 and changed the name to the NRMM (Fatherree 2006). The three main user groups for NRMM are vehicle developers, vehicle procurement and vehicle users (Haley et al. 1979, Ahlvin and Haley 1992).

In this paper we present new algorithms that eliminate the need to iterate for ice thickness and vehicle speed on an ice cover if the water depth is known. Equations are also presented to determine minimum ice thickness and vehicle spacing for a given static load assuming typical midwinter ice properties as well as how to modify them for weaker ice typically found during melting or snow or for longer term loads. These algorithms have been implemented in a research version of NRMM.

2. Minimum Ice Thickness - Short-term Static Load

For short-term static loads on a uniform, clear (black) ice sheet, Fransson (2009) gives the failure load, or maximum vehicle weight, (P_u (N)) as

$$P_u = 1.25(1 + 1.68\alpha)\sigma_f h^2 \quad \alpha = \frac{r}{L}$$
$$L = \sqrt[4]{\frac{D}{\rho_w g}} \qquad D = \frac{Eh^3}{12(1-\upsilon^2)} \tag{1}$$

where h (m) is the ice thickness, σ_f (Pa) is the flexural strength of the ice, r (m) is the load radius, L (m) is the characteristic length, E (Pa) is the elastic modulus of ice, υ is Poisson's ratio, ρ_w (kg/m^3) is the density of water, D(N·m) is the ice flexural rigidity, and g (m/s^2) is gravitational acceleration. Equation (1) is from Panfilov (1960) whose theory is based on a thin plate fully supported by the water below. Using an iterative process and setting the critical load, P_u, equal to the vehicle weight, Eq. (1) was used to determine the ice needed to support a given vehicle for 147 tracked and wheeled vehicles ranging in weight from 18 – 82100 kg. In doing so, the load radius was assumed to be the equivalent radius, $r_e = 0.5\left[{1.3(l \cdot w)^{0.625}}/{(l + w)^{0.25}}\right]$ (Engineering ToolBox 2001) where l (m) and w (m) are the load length and width respectively.

https://doi.org/10.56884/GQVF4190
ISBN 978-1-942112-55-6 (pdf)

It was assumed that g = 9.81 m/s², ρ_w = 1000 kg/m³, σ_f = 0.6 MPa, υ = 0.3, and E = 4500 MPa. Using all of these values and the curve fitting toolbox in *Matlab* (MathWorks) the following equation for minimum ice thickness as a function of vehicle weight was obtained

$$h_{min}(\mathrm{m}) = aW^{0.52}$$
$$\text{where } a = \begin{cases} 2.45\text{x}10^{-3} & W(\mathrm{kg}) \\ 1.62\text{x}10^{-3} & W(\mathrm{lb}) \\ 8.65\text{x}10^{-2} & W(\mathrm{MLC}) \end{cases} \quad (2)$$

where h_{min} is the minimum ice thickness needed to support a vehicle and W is the gross vehicle weight and MLC is military load class (FM 3-34.170 2008). For all fits r^2 = 1.0. Use of Eq. (2) eliminates the need to iteratively determine the ice thickness for a known load of given size.

Current practices are based on Gold's Formula (Gold 1960, 1971), $h_{min} = 4\sqrt{W}$ (h_{min} (in), W (MLC)). Table 1 of the CRREL IEIEB (Ice Engineering Information Exchange Bulletin) (Haynes and Carey 1996) and Table 8-1 of the Engineering Manual EM-1110-2-1612 (US Army 2002) give values for several wheeled and tracked vehicles. These values are more conservative than those proposed using Panfilov's method due to his assumed ice properties. Both sets of data as well as the curve fits for them are shown in Fig. 1. The maximum difference between the two curves is approximately 4.24 cm (1.67 in).

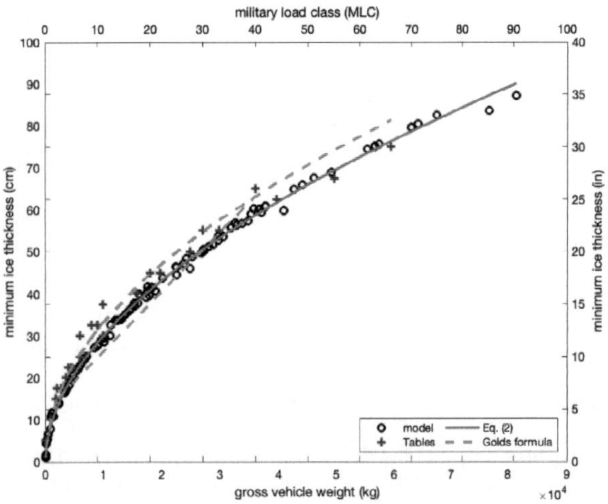

Fig. 1. Data contained in the CRREL bulletin (Haynes and Carey 1996) and US Army EM_110-2-1612 (US Army 2002) as well as model data with the respective curve fits.

3. Vehicle Spacing

Also listed in Table 1 of Haynes and Carey (1996) and Table 8-1 in US Army (2002) are the minimum recommended wheeled and tracked vehicle spacings for stationary vehicles. The rule of thumb for wheeled vehicles is that they should be spaced about 100 times the ice thickness when the ice thickness is at the minimum required. Tracked vehicles that are the same load as their wheeled counterparts can be spaced closer together due to their larger contact areas. Using the data in the tables the following relation was found for the minimum spacing, s_{min} (m), as a function of vehicle weight, W (kg, lb, MLC) ($r^2 \geq 0.975$)

$$s_{min}(\text{wheeled}) = \mathrm{a}_w W^{0.52}$$
$$\text{where } a_w = \begin{cases} 0.28 & W(\mathrm{kg}) \\ 0.19 & W(\mathrm{lb}) \\ 9.55 & W(\mathrm{MLC}) \end{cases}$$
$$s_{min}(\text{tracked}) = \mathrm{a}_t W^{0.47} \quad (3)$$
$$\text{where } a_t = \begin{cases} 0.25 & W(\mathrm{kg}) \\ 0.17 & W(\mathrm{lb}) \\ 0.26 & W(\mathrm{MLC}) \end{cases}$$

The bulletin and engineering manual data along with both curve fits are presented in Fig. 2.

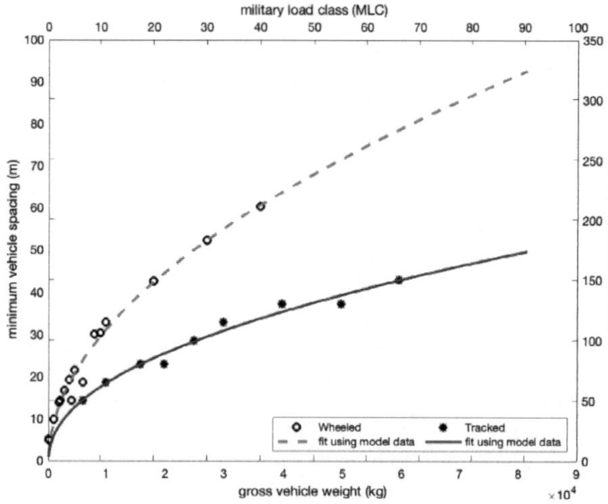

Fig. 2. Minimum stationary vehicle spacing as a function of gross vehicle weight for wheeled and tracked vehicles.

4. Moving Loads

Unlike the minimum ice thickness and minimum spacing calculations for static loads, critical vehicle speed is dependent on the water depth as well as ice thickness. Squire et al. (1996) give the phase speed, c, as

$$c = \sqrt{\frac{\frac{Dk^4}{\rho_w}+g}{k(kh'+\coth kH)}} \quad (4)$$

where $k = {}^{2\pi}/_{\lambda}$ (m⁻¹) is the wave number, λ (m) is the wave length, H (m) is the water depth, and $h' = {}^{h\rho_i}/_{\rho_w}$ (m) is the density weighted ice thickness. The critical vehicle speed occurs where the phase speed, c, is minimized, i.e. where ${}^{\partial c}/_{\partial k} = 0$, or when the vehicle is traveling at the same speed as the water wave underneath the ice. This is also when the largest deflections will occur. Using 37 ice thicknesses ranging from 0.01 – 5.00 m and 3000 evenly spaced water depths from 0.1 – 300.0 m the critical speed, v_{crit} (m/s), was determined using the convergence criteria ${}^{\partial c}/_{\partial k} \leq 0.00005$. The specific ice thicknesses are: 0.01 m; every 0.05 m from 0.05 – 1.00 m; every 0.10 m from 1.10 – 2.00 m; every 0.50 from 2.50 – 5.00 m. The results are shown in Fig. 3.

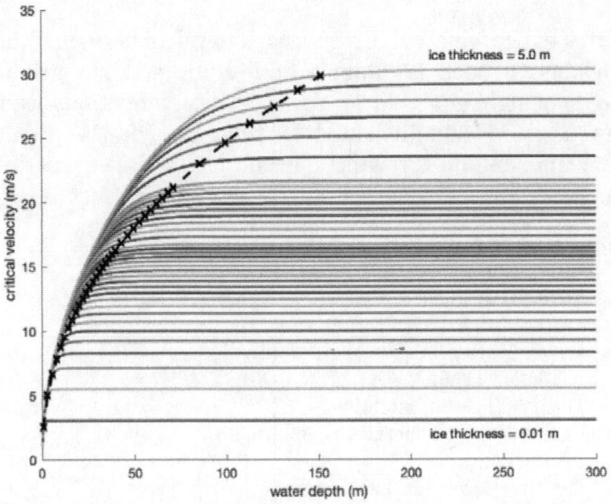

Fig. 3. Critical speed as a function of water depth for 37 ice thicknesses ranging from 0.01 – 5.0 m. Also plotted are the depths at which the asymptotic velocity is considered to be reached.

For each of the 37 ice thicknesses, the critical speed as a function of water depth was plotted. It was noted, that after a certain depth, the curves flatten out and approach a constant, or asymptotic, value. For each ice thickness, the maximum safe speed was defined as the speed at 300 m (the maximum depth we used) - 0.5 m/s (1 knot, 1.2 mph). These are the 'x' points shown in Fig. 4. Using *TableCurve3D* (Inpixon) a maximum speed that applies to all ice thickness and depths, v_{max} (m/s), was determined. It is defined as

$$v_{max} = 16.68\exp\left(-\frac{0.15}{\sqrt{H}}\right) h^{0.37} \tag{5}$$

Again, water depth, *H*, and ice thickness, *h*, are in meters. Equation (5) is plotted in Fig. 4 as the dashed black curve. If the water depth is unknown, the CRREL IEIEB (1996) recommends a maximum vehicle speed of 15 mph (6.7 m/s) which is consistent with *H* = 0.10 m and *h* = 0.31 m, *H* = 1.0 m and *h* = 0.13 m, or *H* = 10.0 m and *h* = 0.10 m.

Once these curves were generated, the 'All Equations' choice in *TableCurve3D* software was employed to determine a single best-fit (r^2 = 0.998) equation for v_{crit} in terms of the water depth and ice thickness

$$v_{crit} = \frac{3.07799981 + 0.84123416 \ln H - 0.5219952 \ln h}{1 - 0.280118 \ln H + 0.02892509 \ln^2 H - 0.0716369 \ln h} - \frac{0.1287891 \ln^2 h + 0.0121051 \ln^3 h}{1 - 0.280118 \ln H + 0.02892509 \ln^2 H - 0.0716369 \ln h} \tag{6}$$

The plot generated with *TableCurve3D* of the 37 individual curves (thick black lines) as well as the fit surface (grided) is shown in Fig. 4.

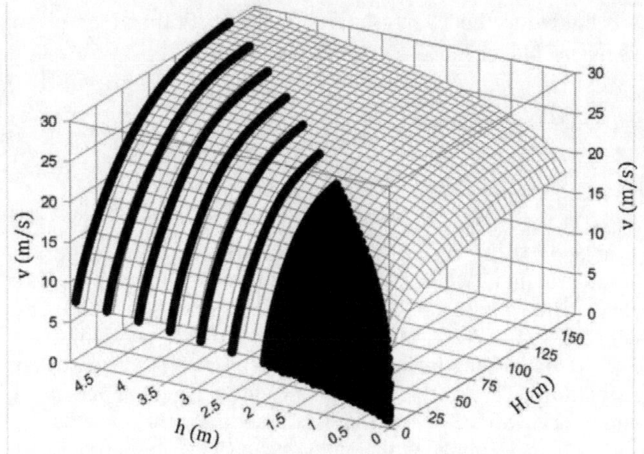

Fig. 4. Plots of the critical velocities for the 37 ice thicknesses over the depths 0 - 300 m (thick black lines) and the resulting best-fit surface (gridded) determined using the 'all equations' mode in *TableCurve3D*.

5. Final Implementation in NRMM

In NRMM3.0_CRRELbeta (Frankenstein and Richmond 2016) if it is indicated that a river or lake is frozen, then Eq. (2) is used to determine if the ice is thick enough to support a single vehicle. If *h* > h_{min}, the ice is thick enough to support the vehicle and Eq. (3) determines the vehicle spacing. If the water depth is known, then maximum vehicle speed $v = \min(v_{crit}, v_{max})$ else *v* = 6.7 m/s (15 mph, 24 kph). Resulting speeds for twelve different water depth are shown in Fig. 5.

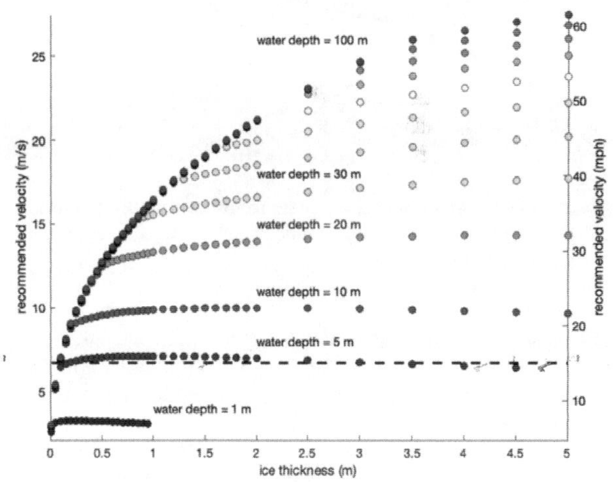

Fig. 5. Recommended vehicle speeds for 12 different water depths (1, 5, 10, 20, 30, 40, 50, 60, 70, 80, 90, 100) meters as a function ofice thickness. The dashed line is the default speed of 6.7 m/s.

6. Adjustments for Weaker Ice and Cracks

The maximum deflection, δ (m), of an infinite ice sheet is given by (Fransson 2009, US Army 2002)

$$\delta = \frac{P}{8\rho_w g L^2} \tag{7}$$

If δ is limited to the ice free-board $\left(1 - {}^{\rho_i}/_{\rho_w}\right)h$ then for a given load the ice thickness can be determined using

$$h = \left(\frac{CW(\text{kg})}{\sqrt{E}}\right)^{2/5} \quad \text{where } C = \frac{\sqrt{3g(1-\upsilon^2)/\rho_w}}{4\left(1-\frac{\rho_i}{\rho_w}\right)} \tag{8}$$

Figure 6 plots the resulting ice thicknesses for three different Young's moduli as well as the Panilov curve fit (Eq. 2) and Gold's formula. The Young's moduli values are from Fransson (2009) and EM_110-2-1612 (2002) for short-term (690 MPa) and long-term (55 MPa) loads. Long-term loads can also be thought of as weaker ice situations. The ice thicknesses calculated using the deflection limited curve with E = 4500 MPa is essentially the same as the thicknesses determined with Panilov's and Gold's methods. As the ice weakens (E decreases) thicker ice is needed to support the load for the same deflection. The thickness needed for weaker ice is Eq. (2) multiplied by the weighting function $w_E = \left({}^{E_0}/_{E}\right)^{0.2}$ where E_o = 4500 MPa.

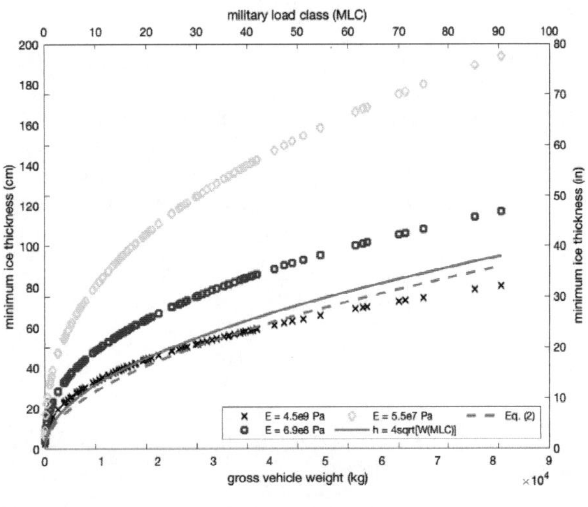

Fig. 6. Ice thickness for deflection limited considerations for three Young's moduli, the curve fit determined from Panfilov's semi-empirical approach, and Gold's formula.

Similarly, the increase in vehicle spacing due to weaker ice Eq. (3) multiplied by the same weighting function w_E. Finally, if wet cracks are present, then the effective load should be doubled (Gold 1971, Haynes and Carey 1996). It is commonly assumed that snow/white ice is weaker than black/clear ice and hence the total ice thickness needs to be adjusted accordingly such that h_{total} = 0.5h_{white} + h_{black} (US Army 2002). This is based on data collected in Russia (Gold 1960). Gow et al. (1978) found that this is untrue if the ice is cold but holds for warmer conditions and is, as such conservative.

7. Declaration of competing interest

We declare that we have no known competing financial interests or personal relationships that could have appeared to influence the work reported in this paper.

8. Conclusions

In this paper we provide algorithms to calculate the minimum ice thickness needed to support both static and moving loads, vehicle spacing, and vehicle speed. These equations can be implemented in mobility models such as NRMM providing valuable information for winter/Arctic logistics and operations planning.

9. Nomenclature

c	Wave celerity	[m/s]
D	Ice flexural rigidity	[N·m]
E	Young's (Elastic) modulus	[Pa]
g	Gravitational acceleration	[m/s²]
h	Ice thickness	[m]
h'	Density corrected ice thickness	[m]
h_{black}	Black/Clear ice thickness	[m]
h_{min}	Minimum ice thickness	[m]
h_{total}	Total ice thickness	[m]
h_{white}	Snow/white ice thickness	[m]
H	Water depth	[m]
H_{min}	Minimum water depth for steady state speed	[m]
k	Wave number	[m⁻¹]
l	Load length	[m]
L	Characteristic length	[m]
MLC	Military Load Class	
P_u	Ultimate load	[N]
r	Load radius	[m]
r_e	Equivalent radius	[m]
s_{min}	Vehicle spacing	[m]
v_{crit}	Critical vehicle speed	[m/s]
v_{max}	Maximum vehicle speed	[m/s]
v_{steady}	Steady state vehicle speed	[m/s]
w	Load width	[m]
w_E	Weaker ice weighting function	
W	Vehicle weight	[kg, lb, MLC]
δ	Ice deflection	[m]
λ	Wave length	[m]
ν	Poisson's ratio	
ρ_i	Ice density	[kg/m³]
ρ_w	Water density	[kg/m³]
σ_f	Ice flexural strength	[Pa]

10. References

Ahlvin, R.B. and P.W. Haley, 1992: NATO Reference Mobility Model Edition II, NRMM II User's Guide. U.S. Army Corps of Engineers Eng. Waterw. Exp. Stn. Tech. Rep. GL-92-19, 286 pp.

Engineering ToolBox. 2001. Available at: https://www.engineeringtoolbox.com/equivalent-diameter-d_205.html

Fatherree, B.H., 2006: The History of Geotechnical Engineering at the Waterways Experiment Station 1932-2000. U.S. Army Corps of Engineers Eng. Res. & Dev. Cen., 362 pp.

Frankenstein, S. and P.W. Richmond. 2016. Code Enhancements, Upgrades and Bug Fixes – NRMM3.0. Proceedings 8th ISTVS Regional Conf., Troy, MI. Woodlands Review, Pulp and Paper Magazine of Canada 61(5), 3-7.

Fransson, L. 2009. Ice Handbook for Engineers, V. 1.2. Luleå University of Technology.

Gold, L.W. 1960. Filed Study on the Load Bearing Capacity of Ice Covers.

Gold, L.W. 1971. Use of Ice Covers for Transportation. Canadian Geotech J. 18(2), 170-181.

Gow, A.J., H.T. Ueda, and J.A. Ricard. 1978. Flexural strength of ice on temperate lakes: Comparative tests of large cantilever and simply supported beams. U.S. Army Corps of Engineers Cold Regions and Eng. Lab. CR-78-9, 18 pp.

Haley, P.W., M.P. Jurkat and P.M. Brady, Jr., 1979: NATO Reference Mobility Model, Edition I Users Guide Volume I. TARADCOM R&D Lab. Tech. Rep.12503, 389 pp.

Haynes, F. D., and K. L. Carey. 1996. Safe Loads on Ice Sheets. Ice Engineering Information Exchange Bulletin MP-96-13. Hanover, NH: ERDC-CRREL.

Hertz, H. 1884. Über das gleichgewicht schwimmender elastischer Platten. Wiedemann's Annalen der Physik und Chemie 22, 449-455.

Jurkat, M.P., C.J. Nuttall and P.W. Haley, 1975a: The U. S. Army Mobility Model (AMM-75). U.S. Army/FRG Army Mobility Sym. Proc., R.W. Siorek, Ed., U.S. Army Tank Automotive Command, Mobility Syst. Lab. Tech. Rep. 12101, 3-50.

Jurkat, M.P., C.J. Nuttall and P.W. Haley, 1975b: The AMC '74 Mobility Model. U.S. Army Tank Automotive Command, Mobility Syst. Lab. Tech. Rep. 11921 (LL149), 430 pp.

Kerr, A. 1975. The Bearing Capacity of Floating Ice Plates Subjected to Static or Quasi-Static Loads. J. Glaciology 17(76), 229-268.

Panfilov, D.F. 1960. Experimental investigation of the carrying capacity of a floating ice plate. In Russian. Izvestia Vsesojuznogo Nauchno-Issledovatelskogo Instituta Gidrotekhniki, 64, 101-115.

Squire, V.A., R.J. Hosking, A.D. Kerr, and P.J. Langhorne. 1996. Moving Loads on Ice Plates. Kluwer Academic Pub., Boston.

US Army. 2002. Ice Engineering, Engineering Manual EM_110-2-1612.

US Army. 2008. FM 3-34.170/MCWP 3-17.4 Military Reconnaissance.

8151 | Study on estimation of traveling states using strain information on chassis of lunar and planetary exploration rovers

Sae Takano [a,*], Kojiro Iizuka [a]

[a] *Department of Machinery and Control Systems, Shibaura Institute of Technology, 307 Fukasaku, Minuma-ku, Saitama City, Saitama, Japan*
** Corresponding author: mf23079@shibaura-it.ac.jp*

ABSTRACT

Lunar exploration is underway around the world. The goals of the investigation are sustainable activity on the lunar surface and technological demonstration for Mars exploration. It is an effective method to land and move for investigation on the lunar surface by rovers. However, there is a possibility of a rover skidding due to the loose soil on the lunar surface. This phenomenon causes the rover to deviate from its intended course. Therefore, it is necessary to obtain the driving state in real time. We focus on the strain generated in the rover's chassis. The strain is caused by some kind of mechanical relationship between the rover's wheels and the ground where it travels. It is therefore expected that the strain has a correlation closely with the driving state. The purpose of this study is to try estimating the driving state of the rover in real time by analyzing the strain value measured with strain gauges mounted on the chassis. In experiment, the data is accumulated with changing voluntarily the slip ratio of a wheel connected to the chassis by changing drawbar mass. By averaging and FFT analyzing of strain values, a correlation is established between the wheel's slip ratio and the strain. The result suggests that the larger the slip ratio of the running wheel, the larger the average value of strain generated the chassis. the larger the slip ratio of the running wheel, the larger the average value of strain generated the chassis. It was also verified that the vibration of certain frequencies was relatively lager when the slip ratio is 0.4~0.8 than one with other slip ratio. This result suggests that the wheel's slip ratio can be assessed by measuring the strain on the chassis. It is concluded that the chassis strain and the rover's driving state are correlated.

Keywords
Planetary exploration
Rover
Moon surface
Strain

1. Introduction

Landing and mobile exploration by rovers is an effective means for the exploration of water resources and geological survey of the moon and planets. The main types of rover movement methods are wheeled, crawler, and legged (Yoshida, 2011). The wheeled type has a simple mechanism and is the most common. On flat terrain, it can move faster and more efficiently than the crawler and walking types. The crawler type has a large ground contact area to distribute the load. Therefore, the crawler type has superior slope climbing performance. On the other hand, the crawler type is difficult to miniaturize because of its complex mechanism around the crawler tracks. When regolith particles penetrate between the crawler and sprockets, the sharp regolith scrapes the sprockets. This can cause the sprockets to come off. The leg type is suitable for traveling on terrain with large bumps and dips in terrain. However, the large number of mechanical components makes reliability and miniaturization issues difficult to overcome. A rover used for unmanned exploration of the lunar and planetary surfaces cannot be repaired in the event of a failure. Therefore, if the rover malfunctions, it may not be able to move, which is the key element of the rover. This would make it impossible to carry out the planned exploration. For these reasons, most of the conventional exploration rovers have adopted a wheeled type with a simple mechanism. However, planetary surfaces such as the moon and Mars have loose soil covered with regolith, which cause the wheels to slip and get stuck. In fact, the Mars Exploration Rover (NASA/JPL) got stuck on the surface of Mars (NASA, NASA's Opportunity Rover Rolls Free on Mars). When a rover runs on a loose ground such as the lunar surface, its wheels repeatedly break and slip. This phenomenon causes wheel spin. However, sensing the running condition is difficult because the process of interaction between the wheel and the ground is non-linear and influenced by many elements. (Ding et al., 2011). To address this problem, methods have recently been proposed to detect rover slippage and other motion hazards using proprioceptive sensing. (Gonzalez et al., 2018; Coloma et al., 2022). On the other hand, in terms of the control of locomotion in vertebrate, in addition to that proprioception is of clear importance, it has been showed that the feedback from muscle spindles regulates adjustments to locomotion disturbance (Santuz et al., 2019). Muscle spindles are sensory organs. They detect changes in muscle stretch and length (Santuz, Akay, 2023). From this information, muscle spindles indirectly sense muscle force, velocity and acceleration (Blum et al., 2020).

This is to be replaced by a rover to detect bending strain on the chassis, the travelling status is presumed to be able to be estimated proprioceptively. Because wheel of the rover directly touches the ground, they influence each other. For this reason, there seems to be some kind of mechanical relationship between the strain which is generated in the rover's chassis. In this study, strain gauges attached on chassis obtain the strain data of chassis. By analyzing the strain value, it is tried to verify the relationship between the slipping state of the rover and strain in the chassis.

2. Measuring bending strain on chassis of rover

The relationship between the ground and the bending strain on the chassis is verified by experiment. Fig. 1 shows the experimental environment. Table 1 shows the experimental conditions. The wheel runs on the loose ground made by Silica sand No. 5. This ground simulates regolith on the lunar surface. The drawbar mass which is connected with rope provide the

https://doi.org/10.56884/QUMN5668
ISBN 978-1-942112-55-6 (pdf)

chassis pulling force. The parameter is drawbar mass. The purpose of changing drawbar mass is to change the slip ratio of running wheel. Drawbar mass are 1.5 kg, 2.5 kg, 3 kg, 4,5 kg, 6 kg and none. The running state is tracked by motion capture system.

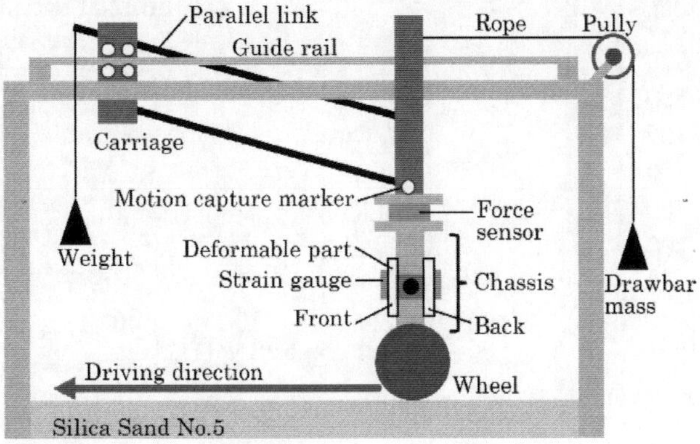

Fig. 1. Experimental environment

Table 1
Experimental conditions

Item	Condition
Ground	Silica sand No. 5
Vertical load [kg]	147
Drawbar mass [kg]	0, 1.5, 2.5, 3.0, 4.5, 6.0
Thickness of deformable part [mm]	3
Rotation speeds [rpm]	5
Number of trials	5

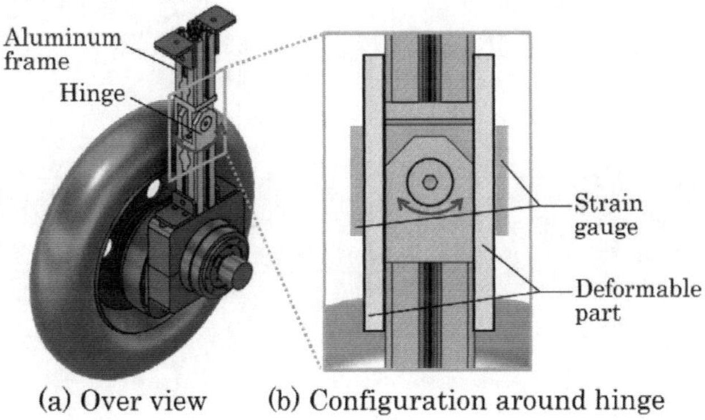

Fig. 2. Proposed chassis

In order to measure bending strain, the chassis should be of a form that maintains rigidity while still allowing for pronounced strain measurement. Therefore, we propose a leg whose motion is restricted in the pitch direction, consisting of an aluminum frame and a hinge, as shown in Fig. 2(a). As shown in Fig. 2(b), only bending strain can be measured by deformable part with a strain gage attached to the hinge. The strain gauge using this experiment is KFGS-5-120-C1 LIM2R made by KYOWA. Strain gauges are attached on two places. As shown in Fig. 1, the strain gage in the direction of wheel travel is "Front" and the opposite side is "Back".

Fig. 3 shows procedure of offset. Before the wheel runs, it is lifted up in the air. After that, it is put on the loose ground without accelerated motion. In this state, the value of strain which is showed at strain logger is offset. The strain data is monitored by data logger while the wheel is running. Fig. 4 shows the direction of strain value. When the chassis bends in the direction in which the wheel runs, the strain logger shows positive value. By contrast, when the chassis bends in the opposite direction, the strain logger shows negative value.

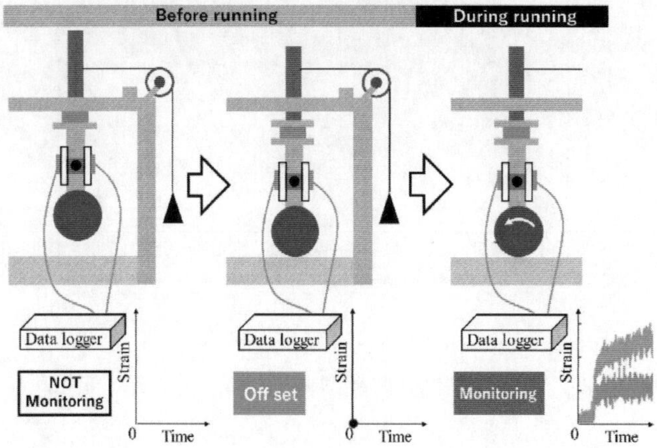

Fig. 3. Procedure of offset of strain value

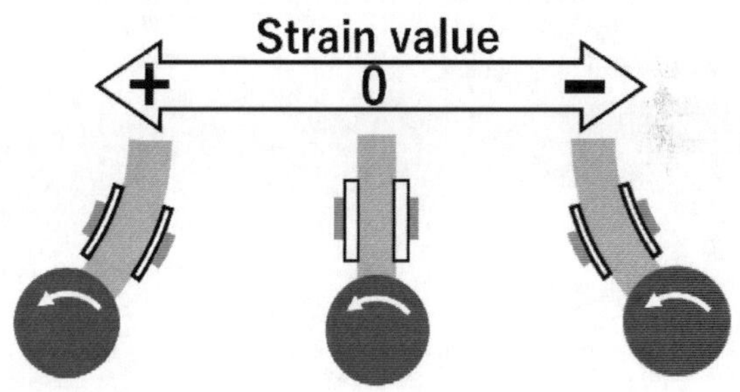

Fig. 4. Direction of strain value

3. Result of experiment

Fig. 5 shows the state of proses of wheel driving. In Fig. 5, the wheel runs from right to left.

Fig. 6 shows the time variation of the strain values when the wheel ran with 4.5kg drawbar mass. After the wheel started to run, the strain value increased rapidly. After that, the value became constant and vibrated fractionally. Under the other terms, the process of changing the value was approximately same. The strain data within 5 seconds which is strain value is extracted from raw strain data. Using this data, three analyses are calculated: average value, slip ratio and FFT analysis.

Fig. 5. State of wheel driving

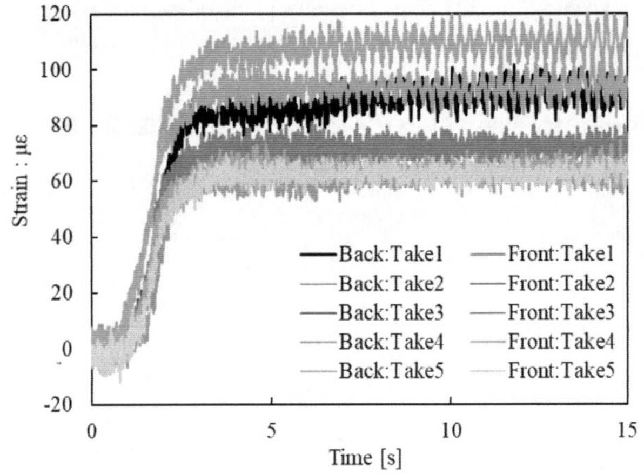

Fig. 6. State of strain value (Drawbar mass : 4.5kg)

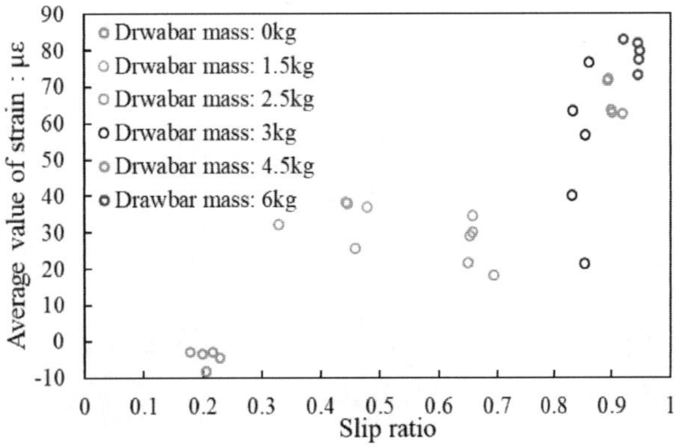

Fig. 7. Relationship between average value of strain of "Front" and slip ratio

The wheel's slip ratio is defined by Equation (1), which is a dimensionless quantity that represents the difference between the wheel peripheral velocity $r\omega$ and the wheel translation velocity v, r is the wheel radius and ω is the wheel angular velocity.

$$s = \frac{r\omega - v}{r\omega} \quad (1)$$

The ratio is calculated using the tracking data by motion capture. Tracking data is extracted for the same time section with strain data.

Fig. 7 shows a graph with the horizontal axis representing the slip ratio and the vertical axis representing the average value of strain of "Front". According to Fig. 8, the larger the slip ratio, the larger the average value of strain. There appears to be a quadratic correlation between slip ratio and average value of strain.

Fig. 8 shows the result of FFT analysis when the wheel ran without drawbar mass. The large power spectrum is shown at around 4 Hz. In addition, some spectrums are appeared from 20 Hz to 25Hz. Fig. 9 shows the result of FFT analysis when the wheel ran with 3 kg drawbar mass. The large power spectrum is shown at around 4 Hz as well as one without drawbar mass. In addition, two large power spectrums are appeared at around 24 Hz and around 27Hz. The results when the wheel ran with 1.5 kg and 2.5 kg drawbar mass resemble to the ones shown Fig. 9. Fig. 10 shows the result of FFT analysis when the wheel ran with 6 kg drawbar mass. A power spectrum is shown at around 4Hz as well as Fig. 8 and Fig. 9. In addition, some spectrums are appeared from 20 Hz to 25Hz, which are smaller than the ones in Fig. 9.

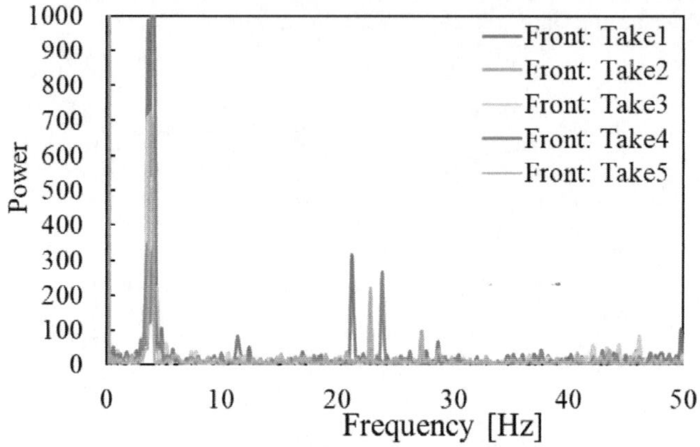

Fig. 8. Results of FFT analysis (without drawbar mass)

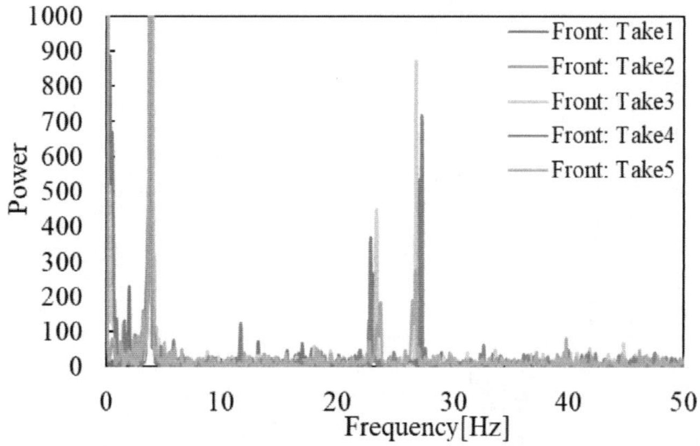

Fig. 9. Results of FFT analysis (with 3kg drawbar mass)

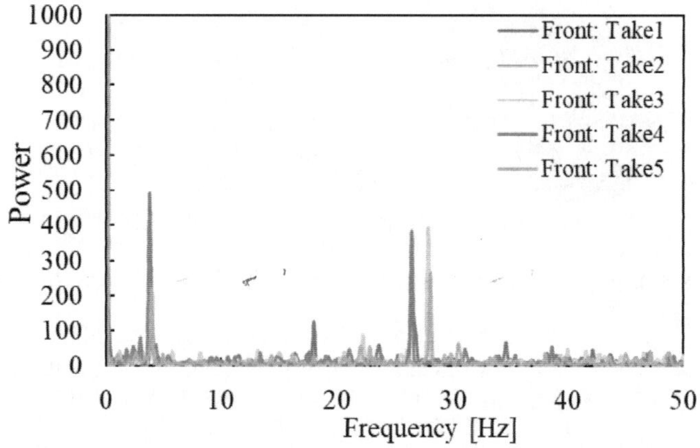

Fig. 10. Results of FFT analysis (with 3kg drawbar mass)

4. Discussion

According to Fig. 7, it seems that the larger the slip ratio, the larger the average value of strain. A larger value of Strain means that the chassis bends more. A bending chassis means that the wheels are moving ahead of the chassis. Therefore, when the wheel slip rate is greater, more wheels are moving ahead of the chassis. This seems to have occurred because of the rope that connects with the upper part of the chassis. The wheel rotates in a forward direction. On the other hand, the upper part of the chassis, which is connected to the rope, is pulled backwards. Consequently, the wheel moves forward more than the chassis. As a result, the chassis bends. The greater the drawbar mass, the greater the force with which the chassis is pulled backwards. Besides, as Fig. 7 shows, the larger the drawbar mass, the larger the slip ratio. Therefore, the larger the slip ratio, the greater the deformation of the chassis, which is considered to result in a larger strain value. This phenomenon may differ from the situation when slipping on a real slope. This is because the wheels are also slanted on the slope, which causes a backward pulling force on not only the chassis but also on the wheels. Therefore, while it is possible to deliberately vary the slip rate in this experiment, it may not fully replicate slipping on a real slope.

When comparing Fig. 8, Fig. 9 and Fig. 10, the spectrum at about 4 Hz was commonly shown under all conditions and had a high value of power. Therefore, the 4 Hz vibration is considered to be the natural vibration of the chassis used in this experiment.

The prominent power spectrums at about 24 Hz and about 27 Hz in Fig. 9 are presumed to be caused by slipping of the wheel. When a wheel runs on a loose soil, the wheel gives shearing force to the ground. This force seems to causes fine vibrations and shear failure of the sand. Repetition of this phenomenon leads to wheel slip. When this phenomenon occurs, some force is exerted on the wheels. It is natural to assume that this force is measured as vibration of the strain in the chassis.

However, as shown in Fig. 10, when the wheel ran with 4.5 kg drawbar mass, the power at around 24 Hz and around 27Hz look smaller than those of Fig. 9 nevertheless the wheel was slipping. The slip rate was approximately 0.9 according to Fig. 7. This fact suggests that if the slip rate is too large, the sand will always move fluidly beyond shear failure. When the sand is moving fluidly, it is not considered to that shear failure would result in higher vibrations at certain frequencies. Therefore, a higher slip rate may have reduced the power at about 24 Hz and 27 Hz.

When installing strain gauges on a lunar rover, the issue of temperature variation is critical and should be well considered since the lunar surface is subject to repeated thermal shocks, with temperatures reaching 120°C during the day and dropping to -180 °C at night (Xiong et al., 2022). The resistive element of a strain gauge is deformed (thermal expansion and contraction) by its own temperature changes. In addition, the object to be measured is also deformed by temperature changes. These two phenomena cause a change in resistance value, which is called an apparent strain. When measuring in high and low temperature environments, such as the lunar surface, one of the best ways to suppress apparent strain is to use a self-temperature compensated gauge. It is also important to know in advance how much apparent strain will actually occur. Research has been reported on strain gauges that can measure accurately even at 800°C (Liu et al., 2019) and fibre-optic strain gauges that function properly in cryogenic environments of 4.2 K (Baxter et al., 2010). On the lunar surface, an effective measure may be to change the type of strain gauge used for measurements between the cooler night time and the warmer day time.

5. Conclusion

The purpose of this study was to try estimating the driving state of the rover in real time by analyzing the strain value measured with strain gauges mounted on the chassis. In the experiments, the strain data was accumulated with changing voluntarily the slip ratio of a wheel connected to the chassis. A correlation was be tried to established between the wheel's slip ratio and the strain by averaging and FFT analyzing of strain values. The result suggests three points. First, the larger the slip ratio of the running wheel, the larger the average value of strain generated the chassis. Second, when the slip ratio is 0.4~0.8, vibrations at frequencies of about 24 Hz and about 27 Hz are larger. This is presumed to be caused by shear failure. Third, when the slip ratio exceeds 0.9, vibrations at frequencies of about 24 Hz and about 27 Hz are smaller. This is presumed to be due to the fact that when the slip rate is very high, the sand moves fluidly, rather than shear failure occurring.

6. Acknowledgements

This work was supported by the Japan Society for the Promotion of Science (JSPS) KAKENHI Grant Number 21K0395.

7. Declaration of competing interest

The authors declare that they have no known competing financial interests or personal relationships that could have appeared to influence the work reported in this paper.

8. References

Baxter, S., Lakrimi, M. H., Thomas, A. M., Gao, Y., Blakes, H., Gibbens, P., Looi, M., 2010. Validation of a novel fiber optic strain gauge in a cryogenic and high magnetic field environment. Cryogenics, 50(10), 700-707.

Blum, K. P., Campbell, K. S., Horslen, B. C., Nardelli, P., Housley, S. N., Cope, T. C., Ting, L. H., 2020. Diverse and complex muscle spindle afferent firing properties emerge from multiscale muscle mechanics. Elife, 9, e55177.

Coloma, S., Martinez, C., Yalçın, B. C., Olivares-Mendez, M. A., 2022. Enhancing Rover Teleoperation on the Moon With Proprioceptive Sensors and Machine Learning Techniques. IEEE Robotics and Automation Letters, 7(4), 11434-11441.

Ding, L., Gao, H., Deng, Z., Nagatani, K., Yoshida, K., 2011. Experimental study and analysis on driving wheels' performance for planetary exploration rovers moving in deformable soil. Journal of Terramechanics, 48(1), 27-45.

Gonzalez, R., Apostolopoulos, D., Iagnemma, K.,2018. Slippage and immobilization detection for planetary exploration rovers via machine learning and proprioceptive sensing. Journal of Field Robotics, 35(2), 231-247.

Liu, H., Mao, X., Yang, Z., Cui, J., Jiang, S., Zhang, W., 2019. High temperature static and dynamic strain response of PdCr thin film strain gauge prepared on Ni-based superalloy. Sensors and Actuators A: Physical, 298, 111571.

NASA, NASA's Opportunity Rover Rolls Free on Mars, <https://www.nasa.gov/vision/universe/solarsystem/mer-060605.html>, (Access : 2022/6/22).

Santuz, A., Akay, T., Mayer, W. P., Wells, T. L., Schroll, A., Arampatzis, A. 2019. Modular organization of murine locomotor pattern in the presence and absence of sensory feedback from muscle spindles. The Journal of physiology, 597(12), 3147-3165.

Santuz, A., Akay, T., 2023. Muscle spindles and their role in maintaining robust locomotion. The Journal of Physiology, 601(2), 275-285.

Xiong, G., Guo, X., Yuan, S., Xia, M., Wang, Z., 2022. The mechanical and structural properties of lunar regolith simulant based geopolymer under extreme temperature environment on the moon through experimental and simulation methods. Construction and Building Materials, 325, 126679.

Yoshida, K., 2011. Development of Mobile Robots for Lunar/Planetary Exploration. Journal of the Japan Society of Precision Engineering, 77(1), 12-15.

8254 | Initial steps towards characterization of a new military cold weather all-terrain vehicle for performance in northern operations

Clifford Witte [a,*], Orian Welling [a], Michael Parker [a], Michael Stearns [a], Zach Tomberg [a]

[a] *Cold Regions Research & Engineering Laboratory, Engineering Resource and Development Center, United States Army Corps of Engineers*

* *Corresponding author: Clifford.F.Witte@erdc.dren.mil*

ABSTRACT

The US Army Corps of Engineers - Cold Regions Research & Engineering Laboratory recently acquired two quad articulated-track cold weather all-terrain vehicle prototypes to begin characterization of their performance on winter surfaces for force projection, modeling, and simulation. The vehicles were previously used in the Cold Weather All-Terrain Vehicle (CATV) program initiated by the United States Army to select a replacement for the aging Bandvagn 206, designated in the US Army fleet as the Small Unity Support Vehicle (SUSV) and in service for over forty years. Researchers at CRREL have extensive experience testing US Army SUSVs and NATO Bv206s in cold regions and have documented the vehicles as one of the most capable in severe weather conditions ranging from flooding, deep mud or thawing 'rasputitsa', or deep snow. At the US Army Mountain Warfare School, the SUSV remains the vehicle of choice for search and rescue in extreme weather. The new prototype vehicles delivered to CRREL are based on the ST Engineering 'Bronco' All-Terrain Tracked Carrier (ATTC) currently in service by Singapore and Thailand militaries and previously deployed by the United Kingdom in Afghanistan. While the US Army ultimately selected the BAE Systems 'Beowulf' as the winner of the CATV program, the tractive performance has been identified by CRREL as comparable between the candidate vehicles with similar articulated quad track design and dimensions, ground pressure, and power delivery. By utilizing the prototype vehicles, researchers at CRREL will gain a head start toward characterizing the CATV in winter conditions before entering service in 2024. This paper will detail the experimental setup, analysis and results for testing conducted in 2023 in Dalton, New Hampshire as well as how the results will be used to begin updating current mobility prediction models.

Keywords
Track models
Track design
Dynamic test
Traction test
Off-road test
Military

1. Introduction

To bolster its capabilities the United States has in recent years emphasized the investigation of equipment shortages which would lead to high failure rates for operations in extremely cold and/or remote regions. As part of a renewed focus on arctic strategy, recent reports to U.S. Congress by the Office of the Under Secretary of Defense (2019), Department of the Army (2021) and the White House (2022) outline an arctic presence requiring significant long lead-time investments in infrastructure and equipment as an anticipatory approach towards changes occurring in the region. One such key area of investment is the enhancement of mobility in the arctic, or specifically the logistics of safely moving personnel and materiel across rugged snow and ice terrain with little to no infrastructure and the most extreme cold temperatures on the planet.

The Bandvagn 206 (Bv206) or Small Unit Squad Support Vehicle (SUSV) as it is known in the United States has been in the U.S. Army's inventory since the 1980s. It is a quad-articulated tracked personal carrier which is designed to operate off-road in the harshest environments. Dwindling inventory has left the US Army with only a few dozen Bv206s left, mostly operating among units in cold regions, wetlands, flooded or muddy terrain. The Bv206 is intended as the stop-gap for both heavily armored vehicles and light wheeled vehicles which struggle on the softest kinds of terrain such as deep snow, wetlands, or highly uneven off-road surfaces requiring a highly articulable vehicle suspension. These are environments, as seen in Fig 1, where the low ground pressure is required to "float" over the soft surface with minimal sinkage.

In 2018, the U.S. Army initiated the Joint All Weather Terrain Support (JAASV) program to begin searching for a replacement for the Small Unit Support Vehicle (SUSV). These are intended to be highly mobile, lightweight vehicles which are capable of traversing over snow, mud, or otherwise off-road terrain which is impassable for most if not all heavily armored tracked vehicles.

By 2021, the program was reclassified as the Cold Weather All-Terrain Vehicle (CATV), further identifying the primary mission of the vehicle as highly mobile in arctic regions, and reduced to a competition between two bidding manufacturers: Oshkosh and BAE Systems. Both models underwent significant test and evaluation by the Cold Regions Test Center in Delta Junction, AK.

Ultimately the U.S. Army selected the BAE Beowulf, a sister variant to the NATO fielded BAE BvS10, as the winner of the competition, and awarded BAE Systems a $278 million dollar contract to produce the CATV, with deliveries planned to begin in September 2023, as noted by a 2023 BAE Systems press release. The BAE Beowulf is seen in Fig. 2.

Fig. 1. Small Unit Support Vehicle, US Army variant of the NATO Bv206

https://doi.org/10.56884/FEDF3359
ISBN 978-1-942112-55-6 (pdf)

2. Background Information

The Cold Regions Research & Engineering Laboratory (CRREL) is one of seven research labs operated by the Engineering Research and Development Center (ERDC), the primary R&D organization for the U.S. Army Corps of Engineers.

The Cold Regions Research & Engineering Laboratory leads scientific research for ERDC in areas of cold expertise and has extensive experience testing vehicles in cold regions.

To jumpstart development of models and simulations for the CATV family of vehicles, CRREL received two prototype Oshkosh 'Bronco' cold weather all-terrain tracked vehicles which previously were being used for evaluation by other organizations. The 'Bronco' is a close variant of the ST Engineering All Terrain Tracked Carrier (ATTC). Although the vehicles are not the final selected 'Beowulf' models for production, the overall characteristics between the two vehicles are similar as they both directly targeted replacement of the SUSV under specifications provided by the U.S. Army.

Fig. 2. Selected BAE 'Beowulf', winner of the CATV competition

Variations to performance between the selected BAE CATV and the CRREL owned 'Bronco' ATTCs are expected and further testing and updates to the model will be performed once the CATV is formally in service. However, some of the most important characteristics to CRREL models are similar between the two models, such as ground pressure and quad articulated track design. Exact vehicle specifications will not be presented in this research due to the sensitive nature of the material.

Fig. 3. Russian tanks immobilized by thawing terrain in Ukraine. Photo released by the Ukrainian Ministry of Defense (2022)

As seen in Fig. 3 and 4, CRREL is involved in updating models which project the mobility of US fleet vehicles into various conditions of snow, thawing, and/or freezing soils, such as in the NATO Reference Mobility Model (NRMM) described by McCullough et al (2017).

The maneuvering characteristics of the CATV are different from other tracked vehicles in the US fleet as the CATV will be the only logistics or combat vehicle with a quad articulated track. CRREL is also working in Army Futures Command (AFC) programs to improve autonomous navigation and control over high-slip terrain. One of the CREEL owned 'Broncos' ATTCs may be used in the future as a testbed for adapting unmanned controls to a quad articulated track type of maneuvering. This will be an important upcoming area of study. As seen in special updates published by NASA Johnson Space Center (2020), work of similar nature is being done for autonomous lunar rovers, which previously have only been driven by remote commands, and will encounter a higher percentage of loss of control or immobilizing situations.

Fig. 4. Peat Soil Terrain Example, highly impassible to tracked and wheeled military vehicles. Picture taken by CRREL researchers in Estonia

CRREL is also investigating an electric vehicle (EV) conversion of one of its units for testing recent advancements in EV technology, especially energy storage.

Both areas of improvement among other areas of basic research will be relevant to the upcoming 'Beowulf' vehicles. This represents a significant cost savings to USACE to not have to procure additional 'Beowulf' vehicles purely for basic research. CRREL intends to construct test kits which can be later deployed on the 'Beowulf' in further developed projects. This will include sensor packages for vehicle perception being tested on the CRREL Mobility Team's fleet of vehicles.

By acquiring the test vehicles, CRREL will be able to move basic research forward from its headquarters in Hanover, NH and field office in Fairbanks, AK.

3. Experimental Setup

CRREL received two variants of the "Bronco" vehicles, a crew cabin variant for hauling passengers and the other a pick-up module for hauling materiel. For this testing, the crew cabin variant was instrumented and tested.

The test vehicle was instrumented for gathering tractive performance data in Drawbar Pull (DBP) and Motion Resistance (MR) testing. These are two of the most common tests that CRREL performs.

Instrumentation includes GPS acquired speed, actual track speeds, an inertial measurement unit (IMU), and an inline tension loadcell. Drawbar Pull is conducted according to SAE J2014 and ASTM F1805-20. However, there are slight variations to make the testing suitable for the offroad terrain. Drawbar testing measures the vehicle's draw-bar traction over a given surface. Gross traction can later be calculated using DBP, MR and Hard Surface Rolling Resistance (HSRR). For this experiment, the Drawbar Pull is done as a constant velocity test at 5 MPH while a hold back vehicle applies its breaks to keep the speed of the test vehicle constant. This method is useful for testing on offroad terrains where it would be expensive or otherwise difficult to set up a stationary drawbar test platform. Fig. 5 shows the test vehicles in DBP configuration with loadcell inline between the two vehicles.

Fig. 5. CRREL 'Bronco' cold weather all-terrain vehicles in Drawbar Pull configuration with holdback vehicle

A load cell is placed between the two vehicles to measure the forward pull force of the test vehicle. Wheel speeds of the vehicle are captured with rotary encoders.

As the test vehicle approaches 8 km/h (or 5 mph), the driver slowly throttles up while the hold-back driver applies the brakes, maintaining 8 km/h. The test vehicle is throttle to maximum power so that the tracks are slipping. In some cases, vehicles with power limitations and high traction do not achieve any slip. In the case of the test vehicles, slip was seen to occur when throttling up indicating no limitations on power.

Motion Resistance is setup as a simple tow as per SAE convention. The test is conducted at 8 km/h as near to the DBP site as possible but in undisturbed, unrutted snow. The value for Motion Resistance is added from Drawbar Pull to calculate Gross Traction (GT).

The experiment is based on the following definition for Gross Traction:

$$Gross\ Traction = DBP + MR$$

4. Results

4.1 Motion Resistance

The values for the Motion Resistance test are reported in Fig. 6. The reported values are much higher than wheeled vehicles owing to the large track surfaces. This reduces fuel economy however the advantages of the reduced ground pressure are worthwhile.

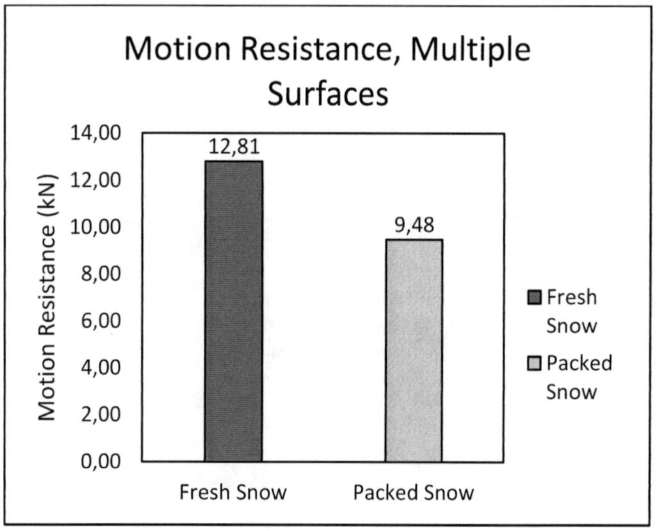

Fig. 6. Motion Resistance Results on Fresh and Packed Snow

Motion Resistance is seen to increase on fresh snow when compared to packed snow. This correlates with the increased force necessary to plow and compress the snow depth.

As the sample size is quite small for this set of data, more testing performed in over winter in 2023-2024 will be added to this dataset. The intention is to produce results for more surfaces as well, with deeper snow depths where the vehicle is floating on a previously uncompressed snow surface. It is likely this future experimentation will include data from sites near Fairbanks, AK.

4.2 Drawbar Pull and Gross Traction

During Drawbar Pull testing, the vehicles presented a maximum load of nearly 50% of its own weight (10,206 kg). This is a common characteristic of ground vehicles on dryer surfaces. A value near 50% of the vehicles weight on the snow surface would be much rarer on other vehicles in the U.S. Army fleet. It is noted that the Gross Traction values were seen to increase when moving to testing over fresh, undisturbed snow as compared to operating on purely packed snow. This could be a 'paddling' effect of the treads, which had less to dig in on the packed snow surface. As this testing is preliminary, the sample size is too small to draw any firm conclusions about the vehicle from this set of data alone. The results for Gross Traction are presented in Fig. 7.

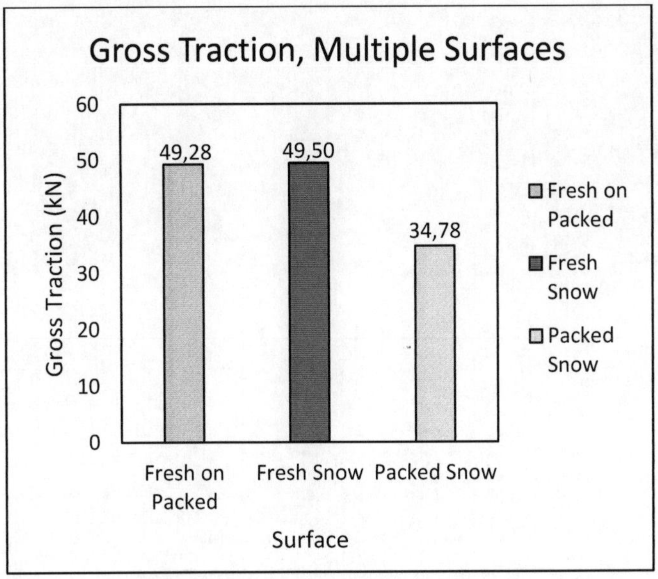

Fig. 7. Motion Resistance Results on Fresh and Packed Snow

Note that the values for fresh undisturbed snow and fresh snow on the packed surface are almost identical. This is likely because the depth of the snow surface was only a couple inches of difference between undisturbed and the fresh over recently packed snow. The snow did not provide a floatation surface because it was not deep enough.

5. Conclusions and future work

Results show promising performance measurements which CRREL intends to be repeated for a larger sample size for integration into updated mobility models.

Among other modernizations such as increased space with shock resistant seating, 4-point seatbelt restraints, and individual heating vents for each soldier inside the crew cab is a welcome sight.

The data presented in this paper is preliminary and represents a very small sample set of tests. CRREL will be conducting more testing on its owned cold weather all-terrain vehicles in the winter seasons to come. CRREL will also utilize the vehicles for supporting other test vehicle experiments. One such example is the development of winter tires for logistics vehicles to avoid circumstances as seen in Fig. 8.

Fig. 8. Immobilized Russian wheeled logistics vehicle in Ukraine (Washington Post, 2022)

CRREL will be able to utilize the prototype vehicles in support roles during other field testing, whether to haul equipment or personal into areas normally only accessible by light high mobility vehicles such as snowmobiles.

The primary purpose of this experiment is to lay the groundwork for further testing on the CRREL owned prototypes. With two vehicles CRREL can stage one vehicle at its main location in Hanover, NH, and another at the CRREL field office in Fairbanks, AK. With daily access to the vehicle, CRREL will be able to develop future systems for continued research on the BAE Systems 'Beowulf' platform as soon as it is available and continue to use the platforms as testbeds for preliminary technology which is intended for the 'Beowulf'.

6. Nomenclature

AFC	Army Futures Command	
ATTC	All-Terrain Tracked Carrier	
Bv206	BAE Systems Bandvagn 206	
BvS10	BAE Systems Bandvagn 210	
CRREL	Cold Regions Research & Engineering Laboratory	
CRTC	Cold Regions Test Center	
CATV	Cold Weather All-Terrain Vehicle	
DBP	Drawbar Pull	[N]
ERDC	Engineer Research & Development Center	
EV	Electric Vehicle	
GT	Gross Traction	[N]
IMU	Inertial Measurement Unit	
JAATS	Joint All Weather Terrain Support	
MR	Motion Resistance	[N]
SUSV	Small Unit Support Vehicle	
USACE	United States Army Corps of Engineers	
USARAK	United States Army Alaska	

7. Acknowledgements

Funding for this effort was provided by the Cold Regions Methodologies for Autonomy and Terramechanics project under the 6.2 BG202 Task: Autonomous Vehicle/Terrain Interactions work package under Modeling and Simulation for Manned Unmanned Teaming, and the Ground Vehicle Snow Mobility project under the 6.3 BL801 Task: Entry and Sustainment in Complex Contested Environments work package under Power Projection in A2AD Environments.

This research acknowledges Team O'Neil Rally School in Dalton, NH for their services in setting up and maintaining test surfaces and usage of their facilities.

8. Declaration of competing interest

The authors declare that they have no known competing financial interests or personal relationships that could have appeared to influence the work reported in this paper.

9. References

American Society for Testing and Materials (ASTM), 2020. Standard Test Method for Single Wheel Driving Traction in a Straight Line on Snow- and Ice-Covered Surfaces American, p11-13, ASTM F1805-20

BAE Systems, 2022. BAE Systems wins U.S. Army's CATV competition, receives $278 Million contract. https://www.baesystems.com/en-us/article/bae-systems-wins-us-armys-catv-competition-receives-278-million-contract/ (accessed July 6 2023)

McCullough,M, Paramsothy, J , Dasch, J, and D. Gorsich. 2017. The Next Generation NATO Reference mobility model development. J Terramechanics, v73 p49-60.

NASA Johnson Space Center, 2020, Lunar Terrain Vehicle (LTV) Special Notice Update - #1. https://sam.gov/opp/9e777623a1f3478296f21f2f0d787113/view (accessed July 6, 2023)

Society of Automotive Engineers (SAE), 2021, Pneumatic Tire/Wheel/Runflat Assembly Qualifications for Military Tactical Wheeled Vehicles, SAE J2014, SAE Int. Surface Vehicle Standards

The White House, 2022. National Strategy for the Artic Region, https://www.whitehouse.gov/wp-content/uploads/2022/10/National-Strategy-for-the-Arctic-Region.pdf/ (accessed September 13, 2023)

Ukraine Ministry of Defense, 2022. Enemy tanks were unearthed in Kharkiv Oblast and handed over to the needs of the Armed Forces. https://armyinform.com.ua/2022/03/29/na-harkivshhyni-vidkopaly-vorozhi-tanky-j-peredaly-na-potreby-zsu/ (accessed September 1, 2023

US Department of the Army, 2021. Regaining Arctic Dominance – The US Army in the Arctic

US Office of the Under Secretary of Defense for Policy. 2019. Report to Congress: Department of Defense Arctic Strategy, 18 p,

8672 | Trajectory optimization for vegetation override in off-road driving

Charles Noren [a,*], Bhaskar Vundurthy [a], Sebastian Scherer [a], Matthew Travers [a]

[a] *The Robotics Institute, School of Computer Science, Carnegie Mellon University, Pennsylvania, United States of America*

* *Corresponding author: cnoren@andrew.cmu.edu*

ABSTRACT

Off-road robotic vehicles often encounter environmental objects while traversing complex terrains. These objects may obstruct the vehicle's motion and must be perceived and reasoned about to ensure safe operation. Historically, limited sensing fidelity for autonomous vehicles led to either overly cautious or unnecessarily risky behaviors when interacting with environmental objects. Some objects must be avoided to ensure operational integrity, but avoiding all environmental objects will likely cause needlessly slow traversals. With contemporary sensing, the capability to tell the difference between these environmental objects is now possible. Given this improved sensing capability, developing controls algorithms which not only reason about avoiding certain classes of objects but can also reason about motions through other classes of objects is now the next step towards robust off-road autonomous driving. In this work, we propose a nonlinear trajectory optimization technique for overriding vegetation that explicitly encodes contact constraints for a subset of vegetative objects found in the operating environment. The subset of objects that the vehicle can safely collide with is derived from existing parameterized collision models, which is estimated by the on-board sensing system, and then characterized by a necessary velocity at the point of collision. These constraints are then enforced during vehicle run-time by the controller. The controller's capabilities are shown in both simulation and used to control a full-sized autonomous utility task vehicle. The vehicle's determination to avoid or override objects is dependent on the specifics of the environment and object geometry.

Keywords
Trajectory Optimization
Vegetation Override
Off-road Driving
Robotics

1. Introduction

"When [I am] going through vegetation and driving through or over it, [I] always have it in the back of my head how I am going to hit it. Typically that is head on with the front bumper as that will be the strongest point on the vehicle and give me the best leverage to run over [the vegetation] and go through it."

-Ryan Arciero, Professional Off-road Driver

The execution of off-road vehicle operations are influenced by the decision of whether to take an action to "avoid" or "strike" environmental objects (Rybansky, 2017). As seen in the quote above, when striking vegetation, professional off-road racing drivers carefully internally model and allow certain collisions for safe travel off-road. It stands to reason that in order to emulate the high performance of professional drivers in off-road conditions, robotic vehicles will also need to consider the same decisions about collisions. However, there is a great imbalance of work in the development of the "avoid" and "strike" actions for robotic platforms, with a majority of work centering on the "avoid" action.

This imbalance may rise from the fact that certain tasks, such as on-road (Paden et al., 2016) and on-trail robotic operations, generally demonstrate the need for avoidance capabilities. Requirements reaching back to the Defense Advanced Research Projects Agency (DARPA) Grand and Urban Challenge(s) (Thurn et al., 2007, Urmson et al. 2009) reflect the need for certain areal or object avoidance behaviors. Thus many methods have evolved in the field robotics, planning, and control domains to address such requirements. Expressing object avoidance as a general constraint in the state space for both state-based planning (LaValle, 2006) and for optimal-control-based approaches (Chen et al. 2017, Howell et al., 2019) are well-established (Paden et al., 2016). Furthermore, using penalty-based methods for object avoidance (Williams et al., 2018) and avoidance-guaranteed "proof-by-construction" techniques have seen renewed attention in recent years (Liu and Tomizuka, 2014; Noren et al., 2021). However, the quote above speaks to a need to traverse objects; and thus only taking avoidance actions or explicitly forbidding object contact may not reflect expert driving behavior.

While works developing the "strike" action set are in fewer numbers, the field robotics, planning, and control communities have all evolved different means to begin reasoning about collisions. The approaches from the field robotics community, such as classifying objects as "strike-able" via visual measures of object geometry or through the use of virtual and physical bumpers (Kelly et al., 2006) draw from the practical needs of robotic platform operation. High-level planning methods, such as those described by Rybansky (2017), develop collision rules dependent on the vehicle configuration and environmental structure to govern expectations (for example, maximum traversal speed through an area) on vehicle operations. Outside of treating contacts as disturbances, control approaches generally fall into two categories: through-contact models and hybrid-contact models. Through-contact models attempt to capture the non-smooth physics of the interaction and directly evaluate those physics during the determination of the control (Howell, 2023; Posa et al., 2013). Hybrid-contact models develop rules-based functional mappings to model the changes of state associated with a discrete event (Hargraves and Paris, 1987) such as a collision. These methods are commonly used in direct collocation trajectory optimization problems for legged locomotion (Kelly, 2017). Unfortunately, many of these methods which allow contact for a "strike" action solely deal with geometric representations of the

https://doi.org/10.56884/NJXE5926
ISBN 978-1-942112-55-6 (pdf)

terrain or object and do not capture all aspects needed for traversing or "overriding" an object.

Yet due to the prevalence and need to interact with vegetation in common area of operations of interest, the off-road mobility and cross-country movement communities have a long history of modeling such non-geometric requirements for vegetation. The U.S. Army Engineer Waterways Experiment Station (Blackmon, 1968) and the U.S. Army Engineer Research and Development Center (Mason et al., 2012) have developed models for the override of different post and vegetation classes. These studies contribute to larger mobility models, such as the North Atlantic Treaty Organization (NATO) Reference Mobility Model (NRMM) (Bradbury et al., 2018), which are used to analyze the capabilities of both crewed vehicles and robotic platforms (Vong et al., 1999). Rybansky (2020) performed a significant study of vegetation overrides from different types of vehicles and developed models for each. However, these mobility models remain confined to the domain of vehicle mobility analysis and have not yet been used in an online capacity for reasoning in robotic platform operations.

The primary contribution of this work is the incorporation of vegetation override models from the off-road mobility and cross-country movement communities in the online planning and control of robotic platforms. Specifically, these collision models are incorporated as collocation and inequality constraints imposed on the vehicle's velocity to ensure the successful override of encountered vegetation. This work begins with a description of the proposed control structure in the second section and controller simulations are shown in the third. Results from hardware experimentation follow the simulation results and the paper concludes with reflections and a discussion of future work.

2. Algorithm Overview

This section begins with a description of direct collocation methods for trajectory optimization and then discusses vegetation and vehicle modeling. The proposed trajectory optimization control technique is then motivated from the description and models.

2.1 *Direct Collocation for Trajectory Optimization*

Trajectory optimization problems determine an optimal state and control sequence that minimizes an objective function (Kelly, 2017). Equation 1 shows a trajectory optimization problem objective function where the boundary term is not explicitly dependent on time:

$$J\big(t_0, t_f, x(t), u(t)\big) = J_f\big(\boldsymbol{x}(t_0), \boldsymbol{x}(t_f)\big) + \int_{t_0}^{t_f} w\big(\tau, x(\tau), u(\tau)\big)d\tau\,, \quad (1)$$

and where decision variables t_0, t_f, $\boldsymbol{x}(t)$, $\boldsymbol{u}(t)$ are, in order: the initial time, final time, and the state and control trajectories. Terms $J_f(\cdot)$ and $w(\cdot)$ are the boundary and integral cost terms. The objective function is then minimized in a mathematical program:

$$\begin{aligned} \min_{t_0,\, t_f,\, x(t),\, u(t)} \quad & J\big(t_0, t_f, x(t), u(t)\big) && (2a)\\ \text{subject to:} \quad & \dot{\boldsymbol{x}}(t) = f\big(t, x(t), u(t)\big)\,, && (2b)\\ & h\big(t, x(t), u(t)\big) \le 0\,, && (2c)\\ & \boldsymbol{g}\big(t_0, t_f, \boldsymbol{x}(t_0), \boldsymbol{x}(t_f)\big) \le \boldsymbol{0}\,, && (2d) \end{aligned}$$

where Eq. 2b are the system dynamics and Eq. 2c and Eq. 2d represents general path and boundary constraints, respectively.

Direct trajectory optimization techniques approximately solve the posed trajectory optimization mathematical program in Eq. 2 by discretizing and transcribing the problem into a more general nonlinear program. A standard-form nonlinear program is

$$\begin{aligned} \min_{z} \quad & R(z) && (3a)\\ \text{subject to} \quad & c(z) = 0\,, && (3b)\\ & d(z) \le 0\,, && (3c) \end{aligned}$$

where all functions are assumed to be at least C^2 smooth. Direct collocation methods represent continuous-time trajectories with a spline of N time-parameterized polynomial segments; and thus the methods discretize the continuous-time trajectory via N+1 knot points. To ensure the solution to the program posed in Eq.3 is feasible with respect to the system dynamics as described in Eq. 2b, an integral form of the dynamics is approximated through numerical quadrature. Through the use of different quadrature rules, classically trapezoidal or Simpson quadrature, different approximating polynomials are recovered. These approximate integrals are thus posed as collocation constraints in Eq. 3b. Additional integral (Eq. 2c) and boundary constraints (Eq. 2d) are posed as constraints in Eq. 3b and Eq. 3c. Finally, a discrete-time representation of the objective function in Eq. 2a must be posed for the nonlinear program (through approximations such as quadrature) and via an augmented state variable $\boldsymbol{z}$ containing all the decision variables at the knot points.

2.2 *Vegetation Override Models*

The vegetation override models developed by the off-road mobility and cross-country movement communities, including Blackmon and Randolph (1968) and Mason et al. (2012), abstract complex collision interactions into useful low computational cost approximations. These approximations generally characterize the required force, work, or velocity a vehicle needs to override a subset of vegetation given some parameterization of the vehicle, environment, or both.

The model presented in Blackmon and Randolph (1968) is regressed from a series of vegetation override tests of different vegetation types in different environments. Blackmon and Randolph (1968) provides unique regressions for force and energy expenditure for these different vegetation types, including: singular coniferous and hardwood trees, arrays of multiple trees struck in unison, and "clumps" of bamboo. From continuous measurements of pushbar force, drivetrain metrics, distance traveled, time, measurements of the impacted trees, and characterizations of the aftermath of the collision, Blackmon and Randolph (1968) constructs a model primarily parameterized by the geometry of individual or multiple trees (for example, the radius of a tree) or the clump diameter.

Mason et al. (2012) presents a model for vertically embedded objects in the ground. These objects consist of posts and small trees. Equations are introduced in Mason et al. (2012) to capture the necessary override force for post-like objects which were then validated primarily through pull tests. The model in Mason et al. (2012) is mainly characterized by vehicle mass and geometry (pushbar height), vegetation and emplacement geometry, soil parameters, and a series of regression coefficients. Mason et al. (2012) then relates a series of energy expenditure and traversal velocity equations from collected data and provides a force model. Equation 10 from Mason et al. (2012) describes the minimum velocity the vehicle is required to travel at in order to override a post, v_{over}, and is given as:

$$v_{over} = \big(2k\alpha\gamma_d D L_t / (m(h + 0.5 * L_t))\big)^{0.5}\,, \quad (4)$$

where L_t is the burial depth, m is the vehicle mass, h is the height from the ground surface at which the override force is applied, D is the post diameter, γ_d is the dry density of soil, and α and k are empirical factors derived as described in Mason et al. (2012).

2.3 Vehicle Modeling

Figure 1 shows the platform used in this work. The platform is a modified Polaris RZR utility task vehicle (UTV) which can travel up to 20 [m/s] in cluttered off-road terrains. The vehicle is ruggedized to collisions and is equipped with an onboard sensor suite that contains monocular and stereo cameras as well as multiple Light Detection and Ranging (LiDAR) sensors. A nonlinear bicycle model was used to represent the vehicle dynamics, with the vehicle state given as: $\boldsymbol{x} = [p_x, p_y, \psi, \delta, v]$. In order, the elements of this state vector are the vehicle: x-position, y-position, heading, steering angle, and velocity. The vehicle controls include acceleration and steering rate: $\boldsymbol{u} = [a, \dot{\delta}]$. The continuous time vehicle dynamics are:

$$\dot{p_x} = v * \cos(\psi + \operatorname{atan}\left(\frac{L_f}{L} * \delta\right)), \quad (5)$$

$$\dot{p_y} = v * \sin(\psi + \operatorname{atan}\left(\frac{L_f}{L} * \delta\right)), \quad (6)$$

$$\dot{\psi} = \frac{v}{L} * \cos\left(\operatorname{atan}\left(\frac{L_f}{L} * \delta\right)\right) * \tan(\delta), \quad (7)$$

$$\dot{\delta} = \dot{\delta}, \quad (8)$$

$$\dot{v} = a, \quad (9)$$

where the parameters may be found in Table 1. The vehicle's total length is L_T, wheelbase length is L, front axle to center of mass distance is L_f, front axle to nose distance is L_{nose}, lower bull-bar strut height is h_{nose}, width is w, and mass is m.

2.4 Direct Collocation for Vegetation Override

One central aspect of the collision models presented in the previous section is that each model captures a loss of kinetic energy due to the collision corresponding to the failing of the vegetation or post. Yet the structure of the trajectory optimization problem and the nonlinear program in Eq. 2 and Eq. 3 require at least C^2 smooth functions to accurately represent the dynamics in the collocation constraints. Thus to incorporate the prior mentioned vegetation models in the trajectory optimization, the impact dynamics must be accounted for. From the formulation, the collocation constraints are evaluated at specific collocation points, which need not be the same as the knot points described earlier. This realization gives rise to the idea of introducing a pointwise-in-time discontinuity in the form of a functional map, commonly described as a "jump" map, between the two states which represents the collision. This concept exists at the center of hybrid-contact representations of collisions and multi-phase direct collocation methods, and it is discussed in more detail in Hargraves and Paris (1987) and Kelly (2017). These approaches have not been implemented using the aforementioned vegetation models.

Table 1
Platform Parameters

Property	Unit	Magnitude
L_T	[m]	3.785
L	[m]	2.972
L_f	[m]	1.412
L_{nose}	[m]	0.915
h_{nose}	[m]	0.533
w	[m]	1.828
m	[kg]	901.0

Fig. 1. The modified Polaris RZR UTV

To capture the effects of the collision at a knot point, a mapping must be defined to transition the state at the time of collision, $\boldsymbol{x}(t_{col})$, to a new post-collision state. That knot point representing the pre-collision state, $\boldsymbol{x_{col}} = \boldsymbol{x}(t_{col})$, is mapped to the next knot point via the mapping:

$$\boldsymbol{x_{k+1}} = \boldsymbol{f_{col}}(\boldsymbol{x_k}), \quad (10)$$

where subscript k is a general indexing of the knot points, instead of via the selected quadrature rule. While the technique is compatible with the different collision models as described in the vegetation modeling section, this work represented the collision as a loss of velocity at the point of collision. More specifically,

$$\boldsymbol{x_{k+1}} = \boldsymbol{f_{col}}(\boldsymbol{x_k}) = \boldsymbol{x_k} - [0; 0; 0; 0; v_{over}], \quad (11)$$

where v_{over} could be defined (for example, via Mason's (2012) model) in Eq. 4. and the last element of $\boldsymbol{x_k}$ corresponds to the velocity (Eq. 9). However, enforcing such a mapping at the point of collision does not solely account for the effects of the collision. For the simple jump map in Eq. 11, collisions could propel the system in reverse at low speeds, so either a guard function or a lower-bound on the minimal allowable velocity could be enforced to ensure compliance with physical laws. A natural requirement to ensure the vegetation is overridden is thus to enforce the velocity reaches at least v_{over} at the time of collision. Just as the velocity constraint is imposed at the point of collision additional constraints or allowances, may be associated with knot point $\boldsymbol{x_{col}}$.

Unfortunately, specifying the time of contact or the contact sequence is a non-trivial matter. Similarly for the collision problem, while the time of collision is not known, the point in space at which the contact occurs is. However, using fixed time-stepping methods which discretized the problem posed in Eq. 2 with a constant time step, specifying which particular knot point (if any) will represent $\boldsymbol{x_{col}}$ is equivalent to knowing the specified contact sequence at best and may result in an unsolvable problem at worst. However, by adding the time step used during the determination of the collocation constraints into the decision variables and by allowing the solver determine a unique time-step for each segment, then the knot point which specifies the collision takes new meaning. Instead of representing a specific time of collision, now the knot point $\boldsymbol{x_{col}}$ may be enforced as a specific state at the point of collision through an equality constraint:

$$\boldsymbol{x_{col}} = \boldsymbol{x_{obj}}, \quad (12)$$

where $\boldsymbol{x_{obj}}$ is a positional representation of the object in the state space. Equation 12 is then added as a general equality constraint in Eq. 3b. Selecting the specific index of $\boldsymbol{x_{col}}$ is a method hyperparameter. Associated with this additional time decision variable includes the need to provide bounds on sizes of the time steps. Simple inequality bounds may be posed as $h_{low} \leq h \leq h_{high}$ and $h_{low} > 0$. This constraint on the time step lower bound ensures the solver does not take un-physical actions.

Finally, if no feasible trajectories are found during the solution, such as if the needed velocity to override given a particular model was not obtainable before collision, the platform is commanded to remain stationary or to take an emergency avoidance action. While no uncertainty in the position of the object was considered in this work, uncertainty in the measurements of the vegetation was addressed by taking the maximum v_{over} as prescribed by the vegetation model during the time the vegetation was observed.

3. Simulation Results

Before running on the vehicle hardware, a series of tests were conducted in simulation. For all the simulation tests and the following hardware tests, the Mason et al. (2012) model was utilized by modifying only the vehicle and post parameters (L_t, m, h, D). The choice to use posts instead of override local vegetation was made in order to lower the studies impact on the local environment.

An example override for a 31.75 [mm] post that is embedded 0.3048 [m] into the ground is shown in Fig. 2. The post is placed 20 [m] in front of the robot in the x-direction. In the figure, the top plot represents distance from the goal and the bottom represents the simulated velocity. Of particular interest in this simulation is the collision that occurs at the 20 [m] mark, where the velocity drops substantially due to the collision with the simulated post. In order to increase the trajectory optimization problem solve times in simulation, the problem was initialized by dissecting the trajectory into m segments, where m is the number of expected collisions plus one, and then by linearly interpolating between the start, goal, and all object locations ($\boldsymbol{x_{obj}}$). While the implementation is not overly optimized, for the problems considered in this work, the initialization dropped the number of

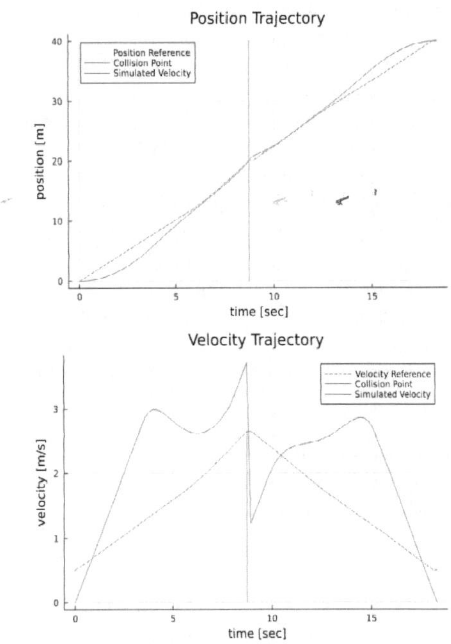

Fig. 2. Simulated vegetation override for 31.75 [mm] post. The controller achieves an actual velocity (green) above the required threshold

solver iterations by around a third. Note the one exception to this statement is in the initialization of the control trajectory, in which the initialization held constant values.

4. Hardware Results

Two hardware results are presented in this section, both dealing using the Mason et al. (2012) post override models as described in the Section 2. In both experiments, the vehicle was commanded to travel at a nominal speed of 5 [m/s]. A model predictive controller, similar to the unconstrained version of the work provided in Chen et al. (2017) was implemented as a tracking controller. The results provided in Fig. 3 and Fig. 4 reflect this tracking controller following the trajectory as determined by the trajectory optimization controller described in Section 2.

4.1 Experiment 1 – Straight Line Test

The first experiment that was conducted on hardware was a straight-line test. The object that the platform collided with was a 31.75 [mm] pine dowel rod that stood 0.914 [m] above the ground and was embedded around 0.305 [m] below the ground. The soil surrounding the dowel rod was compacted by hand. It had not rained in over a week and thus soil conditions were dry, even at a depth of 0.305 [m]. The minimum required override velocity computed using the model by Mason et al. (2012) was 2.7118 [m/s]. The post was placed around 11.70 [m] from the front of the robotic platform. The red arrow in Fig. 4 marks the position of the object as seen by the robotic platform's perception system. In Fig. 3, the vehicle began its trajectory at viewpoint zero. GPS had the vehicle localized at the centroid of the blue circle in Fig. 3 at the time of collision. The vehicle's end goal is depicted as a red square.

In order to constrain the platform into overriding the vegetation, keep-out zones were enforced around the vehicle. These zones are all areas shown in bright pink in Fig. 3. The gray areas in Fig. 3 indicate all the surrounding objects that are above a height limit of 0.5 [m]. The planned vehicle trajectory is shown in green. Platform viewpoints are provided in Fig. 3, which are captured along the executed trajectory. A subset of key viewpoints correspond to the numbers located on the obstacle and collision map subfigure in Fig. 3. Note that the vegetation in each photo is highlighted with a bounding box for better visibility of the extremities of the vegetation.

The vehicle reached a speed greater than the necessary threshold in order to impact the vegetation. The collision with the pine dowel rod occurred at around 4.5 [m/s]. The registered collision time was earlier than expected, but this is likely due to the fact that the point of time of collision is calculated when the vehicle first makes contact with the vegetation and the vehicle models used in the collision planning did not include the length of the nose of the vehicle. The vehicle suffered no damage during the test, but the post was almost completely failed. After collision, the post remained at an angle, with the end of the post that was suspended in mid-air sitting around 0.20 [m] over the ground surface. The post had displaced some soil, as evident in the post-collision subfigure in Fig. 3. While the soils were compacted by the experimenters before the tests, this displaced soil may be evident a lack of strong compaction at the surface. The post failed roughly 0.271 [m] from its bottom-most point, a little below the ground plane.

4.2 Experiment 2 – Post Override

The second experiment used the model from Mason et al (2012) to represent the post for override. This post is 25.4 [mm] in diameter, stands 0.914 [m] above the ground, and is embedded around 0.305 [m] below the ground. The parameterization of the soil was assumed consistent with the results found in Mason et al. (2012). This yielded an override minimum velocity of 2.4255 [m/s]. The post's position is marked with indicator "B" in Fig. 4. Additionally, the electrical pole in Fig. 4 that may be seen in viewpoint zero is indicated with an "A" in the collision map for localization. The vehicle began its trajectory at viewpoint zero. GPS had the vehicle localized at the centroid of the blue circle in Fig. 4 at the time of collision.

The vehicle reached the target speed and x-coordinate position at the correct collision time. Again, note the vehicle's nose is making contact with the pole at the time of the collision. Another thing to note is vehicle did strike the post on the passenger-side of the front bull-bar. This is a near-head-on collision with the post, but the alignment was meant to be towards the center of the bull-bar not the side.

After the collision, the post was completely failed by the platform. Additionally, the post had been slightly pulled out of the ground by around 50 [mm]. Figure 4 contains an image of the failed post. As in Experiment 1, the platform was not harmed.

5. Conclusions and future work

This work presents a trajectory optimization method that combines hybrid-dynamics, a free-time formulation, and existing parameterized vegetation models from the off-road mobility and cross-country movement literature in order to override environmental vegetation. The method works by enforcing minimum velocity constraints such that the collisions with the vegetation occur at a minimum override speed.

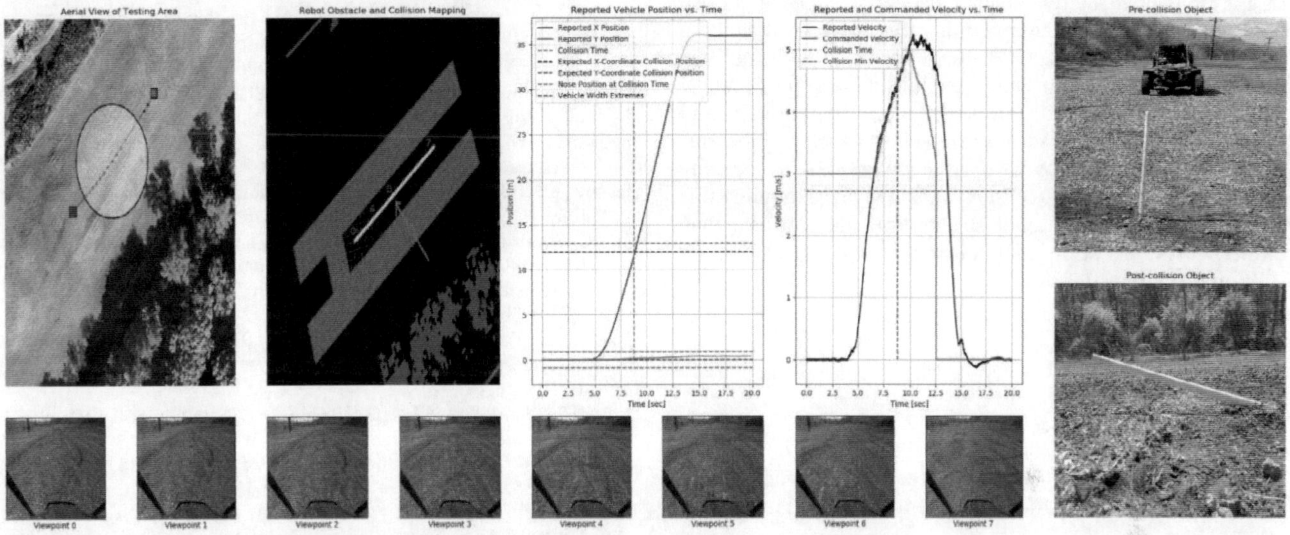

Fig 3. Hardware results from the first hardware experiment: a straight line trajectory through an embedded 31.75 [mm] post

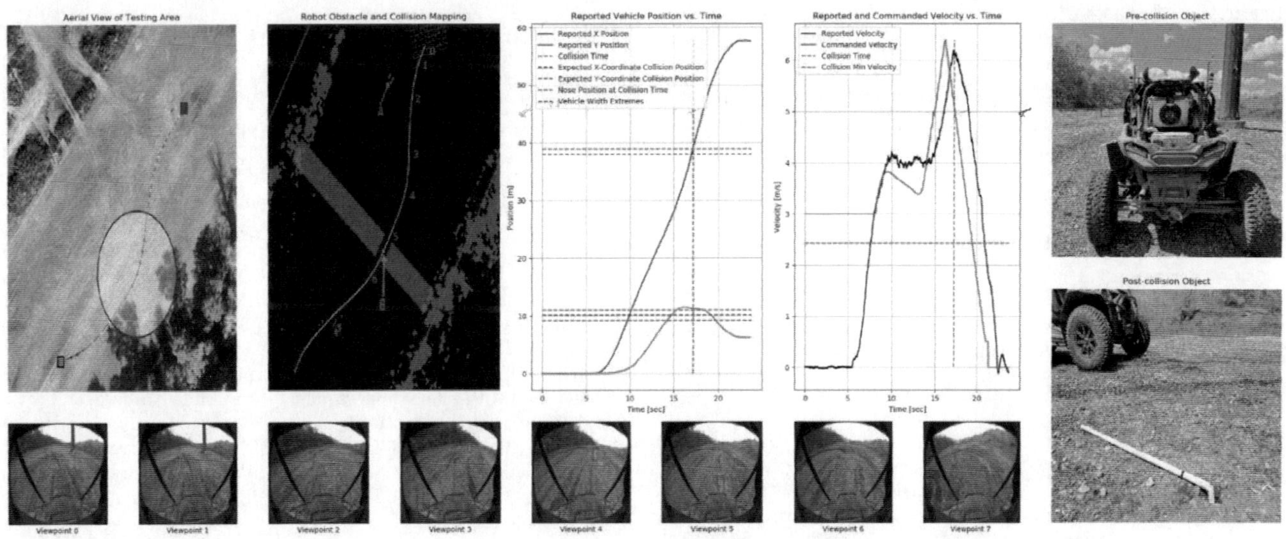

Fig 4. Results from the second hardware experiment: a turning trajectory through an embedded 25.4 [mm] post

These collisions occur at a designer-specified index of collision, but the designer need not select the time of collision as that is handled directly by the solver.

While both simulation and hardware results are expected to show losses in kinetic energy due to the collision, surprisingly, no large loss of kinetic energy is experienced during the collision with the vegetation. While the vehicle engine was in use during the impacts presented in this work and the bull-bar has a compliant mount, the loss of energy associated with both experiments should have resulted in a more identifiable drop in the kinetic energy. These effects could be more pronounced during slower overrides, so running experiments that supports characterizing the behavior at these slower speeds should take priority.

While the collisions modeled in this work occur at a unique instance in time, future studies should also look into addressing extended collisions with distributed objects, such as dense foliage.

General improvements to the algorithm include changing the vehicle model to capture the frontal geometry, better representations of vegetative objects (the representation of bush as a singular point such as the center of a tree showed difficulty in determining collision time), and better integration with the low-level vehicle interface. Furthermore, exploring the implications of the designer-selected collision index hyperparameter and developing automated approaches to how best to select that index should be made a priority. Work in adaptive gridding or meshing could assist in selecting the collision index. Finally, extending this algorithm to address additional objects that the platform may encounter will build a holistic solution to the challenge of obstacle interaction in off-road driving.

6. Acknowledgements

This research was sponsored by DARPA (#HR001121S0004). The views, opinions, and/or findings expressed are those of the author(s) and should not be interpreted as representing the official views or policies of the Department of Defense or the U.S. Government.

Distribution Statement "A" (Approved for Public Release, Distribution Unlimited)

The authors of this publication would like to acknowledge the contributions of Scott Hurley and Jacob Lammott, for their assistance in the field experimentation. Additional thanks to Robert Bittner, Norman Papernick, and Dr. Herman Herman. We would also like to give our thanks to Ryan Arciero for providing his thoughts on vegetation override. This work would not have been possible without the efforts of Team DEAD Fast.

7. Declaration of competing interest

The authors declare that they have no known competing financial interests or personal relationships that could have appeared to influence the work reported in this paper.

8. References

Blackmon, C. A., Randolph, D.D., 1968. An analytical model for predicting cross-country vehicle performance, vol. II: longitudinal obstacles. Appendix B: vehicle performance in lateral and longitudinal obstacles (vegetation). TR No. 3-783. U.S. Army Engineer Waterways Experiment Station, Vicksburg, MS.

Bradbury, M., Dasch, J., Gonzalez-Sanchez, R., Hodges, H., Iagnemma, K., Jain, A., Jayakumar, P., Letherwood, M., McCullough, M., Priddy, J., Wojtysiak, B., 2018. Next Generation NATO Reference Mobility Model (NRMM) Development. AC/323(AVT-ET-148)TP/728. North Atlantic Treaty Organization Science and Technology Organization.

Chen, J., Zhan W., Tomizuka M., 2017. Constrained iterative LQR for on-road autonomous driving motion planning. 2017 IEEE 20th International Conference on Intelligent Transportation Systems (ITSC). Yokohama, Japan, pp. 1-7.

Hargraves, C. R., Paris, S. W., 1987. Direct trajectory optimization using nonlinear programming and collocation. J. Guid., Ctrl, and Dyn. 10 (4), pp. 338-342.

Howell, T. A., Jackson B. E., Manchester Z., 2019. ALTRO: A Fast Solver for Constrained Trajectory Optimization. 2019 IEEE/RSJ International Conference on Intelligent Robots and Systems (IROS), Macau, China, pp. 7674-7679.

Howell, T.A., Tracy, K., Le Cleac'h, S., Manchester, Z., 2023. CALIPSO: A Differentiable Solver for Trajectory Optimization with Conic and Complementarity Constraints. In: Billard, A., Asfour, T., Khatib, O. (eds) Robotics Research. ISRR 2022. Springer Proceedings in Advanced Robotics. 27, pp 1-25.

Kelly, A., Stentz, A., Amidi, O., Bode, M., Bradley, D., Diaz-Calderon, A., Happold, M., Herman, H., Mandelbaum, R., Pilarski, T., Rander, P., Thayer, S., Vallidis, N., Warner, R., 2006. Towards Reliable Off Road Autonomous Vehicles Operating in Challenging Environments. The Int. J. Rob. Res. 25, pp. 449-483.

Kelly, M., 2017. An introduction to trajectory optimization: How to do your own direct collocation," SIAM Review. 59, pp. 849–904.

LaValle, S. M., 2006. Planning Algorithms, University of Illinois, Cambridge University Press (Online, accessed 05/15/2023).

Liu, C., Tomizuka,, M., 2014. Control in a Safe Set: Addressing Safety in Human-Robot Interactions. Proceedings of the ASME 2014 Dynamic Systems and Control Conference. San Antonio, Texas, USA. October 22–24, 2014. 3, pp. 1-10.

Mason, G. L., Gates, B. Q., Moore, V. D., 2012. Determining forces required to override obstacles for ground vehicles. J. Terramech. 49 (3-4), pp. 191-196.

Noren, C., Zhao, W., and Liu, C., 2021. Safe Adaptation with Multiplicative Uncertainties Using Robust Safe Set Algorithm. Proceedings of the 2021 Modeling Estimation and Controls Conference (MECC), 54, pp. 360-365.

Paden, B., Čáp, M., Yong, S. Z., Yershov, D., and Frazzoli, E., 2016. A Survey of Motion Planning and Control Techniques for Self-Driving Urban Vehicles. IEEE Trans. Intell. Veh. 1, pp. 33-55.

Posa, M., Cantu, C., Tedrake, R., 2013. A direct method for trajectory optimization of rigid bodies through contact. The Int. J. Rob. Res. 33(1), pp. 69-81.

Rybansky, M., 2017. Trafficability analysis through vegetation. Proceedings of the International Conference on Military Technologies (ICMT). Brno, Czech Republic, pp. 207-210.

Rybansky, M., 2020. Determination the ability of military vehicles to override vegetation. J. Terramech. 91, pp. 129-138.

Thrun, S., et al., 2007. Stanley: The Robot That Won the DARPA Grand Challenge. The 2005 DARPA Grand Challenge. Springer Tracts in Advanced Robotics. Berlin, Heidelberg. 36, pp. 1-44. https://doi.org/10.1007/978-3-540-73429-1_1.

Urmson, C., et al. 2009. Autonomous Driving in Urban Environments: Boss and the Urban Challenge. The DARPA Urban Challenge. Springer Tracts in Advanced Robotics, Springer, Berlin, Heidelberg. 56, pp. 1-59 https://doi.org/10.1007/978-3-642-03991-1_1.

Vong, T. T., Hass, G. A., Henry, C.L., 1999. NATO Reference Mobility Model (NRMM) Modeling of the DEMO III Experimental Unmanned Ground Vehicle (XUV). Final Report. Army Research Lab, Aberdeen Proving Ground, Maryland. pp. 1-69.

Williams, G., Drews P., Goldfain, B., Rehg, J. M., Theodorou E. A., 2018 Information-Theoretic Model Predictive Control: Theory and Applications to Autonomous Driving, IEEE Transactions on Robotics, 34, 6, pp. 1603-1622.

9268 | Experiments of singkage characters of wire mesh wheel under different slip ratio

Yang Wang [a], Meng Zou [a,]*, Jianqiao Li [a]

[a] Key Lab for Bionics Engineering of Education Ministry, Jilin University
* Corresponding author: zoumeng@jlu.edu.cn

ABSTRACT

For researching the sinkage of the lunar rover on Earth, a soil bin test system under light-load conditions was developed, which was used for researching the soil-wheel interactions between the wire mesh wheel of rover and the lunar soil simulant. Results of the experiments of sinkage characters of wire mesh wheel under different slip ratio shown that: The sinkage increased with the increase of slip ratio, the height of rut decreased with the increase of slip ratio, while lunar soil simulant was under different prepared conditions; The sinkage and the height of rut both increased, when lunar soil simulant changed from compact to soft. Experiments of rut surface scan of wire mesh wheel shown that, the maximum sinkage of rut increased firstly and then decreased, the maximum negative sinkage of rut increased, the shearing action of terrain increased, the surface of rut was8 destroyed gradually, with the increase of slip ratio. The results provided basic experimental data for evaluating the sinkage of lunar rover and the physical and mechanical characteristics of lunar soil simulant through the rut of wire mesh wheel.

Keywords
Wire mesh
Sinkage
Wheel rut
Slip ratio
Lunar soil simulant

1. Introduction

As the trafficability of lunar terrain was low, rover sank deeply and slipped seriously when moving on lunar terrain (Dang et al., 2022, Arvidson et al., 2002, Baichao, 2009). In order to improve the grip ability of the wheels, avoid serious slippage and sagging of the wheels in the soft lunar soil environment, and meet the requirements of lightweight lunar rover, the lunar rover is suitable for using spring griddle net wheels with good grip performance and light weight (Hao, 2013, Wong, 2009).

For researching the sinkage of the lunar rover driving on the lunar surface on Earth, considering the low gravity environment of lunar, developed a soil bin test system under light-load conditions according to the principle of equivalent load, and the soil bin was used for researching the soil-wheel interactions between the wire mesh wheel of rover and the lunar soil simulant. The soil bin could obtain the 3-D graphics of rut' surface of wire mesh wheel with a laser scanner.

2. Materials and methods

2.1 Wire mesh wheel

Wire mesh wheel generally had a solid hub with a woven wire carcass, which made it had the advantages of low mass. The wire mesh tread also helped clear the soil on the wheel and mitigate soil buildup in front of the wheel. Wire mesh wheel developed more drawbar pull than the hoop spring wheel on the soft terrain, which provides a technical basis for developing wheel for manned lunar roving vehicle (Yang, 2014, Haibo et al., 2008).

The test wire mesh wheel, as shown in Fig.1, diameter dimension is 300mm, width dimension is 150mm. The tread was weaved of titanium alloy wire with diameter 0.6mm, and the dimension of the square hole on the tread is 3mm x 3mm. Twenty-four lugs were divided into two rows and installed between each adjacent rims, respectively, each lug with dimensions of 75mm (length) × 10mm (height) × 1.5mm (thickness) (Haibo et al., 2008).

Fig.1. Wire mesh wheel

https://doi.org/10.56884/OVJS8996
ISBN 978-1-942112-55-6 (pdf)

2.2 *Soil bin test system under light-load*

Figure. 2 shows the soil bin test system under light-load, which was consisted of a soil bin testbed and a laser scanner. The overall dimension of the soil bin is 1500mm (length) × 800mm (width) × 400mm (depth), spread the JLU-2 lunar soil simulant, and the maximum wheel load of the test system is 100N.

Fig.2. Soil bin test system

The testbed was consisted of soil bin, trolley, metal frame. The trolley, as shown in Fig. 3, mainly consisted of vertical displacement sensor, horizontal displacement sensor, rotation sensor, force sensor and torque sensor, test wheel. A cable-pull displacement sensor (XPS-S-5000-11R0) was used to measure the sinkage of wheel, measurement range was 0m to 5m and measurement resolution was 0.1mm. A roller encoder (D38H-8G05L-360BM-5M)) was applied to measure horizontal displacement of wheel, measurement range is 0m to 5m and measurement resolution was 0.1mm (Han et al., 2016).

Fig.3. Trolley

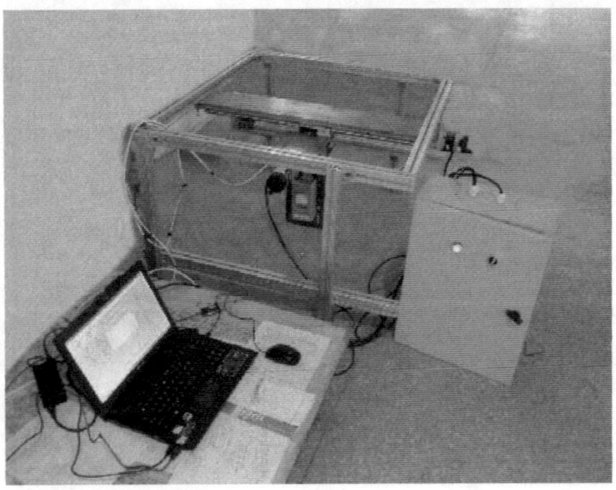

Fig.4. Laser scanner system

In order to obtain the digit parameters of wheel rut in the soil bin, a laser scanner was developed (Fig.4 shows), which was consisted of a laser sensor, a 2-dimensional traversing frame, the maximum scanning area of the laser scanner system was 650mm (length) × 400mm (width), measurement range was 0.05m to 65m and scanning resolution was 0.1mm. The scanner could obtain the 3-D graphics of the surface of rut by using contactless measuring method without disturbing the wheel rut.

3. Results and discussion

3.1 *Sinkage of wheel and height of rut under different slip ratio*

The basic conditions of experiments were as follows: lunar soil simulant was prepared into three different states, soft state, natural state, compact state; Wheel load and speed was set as 40N and 50m/h, respectively; Slip ratio were achieved by different traction forces.

Measured the sinkage of wheel through trolley, calculated out the height of rut surface by using the line scan model of laser scanner.

Rut of wire mesh wheel under different slip ratio, while lunar soil simulant under natural state, as shown in Fig. 5.

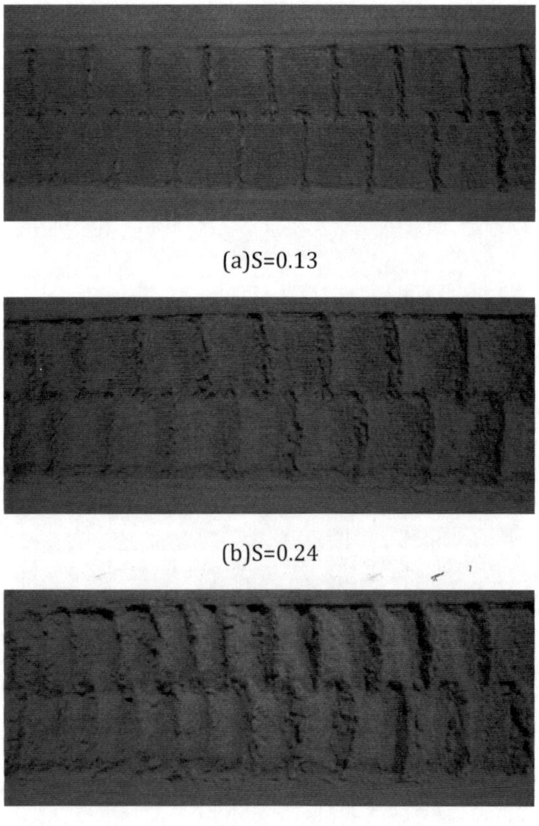

(a)S=0.13

(b)S=0.24

(c)S=0.40

Fig. 5. Rut of wire mesh wheel under different slip ratio, lunar soil simulant under natural state

Sinkage of wheel under different slip ratio and different lunar soil simulant states, as shown in Fig. 6. The sinkage of wheel increased with slip ratio under three lunar soil simulant states, as the digging effect of the lugs of wire mesh wheel on lunar soil simulant increased with slip ratio, slip sinkage of wheel increased with slip ratio.

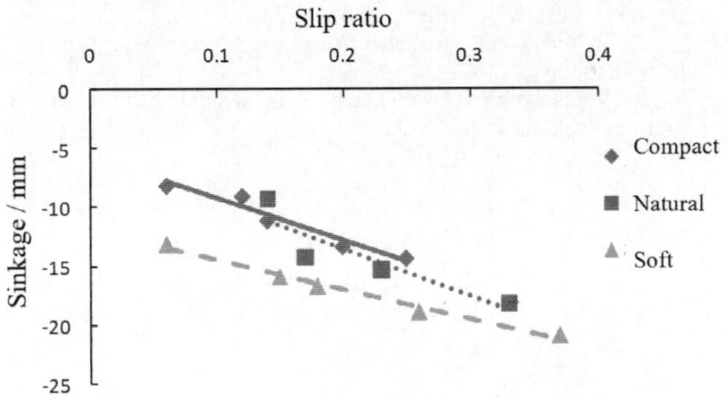

Fig. 6. Sinkage of wheel under different slip ratio and different lunar soil simulant states

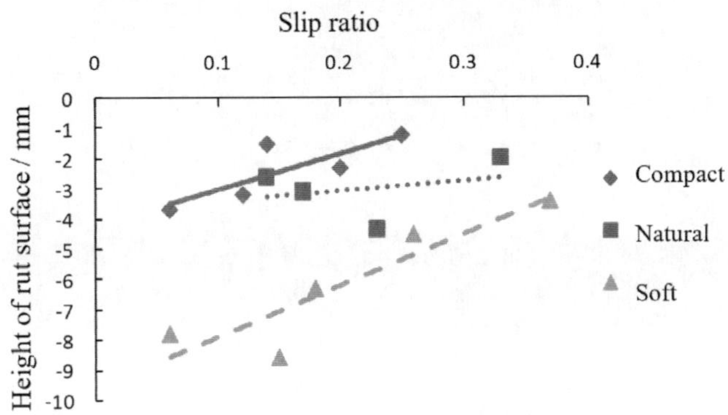

Fig. 7. Height of rut surface under different slip ratio and different lunar soil simulant states

Height of rut surface under different slip ratio and different lunar soil simulant states, as shown in Fig. 7. Height of rut surface decreased with slip ratio under three lunar soil simulant states, as the digging effect of the lugs of wire mesh wheel on lunar soli simulant increased with slip ratio, lunar soil simulant was spilled back by the lugs of wire mesh wheel, spread on the surface of rut, and renewed the rut.

Besides, sinkage of wheel and height of rut surface both increased, when lunar soil simulant changed from compact state to soft state. The sinkage of wheel increased with slip ratio under three lunar soil simulant states, but the height of rut surface decreased with slip ratio under three lunar soil simulant states.

3.2 Surface scan of rut

The basic conditions of experiments were as follows: lunar soil simulant was prepared into natural state; Load was 40N, speed of wheel was 50m/h; Slip ratio were achieved by different traction forces. Scanning area was 200mm by 222 mm, scanning spacing was 2mm.

Surface scan of rut of wire mesh wheel under different slip ratio, as shown in Fig. 8. The recurring cycle along the length of rut decreased with slip ratio, the shearing action between wheel and lunar soil simulant increased with slip ratio, the integrity of rut surface decreased with slip ratio.

Scan sinkage of rut of wire mesh wheel under different slip ratio, as shown in Fig. 9, the negative sinkage of rut was used the highest point of rut to subtract the reference surface of lunar soil simulant.

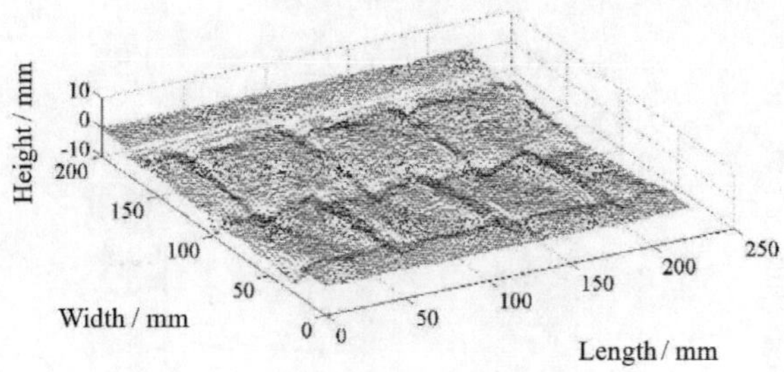

(a)S=0.08

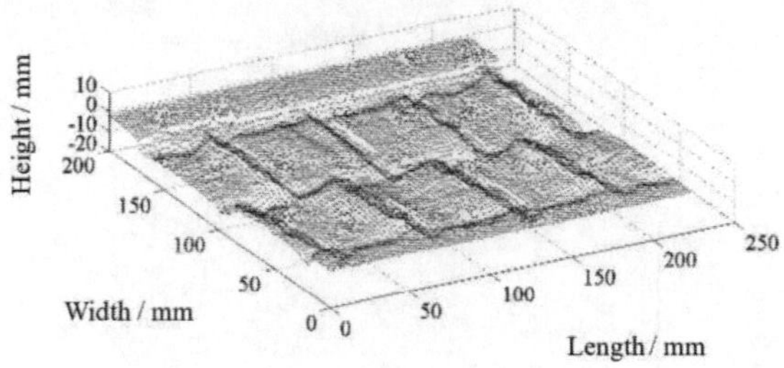

(b)S=0.22

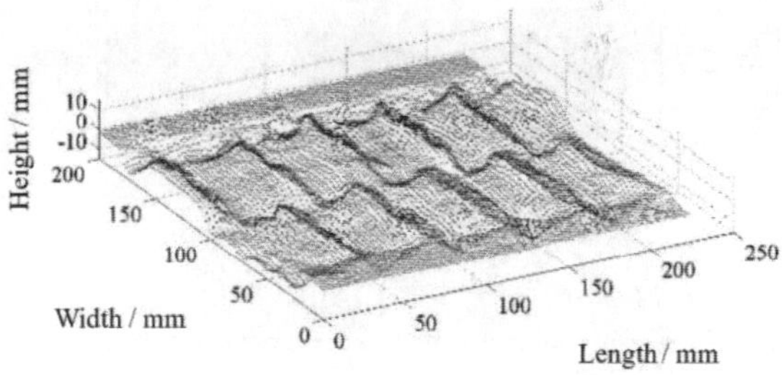

(c)S=0.33

Fig. 8. Surface scan of rut of wire mesh wheel under different slip ratio

The maximum sinkage of rut increased with slip ratio firstly, and then decreased with slip ratio, as the time of wheel compact on lunar soil simulant was longer under low slip ratio, lunar soil simulant was spilled back by the lugs of wire mesh wheel more seriously under high slip ratio. The maximum negative sinkage of rut increased with slip ratio.

4. Conclusions

Results of the experiments of sinkage characters of wire mesh wheel under different slip ratio shown that, the sinkage increased with slip ratio, the height of rut decreased with slip ratio, while lunar soil simulant was under different prepared conditions; The sinkage and the height of rut both increased, when lunar soil simulant changed from compact to soft.

Experiments of rut surface scan of wire mesh wheel shown that, the maximum sinkage of rut increased firstly and then decreased, the maximum negative sinkage of rut increased, the shearing action of terrain increased, the surface of rut was destroyed gradually, with the increase of slip ratio. The results provided basic experimental data for evaluating the sinkage of lunar rover and the physical and mechanical characteristics of lunar soil simulant through rut of wire mesh wheel.

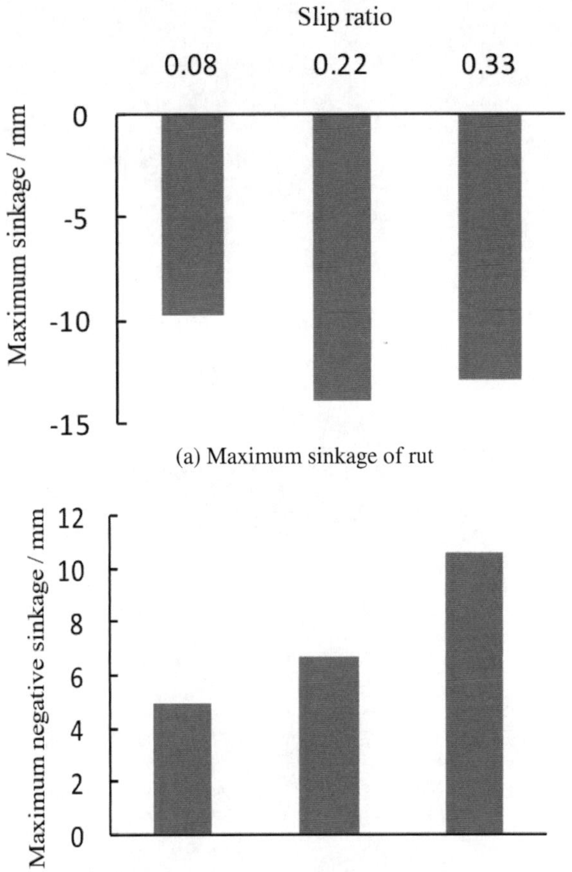

(a) Maximum sinkage of rut

(b) Maximum negative sinkage of rut

Fig. 9. Scan sinkage of rut of wire mesh wheel under different slip ratio

5. Acknowledgements

This work was supported by the National Natural Science Foundation of China (grant numbers 52075217, 51775233) and Key Funding of Science and Technology Department of Jilin Province (grant numbers 20200401144GX).

6. Declaration of competing interest

The authors declare that they have no known competing financial interests or personal relationships that could have appeared to influence the work reported in this paper.

7. References

Arvidson, E. R. et al. 2002. FIDO prototype Mars rover field trials, Black Rock Summit, Nevada, as test of the ability of robotic mobility systems to conduct field science. J. Geophys. Res. 107(E11), 8002.

Baichao, C., 2009. Design of a new locomotion system for lunar rover." Jilin University, China.

Dang, Z., Zhang, J., Chen, B., 2022. Field Validation of Egress Process for Planetary Rover. Proc. 11th Asia-Pacific Regional Conference of the ISTVS.

Haibo, G., Yang, J., Zongquan, D., 2008. Development on spring griddle net wheel of lunar rover. J. Harbin Institute of Technology 40(11), 1712-1716.

Han, H., Jianqiao L., Baichao, C., 2016. Performance evaluation of a wire mesh wheel on deformable terrains. J. Terramech. 68, 9–22.

Hao, L., 2013. The JLU-B series Lunar soil simulant and the interaction with the rigid grouser wheel. Jilin University, China.

Wong, J.Y., 2009. Terramechanics and off-road vehicle engineering (2nd ed). Elsevier, Oxford.

Yang, W., 2014. Research on travelling environment simulation and ruts of lunar rover. Jilin University, China.

9992 | Soil shear strength values obtained from its color

Alaa El Hariri [a,*], Péter Kiss [b]

[a] *Mechanical Engineering Doctoral School – Hungarian University of Agriculture and Life Sciences (MATE), 2100 Gödöllő, Páter Károly u. 1., Hungary*

[b] *Institute of Technology, Department of Vehicle Technology- Hungarian University of Agriculture and Life Sciences (MATE), 2100 Gödöllő, Páter Károly u. 1., Hungary*

* *Corresponding author: Alaahariri123@outlook.com*

ABSTRACT

The soil shear strength and its parameters are important values that reflect the resistance of the soil to shearing. The soil shear strength relates the shear cohesion, internal friction angle, and the applied consolidation pressure through the linear Coulomb model. The paper embeds shear strength, shear cohesion, and internal friction angle results of a sandy loam soil and a clay soil at different moisture contents. The shear strength values were obtained using the direct shear test, also by drawing the Mohr-Coulomb line. Beside the mentioned values, the article contains color results of the tested soil types at different moisture contents, measured using the spectrophotometric technology in the visible band (spectrum; 400-700 nm). Relating the color of the soil to its shear strength values through its moisture content, might be helpful in the field for identifying the shear strength and its parameters based on the color of the soil, and this target is achieved by laboratory empirical work. The process of correlating the soil color and its shear strength values at different moisture contents is presented in the paper. The measurements were carried out at the Hungarian University of Agriculture and Life Sciences (MATE - Department of Vehicle Technology Laboratory).

Keywords
Soil shear strength
Sandy loam soil
Clay soil
Direct shear test
Shear cohesion
Internal friction angle
Spectrophotometer
Soil color reflectance

1. Introduction

The change in the moisture content leads to changing the mechanical properties (load bearing capacity and shear strength) and also the physical behaviour of the soil (El Hariri et al., 2023) (Ahmed et al., 2023). The forces acting from the vehicle's tractive element on the soil, when in motion, are the normal, tangential, and the lateral forces, thus if the soil is not able to withstand these forces (stresses), the soil fails forming drawbacks to the vehicle's motion (in some cases the vehicle might get stuck totally in the soil). In terramechanics science, it is crucial to take into account the shear strength values of the soil for being inserted in a model such as the model proposed by (Janosi and Hanamoto, 1961), thus predicting the required performance parameter (as the traction force) of a vehicle on soil at a moisture value.

The question is how to identify the shear strength parameters in the field, knowing that in field the soil is inhomogeneous. In most of the terramechanics researches, the shear cohesion and the internal friction angle are obtained by laboratory work, such as using the direct shear machine or the triaxial test (Shoop, 1993). Even though the shear strength values are obtained in the laboratory by having the soil homogeneous (sieved) for simplifying the study case, but transferring this data to be used in the field is the solution for a real time terramechanics case, so in the field an indication is needed to know which strength values from the laboratory work (for a soil type) to be used in the models.

Since the soil color is an apparent useful property (Baumgardner, 1985), it might be an indication to detect the moisture of the soil in the field. The soil color is used as a chip for identifying the kind of the soil (soil survey staff, 1999). The organic matter of the soil has been related to its color by researchers (Shields, 1968) (Al-Abbas, 1972) (Cihlar and Protz, 1973) (Page, 1974) (Henderson and Szilagyi, 1989) (Dobos and Ciolkosz, 1990) (Ivey and McBride, 1999). Different methods have been used for measuring the soil color and its reflectance aiming to find the field conditions (Melville and Atkinson, 1985) (Post et al., 1993). Identifying the moisture content of the soil using its spectral data is important in agriculture (Muller and Decamps, 2001).

The spectrophotometric technology is one of the methods used for measuring the soil color. Using a visible range spectrophotometer, soil reflectance results will be obtained at different moisture contents. Ending up with a reflectance curve of the soil color, serves in predicting the moisture of the soil upon having input color value inserted into the curve's equation, thus the ability to obtain the moisture of the soil in the field (field of same soil type as the laboratory tested soil).

Dealing with the soil in the laboratory will facilitate the engineering study case, thus for a specific type of soil, the color will be measured using the visible range spectrophotometer. The shear strength and its parameters will be measured using the direct shear test at different moisture contents, since based on researchers' literatures, this test closely emulates the shearing mechanism of the soil beneath the tractive element (Shoop, 1993). Plotting the results of the shear strength and its parameters as function of moisture content, will result in curves that based on their equations, the shear strength values can be predicted when needed for any input moisture value (curves will be prepared by laboratory work, before usage in field).

This idea is beneficial to be implemented in autonomous off-road engineering designs - where the machine is not controlled by an operator -, since moving on soil zone that is of high moisture content might lead to the slippage or sinkage of the vehicle, and maybe immobilization might occur. A remote sensor for measuring the color of the soil can be used, either installed on the

https://doi.org/10.56884/PQZU1454
ISBN 978-1-942112-55-6 (pdf)

vehicle or on a drone, so before moving on the doubted soil zone, the color can be detected from the remote sensor (spectrophotometric technology), and based on the provided soil color reflectance, the moisture can be predicted (from color curve - of the soil type - that is prepared in the laboratory). The moisture obtained having it inserted in the prepared (laboratory) shear strength curves, will end up with the required soil shear strength values.

2. Measuring the soil shear strength values

2.1 Measuring the shear strength of tested soil types

Based on scientific references showing that the direct shear test closely resembles the shearing that is resulting from the tractive element on the soil, thus its results (shear strength values) can be used for studying the mobility case of a vehicle on soil. The soil types that were tested are the clay and the sandy loam soil. The sandy loam soil is of mechanical composition (90.50 % sand, 3.20 % silt, and 6.30 % clay), and the clay soil is 100 % clay. Having the soil initially prepared in the laboratory and at ambient conditions, the soils were sieved for removing the impurities, and thus dealing with the soil as homogeneous material for the sake of simplifying the study case. Initially having the soils at ambient conditions, samples were placed in boxes for getting moistened. Water was added to each sample (in each box), for increasing its moisture content. The water was added to each box in an increment way, aiming to study the influence of the increase in the moisture on the shear strength. In each box after adding water, the soil was stirred till becoming homogeneous in phase (uniform).

Having the soils moistened and prepared in the boxes, the moisture of each sample was measured using the analyzer, that works based on the principle of drying. The used analyzer is a gravimetric method, where an amount of soil is placed on a tray in the analyzer, and the heating plates start to dry this sample. Upon reaching the complete dryness of the placed sample (zero moisture; no decrease in its mass), the moisture percentage is calculated by the anyalzer through deducting the final mass (remaining mass) from the initial mass and then dividing the result by the initial mass. This measurement is wet based, since dividing the difference by the initial wet mass of the soil (placed in the analyzer). The moisture amount is obtained for each of the soil samples, thus identifying the moisture of each prepared soil sample using the precise gravimetric method.

The samples of the two soil types are prepared (boxes covered to prevent the decrease in the moisture), then using the direct shear test, the shear strength of each sample will be measured. The direct shear test used is shown in Fig 1. In the direct shear box (shown in Fig. 2), place an amount of soil (110 grams), then apply a consolidation load equal to 30 kg. This consolidation load (30 kg) is used, since when the resulting gravitational force is divided by the area of the pressing plate (0.002743246 square meter of the shear box), the resulting applied stress will be equal to 1.073 bar (real case stress applied from wheel on ground). Since the soil in the field is undisturbed (real case), soil in laboratory is disturbed, the soil placed in the direct shear box was pressed for 5 minutes (by the 30 kg) before running the machine. Five minutes was chosen as a compaction time, longer compaction time means that the soil compacts more, thus different shear strength result. After compacting the soil for 5 minutes, the consolidation load was removed and the thickness of the soil in the direct shear box was measured. Having the soil mass (110 grams) and the volume of the soil sample in the shear box (area of the shear box multiplied by the soil thickness), the wet density of each soil sample can be calculated (dividing mass over the volume). Figures 3 and 4 show the density results of the two tested soil types.

For the sandy loam soil, the density was calculated at the following moisture contents, 1.41, 4.36, 6.24, 8.18, 10.15, 13.36, 18.28 % (unit of moisture content is percentage; symbol %). For the moisture value (18.28 %) the soil was not compacted for 5 minutes, since the moisture is high, for not draining the moisture content (just load then de-load directly, and record the thickness of soil).

For the clay soil, the density was calculated at the following moisture contents, 4.61, 8.58, 11.54, 14.43,18.22, 23.27, 26.14 %. For the high moisture values (beyond 20 %) the soil was not compacted for 5 minutes, for not draining the moisture content.

The density results at different moisture contents are shown in table 1.

Starting with the first soil sample (at ambient conditions, for the 2 soils), and after compacting the soil in the direct shear box for 5 minutes and measuring the soil thickness for the density, the pressing plate was placed back on the soil in the shear box, and the 30 kg was applied again for starting the shear test. The direct test was set on 9 mm/min shearing speed (highest in the machine), then the machine was activated.

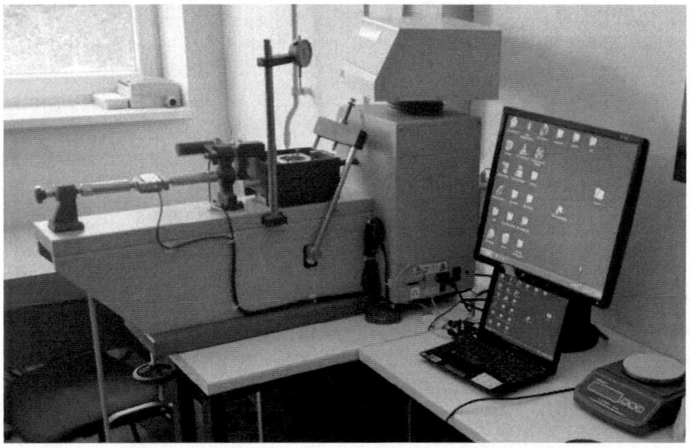

Fig. 1. Direct shear test machine (ELE 26–2112/01; MATE university - Hungary)

Fig. 2. Shear box of the shear machine

The same test was repeated on the other soil samples (the two types), and the shear strength results were recorded. Force and displacement sensors are installed on the direct shear machine, thus ending up with shear force – displacement curve. The maximum point or the peak point on the curve is the maximum applied force, which beyond this force the shearing failure of the soil occurs. The maximum force divided by the area of the shear box, results in the maximum shear stress or the shear strength of the tested soil.

Table 1
The density results of the tested soils at different moisture contents

Sandy loam soil		Clay soil	
Moisture [m/m %]	Density [g/cm^3]	Moisture [m/m %]	Density [g/cm^3]
1.41	1.5	4.61	1.28
4.36	1.58	8.58	1.36
6.24	1.6	11.54	1.39
8.18	1.63	14.43	1.4
10.15	1.76	18.22	1.5
13.36	1.82	23.27	1.87
18.28	2.08	26.14	1.94

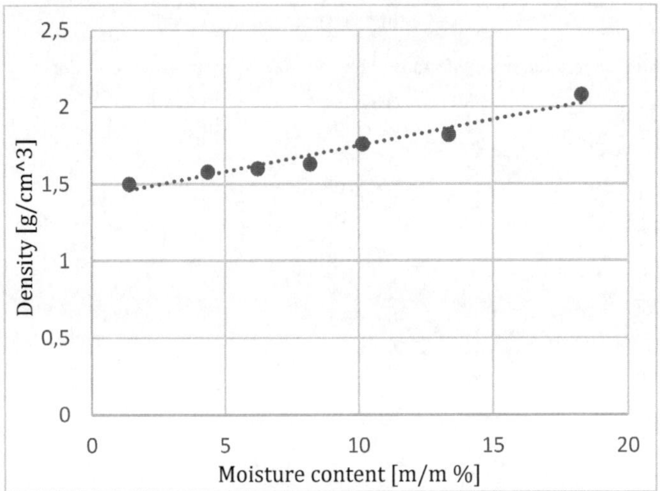

Fig. 3. Sandy loam soil wet density as function of moisture content

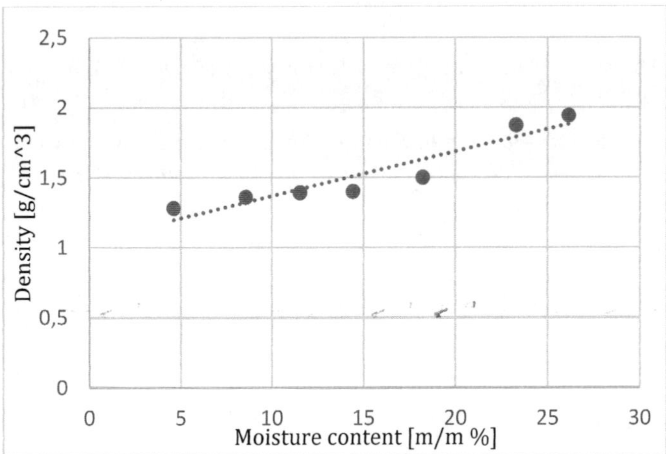

Fig. 4. Clay soil wet density as function of moisture content

With the increase in the moisture content for both types of soil, the density increases. Across the moisture content range (in both sandy loam and clay), the density under the same compaction load (30 kg) is increasing, showing that for the two tested soils, the moisture makes the soil more compactible. The increase in the moisture content, increases the compressibility of the tested soils under the same compaction load.

The shear strength results of sandy loam soil were measured at the following moisture values, 1.41, 2.23, 3.21, 4.36, 5.88, 6.24, 7.55, 8.19, 9.33, 10.15, 11.47, 12.45, 13.36, 14.12, 15.16, 18.28 (% is the unit).

The shear strength results of clay soil were measured at the following moisture values, 3.59, 4.61, 5.5, 8.45, 9.23, 11.69, 15.89, 17.11, 20.38, 23.04 %.

The obtained shear strength results of the two tested soil types at different moisture contents are shown in Fig 5 and Fig 6 (table 2 shows all the obtained values).

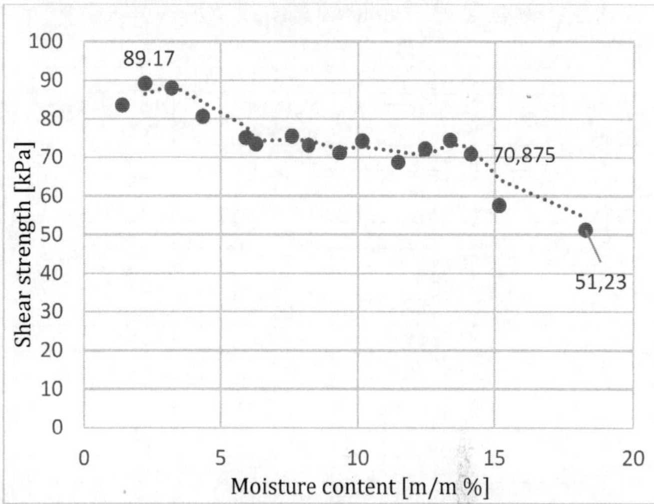

Fig. 5. Sandy loam soil shear strength as function of moisture content

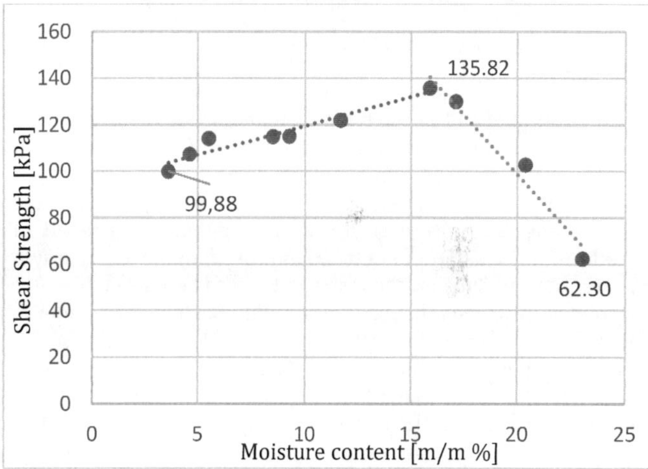

Fig. 6. Clay soil shear strength as function of moisture content

For the sandy loam soil, with the increase in the moisture content, the shear strength increases to a maximum value equal to 89.17 kPa at 2.23 % moisture content. Beyond the maximum shear strength value, with the increase in the moisture, the shear strength decreases till reaching a constant level of a shear strength value around 70 kPa in the moisture range (5.88 till 14.12 %). Beyond the 14.12 %, with the increase in the moisture, the shear strength decreases, reaching a value 51.23 kPa at 18.28 % moisture content. In the range where the shear strength is approximately constant, based on Mohr-Coloumb equation, since the same consolidation load is applied, thus the change in the cohesion and the internal friction angle in this range was ending up with the same shear strength value. At high moisture contents the shear strength decreases, since the water acts as lubrication at the particle's level, so decreasing the internal friction, and breaking cohesive bonds, leading to a low shear strength result (Salman and Kiss, 2018).

For the clay soil, with the increase in the moisture content, the shear strength increases till reaching a maximum value 135.82 kPa at 15.86 % moisture value, and beyond this value, the strength starts to decrease with the increase in the moisture, till reaching 62.30 kPa at 23.04 % moisture content.

Table 2
The shear strength results of the tested soils at different moisture contents.

Sandy loam soil		Clay soil	
Moisture [m/m %]	Shear strength [kPa]	Moisture [m/m %]	Shear strength [kPa]
1.41	83.57	3.59	99.88
2.23	89.17	4.61	107.28
3.21	88.04	5.5	114.08
4.36	80.65	8.45	114.86
5.88	75.14	9.23	115.03
6.24	73.50	11.69	121.99
7.55	75.54	15.89	135.82
8.19	73.21	17.11	129.98
9.33	71.25	20.38	102.64
10.15	74.29	23.04	62.30
11.47	70.72	---	---
12.45	72.32	---	---
13.36	74.51	---	---
14.12	70.87	---	---
15.16	57.56	---	---
18.28	51.23	---	---

2.2 *Measuring the shear strength parameters of the tested soil types*

The cohesion and the internal friction angle of the tested soils at different moisture contents were obtained from the Coloumb line. The Coloumb's equation based on Mohr's principle, relates the shear strength parameters and the normal applied stress (consolidation stress) linearly for calculating the shear strength of the soil (Mohr, 1900) (Coulomb, 1776). The Coloumb line is drawn by fitting a trendline into shear strength plotted points at different consolidation stresses, having the soil at a fixed moisture content. So, taking a soil sample at a moisture content, and measuring its shear strength under different consolidation stresses. Using the masses 10, 20, 30, 50 and 70 as normal loads, the shear strength results of the tested soil (at a moisture content) were obtained. The same procedure was repeated on the soils, but at different moisture contents, using the same applied stresses. The intercept of the Coloumb line with the y-axis (vertical axis) is the cohesion, and the angle that the line is making with the horizontal is the internal friction angle, thus obtaining the cohesion and the internal friction angle of the soil at a moisture content.

The cohesion and internal friction angle of the sandy loam soil were obtained at moisture contents 2.45, 4.61, 6.41, 9.4, 11.19, 12.7, 14.43, 15.47 %.

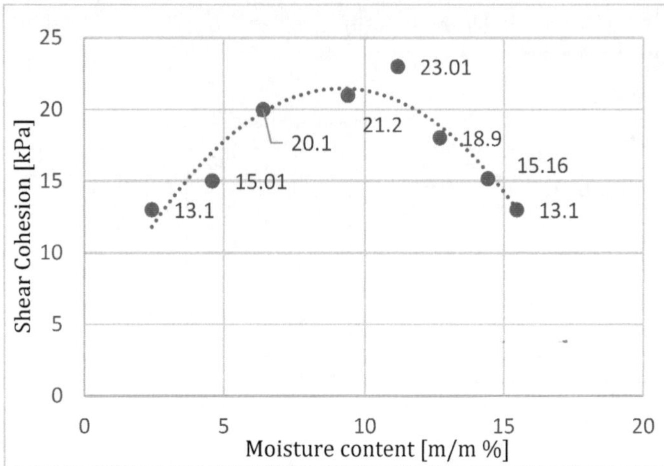

Fig.7. Shear Cohesion of sandy loam soil at different moisture contents

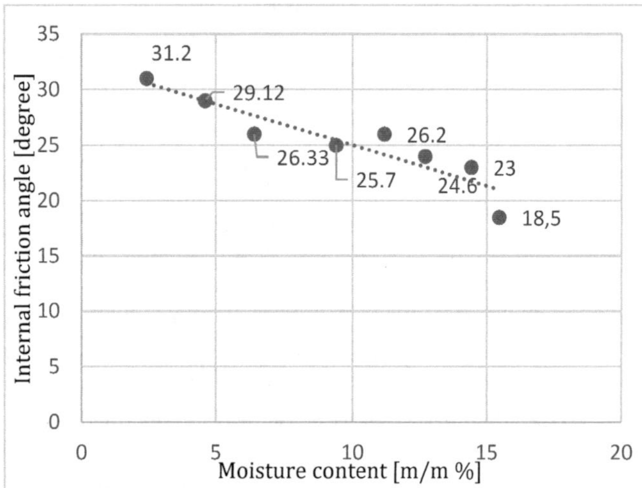

Fig. 8. Internal friction angle of sandy loam soil at different moisture contents

The cohesion and internal friction angle of the clay soil were obtained at 3.9, 7.91, 11.99, 16.35, 20.2, 23.66 % moisture values.

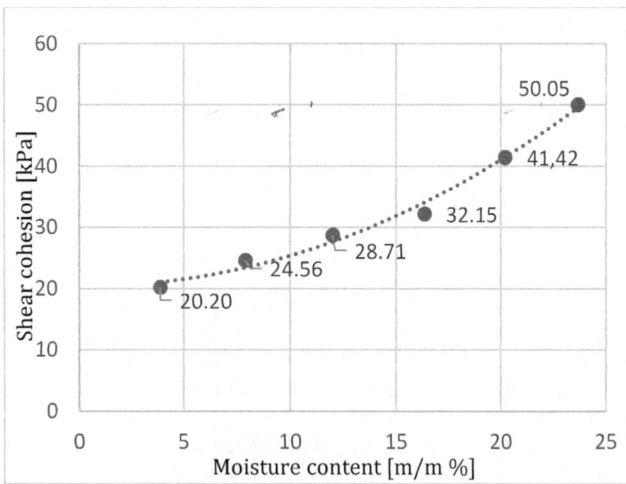

Fig. 9. Shear cohesion of clay soil at different moisture contents

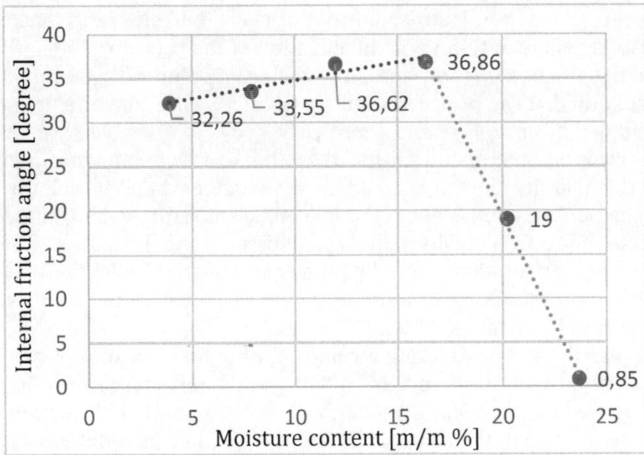

Fig. 10. Internal friction angle of clay soil at different moisture contents

For the sandy loam soil, with the increase in the moisture content, the cohesion increases till reaching a maximum value of 23.01 kPa at moisture content 11.19 %. Beyond this point, with the increase in the moisture, the cohesion starts to decrease till reaching a cohesion value 13.1 kPa at 15.47 %. The saturation level of the sandy loam soil is at moisture content around 26 %. Figure 8 shows that with the increase in the moisture content in the sandy loam soil, the internal friction angle decreases.

For the clay soil, with the increase in the moisture content, the shear cohesion increases reaching a maximum value of 50.05 kPa at the maximum tested moisture content, knowing that the saturation level of the tested clay soil is around 37 % moisture content. The internal friction increases reaching a maximum value of 36.86 degree at moisture content 16.35 %, then after this point, the value starts to decrease reaching 0.85 at moisture value 23.66 %.

At moisture content close to the saturation level, the cohesion and the internal friction angle are very low, since the water films between the soil particles break the cohesive bonds, and the water acts as a lubricant.

3. Measuring the color reflectance

The soil color was measured using a visible range spectrophotometer. The visible range spectrophotometer was used for two reasons: 1-since in field, the influence of the moisture on the soil color can be detected by the human eye, 2- in order to see if in fact the soil reaches a level where there is no more change in its color (most of the soils), that is why the eye cannot recognize a change in the soil color at high moisture contents. The visible range spectrophotometer works based on the principle of sending light waves (in the visible spectrum 400 – 700 nm), thus the color of the soil is the reflected wavelength. The color of the soil is decoded by a number (quantity) which is the reflected wavelength, the reason behind making this measuring method (spectrophotometric technology) a quantitative precise method. The visible range spectrophotometer used is (CM-700d), shown in Fig 11.

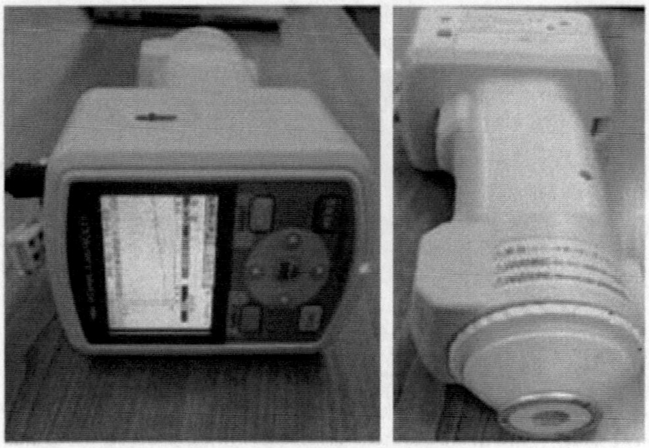

Fig. 11. The visible range spectrophotometer used (CM-700d)

Plotting the reflectance results at 700 nm as function of the moisture values for the two soils. Figures 12 and 13 show the color reflectance at 700 nm as function of moisture content (table 3 embeds all the reflectance results).

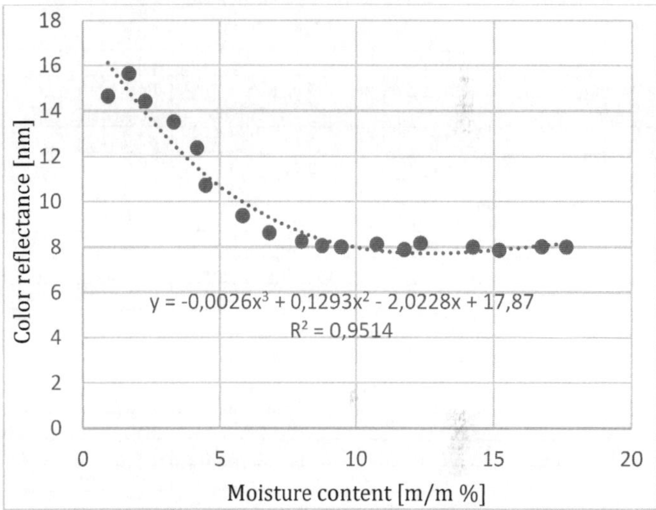

Fig. 12. Sandy loam soil color reflectance at 700 nm as function of moisture content

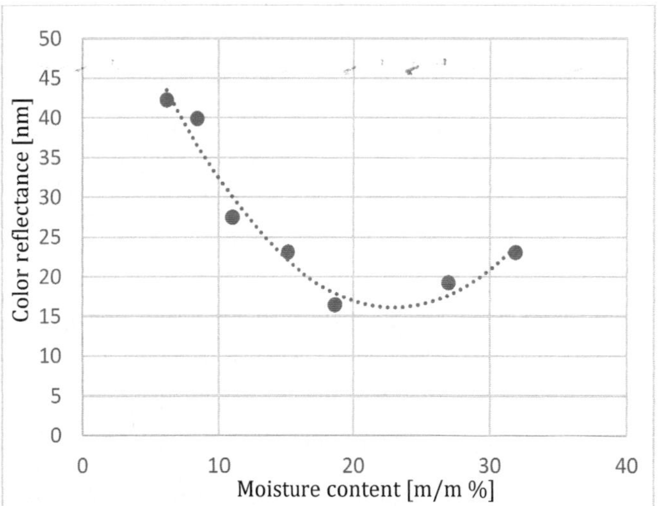

Fig. 13. Clay soil color reflectance at 700 nm as function of moisture content

For the sandy loam, with the increase in the moisture content, the color reflectance decreases, till reaching a level where the reflectance become approximately constant (beyond the moisture value 9.46 %). The constant stage in the reflectance reveals either that there is no more change in the color of the soil, beyond 9.46 % moisture, or the visible range spectrophotometer is not able to detect the change in the color at this level (maybe another range spectrophotometer can detect the change; such as infrared); the constant complies with what the eye sees (no change in the soil color at high moisture in most soils).

For the clay soil, with the increase in the moisture content, the reflectance decreases till reaching a minimum point, at moisture value 18.67 %. Beyond 18.67 %, with the increase in the moisture, the reflectance increases. With increasing the moisture beyond 18.67 %, the clay soil becomes paste like material, and thus not more particulate material, which is the reason behind the increase in the color reflectance. With sending a light wave to the soil as a particulate material, a part of the sent wavelength gets scattered at the particle's level, thus having the soil as a paste like material will end up with increasing the reflectance.

Table 3
The color reflectance results of the tested soils at different moisture contents

Sandy loam soil		Clay soil	
Moisture [m/m %]	Reflectance at 700 [nm]	Moisture [m/m %]	Reflectance at 700 [nm]
0.92	14.64	6.16	42.24
1.68	15.62	8.45	39.9
2.27	14.42	10.97	27.44
3.31	13.5	15.12	23.09
4.17	12.36	18.67	16.41
4.49	10.71	26.95	19.2
5.8	9.36	31.88	23.04
6.78	8.60	---	---
7.99	8.23	---	---
8.77	8.03	---	---
9.46	7.98	---	---
10.76	8.11	---	---
11.75	7.87	---	---
12.34	8.14	---	---
14.24	7.98	---	---
15.2	8.11	---	---
16.75	7.87	---	---
17.65	8.14	---	---

4. Relating the color to the shear strength values

The main target behind the article is predicting the strength values of the soil from its color through the moisture content, which means that the moisture will be predicted from the color at first. The intercept of the moisture value with any of the strength curves (in the above strength figures, 5-10), leads to the strength value (graphically; moisture - curve intercept - projection on y-axis). Also, by obtaining the equations of the strength curves (might be divided into ranges in some curves), and substituting the moisture in it, the strength value can be calculated.

Since the material being studied is a particulate material, then there should be a change in the color reflectance results upon repeating the measurement (at the same moisture content). The difference in the obtained results, leads to the estimation of the moisture content, thus using the term predicting the moisture content, and not giving the exact value as when measured using the gravimetric method.

For the sandy loam soil, fitting a curve into the reflectance results, will lead to finding the equation of the color (at 700 nm) change with moisture content. Thus, ending up with a curve equation that can be used in field for predicting the moisture upon having a color value measured by a visible range spectrophotometer and substituted in the equation (knowing that in the field the moisture should be in the range where the color is changing; equation is not applicable at high moisture values, since same reflectance results in the high moisture range).

Another solution for finding the color equation, is by fitting a curve into the reflectance points in the range where the color is changing (moisture < 9.46 %).

For the clay soil, fitting a curve into the first moisture range (below 20.63 %), where there is decay in the reflectance with the increase in the moisture content, will lead to finding the equation of the change in the color (at 700 nm) with moisture content. As mentioned, that beyond the 20.63 % moisture content, the clay becomes a paste like material, and such material doesn't exist in real time terramechanics case, so it is not necessary to deal with the color change in this range. Ending up a curve equation in the first range, can be used in field for predicting the moisture upon having a color value measured by a visible range spectrophotometer and substituted in the equation.

Having the moisture predicted from the color, and by substituting it in any of the shear strength curves (shear strength, cohesion, and internal friction angle), the shear strength value at the specified moisture content can be obtained. Thus, the shear strength value of the soil at a moisture content is ready to be substituted in the used model (proposed by researchers), for calculating the term needed for studying the mobility of the vehicle on the soil.

5. Soil moisture and strength values obtained from its color

The soil color will be used as an indication for identifying its moisture amount, so being able to know the strength value relying on already prepared laboratory results (for the soil type). Therefore, the color of the soil is linked to its strength values through the moisture content.

Giving an example (on sandy loam soil) about the above work, which is finding the moisture of the soil from its color, and then predicting the strength values from its strength curves.

For the sandy loam soil, the equation of the curve that is fitting the plotted color reflectance points (at 700 nm), and of high regression analysis is (Equation 1; shown in Fig. 12):

$$y = -0.0026x3 + 0.1293x2 - 2.0228x + 17.87 \quad (1)$$

For another sample (sandy loam; different moisture), the color was measured by the spectrophotometer, and is of reflectance value (at 700 nm) equal to 11.72 %; percentage is the unit of color reflectance.

Substituting the color value in the equation, gives 3 solutions, since it is a third-degree polynomial equation. The results are:

$$X_1= 4.03 \quad (2)$$

$$X_2= 22.88 + 7.99\mathbf{i} \text{ (complex value)} \quad (3)$$

$$X_3= 22.88 - 7.99\mathbf{i} \text{ (complex value)} \quad (4)$$

X_2 and X_3 are rejected, since they are complex values, and the moisture value should be a rational number (X_1).

Using the moisture analyzer (gravimetric method), the soil moisture was measured, and is of a value 4.28 percent. The two moisture values are close to each other.

For the shear strength value at the moisture 4.03 percent (obtained from color), the shear strength of the soil at this moisture is around 80 kPa (approximate value) (x intercept with curve in Fig. 5, then project on the y-axis, 'shear strength axis').

For obtaining the exact value, the shear strength curve can be divided into more than one part, having each in a moisture interval and of an equation (shear strength as function of moisture). The moisture is then substituted in the interval equation, thus ending with the shear strength value mathematically.

Using the curves in the Fig. 7 and 8, the shear cohesion and the internal friction angle can be estimated at the given moisture value (4.03 %). The shear cohesion is about 15 kPa, and the internal friction angle is about 29 degrees (from the curves). For obtaining the precise shear cohesion value, the parabolic equation (with best regression analysis) of the curve in Fig. 7 should be obtained, and then finding the result at the given moisture. The same procedure is to be followed for obtaining the internal friction angle, through obtaining the equation of the curve (straight line) in Fig. 8.

6. Conclusion and future work

The article's work shows that the idea of relating the shear strength of soil to its color is applicable. Studying the soil mechanical properties and its color property in the field is a complicated study case, since in field the soil is mixed with many impurities (ashes, stones ...; also, there are holes and roots), where the soil is not homogeneous. For simplifying the soil study case, the laboratory work is the solution, by dealing with the soil as a homogeneous material, thus measuring its shear strength values and its color using already existing technologies (direct shear test, spectrophotometer). Different models are proposed by researchers for studying the performance of vehicles over soil in field, but for the applicability of these models, values are needed to be substituted in them, such as the cohesion and the internal friction angle. What mostly influences the strength of the soil is the moisture content. Studying the shear strength parameters of the soil at different moisture contents using a laboratory machine (direct shear test since closely resembles the terramechanics case) is a crucial step for having the shear strength values ready to be used in any terramechanics model. Plotting the shear strength values at different moisture contents, will help in ending up with the curve's equations (fitting the plots), thus in field, for any given moisture value, the required shear strength value can be obtained (shear strength, cohesion, and internal friction angle).

The spectrophotometric technology is a quantitative method for measuring the color of a body, thus it was used for measuring the color of the tested soils at different moisture contents. The visible range spectrophotometer meter was used, by sending light waves in the spectrum band (400-700 nm), thus receiving reflectance to the sent wavelengths. Plotting the color results at 700 nm of each soil at different moisture contents, will end up with a curve showing the color change tendency with moisture, thus fitting a curve into the plots, will result in an equation of the color change with moisture. The obtained equation is used to solve for calculating the moisture content at any measured color value (700 nm; input). Thus, with having the color result at 700 nm, the moisture can be predicted, therefore by substituting the moisture in the shear strength curves, the demanded value is obtained.

As future work, the strength value results will be checked by field measurements, since there are no standardized results to compare the obtained results to. The results will be tested by substitution in models, and then comparing the predicted results to measured ones (such as net traction results; predicted by a model compared to sensor result). Validating the results by field measurements, will lead to precise shear strength values equations as function of moisture content (curve tendency). Also, the applicability of the color curves will be checked in field, for predicting the moisture from the color.

7. Acknowledgements

Project no. 2022-2.1.1-NL-2022-00012 has been implemented with the support provided by the Ministry of Culture and Innovation of Hungary from the National Research, Development and Innovation Fund, financed under the 2022-2.1.1-NL funding scheme.

8. Declaration of competing interest

The authors declare that they have no known competing financial interests or personal relationships that could have appeared to influence the work reported in this paper.

9. References

Ahmed, A.E.E.; El Hariri, A.; Kiss, P. , 2021. Soil strength and load bearing capacity measurement techniques. HAE 2021, 40, 16–27.

Al-Abbas, A.H., Swain, P.H., Baumgardner, M.F., 1972. Relating organic matter and clay content to the multi- spectral radiance of soils, Soil Sci., 114, 477 – 485.

Baumgardner, M.F., Sylva, L.F., Biehl, L.L., Stoner, E.R., 1985. Reflectance Properties of Soil, Advances in Agronomy. 1985, 38, 1-44.

Coulomb, C.A., 1776. Essai sur une Application des Regles de Maximis et Minimis Quelques Problemes de Statique, Relatits a L'architecture; Memoires de Mathematique de l'Academie Royale de Science: Paris, France, p. 7.

Cihlar, J.; Protz, R., 1973. Surface characteristics of mapping units related to aerial imaging of soils, Can. J. Soil Sci., 53, 249 – 257.

Ciolkosz, E.J., 1993. Soil Color. Soil Science Society of America, Madison, WI, pp. 35 – 50.

Dobos, R.R., Ciolkosz, E.J., Waltman, W.J., 1990. The effect of organic carbon, temperature, time, and redox conditions on soil color, Soil Sci., 150, 506 – 512.

El Hariri, A.; Ahmed, A.E.E.; Kiss, P., 2023. Review on soil shear strength with loam sand soil results using direct shear test. J. Terramech. 107, 47-59.

Henderson, T.L., Szilagyi, A., Baumgardner, M.F., Chen, C.T., Landgrebe, D.A., 1989. Spectral band selection for classification of soil organic matter content, Soil Sci. Soc. Am. J. 53, 1778 – 1784.

Ivey, J.L., McBride, R.A., 1999. Delineating the zone of topsoil disturbance around buried utilities on agricultural land, Land Degrad. Dev., 10, 531 – 544.

Janosi, Z., Hanamoto, B., 1961. The analytical determination of drawbar pull as a function of slip for tracked vehicles in deformable soils, The 1st International Conference on the Mechanics of Soil-Vehicle Systems, Torino: Edizioni Minerva Tecnica, (p. 22).

Mohr, O., 1900. Welche Umstände bedingen die Elastizitätsgrenze und den Bruch eines Materials? Z. Ver. Dtsch. Ing., 44, 1524–1530.

Muller, E., Decamps, H., 2001. Modeling soil moisture-reflectance. Remote Sensing of Environment, vol. 76 (n°2), pp. 173-180.

Melville, M.D., Atkinson, G., 1985. Soil colour: its measurement and its designation in models of uniform colour space, J. Soil Sci., 36, 495 – 512.

Page, N.R., 1974. Estimation of organic matter in Atlantic coastal plain soils with a color difference meter, Agron. J., 66, 652 – 653.

Post, D.F., Levine, S.J., Bryant, R.B., Mays, M.D., Batchily, A.K., Escadafal, R., Huete, A.R., 1993. Correlations between field and laboratory measurements of soil color In: Bigham, J.M., Ciolkosz, E.J. (Eds.), Soil Color. Soil Science Society of America, Madison, WI, pp. 35 – 50.

Salman, N.D., Kiss, P., 2018. Survey: Effect of bulk density and moisture content of soil on the penetration resistance and penetration depth, Mechanical Engineering Letters: R and D. Research and Development 17, 109.

Soil Survey Staff, 1999. Soil Taxonomy: A Basic System of Soil Classification for Making and Interpreting Soil Surveys, 2nd ed.; U.S. Department of Agriculture, Natural Resources Conservation Service: Washington, WA, USA.

Shields, J.A., Paul, E.A., St. Arnaud, R.J., Head, W.K., 1968. Spectrophotometry measurement of soil color and its relationship to moisture and organic matter. Can. J. Soil Sci., 48, 271 – 280.

ISTVS

Abstract-only submissions

1166 | Selection of energy saving engine mode based on the power delivery and fuel consumption of a 95-kW tractor during agricultural operations

Md Abu Ayub Siddique [a,*], Seung-Min Baek [a], Mo-A Son [a], Ye-In Song [a], Yong-Joo Kim [a,*]

[a] *Department of Agricultural Machinery Engineering, Chungnam National University, Daejeon 34134, Republic of Korea*

* *Corresponding author: babina@cnu.ac.kr*

The objective of this article is to estimate the power delivery efficiency and fuel consumption based on engine modes. In this study, 95-kW powershift tractor was used to analyze the power delivery and estimate the fuel consumption during agricultural operations. The asphalt driving, plow and rotary tillage were conducted for the field the experiment at the engine modes of conventional, APS (Auto power shift) power, and APS ECO. To analyze the field condition, soil hardness, and soil water content were measured as well as soil sample was collected from the experimental site to analyze the soil texture using the USDA soil texture triangle. It is observed that the engine required powers at various engines have a significant different of the asphalt driving, plow, and rotary tillage. However, the axle powers at all engine modes for each operation showed almost similar due to the similar forward speed of the tractor. In case of fuel consumption, it was found that the fuel consumption at each engine mode has a significant different. It was also observed the APS ECO mode was the most economic compared to the conventional and APS power for the asphalt driving, plow and rotary tillage. Therefore, we recommend by this study to the users to perform the tractor at APS ECO engine mode for asphalt driving, plow, and rotary tillage considering fuel economic and high working load.

2297 | Traction effects on vertical and horizontal contact stresses and the risk of soil structure deformation

Loraine ten Damme [a,*], Lars J. Munkholm [b], Per Schjønning [b], Thomas Keller [b], Alvaro Calleja Huerta [b], Mathieu Lamandé [b]

[a] *Aarhus University, Denmark; Swedish University of Agricultural Sciences*

[b] *Aarhus University, Denmark*

* *Corresponding author: ltd@agro.au.dk*

Arable soil is often exposed to high mechanical stresses during field operations, which bears a great risk of soil structure deformation. The stress state under the wheels determines whether a risk of soil compression and/or soil distortion prevails. However, soil compaction risk assessments typically consider the risk of compression from the applied load but neglect tractive forces and the risk of distortion. We investigated traction effects on both the vertical and horizontal (longitudinal) stress distributions near the tyre-soil interface as a next step in advancing more accurate predictions of soil structure deformation. Vertical and horizontal stress measurements were made during wheeling in a field experiment with stress transducers installed at 0.1 m depth. We studied the stress distributions of a tractor's rear wheel (6.7 Mg, 650/60 R38, 80 kPa) that was towed (i.e., no traction) or pulling (i.e., with traction, about 42 kN drawbar pull) and compared these with estimated soil strengths. We show how traction influenced tyre-soil interaction and the risk of soil structure deformation. Among others, with traction the tyre-soil contact area was larger and the magnitude of vertical stress lower. This reduced the risk of soil compression noticeably. Soil shear strength was reduced due to reduced constraining normal load. The latter, in combination with higher horizontal stresses, increased the risk of soil distortion considerably. Our results highlight the importance of considering traction in soil compaction risk assessments. However, reliable assessments require incorporation of the risk of soil distortion in addition to the risk of soil compression.

3815 | Data measurement and performance evaluation of an electric all-wheel-drive tractor during plow tillage

Seung-Yun Baek [a], Yong-Joo Kim [a,*], Seung-Min Baek [a], Ji-Won Choi [a], Min-Jong Park [a]

[a] *Chungnam National University, Republic of Korea*

* *Corresponding author: babina@cnu.ac.kr*

This study was conducted for basic research on AWD (all wheel drive) systems applicable to tractors as the need for research on electric AWD systems increases. The platform consists of an electric power transmission system consisting of 4 sets of electric motors, helical reducers and planetary reducers, wheels, and an electric system including chargers, batteries (LiFePO4), converters and inverter (motor controller). The data measurement system consisted of analog (current, traction force) and digital (battery voltage, current, motor rotation speed, SOC (state of charge) level, travel speed) components using a CAN (controller area network) bus. Data measured during plow tillage were analyzed and used to calculate the data such as motor power, motor torque, traction power, and driving power. The traction performance of the electric AWD tractor was evaluated using tractive efficiency (TE) using calculated values (traction power and driving power) and dynamic ratio (DR). This study presented limitations and improvement methods of electric AWD tractors through data analysis and comparison with conventional tractors. In future study, we plan to conduct research on optimization and performance verification of the electric AWD tractor that reflect improvements.

3859 | Analysis for dynamic characteristics of a tractor cabin during agricultural operations

Jong-Dae Park [a], Hyeon-Ho Jeon [a], Su-Young Yoon [a], Min-Jong Park [a], Yong-Joo Kim [a,*]

[a] *Chungnam National University, Republic of Korea*

* *Corresponding author: babina@cnu.ac.kr*

This study was conducted to develop a dynamic characteristic control system for improving the cabin environment during agricultural operations. The specification of platform used in the study was a 67 kWclass tractor with a weight of 4,000 kg and dimensions of 4,020 × 2,270 × 2,790 mm (lenth × width × height), respectively. For measuring data of dynamic characteristic of the cabin, plow tillage and rotary tillage, which account for the largest proportion of the tractor operations, were selected for major operations. To measure data of dynamic characteristic of the cabin, which continuously changes during operations, an Ellipse Series INS with a built-in IMU was attached to the center of gravity of the cabin. During field test, the gear stages of plow tillage were B2 (5.1 km/h) and B3 (7.6 km/h), and the gear stages of rotary tillage were A3 (2.67 km/h) and A4 (3.66 km/h), which are the most commonly used. To analyze dynamic characteristic such as roll and pitch during operations according to the gear stages, the data showed by box plot. Roll and pitch average angles were -0.7° and 0.4°, respectively, for the plow tillage. Roll and pitch average angles were -0.8° and -0.1°, respectively, for the rotary tillage. As operations rate increased, the roll and pitch angles became larger, varying within a maximum range of 2°. As a result, it was determined that the rotary tillage showed a larger change compared to the plow tillage in the case of roll. Both operations showed a similar change in the case of pitch. Also, roll and pitch showed a larger change as the travel speed of each operation increased. In the future, we plan to conduct additional research on the hydraulic control system related to the cabin to provide a more comfortable workspace for operators.

3893 | Analysis of load factor and performance evaluation for developing an electric driven tractor

Min-Jong Park [a], Su-Young Yoon [a], Ji-Won Choi [a], Jong-Dae Park [a], Yong-Joo Kim [a,*]

[a] *Chungnam National University, Republic of Korea*

* *Corresponding author: babina@cnu.ac.kr*

This study aims to develop a 44 kW electric driven tractor consisting of an E-driving system. Because electric tractor is in the design process, we installed a measurement system on a conventional agricultural tractor. Torquemeters were installed between wheels and axles to measure wheel torque, proximity sensors were installed to measure wheel rotation speed and GPS (Global Positioning System) was installed to measure driving speed of the tractor. Antenna, amplifier, and DAQ (Data Acquisition) were installed to transfer and store the measured data. Workload data were measured during plow tillage, rotary tillage, loader operations, asphalt and field driving. Measured data were converted for validation of designing a 44 kW electric driven tractor using an engine load factor. The average engine load factor was 0.51 for rotary tillage, 0.46 for plow tillage, 0.38 for asphalt driving, 0.17 for field driving and 0.09 for load operation. For gear stage L, plow tillage and loader operations were conducted within the maximum torque and rotary tillage, field and asphalt driving were performed within the rated torque. For gear stage M, the plow tillage and loader operations were conducted outside of the maximum torque, which were impossible to perform. For gear stage H, all operations except the asphalt driving were conducted outside of the maximum torque, which were impossible to perform. Thus, high-load operations such as plow tillage and loader operations can be conducted in gear stage L, and high-speed operations such as asphalt driving can be conducted with motor rated speed in gear stage H. Consequently, it is judged that an E-driving system consisting of a motor and range shift can perform the major operations of a 44 kW tractor depending on gear stages.

4336 | High-fidelity 3D-DEM simulation for wheel mobility in low gravity environment

Takuya Omura [a,*] Genya Ishigami [a]

[a] *Keio University, Japan*

* *Corresponding author: ta_hu823tyann@keio.jp*

The effect of gravity on wheeled rover mobility on loose soil remains a challenging issue. Among several experimental techniques, such as parabolic flight or reduced-weight tests simulating low gravity, the discrete element method (DEM) is one of the most promising methods for analyzing the wheel-soil interaction in various gravity environments. However, the mechanical parameters used in the DEM are often identified ambiguously or empirically, resulting in low-fidelity simulation results. In this study, we comprehensively calibrate the mechanical parameters of a lunar regolith simulant applicable to the 3D-DEM to simulate the effect of gravity on wheel mobility. The DEM parameters were determined from the experimental results of two tests: the direct shear test and the minimum density test. We then tested a rigid wheel on the lunar simulant using the DEM with calibrated parameters in the following conditions: (1) Earth gravity (1 G) with the wheel mass of M, (2) moon gravity (1/6 G) with the wheel mass of M, and (3) Earth gravity (1 G) with the reduced-mass of 1/6M. The simulation results show that the wheel sinkage and the slip ratio on the moon are almost equivalent to those on Earth. On the other hand, in the reduced-weight test, the wheel sinkage and the slip ratio were smaller than the moon's results. This result implies that the reduced-weight test might overestimate the wheel performance.

4508 | Semi-empirical terramechanics model for variable terrain height

Eric Karpman [a,*], Jozsef Kovecses [a], Marek Teichmann [b]

[a] McGill University, Canada

[b] CM Labs Simulation, Inc., Canada

* Corresponding author: eric.karpman@mail.mcgill.ca

Dynamic simulations of various types of off-road vehicles, from planetary rovers to agricultural equipment, have long relied on well-established semi-empirical terramechanics models. While these models do have drawbacks and reliability issues that have been addressed by numerous works in the decades since the models were first introduced, semi-empirical approaches remain one of the few ways to simulate realistic wheel-soil interaction in real-time. One of the drawbacks of these methods is that they are constrained by the assumption that the terrain is a flat plane. The models work by integrating normal and shear stresses along the wheel-terrain contact patch. The normal stress at each point along the contact patch is determined based on an equation that computes soil pressure based on semi-empirical parameters, the dimensions of the wheel and the sinkage of the wheel at that point. The sinkage at any given point along the contact patch is generally determined based on the distance between the point and the flat plane that defines the terrain. This means that the sinkage of each point along the contact patch is computed based on one constant initial terrain height. In this work, we propose a modified version of the semi-empirical model in which the terrain is defined as a height-field rather than a flat plane. Thus, more complex terrain shapes can be used in simulation. Of particular interests are terrains with varying slopes, steps, and terrain shapes that would result in gaps in the wheel-soil contact patch rather than a continuous patch.

4727 | Analysis of gear durability for agricultural tractor transmission

Seung-Min Baek [a,*], Seung-Yun Baek [a], Hyeon-Ho Jeon [a], Yong-Yoo Kim [a]

[a] Chungnam National University, Republic of Korea

* Corresponding author: babina@cnu.ac.kr

The purpose of this study was to analyze gear durability of tractor transmission using simulation analysis. A field test was performed using a 86 kW class tractor and the type of gear failure was analyzed. A simulation model of the tractor transmission was developed and analyzed for the safety factor and lifespan by changing the material of the damaged gear. The field test was conducted for plow and rotary tillage and was completed after about 107 hours due to operational problems. After field test, it was found that range shift A and B gears were damaged. In order to advance durability of the transmission, four materials of alloy steel for machine structural use such as SCr420, SNCM220, SCM822, and SNC815 were selected, and the safety factor and service life of the damaged range shift gears were analyzed according to gear materials. As a result of simulation analysis, SCM822 material satisfied the target life and was selected as a material for change. Therefore, it is considered that the durability of the transmission has been secured by satisfying the target life.

4895 | Theoretical analysis and FEM simulation of the interfacial forces between tracked vehicle and snow soil

Yan Qingdong [a], Zhu Ming [a], Wei Wei [a,*], Liu Cheng [a], Liu Jianfen [a], Meng Qingkai [a]

[a] Beijing Institute of Technology, China

* Corresponding author: weiweibit@bit.edu.cn

Based on the pressure-sinkage characteristics and the shear force-shear displacement characteristics of snow under loading, unloading and reloading, theoretical models of vertical displacement distribution and vertical pressure distribution for snow and interfacial forces between track and snow under different driving conditions were built. A multi-body dynamics model of track and plastic material models of snow were built to simulate the interfacial forces of the track on snow. The results of plastic deformation and vertical pressure distribution of snow were extracted from the simulation results to verify the theoretical models. A muti-body dynamics model of articulated tracked vehicles was built, and the driving performance of it on snow was evaluated. An experimental study on articulated tracked vehicle driving in snow is carried out. By comparing the experimental results and the simulation results of the plastic deformation characteristics of snow under tracked vehicle loading, the correctness of the coupling simulation model and the theoretical model is verified.

4976 | Performance evaluation of autonomous driving lateral control simulation model of agricultural tractor

Mo-A Son [a], Hyeon-Ho Jeon [a], Seung-Yun Baek [a], Ye-In Song [a], Yong-Yoo Kim [a,*]

[a] Chungnam National University, Republic of Korea

* Corresponding author: babina@cnu.ac.kr

Interest in automation of agricultural machinery is increasing. Recently, performance evaluation studies using simulation models are being conducted to reduce time and cost. Therefore, in this study, the performance evaluation of the autonomous control model of an agricultural tractor was performed. The simulation model was developed using the specifications of a 78 kW tractor and the software RecurDyn. The input data was set as traction data from the six-component load cells, the rotational speed of each axle, and the position of the 3-point hitch. Validation of the simulation model was conducted by comparing the error between the measured data and simulation results. The performance evaluation of the steering control model was performed on the road model (friction coefficient conversion). The steering control model was configured to perform steering control by receiving the lateral error in real time from the dynamics model. A trial-and-error method was used to select coefficients for the autonomous control model. The RMS of the measured data was 3.29 and 3.44 for the anterior left and right sides, respectively, and 6.98 and 7.41 for the posterior left and right sides, respectively. The RMS of the simulation results were 3.24 and 3.53 for the anterior left and right side, respectively, and 8.26 and 8.55 for the posterior left and right side, respectively. The errors were 1% and 2% for the anterior left and right side, respectively, and 15% and 13% for the posterior left and right side, respectively. The error occurred within the range of about 10%, and it was judged appropriate to conduct research on performance evaluation using this model. The RMS according to the friction coefficient conversion was 3.7 and 4.5, respectively, showing a difference of about 1.

5356 | Development and validation of head-feeding combine harvester multibody dynamics simulation model

Cheol-Woo Yang [a], Hyeon-Ho Jeon [a], Min-Jong Park [a], Ji-Won Choi [a], Yong-Joo Kim [a,*]

[a] *Chungnam National University, Republic of Korea*

* *Corresponding author: babina@cnu.ac.kr*

Head-feeding combine harvester are expensive and time-consuming to produce prototype, performance evaluation using simulation should be performed in the design stage. Recently, research has been conducted to solve these problems by replacing the driving test of agricultural machinery with simulation. In this study, a head-feeding combine harvester dynamics simulation model was developed and verified. The simulation model was developed using commercial software (Recurdyn, V9R4, Korea). The model development was performed using the specifications of a 6-row head-feeding combine harvester (CF690G, TYM, Korea). The speed condition of the simulation was set to 1.5 m/s. The simulation result set as the torque of the sprocket. For the development and validation of the simulation model, a measurement system was developed and installed on the head-feeding combine harvester. The driving test was conducted at the paddy field located in Dangjin, Chungnam (36°55'49.1"N, 126°37'57.3"E). The simulation was verified by comparing the measured data with the simulation results. The simulation results showed an average sprocket torque of approximately 1241.17 Nm. The measurement resulted in an average sprocket torque of approximately 1143.49 Nm. The error between the measured and simulated values is around 7%, and the measured and simulated values are similar. In the future research, the study to measure the load data during agricultural work and improve the simulation model using the load data will be conducted.

6059 | Comparison of multi-tracked running gears for light UGVs in terms of terrain obstacles overcoming abilities

Daniela Szpaczyńska [a,*], Marian Łopatka [a], Piotr Krogul [a]

[a] *Military University of Technology, Poland*

* *Corresponding author: daniela.szpaczynska@wat.edu.pl*

Rubber tracked running gears are widely used in high-mobility unmanned ground vehicle to increase the ability to overcome terrain obstacles that are often found in off-road terrain. The paper investigates the process of overcoming an obstacle by light unmanned ground vehicle using multibody dynamics simulations. The two-dimentional model of running gear was assumed and created in the ADAMS program environment. As well as the kinematics of the running system components, the flexibility of the elastomer belt and also the reactions of the tractive force in the track-ground contact surface were taken into account. On the basis of the simulations, the functionality of multi-track systems was compared with classic tracked systems. Properties that could be improved through the use of multi-tracked running gears in light land platforms were determined.

6143 | Flexible tire-terrain interaction model for real time simulations

Mahdi Maleki [a,*], Jozsef Kovecses [a]

[a] *McGill University, Canada*

* *Corresponding author: mahdi.maleki@mail.mcgill.ca*

The analysis of tire dynamics is important in the simulation of vehicle behaviours. The forces exerted on a tire depend on the tire and the terrain interaction at the contact patch, and the tire structure directly influences this interaction. Complex models with a high number of degrees of freedom, such as lumped parameter models or finite element models, are typically required to represent tire flexibility appropriately. Employing these models, however, can lead to high computational costs that can be a significant challenge for real-time simulation.

In this work, we developed a reduced flexible tire model capable of representing the effect of tire flexibility to reduce the computational costs of the simulation. This model reduction is done by defining an effective stiffness of the flexible tire (which accounts for the effects of tire deformation at the contact patch) and augmenting it with the rigid wheel to represent the effect of stiffness at the contact patch. The flexible tire model, which has been used as a base, is a lumped parameter model in which the tire sidewall and belt are represented by a set of lumped point masses connected together and to the rigid wheel rim with sets of parallel spring and damper elements.

In addition to reducing the degrees of freedom of the system, the proposed method makes it possible to perform the simulation with relatively large time steps, which significantly reduces the simulation run time. By this approach, we could achieve up to 25 times faster simulations compared to the ordinary model while maintaining adequate accuracy.

6211 | Wheel-soil terramechanics model augmentation using machine learning across multiple soil types

Eric Karpman [a,*], Jozsef Kovecses [a], Marek Teichmann [b]

[a] *McGill University, Canada*

[b] *CM Labs Simulation, Inc., Canada*

* *Corresponding author: eric.karpman@mail.mcgill.ca*

Terramechanics simulations for real time applications have primarily made use of traditional semi-empirical models since they are the only ones computationally efficient enough to run at real time interactive rates. A drawback to these models is that they are formulated based on an assumption that the wheel is operating under steady state conditions. This can adversely affect their accuracy or even their compatibility in dynamic simulations. Computationally demanding approaches such as the Finite Element Method (FEM) and the Discrete Element Method (DEM) address this issue, but the time required to run simulations is prohibitive for certain applications. We have previously proposed a method for real time terramechanics simulation that addressed the issue of the steady state assumption in the traditional semi-empirical models by using machine learning (ML) to augment the forces predicted by existing semi-empirical models in an effort to improve accuracy in dynamic simulations. By augmenting the semi-empirical predictions with ML rather than using ML to predict the total force, we aimed to preserve the insights provided by the semi empirical model. DEM experiments were run to establish a ground truth for the soil reaction forces, and at each time step we compared predictions of the semi-empirical model to those recorded in DEM simulations. We then train a ML model to augment the semi-empirical model to match DEM results. Our past work showed how this approach can work for a single soil type. Now, we will present the work with an expanded scope in which multiple soil types are used to train the ML algorithm - resulting in a single trained model that can be used for a variety of different soil types without the need to re-train the ML portion of the model each time.

6955 | Development of dynamics simulation model for agricutural tractor according to agricultural work

Hyeon-Ho Jeon [a], Seung-Yun Baek [a], Seung-Min Baek [a], Mo-A Son [a], Yong-Yoo Kim [a,*]

[a] *Chungnam National University, Republic of Korea*

* *Corresponding author: babina@cnu.ac.kr*

The behavior of tractor is affected by the load. Especially, the behavior is significantly changed, an accident, such as rollover, is occurring. Therefore, the behavior of the vehicle affects work stability. To increase the work stability, it is necessary to predict the behavior of the vehicle by predicting the load occurred in a vehicle. This study was conducted as a basic study for the development of a tractor dynamics model. The simulation model was developed using a multi-body dynamics software and a steering controller for a simulation model. The specification of the tractor was 4,225 mm in length, 2,140 mm in width, 2,830 mm in height, and 3,985 kg. The center of gravity was 984 mm from the rear axle, 973 mm above ground, and 986 mm from the left side of the vehicle. Simulations were conducted during conditions of Plough tillage and condition of driving on a field. A field test was conducted to make and validate the simulation model. The measurement system is configured to measure traction, rotation speed, axle torque and behavior of a tractor. Traction and axle rotation speed were used as input data for simulation model. The load on axles and vehicle behavior were used as validation data for simulation. Field tests were conducted to measure input value and verify the simulation model. Field tests were conducted at M2 high (5.2 km/h ~ 6.7 km/h). Analysis between simulation result and measurement data showed that the difference in axle load was less than 10%. The load on axles can be predicted by using the developed model and evaluating the steering controller model can be conducted using this dynamics simulation model. In the next study, it is necessary to evaluate the steering controller model thorough the dynamics model.

7994 | Longitudinal friction coefficient comparison on actual and laboratory surfaces

Carl Becker [a,*], Schalk Els [a]

[a] *University of Pretoria, South Africa*

* *Corresponding author: carl.becker@up.ac.za*

Tyre models used in soft soil simulation analysis requires tyre parameters in the form of stiffnesses in multiple di-rections. These parameters are obtained from measurements on hard terrain as these parameters are a function of the tyre carcass construction. Vehicles used on soft terrain are also used on hard terrain. Tyre testing is not a trivial or inex-pensive exercise and outdoor testing has limitations on repeatability. This is the case for testing on soft terrain/soil and hard terrain, thus it is preferred to conduct laboratory tests when characterizing tyres. This study investigates the longi-tudinal friction coefficient measured on surfaces with different surface roughness. The surface roughness has a direct influence on the friction coefficient generated by a tyre during tests.

Tests are conducted on the actual concrete surface of interest, typically used in field tests. A mold is cast from the actual surface in order to replicate the same surface in laboratory conditions. Tests are then conducted on the replicat-ed surface and different sandpaper surfaces to determine which surface is the most representative in the laboratory. Multiple tyres are tested on the different surfaces and results compared.

8754 | Characterization of lunar simulant (FJS-1), toyoura sand, and silica sand with physical experiments and dem simulation

Isabel Casasbuenas [a,*], James Hurrell [b], Keisuke Takehana [b], Kazuya Yoshida [b]

[a] *Colorado School of Mines, USA*

[b] *Tohoku University, Japan*

* *Corresponding author: yicasasbuenasc@unal.edu.co*

Geotechnical properties of the soils allow understanding of the mechanical behavior of soil under different stress conditions. However, the geomechanical properties of soils under Earth's gravity are liable to be different from those under lunar gravity, making direct comparisons with experiments difficult. Therefore, Discrete Element Method (DEM) simulations can be used alongside physical experiments to study the geotechnical properties of the soil in a lunar environment. Understanding the mechanical behavior of the lunar soil will aid in performing future engineering operations such as mobility, construction, mining, and foundation design. Angle of Repose and Triaxial compression test simulations in lunar and terrestrial gravity environments were performed to understand soil behavior under these two environments. Angle of repose (AoR) testing used the fixed funnel method. A set volume and mass of material are poured through a funnel forming a pile below. A value is then measured for AoR which is related to the internal friction angle. Laboratory triaxial tests were performed on three specimens to simulate stress conditions and determine shear strength and other geotechnical properties under Earth's gravity conditions. Vertical loads were applied at a 0.5 mm/min rate and confinement pressure from 10 to 60 kPa. Cohesion, soil stiffness, shear stress, and friction angle obtained from the triaxial and angle of repose tests are compared with the results obtained in DEM simulations. Exploring the difference between static and dynamic tests could contribute to understanding where physical properties of the lunar soil play a major role in the wheel soil interaction. The result of the present work will contribute to ongoing research in wheel-soil interaction.

8859 | An efficient and high-fidelity track model for dynamic simulation of off-road tracked vehicles

Oz Ben-Yosef [a,*], Dror Rubinstein [a,*]

[a] *Ariel University, Department of Mechanical Engineering and Mechatronics, Israel*

* *Corresponding author: ozby@ariel.ac.il, drorru@ariel.ac.il*

A high-fidelity simulation model of tracked vehicles is required for proper prediction of the mobility of tracked vehicles traveling over soft soils. The vehicle components can be modelled using standard tools of multi-body programs. A track model was developed and successfully worked together with Altair's MotionSolve multi-body program. The model based on classic soil mechanics equations. The grousers, which are a significant part of many types of track-links, are taking into the account. The plasticity and viscosity properties of the soil are considered in the model. Verification tests were conducted in an agricultural field in the Jezreel Valley. The tests were run over soils with varying mechanical prop-erties, achieved through irrigation and tillage. The chosen test vehicle was an M113 armoured carrier. Several drawbar pull loads were applied on each soil condition. Reasonable correlation between the tests and simulation results were achieved.

However, this model is not efficient and require consumption of large amount of CPU time. On the other hand, the more efficient models are based on simplifying assumptions with lack of accuracy. This work proposes a method for creating an efficient high-fidelity model. This can be implemented by representing the interaction between the track and the ground according to the previous work. The solution for the track-links will be done independently to the solu-tion of the entire vehicle. The solution of the track link will be obtained through a solver that will be developed for this purpose. The solution of the entire vehicle will be done by the solver of the multibody program (Altair MotionSolve). Good correlation between the efficient model and the previous model were achieved.

8963 | Analysis of the influence of terrain unevenness and velocity on the ground pressure exerted by demining rollers

Dariusz Kalinko [a,*], Marian Janusz Łopatka [a,*], Arkadiusz Rubiec [a,*], Mirosław Przybysz [a,*], Piotr Krogul [a,*]

[a] *Military University of Technology in Warsaw, Poland*

* *Corresponding author: dariusz.kalinko@wat.edu.pl, marian.lopatka@wat.edu.pl, arkadiusz.rubiec@wat.edu.pl, miroslaw.przybysz@wat.edu.pl, piotr.krogul@wat.edu.pl*

The effectiveness of demining rollers heavily depends on their performance in different terrains and operating conditions. Conventional demining roller structures with rigid wheels are widely used in demining operations, but their performance in real work conditions is still a significant scientific aspect. In this paper, the results of simulation tests conducted on demining roller systems with rigid wheels, without additional ground pressure systems are presented. A multi-body numerical model of a vehicle with a single demining section is described. The influence of terrain unevenness and velocity on the ground pressure exerted by roller is discussed.

9504 | Soil deformation model for efficient simulation of off-road vehicles and compaction force analysis

Yang Jiao [a,*], Jozsef Kovecses [a,*]

[a] McGill University, Canada

* Corresponding author: yangjiao0706@gmail.com, jozsef.kovecses@mcgill.ca

The performance evaluation of off-road vehicles is important since their operating environment is always complex. There are various types of models developed to predict vehicle soil interaction, ranges from semi-empirical models to discrete element(DEM) finite element based representations. For scenarios that require fast simulation, semi-empirical model is a better choice since it has the lowest computational cost. However, it does not incorporate the influence of soil deformation and has a quasi-static nature. To incorporate the influence of soil deformation, a new soil deformation model is proposed and developed. The geometry of the soil in this model is represented by using a height map with vertical layers. The soil deformation is separated into two parts and the total soil mass is conserved. The soil mass transmission resulted by vertical soil compaction is the main focus of this work, and it is captured by using a soil mass prediction neural network. The change in density and void ratio on each soil layer is then computed based on the mass transmission. The other part that resulted by tangential movements would also be briefly introduced. The data gathering is done by conducting Bevameter tests in DEM simulation environment by using EDEM. Various compaction tool widths, speeds and angles are used to generate the data for training and validation purposes. The influence of the compaction speed and angle on soil compaction force and soil skeleton states (e.g. density and void ratio at each layer) are also analyzed to investigate the possibility of optimizing the prediction of the Bekker pressure sinkage equation and the soil deformation model. The relation between Bekker parameters and general soil mechanics parameters is also discussed in this work.

9716 | Non-pneumatic tire for special vehicle – a review of selected functional properties

Marcin Żmuda [a,*], Jerzy Jackowski [a,*], Marcin Wieczorek [a,*]

[a] Wojskowa Akademia Techniczna im. Jarosława Dąbrowskiego (Military University of Technology), Poland

* Corresponding author: marcin.zmuda@wat.edu.pl, jerzy.jackowski@wat.edu.pl, marcin.wieczorek@wat.edu.pl

The wheel is an important element of the vehicle, as it is the only one that has direct contact with the road. Pneumatic tires are the most commonly used on vehicles nowadays. The tires transfer traction forces and directional control of the vehicle, they affect on parameters of the contact patch and the enveloping properties. Those functions of tires require maintaining compressed air at a certain pressure inside the pneumatic tire. Losses of compressed air, as a result of damage to a pneumatic tire, makes it difficult or impossible to use it further. High reliability of tires and resistance to damage is particularly important in the case of special vehicles that move mainly in difficult terrain, e.g. soft soil, natural and artificial obstacles, objects with sharp edges. In those conditions, an alternative solution are non-pneumatic-tires, in which the function of compressed air has been taken over by the so-called supporting structure. Type of supporting structure, including shape and materials used, influence on the load -carrying mechanism, directional stiffness and pressures in the contact patch. The paper analyzes the method of shaping selected properties of non-pneumatic tires and the influence of the applied supporting structure on selected operational characteristics of non-pneumatic tires.

Author index

Name	Affiliation	Submissions
Ahmed, Ahmed Elawad Eltayeb	Hungarian University of Agriculture and Life Sciences (MATE), Hungary	2936 | Moisture content impacts on soil load-bearing capacity and its spectral behaviour
Arvidsson, Erik	Luleå University of Technology, Sweden	6027 | Digital Precision Planning Tool for Autonomous Forest Regeneration of Mixed Tree Species
Baek, Seung-Min	Department of Agricultural Machinery Engineering, Chungnam National University, Daejeon, Republic of Korea	1166 | Selection of energy saving engine mode based on the power delivery and fuel consumption of a 95-kW tractor during agricultural operations *(abstract only)*
		3815 | Data measurement and performance evaluation of an electric all-wheel-drive tractor during plow tillage *(abstract only)*
		4727 | Analysis of gear durability for agricultural tractor transmission *(abstract only)*
		6955 | Development of dynamics simulation model for agricutural tractor according to agricultural work *(abstract only)*
Baek, Seung-Yun	Department of Agricultural Machinery Engineering, Chungnam National University, Daejeon, Republic of Korea	3815 | Data measurement and performance evaluation of an electric all-wheel-drive tractor during plow tillage *(abstract only)*
		4727 | Analysis of gear durability for agricultural tractor transmission *(abstract only)*
		4976 | Performance evaluation of autonomous driving lateral control simulation model of agricultural tractor *(abstract only)*
		6955 | Development of dynamics simulation model for agricutural tractor according to agricultural work *(abstract only)*
Becker, Carl	University of Pretoria, South Africa	7994 | Longitudinal friction coefficient comparison on actual and laboratory surfaces *(abstract only)*
Ben-Yosef, Oz	Ariel University, Department of Mechanical Engineering and Mechatronics, Israel	8859 | An efficient and high-fidelity track model for dynamic simulation of off-road tracked vehicles *(abstract only)*
Biro, Norbert	UNI-MATE, Hungary	7477 | Evaluating the pressure performance of DPF filters using engine bench analysis
Budzyński, Piotr	Lublin University of Technology, Poland	3835 | Effect of steer angle rate upon tire lateral force generation on two different soils
Calleja Huerta, Alvaro	Aarhus University, Denmark	2297 | Traction effects on vertical and horizontal contact stresses and the risk of soil structure deformation *(abstract only)*

Name	Affiliation	Submissions
Cao, Hongtao	Key Lab for Bionics Engineering of Education Ministry, Jilin University, China	2145 | Obstacle performance and wheel failure test analysis of Zhurong Rover
Card, Angela	CAVS, Mississippi State University, USA	1522 | Tractive performance of rigid wheel in granular media using coarse-scale DEM models
Casasbuenas, Isabel	Colorado School of Mines, USA	8754 | Characterization of lunar simulant (FJS-1), toyoura sand, and silica sand with physical experiments and dem simulation *(abstract only)*
Cheng, Liu	Beijing Institute of Technology, China	4895 | Theoretical analysis and FEM simulation of the interfacial forces between tracked vehicle and snow soil *(abstract only)*
Choi, Ji-Won	Department of Agricultural Machinery Engineering, Chungnam National University, Daejeon, Republic of Korea	3815 | Data measurement and performance evaluation of an electric all-wheel-drive tractor during plow tillage *(abstract only)*
		3893 | Analysis of load factor and performance evaluation for developing an electric driven tractor *(abstract only)*
		5356 | Development and validation of head-feeding combine harvester multibody dynamics simulation model *(abstract only)*
Cole, Michael	U.S. Army Ground Vehicle Systems Center, USA	6630 | Real-time measurement of tire sinkage using stereo cameras
Czajka, Aleksander	Lublin University of Technology, Poland	3507 | Determination of the tractive forces of a tractor based on the reconstruction of tire – soil contact surface obtained by means of photogrammetry
Driggs-Campbell, Katherine	University of Illinois at Urbana-Champaign, Department of Electrical and Computer Engineering, USA	2801 | In Situ Soil Property Estimation for Autonomous Earthmoving Using Physics-Infused Neural Networks
El Hariri, Alaa	Hungarian University of Agriculture and Life Sciences (MATE), Hungary	9992 | Soil shear strength values obtained from its color
Els, P. Schalk	University of Pretoria, South Africa	2437| Improving predictive control methods on off-road vehicles with realistic steering preview inputs
		6736 | Ride comfort comparison between 4-poster and straight-line driving simulations
		7994 | Longitudinal friction coefficient comparison on actual and laboratory surfaces *(abstract only)*
Eto, Ryosuke	National Defense Academy, Japan	1638 | Analysis of tire characteristics on road surface with volcanic ash fall
Frankenstein, Susan	CRREL, USA	7577 | Implementation of moving loads on ice in NRMM
Fujimura, Kanta	Shibaura Institute of Technology, Japan	0709 | Proposal of Swarm Rovers' Collaborative Locomotion with Expansion and Contraction Mechanism for Driving a Loose Slope

Name	Affiliation	Submissions
Fujiwara, Daisuke	Suwa University of Science, Japan	0709 | Proposal of Swarm Rovers' Collaborative Locomotion with Expansion and Contraction Mechanism for Driving a Loose Slope
		0889 | Skidding suppression method using "Discrete 4-Wheel-Drive typed rover" considering steering angle at traveling across loose slope
		1453 | Suppressing the reduction of the traveling displacement on loose soil for rovers with function of a wheel-walking
Gardyński, Leszek	Lublin University of Technology, Poland	3507 | Determination of the tractive forces of a tractor based on the reconstruction of tire – soil contact surface obtained by means of photogrammetry
Gorsich, David	Department U.S. Army DEVCOM Ground Vehicle Systems Center, USA	3493 | Numerical modeling of a tire on undrained saturated clay using FEM, ALE, and SPH
		6630 | Real-time measurement of tire sinkage using stereo cameras
Grebner, Karl	CAVS, Mississippi State University, USA	1522 | Tractive performance of rigid wheel in granular media using coarse-scale DEM models
He, Qingze	Shibaura Institute of Technology, Japan	1453 | Suppressing the reduction of the traveling displacement on loose soil for rovers with function of a wheel-walking
Higashiyama, Ryoma	Shibaura Institute of Technology, Japan	1098 | Simulation of change in supporting force when imparting vibration by distinct element method
Hjelm, Karin	Swedish University of Agricultural Sciences, Alnarp, Sweden	6027 | Digital Precision Planning Tool for Autonomous Forest Regeneration of Mixed Tree Species
Hurrell, James	Tohoku University, Japan	0448 | Lunar rover discrete element method study and calibration
		8754 | Characterization of lunar simulant (FJS-1), toyoura sand, and silica sand with physical experiments and dem simulation *(abstract only)*

Name	Affiliation	Submissions
Iizuka, Kojiro	Shibaura Institute of Technology, Japan	0709 | Proposal of Swarm Rovers' Collaborative Locomotion with Expansion and Contraction Mechanism for Driving a Loose Slope
		0889 | Skidding suppression method using "Discrete 4-Wheel-Drive typed rover" considering steering angle at traveling across loose slope
		1098 | Simulation of change in supporting force when imparting vibration by distinct element method
		1453 | Suppressing the reduction of the traveling displacement on loose soil for rovers with function of a wheel-walking
		8151 | Study on estimation of traveling states using strain information on chassis of lunar and planetary exploration rovers
Ishigami, Genya	Keio University, Japan	4336 | High-fidelity 3D-DEM simulation for wheel mobility in low gravity environment *(abstract only)*
Jackowski, Jerzy	Wojskowa Akademia Techniczna im. Jarosława Dąbrowskiego (Military University of Technology), Poland	9716 | Non-pneumatic tire for special vehicle - a review of selected functional properties *(abstract only)*
James, Thomas	CAVS, Mississippi State University, USA	1522 | Tractive performance of rigid wheel in granular media using coarse-scale DEM models
Jelinek, Bohumir	CAVS, Mississippi State University, USA	1522 | Tractive performance of rigid wheel in granular media using coarse-scale DEM models
Jeon, Hyeon-Ho	Department of Smart Agriculture Systems, Chungnam National University, Daejeon, Republic of Korea	3859 | Analysis for dynamic characteristics of a tractor cabin during agricultural operations *(abstract only)*
		4727 | Analysis of gear durability for agricultural tractor transmission *(abstract only)*
		4976 | Performance evaluation of autonomous driving lateral control simulation model of agricultural tractor *(abstract only)*
		5356 | Development and validation of head-feeding combine harvester multibody dynamics simulation model *(abstract only)*
		6955 | Development of dynamics simulation model for agricutural tractor according to agricultural work *(abstract only)*
Jianfen, Liu	Beijing Institute of Technology, China	4895 | Theoretical analysis and FEM simulation of the interfacial forces between tracked vehicle and snow soil *(abstract only)*

Name	Affiliation	Submissions
Jiao, Yang	McGill University, Canada	9504 | Soil deformation model for efficient simulation of off-road vehicles and compaction force analysis *(abstract only)*
Jones, Kathleen	USA	7577 | Implementation of moving loads on ice in NRMM
Jozwik, Jerzy	Lublin University of Technology, Faculty of Mechanical Engineering, Department of Production Engineering, Poland	7198 | Analysis of aircraft dynamics while overcoming terrain obstacles
Kalinko, Dariusz	Military University of Technology in Warsaw, Poland	8963 | Analysis of the influence of terrain unevenness and velocity on the ground pressure exerted by demining rollers *(abstract only)*
Kamprath, Nathan	GSVC, USA	3280 | Comparison of tweels and pneumatic tires on LTATV military vehicle
Karlberg, Magnus	Luleå University of Technology, Sweden	2765 | Obstacle detection vision system enabling autonomous mounding on clearcuts
		4141 | Modeling and verification of a full-scale forestry vehicle real-time multi-physics digital twin
		6027 | Digital Precision Planning Tool for Autonomous Forest Regeneration of Mixed Tree Species
Karpman, Eric	McGill University, Canada	4508 | Semi-empirical terramechanics model for variable terrain height *(abstract only)*
		6211 | Wheel-soil terramechanics model augmentation using machine learning across multiple soil types *(abstract only)*
Kat, Cor-Jacques	University of Pretoria, South Africa	6736 | Ride comfort comparison between 4-poster and straight-line driving simulations
Keller, Thomas	Swedish University of Agricultural Sciences; Agroscope, Sweden	2297 | Traction effects on vertical and horizontal contact stresses and the risk of soil structure deformation *(abstract only)*

Name	Affiliation	Submissions
Kim, Yong-Joo	Department of Agricultural Machinery Engineering, Chungnam National University, Daejeon, Republic of Korea	1166 | Selection of energy saving engine mode based on the power delivery and fuel consumption of a 95-kW tractor during agricultural operations *(abstract only)* 3815 | Data measurement and performance evaluation of an electric all-wheel-drive tractor during plow tillage *(abstract only)* 3859 | Analysis for dynamic characteristics of a tractor cabin during agricultural operations *(abstract only)* 3893 | Analysis of load factor and performance evaluation for developing an electric driven tractor *(abstract only)* 4727 | Analysis of gear durability for agricultural tractor transmission *(abstract only)* 4976 | Performance evaluation of autonomous driving lateral control simulation model of agricultural tractor *(abstract only)* 5356 | Development and validation of head-feeding combine harvester multibody dynamics simulation model *(abstract only)* 6955 | Development of dynamics simulation model for agricutural tractor according to agricultural work *(abstract only)*
Kiss, Peter	Hungarian University of Agriculture and Life Sciences (MATE), Hungary	2936 | Moisture content impacts on soil load-bearing capacity and its spectral behavior 4681 | Simplified models of terrain-vehicle interaction for real-time applications 4816 | Effects of humidity on the emissions of the diesel engines 7477 | Evaluating the pressure performance of DPF filters using engine bench analysis 9992 | Soil shear strength values obtained from its color
Kizaki, Shino	Tohoku University, Japan	4595 | Evaluation and Comparison of Driving Performance of Lunar Exploration Rover Wheel in Different Soils
Klepka, Mateusz	Lublin University of Technology, Poland	6861 | Researches on airplane performance on a grass airfield
Korga, Sylwester	Lublin University of Technology, Faculty of Electrical Engineering and Computer Science, Department of Computer Science, Poland	7198 | Analysis of aircraft dynamics while overcoming terrain obstacles

Name	Affiliation	Submissions
Körmöczi,Dávid	MATE Doctoral School of Mechanical Engineering, Hungary	4681 | Simplified models of terrain-vehicle interaction for real-time applications
Kosiuczenko, Krzysztof	Military Institute of Armour and Automotive Technology, Sulejówek, Poland	3796 | The issue of assessing the suitability of own simulation models for testing off-road vehicles
Kovecses, Jozsef	McGill University, Canada	4508 | Semi-empirical terramechanics model for variable terrain height *(abstract only)*
		6143 | Flexible tire-terrain interaction model for real time simulations *(abstract only)*
		6211 | Wheel-soil terramechanics model augmentation using machine learning across multiple soil types *(abstract only)*
		9504 | Soil deformation model for efficient simulation of off-road vehicles and compaction force analysis *(abstract only)*
Kravcov, Alexander	Czech Technical University, Prague, Czechia	3796 | The issue of assessing the suitability of own simulation models for testing off-road vehicles
Krogul, Piotr	Military University of Technology, Poland	6059 | Comparison of multi-tracked running gears for light UGVs in terms of terrain obstacles overcoming abilities *(abstract only)*
		8963 | Analysis of the influence of terrain unevenness and velocity on the ground pressure exerted by demining rollers *(abstract only)*
Kuszneruk, Michał	Lublin University of Technology, Poland	2177 | Effects of road wheel load, driving speed and track slip upon stress state in sandy soil under tracked vehicle loading
		3835 | Effect of steer angle rate upon tire lateral force generation on two different soils
L'Afflitto, Andrea	Virginia Tech, USA	6630 | Real-time measurement of tire sinkage using stereo cameras
Lamandé, Mathieu	Aarhus University, Denmark	2297 | Traction effects on vertical and horizontal contact stresses and the risk of soil structure deformation *(abstract only)*
Lehto, Mattias	Luleå University of Technology, Sweden	4141 | Modeling and verification of a full-scale forestry vehicle real-time multi-physics digital twin
Li, Jianqiao	Key Lab for Bionics Engineering of Education Ministry, Jilin University, China	2145 | Obstacle performance and wheel failure test analysis of Zhurong Rover
		9268 | Experiments of sinkage characters of wire mesh wheel under different slip ratio
Li, Songyu	Luleå University of Technology, Sweden	2765 | Obstacle detection vision system enabling autonomous mounding on clearcuts

Name	Affiliation	Submissions
Lideskog, Håkan	Luleå University of Technology, Sweden	2765 | Obstacle detection vision system enabling autonomous mounding on clearcuts 4141 | Modeling and verification of a full-scale forestry vehicle real-time multi-physics digital twin 6027 | Digital Precision Planning Tool for Autonomous Forest Regeneration of Mixed Tree Species
Lin, Tony	Goodyear Tire and Rubber company, USA	7523 | Assessing Performance of Light Wheeled Vehicles on GRC-1 using 3D Scanned Footprint Measurements
Lindbäck, Torbjörn	Luleå University of Technology, Sweden	4141 | Modeling and verification of a full-scale forestry vehicle real-time multi-physics digital twin
Longwic , Rafał	Lublin University of Technology, Poland	2177 | Effects of road wheel load, driving speed and track slip upon stress state in sandy soil under tracked vehicle loading
Łopatka, Marian	Military University of Technology, Poland	6059 | Comparison of multi-tracked running gears for light UGVs in terms of terrain obstacles overcoming abilities *(abstract only)* 8963 | Analysis of the influence of terrain unevenness and velocity on the ground pressure exerted by demining rollers *(abstract only)*
Łukaszewicz, Andrzej	Bialystok University of Technology, Poland	7198 | Analysis of aircraft dynamics while overcoming terrain obstacles
Maleki, Mahdi	McGill University, Canada	6143 | Flexible tire-terrain interaction model for real time simulations *(abstract only)*
Mason, George	CAVS, Mississippi State University, USA	1522 | Tractive performance of rigid wheel in granular media using coarse-scale DEM models
Ming, Zhu	Beijing Institute of Technology, China	4895 | Theoretical analysis and FEM simulation of the interfacial forces between tracked vehicle and snow soil *(abstract only)*
Mukherjee, Jyotirmoy	Virginia Tech, USA	6630 | Real-time measurement of tire sinkage using stereo cameras
Munkholm, Lars J.	Aarhus University, Denmark	2297 | Traction effects on vertical and horizontal contact stresses and the risk of soil structure deformation *(abstract only)*
Noren, Charles	Carnegie Mellon University, USA	8672 | Trajectory optimization for vegetation override in off-road driving
Nottage, Dustin	Engineer Research and Development Center - Construction Engineering Research Laboratory, USA	2801 | In Situ Soil Property Estimation for Autonomous Earthmoving Using Physics-Infused Neural Networks
Nowakowski, Marek	Military Institute of Armour and Automotive Technology, Sulejówek, Poland	3796 | The issue of assessing the suitability of own simulation models for testing off-road vehicles

Name	Affiliation	Submissions
Omura, Takuya	Keio University, Japan	4336 | High-fidelity 3D-DEM simulation for wheel mobility in low gravity environment *(abstract only)*
Pandit, Rashna	Virginia Tech, USA	3493 | Numerical modeling of a tire on undrained saturated clay using FEM, ALE, and SPH
Park, Jong-Dae	Department of Biosystems Machinery Engineering, Chungnam National University, Daejeon, Republic of Korea	3859 | Analysis for dynamic characteristics of a tractor cabin during agricultural operations *(abstract only)*
		3893 | Analysis of load factor and performance evaluation for developing an electric driven tractor *(abstract only)*
Park, Min-Jong	Department of Agricultural Machinery Engineering, Chungnam National University, Daejeon, Republic of Korea	3815 | Data measurement and performance evaluation of an electric all-wheel-drive tractor during plow tillage *(abstract only)*
		3859 | Analysis for dynamic characteristics of a tractor cabin during agricultural operations *(abstract only)*
		3893 | Analysis of load factor and performance evaluation for developing an electric driven tractor *(abstract only)*
		5356 | Development and validation of head-feeding combine harvester multibody dynamics simulation model *(abstract only)*
Parker, Michael	ERDC-CRREL, USA	3280 | Comparison of tweels and pneumatic tires on LTATV military vehicle
		8254 | Initial steps towards characterization of a new military cold weather all-terrain vehicle for performance in northern operations
Parmar, Sumit	Indian Institute of Technology Kharagpur, Kharagpur, India	4456 | Modeling soil-tool interaction of a cultivator sweep using DEM
Patidar, Prakhar	Indian Institute of Technology Kharagpur, Kharagpur, India	4456 | Modeling soil-tool interaction of a cultivator sweep using DEM
Peenze, Andries	University of Pretoria, South Africa	2437| Improving predictive control methods on off-road vehicles with realistic steering preview inputs
Pillinger, Gyorgy	Hungarian University of Agriculture and Life Sciences (MATE), Hungary	2936 | Moisture content impacts on soil load-bearing capacity and its spectral behaviour
Priddy, Jody	U.S. Army Engineer Research and Development Center, Geotechnical and Structures Laboratory, USA	1522 | Tractive performance of rigid wheel in granular media using coarse-scale DEM models
Przybysz, Mirosław	Military University of Technology in Warsaw, Poland	8963 | Analysis of the influence of terrain unevenness and velocity on the ground pressure exerted by demining rollers *(abstract only)*

Name	Affiliation	Submissions
Pytka, Jarosław	Lublin University of Technology, Poland	2177 | Effects of road wheel load, driving speed and track slip upon stress state in sandy soil under tracked vehicle loading 3507 | Determination of the tractive forces of a tractor based on the reconstruction of tire – soil contact surface obtained by means of photogrammetry 3835 | Effect of steer angle rate upon tire lateral force generation on two different soils 6861 | Researches on airplane performance on a grass airfield 7198 | Analysis of aircraft dynamics while overcoming terrain obstacles
Qingdong, Yan	Beijing Institute of Technology, China	4895 | Theoretical analysis and FEM simulation of the interfacial forces between tracked vehicle and snow soil *(abstract only)*
Qingkai, Meng	Beijing Institute of Technology, China	4895 | Theoretical analysis and FEM simulation of the interfacial forces between tracked vehicle and snow soil *(abstract only)*
Ravichandran, Nikhil	Virginia Polytechnic Institute and State University, USA	7523 | Assessing Performance of Light Wheeled Vehicles on GRC-1 using 3D Scanned Footprint Measurements
Rizzo, Denise M.	Department U.S. Army DEVCOM Ground Vehicle Systems Center, USA	3493 | Numerical modeling of a tire on undrained saturated clay using FEM, ALE, and SPH
Rubiec, Arkadiusz	Military University of Technology in Warsaw, Poland	8963 | Analysis of the influence of terrain unevenness and velocity on the ground pressure exerted by demining rollers *(abstract only)*
Rubinstein, Dror	Ariel University, Department of Mechanical Engineering and Mechatronics, Israel	8859 | An efficient and high-fidelity track model for dynamic simulation of off-road tracked vehicles *(abstract only)*
Sakata, Takaaki	Shibaura institute of technology	0889 | Skidding suppression method using "Discrete 4-Wheel-Drive typed rover" considering steering angle at traveling across loose slope
Sandu, Corina	Virginia Polytechnic Institute and State University, USA	3493 | Numerical modeling of a tire on undrained saturated clay using FEM, ALE, and SPH 6630 | Real-time measurement of tire sinkage using stereo cameras 6690 | Benchmarking of methods to evaluate penetration resistance for Clegg hammer and cone penetrometer 7523 | Assessing Performance of Light Wheeled Vehicles on GRC-1 using 3D Scanned Footprint Measurements
Sebeck, Katherine	Department U.S. Army DEVCOM Ground Vehicle Systems Center, USA	3493 | Numerical modeling of a tire on undrained saturated clay using FEM, ALE, and SPH

Name	Affiliation	Submissions
Scherer, Sebastian	Carnegie Mellon University, USA	8672 | Trajectory optimization for vegetation override in off-road driving
Schjønning, Per	Aarhus University, Denmark	2297 | Traction effects on vertical and horizontal contact stresses and the risk of soil structure deformation *(abstract only)*
Shenvi, Mohit Nitin	Virginia Tech, USA	6690 | Benchmarking of methods to evaluate penetration resistance for Clegg hammer and cone penetrometer
Shoop, Sally	ERDC-CRREL, USA	3280 | Comparison of tweels and pneumatic tires on LTATV military vehicle
Siddique, Md Abu Ayub	Department of Agricultural Machinery Engineering, Chungnam National University, Daejeon, Republic of Korea	1166 | Selection of energy saving engine mode based on the power delivery and fuel consumption of a 95-kW tractor during agricultural operations *(abstract only)*
Simiński, Przemysław	Military Institute of Armour and Automotive Technology, Sulejówek, Poland	3796 | The issue of assessing the suitability of own simulation models for testing off-road vehicles
Skonieczny, Krzysztof	Concordia University, Canada	1805 | Accelerating Graph Networks for Real Time Physics Simulation
Skorupa, Thomas	U.S. Army Ground Vehicle Systems Center, USA	1522 | Tractive performance of rigid wheel in granular media using coarse-scale DEM models
Śliczniak, Tomasz	Akademia Zamojska, Poland	3507 | Determination of the tractive forces of a tractor based on the reconstruction of tire – soil contact surface obtained by means of photogrammetry
Son, Mo-A	Department of Agricultural Machinery Engineering, Chungnam National University, Daejeon, Republic of Korea	1166 | Selection of energy saving engine mode based on the power delivery and fuel consumption of a 95-kW tractor during agricultural operations *(abstract only)*
		4976 | Performance evaluation of autonomous driving lateral control simulation model of agricultural tractor *(abstract only)*
		6955 | Development of dynamics simulation model for agricutural tractor according to agricultural work *(abstract only)*
Sonalkar, Chaitanya	Virginia Polytechnic Institute and State University, USA	7523 | Assessing Performance of Light Wheeled Vehicles on GRC-1 using 3D Scanned Footprint Measurements
Song, Ye-In	Department of Agricultural Machinery Engineering, Chungnam National University, Daejeon, Republic of Korea	1166 | Selection of energy saving engine mode based on the power delivery and fuel consumption of a 95-kW tractor during agricultural operations *(abstract only)*
		4976 | Performance evaluation of autonomous driving lateral control simulation model of agricultural tractor *(abstract only)*

Name	Affiliation	Submissions
Soni, Peeyush	Indian Institute of Technology Kharagpur, Kharagpur, India	4456 | Modeling soil-tool interaction of a cultivator sweep using DEM
Soylemezoglu, Ahmet	Engineer Research and Development Center - Construction Engineering Research Laboratory, USA	2801 | In Situ Soil Property Estimation for Autonomous Earthmoving Using Physics-Infused Neural Networks
Stearns, Mike	USACE-ERDC-CRREL, USA	5552 | New approach to controlled vehicle drawbar pull testing
		8254 | Initial steps towards characterization of a new military cold weather all-terrain vehicle for performance in northern operations
Suda, Gentaro	Tohoku University, Japan	4595 | Evaluation and Comparison of Driving Performance of Lunar Exploration Rover Wheel in Different Soils
Suga, Tatsuhiko	Shibaura institute of technology Kojiro Iizukaa,*, a, Taka-aki Sakataa and Daisuke Fujiwara	0889 | Skidding suppression method using "Discrete 4-Wheel-Drive typed rover" considering steering angle at traveling across loose slope
Swamy, Varsha S.	Virginia Tech, USA	3493 | Numerical modeling of a tire on undrained saturated clay using FEM, ALE, and SPH
Szőllősi, Dániel	Hungarian University of Agriculture and Life Sciences (MATE), Hungary	4816 | Effects of humidity on the emissions of the diesel engines
Szpaczyńska, Daniela	Military University of Technology, Poland	6059 | Comparison of multi-tracked running gears for light UGVs in terms of terrain obstacles overcoming abilities *(abstract only)*
Takehana, Keisuke	Tohoku University, Japan	4595 | Evaluation and Comparison of Driving Performance of Lunar Exploration Rover Wheel in Different Soils
		8754 | Characterization of lunar simulant (FJS-1), toyoura sand, and silica sand with physical experiments and dem simulation *(abstract only)*
Tarkowski, Sławomir	Lublin University of Technology, Poland	3507 | Determination of the tractive forces of a tractor based on the reconstruction of tire – soil contact surface obtained by means of photogrammetry
Tatarynow, Dawid	Lublin University of Technology, Poland	2177 | Effects of road wheel load, driving speed and track slip upon stress state in sandy soil under tracked vehicle loading
		3835 | Effect of steer angle rate upon tire lateral force generation on two different soils
Takano, Sae	Shibaura Institute of Technology, Japan	8151 | Study on estimation of traveling states using strain information on chassis of lunar and planetary exploration rovers
Takehana, Keisuke	Tohoku University, Japan	0448 | Lunar rover discrete element method study and calibration
Tanaka, Tomomi	Tohoku University, Japan	4595 | Evaluation and Comparison of Driving Performance of Lunar Exploration Rover Wheel in Different Soils
Tarkowski, Sławomir	Lublin University of Technology, Poland	6861 | Researches on airplane performance on a grass airfield

Name	Affiliation	Submissions
Teichmann, Marek	CM Labs Simulation, Inc., Canada	4508 | Semi-empirical terramechanics model for variable terrain height *(abstract only)* 6211 | Wheel-soil terramechanics model augmentation using machine learning across multiple soil types *(abstract only)*
ten Damme, Loraine	Aarhus University, Denmark; Swedish University of Agricultural Sciences	2297 | Traction effects on vertical and horizontal contact stresses and the risk of soil structure deformation *(abstract only)*
Tikanmäki, Antti	University of Oulu, Finland	4141 | Modeling and verification of a full-scale forestry vehicle real-time multi-physics digital twin
Tomberg, Zach	USACE-ERDC-CRREL, USA	5552 | New approach to controlled vehicle drawbar pull testing 8254 | Initial steps towards characterization of a new military cold weather all-terrain vehicle for performance in northern operations
Tomiło, Paweł	Lublin University of Technology, Poland	0274 | Deep Learning Method for IMU-based tracking of Martian rover 2177 | Effects of road wheel load, driving speed and track slip upon stress state in sandy soil under tracked vehicle loading 3835 | Effect of steer angle rate upon tire lateral force generation on two different soils 6861 | Researches on airplane performance on a grass airfield
Travers, Matthew	Carnegie Mellon University, USA	8672 | Trajectory optimization for vegetation override in off-road driving
Uno, Kentaro	Tohoku University, Japan	0448 | Lunar rover discrete element method study and calibration 4595 | Evaluation and Comparison of Driving Performance of Lunar Exploration Rover Wheel in Different Soils
Untaroiu, Costin	Virginia Tech, USA	6690 | Benchmarking of methods to evaluate penetration resistance for Clegg hammer and cone penetrometer
Usmani, Mohammad Jasim	Concordia University, Canada	1805 | Accelerating Graph Networks for Real Time Physics Simulation
van Westendorp, Ruben	Bracke Forest AB, Sweden	2765 | Obstacle detection vision system enabling autonomous mounding on clearcuts
Vennik, Kersti	Estonian Military Academy, Estonia	3797 | Disturbed and naturally recovered soil surface as a ground for subsequent vehicle mobility operations
Vilsan, Alexandru	Virginia Polytechnic Institute and State University, USA	7523 | Assessing Performance of Light Wheeled Vehicles on GRC-1 using 3D Scanned Footprint Measurements
Vundurthy, Bhaskar	Carnegie Mellon University, USA	8672 | Trajectory optimization for vegetation override in off-road driving

Name	Affiliation	Submissions
Wagner, W. Jacob	Engineer Research and Development Center-CERL, and University of Illinois at Urbana-Champaign, Dept. of ECE, USA	2801 | In Situ Soil Property Estimation for Autonomous Earthmoving Using Physics-Infused Neural Networks
Watanabe, Dai	Shibaura Institute of Technology, Japan	1098 | Simulation of change in supporting force when imparting vibration by distinct element method
Watanabe, Tomohiro	Niigata University, Japan	1098 | Simulation of change in supporting force when imparting vibration by distinct element method
Wei, Wei	Beijing Institute of Technology, China	4895 | Theoretical analysis and FEM simulation of the interfacial forces between tracked vehicle and snow soil *(abstract only)*
Welling, Orian	USACE-ERDC-CRREL, USA	5552 | New approach to controlled vehicle drawbar pull testing
		8254 | Initial steps towards characterization of a new military cold weather all-terrain vehicle for performance in northern operations
Wheeler, Allan	ERDC-CRREL, USA	3280 | Comparison of tweels and pneumatic tires on LTATV military vehicle
White, Hannah	Virginia Tech, USA	6630 | Real-time measurement of tire sinkage using stereo cameras
Wieczorek, Marcin	Wojskowa Akademia Techniczna im. Jarosława Dąbrowskiego (Military University of Technology), Poland	9716 | Non-pneumatic tire for special vehicle – a review of selected functional properties *(abstract only)*
Witte, Clifford	ERDC-CRREL, USA	3280 | Comparison of tweels and pneumatic tires on LTATV military vehicle
		8254 | Initial steps towards characterization of a new military cold weather all-terrain vehicle for performance in northern operations
Yamakawa , Junya	National Defense Academy, Japan	1638 | Analysis of tire characteristics on road surface with volcanic ash fall
Yang, Cheol-Woo	Department of Biosystems Machinery Engineering, Chungnam National University, Republic of Korea	5356 | Development and validation of head-feeding combine harvester multibody dynamics simulation model *(abstract only)*
Yang, Wang	Key Laboratory of Bionic Engineering, Ministry of Education, Jilin University, Changchun, China	9268 | Experiments of sinkage characters of wire mesh wheel under different slip ratio
Yerro, Alba	Virginia Tech, USA	3493 | Numerical modeling of a tire on undrained saturated clay using FEM, ALE, and SPH
Yoon, Su-Young	Department of Agricultural Machinery Engineering, Chungnam National University, Daejeon, Republic of Korea	3859 | Analysis for dynamic characteristics of a tractor cabin during agricultural operations *(abstract only)*
		3893 | Analysis of load factor and performance evaluation for developing an electric driven tractor *(abstract only)*

Name	Affiliation	Submissions
Yoshida, Kazuya	Tohoku University, Japan	0448 | Lunar rover discrete element method study and calibration 4595 | Evaluation and Comparison of Driving Performance of Lunar Exploration Rover Wheel in Different Soils 8754 | Characterization of lunar simulant (FJS-1), toyoura sand, and silica sand with physical experiments and dem simulation *(abstract only)*
Zalewska-Tytłak, Anna	Lublin University of Technology, Poland	6861 | Researches on airplane performance on a grass airfield
Żmuda, Marcin	Wojskowa Akademia Techniczna im. Jarosława Dąbrowskiego (Military University of Technology), Poland	9716 | Non-pneumatic tire for special vehicle – a review of selected functional properties *(abstract only)*
Żyła, Kamil	Lublin University of Technology, Faculty of Electrical Engineering and Computer Science, Department of Computer Science, Poland	7198 | Analysis of aircraft dynamics while overcoming terrain obstacles
Zou, Meng	Key Lab for Bionics Engineering of Education Ministry, Jilin University, China	2145 | Obstacle performance and wheel failure test analysis of Zhurong Rover 9268 | Experiments of sinkage characters of wire mesh wheel under different slip ratio